Wavelets, Frames and Operator Theory

CONTEMPORARY MATHEMATICS

345

Wavelets, Frames and Operator Theory

Focused Research Group Workshop on Wavelets, Frames and Operator Theory

January 15–21, 2003
University of Maryland
College Park, Maryland

Christopher Heil
Palle E.T. Jorgensen
David R. Larson
Editors

American Mathematical Society
Providence, Rhode Island

This volume contains papers from an AMS Special Session on "Wavelets, Frames and Operator Theory" from the 2003 Annual Meeting of the AMS in Baltimore, Maryland, January 15–18, 2003, and an NSF-sponsored workshop at the University of Maryland, January 19–21, 2003. Both events were associated with the NSF Focused Research Group (FRG) with support from NSF grant DMS-0139759.

2000 *Mathematics Subject Classification.* Primary 20C20, 41A17, 42C15, 42C40, 43A85, 46C05, 46C99, 46E25, 47C05, 65T60.

Library of Congress Cataloging-in-Publication Data

Wavelets, frames and operator theory: Focused research group workshop on wavelets, frames and operator theory, January 19–21, 2003 / Christopher Heil, Palle E. T. Jorgensen, David R. Larson, editors.

p. cm.– (Contemporary mathematics, ISSN 0271-4132; 345)

Includes bibliographical references.

ISBN 0-8218-3380-4 (alk. paper)

1. Wavelets (Mathematics)–Congresses. 2. Frames (Combinatorial analysis)–Congresses. 3. Operator theory–Congresses. I. Heil, Christopher E., 1960– II. Jørgensen, Palle E. T., 1947– III. Larson, David R., 1942– IV. Contemporary mathematics (American Mathematical Society); v. 345.

QA403.3.W392 2004

515′.2433–dc22 2004041027

∞ The paper used in this book is acid-free and falls within the guidelines
established to ensure permanence and durability.
Visit the AMS home page at http://www.ams.org/

10 9 8 7 6 5 4 3 2 1 09 08 07 06 05 04

Contents

Preface

This book grew out of the Special Session on *Wavelets, Frames, and Operator Theory* that we organized at the 2003 Annual Meeting of the AMS in Baltimore, January 15–18, 2003, and an immediately following NSF-sponsored workshop organized by John Benedetto at The University of Maryland, January 19–21, 2003. Both events were associated with the NSF Focused Research Group (FRG) of which we are a part, and whose other members are Akram Aldroubi, Lawrence W. Baggett, John J. Benedetto, Gestur Ólafsson, and Yang Wang. The speakers in the Special Session and the Maryland workshop were invited to contribute papers, and this volume is the very pleasant result.

We hope that those events and more like them that have since taken place or are planned for the future, and the present book itself, will act as a catalyst, encouraging members of our community to work on one or more of the many facets of the intertwined subjects of wavelets, frames, and operator theory. Some of the papers included here focus more on one of the three areas than the other two, but all hint at exciting connections and interrelationships. They stand at the crossroads of harmonic analysis, operator theory, and applied mathematics. Some papers are theoretical, some applied, but most are a mix of theory and applications, each inspiring the other. Wavelets and frames have emerged as significant tools in mathematics and in technology over the past two decades. They interact with harmonic analysis, with operator theory, and with a host of applications. In their primitive form, both wavelets and frames originate with the vector space notion of a basis. They are used in the analysis of functions, and the functions make up infinite-dimensional spaces, typically Hilbert spaces. While many wavelet constructions yield orthonormal bases, frames by their very nature, including many interesting classes of wavelets, satisfy conditions which are more general than the familiar orthogonality relations. Historically, operator theory has played a big part in the analysis of both wavelets and frames, and we hope to highlight this feature in our collection of papers.

The workshops, the research, and the publication of this volume were supported in part by our FRG grant from the National Science Foundation.[1] It is also a pleasure to thank Brian Treadway, whose assistance was essential to the smooth

[1]DMS-0139759 Collaborative Research: Focused Research on Wavelets, Frames, and Operator Theory. Description: In this project, fundamental problems are addressed in wavelet theory, non-uniform sampling, frames, and the theory of spectral-tile duality. These problems are inextricably interwoven by concept and technique. Operator theory provides the major unifying framework, combined with an integration of ideas from a diverse spectrum of mathematics including classical Fourier analysis, noncommutative harmonic analysis, representation theory, operator algebras, approximation theory, and signal processing. For example, the construction, implementation, and ensuing theory of single dyadic orthonormal wavelets in Euclidean space requires significant input from all of these disciplines as well as deep spectral-tile results.

completion of this volume. Brian managed the correspondence with referees and authors, organized the many drafts of papers, helped bring them into the TEX format for the book series, and managed and assisted us in numerous typesetting issues.

Christopher Heil, Palle E. T. Jorgensen, and David R. Larson
September 21, 2003

Speakers at the two sessions of the 2003 Annual Meeting of the American Mathematical Society, held in Baltimore, and the Wavelet Workshop held in College Park, Maryland, out of which the papers in this volume arose

Symbols are used in the list to designate the specific sessions, as follows:

(S): AMS Special Session on "Wavelets, Frames and Operator Theory", at the 2003 Annual Meeting of the American Mathematical Society (AMS–MAA–ASL–AWM–NAM–SIAM Joint Mathematics Meetings, Baltimore, MD, January 15–18, 2003);
(C): AMS Session on "Operator Theory", at the same Annual Meeting;
(W): NSF Focused Research Group Workshop on the theme "Modulation Spaces and the Continuous Wavelet Transform" (University of Maryland, College Park, MD, January 19–21, 2003).

The ten speakers marked (C) had requested to be part of our Special Session, but unfortunately could not be included due to lack of space. The AMS incorporated these speakers into a related session of contributed talks. We thank the AMS for kindly honoring our request to include them in this way, and we consider them informally included in our session for the purpose of this volume.

Affiliations follow the session symbols for each speaker. Where more than one affiliation is listed, the first is from the time of the sessions, and the others are later locations.

Akram Aldroubi (S), Vanderbilt University
http://www.math.vanderbilt.edu/~aldroubi/
"Non uniform sampling and reconstruction in irregular spaces"
Radu Balan (S), Siemens Corporate Research
"Measure function and redundancy of Weyl-Heisenberg multiframes and superframes"
Robert Benedetto (S), Amherst College
http://www.cs.amherst.edu/~rlb/
"Wavelets on p-adic fields and related groups"
Ola Bratteli (S), University of Oslo
http://www.math.uio.no/~bratteli/
"Global structure of the scaling-wavelet variety"
Peter G. Casazza (S), University of Missouri, Columbia
http://www.math.missouri.edu/~pete/
"Existence and construction of finite frames with a given frame operator"

Ingrid Daubechies (S), Princeton University
http://www.princeton.edu/~icd/
"An iterative algorithm for ill-posed inverse problems where the object has a sparse wavelet expansion"
Dorin Dutkay (S), University of Iowa
http://www.math.uiowa.edu/~ddutkay/
"The local trace function of shift invariant subspaces"
Hans G. Feichtinger (S, W—plenary talk), University of Vienna
http://www.univie.ac.at/NuHAG/FEI/
"Approximation of linear operators by Gabor multipliers" (S)
Matthew C. Fickus (S, W), Cornell University
http://www.math.cornell.edu/People/Postdocs/fickus.html
"Frames in communications" (S)
"A physical interpretation for finite tight frames" (W)
Yevgeniy V. Galperin (C), Sacred Heart University
"Embeddings of Fourier-Lebesgue spaces into modulation spaces: Optimality of sufficient conditions"
Joel K. Glenn (C), Colorado College
"Frequency estimation and vortex analysis using wavelet coefficients"
Karlheinz Gröchenig (S, W—plenary talk), University of Connecticut
http://www.math.uconn.edu/~groch/
"Localization of frames" (S, W)
Christopher Hammond (C), University of Virginia
"On the norm of a composition operator with linear fractional symbol"
Deguang Han (S), University of Central Florida
http://pegasus.cc.ucf.edu/~dhan/main.html
"Operator parametrization and tight frame approximations"
Doug Hardin (S), Vanderbilt University
http://math.vanderbilt.edu/~hardin/
"Continuous orthogonal wavelets on semi-regular triangulations"
Denise Jacobs (C), United States Military Academy
http://www.dean.usma.edu/math/people/jacobs/
"Orthogonal wavelets in higher dimensions"
Brody Johnson (W), Georgia Institute of Technology; Saint Louis University
http://euler.slu.edu/Dept/Faculty/johnson/
"Oversampling wavelet frames"
Norbert Kaiblinger (S), University of Vienna
http://www.mat.univie.ac.at/~kaib
"Varying the lattice of Gabor frames"
Keri Kornelson (S), Texas A & M University
http://www.math.tamu.edu/~keri.kornelson/
"Ellipsoidal tight frames"
Gitta Kutyniok (S, W), University of Paderborn
http://www-math.uni-paderborn.de/~gittak/
"Density of weighted wavelet frames" (S)
"A qualitative uncertainty principle for functions generating a Gabor frame on LCA groups" (W)

Demetrio Labate (S, W), Washington Univ., St. Louis; North Carolina State Univ.
http://www4.ncsu.edu:8030/~dlabate/
"A unified theory of reproducing function systems" (S)
"Oversampling of affine systems" (W)
Jeffrey C. Lagarias (S), AT&T Labs—Research
http://www.research.att.com/~jcl/
"A family of piecewise-linear plane maps and associated nonlinear difference operators of Schrödinger type"
Mark Lammers (S, W), Western Washington U.; U. of North Carolina, Wilmington
http://people.uncw.edu/lammersm/
"Wilson bases and convolution" (S)
Zeph Landau (S, W), Mathematical Sciences Research Institute; Microsoft Corp.
"Densities of frames" (S)
"Measuring sequences, subspaces, and frames" (W)
Ursula Molter (S), Universidad de Buenos Aires
http://mate.dm.uba.ar/~umolter/
"Optimal shift-invariant spaces"
Krzysztof Nowak (C), Drexel University
"Best projections of Gabor multiplier type"
Kasso Okoudjou (S, W), Georgia Institute of Technology; Cornell University
http://www.math.cornell.edu/~kasso/
"Bilinear pseudodifferential operators on modulation spaces" (S)
"Gabor analysis in amalgam spaces" (W)
Gestur Ólafsson (S), Louisiana State University
http://www.math.lsu.edu/~olafsson/
"Frames and groups"
Judith Packer (S), University of Colorado, Boulder
http://spot.colorado.edu/~packer/
"An analogue of the Bratteli-Jorgensen loop group action for m-systems in the GMRA setting"
Manos Papadakis (S), University of Houston
http://www.math.uh.edu/~mpapadak/
"Symmetric univariate QM filters with Gaussian decay"
Alexander Powell (S), University of Maryland; Princeton University
"A (p, q) weighted version of a theorem of J. Bourgain"
T. Gabriel Prajitura (C), State University of New York, Brockport
"Approximation by countably hypercyclic operators"
Dmitry Ryabogin (C), University of Missouri, Columbia
http://www.math.missouri.edu/~ryabs/
"The Calderón reproducing formula and rough singular integrals"
Ziemowit Rzeszotnik (S), University of Texas, Austin
http://www.ma.utexas.edu/users/zioma/
"Unitary operators preserving wavelets"
Songkiat Sumetkijakan (W), University of Maryland; Chulalongkorn University
http://pioneer.netserv.chula.ac.th/~ssongkia/
"On the neighborhood-mapping construction of wavelet sets in $\mathbb{R}^d$"

Qiyu Sun (S), University of Houston; University of Central Florida
http://gauss.math.ucf.edu/~qsun/
"Symmetric univariate QM filters with Gaussian decay II"
David Walnut (S), George Mason University
"Local reconstruction from averages"
Ying Wang (C, W), Marywood University
"On joint perturbations of Gabor frames" (C)
"On perturbations of irregular Gabor frames" (W)
Eric Weber (S, W), University of Wyoming; Iowa State University
http://www.math.iastate.edu/esw/
"Superwavelets and generalized multiresolution analysis" (S)
"Orthogonal frames of translates" (W)
Guido Weiss (S, W—plenary talk), Washington University, St. Louis
http://www.math.wustl.edu/~guido/
"On the connectivity of wavelets" (S)
"A unified theory for the characterization of reproducing systems" (W)
Janine Wittwer (C), Williams College
http://www.williams.edu/Mathematics/jwittwer/
"Wavelets and Bellman functions"
Richard A. Zalik (S), Auburn University
http://www.auburn.edu/~zalikri/zalikri.html
"On MRA Riesz wavelets"
Shijun Zheng (C, W), University of Maryland; Louisiana State University
"Littlewood-Paley theory associated with Schrödinger operators with hyperbolic secant potentials" (C)
"Schrödinger operator, Besov spaces, and wavelet computations for thin film image processing" (W)

Members of the NSF Focused Research Group on "Wavelets, Frames and Operator Theory"

http://www.math.uiowa.edu/~jorgen/waveletFRG.html

Akram Aldroubi, Vanderbilt University
http://www.math.vanderbilt.edu/~aldroubi/
Lawrence W. Baggett, University of Colorado
http://spot.colorado.edu/~baggett/
John J. Benedetto, University of Maryland
http://www.math.umd.edu/~jjb
Christopher E. Heil, Georgia Institute of Technology
http://www.math.gatech.edu/~heil/
Palle E.T. Jorgensen, University of Iowa
http://www.math.uiowa.edu/~jorgen/
David R. Larson, Texas A & M University
http://www.math.tamu.edu/directory/faculty/Larson_formalpage.html
Gestur Ólafsson, Louisiana State University
http://www.math.lsu.edu/~olafsson/
Yang Wang, Georgia Institute of Technology
http://www.math.gatech.edu/~wang/

Contemporary Mathematics
Volume **345**, 2004

How to construct wavelet frames on irregular grids and arbitrary dilations in $\mathbb{R}^d$

Akram Aldroubi, Carlos Cabrelli, and Ursula M. Molter

ABSTRACT. In this article, we present a method for constructing wavelet frames of $L^2(\mathbb{R}^d)$ of the type $\{|\det A_j|^{1/2}\psi(A_j x - x_{j,k}) : j \in J,\ k \in K\}$ on irregular lattices of the form $X = \{x_{j,k} \in \mathbb{R}^d : j \in J,\ k \in K\}$, and with an arbitrary countable family of invertible $d \times d$ matrices $\{A_j \in GL_d(\mathbb{R}) : j \in J\}$. Possible applications include image and video compression, speech coding, image and digital data transmission, image analysis, estimations and detection, and seismology.

1. Introduction

In this article we present a general method for constructing well-localized wavelet frames $\{|\det A_j|^{1/2}\psi(A_j x - x_{j,k}) : j \in J,\ k \in K\}$ of $L^2(\mathbb{R}^d)$ on arbitrary grids $X = \{x_{j,k} \in \mathbb{R}^d : j \in J,\ k \in K\}$, and with arbitrary dilation matrices $\{A_j\}_{j \in J}$. The construction presented here is a special case of a more general method for constructing time-frequency frame atoms in several variables discussed in [**ACM03**]. Although there has been considerable interest in trying to obtain wavelet frame decompositions of the space $L^2(\mathbb{R}^d)$, on irregular grids and with unstructured dilation matrices (see [**Bal97**], [**BCHL03**], [**Chr96**], [**Chr97**], [**CH97**], [**CDH99**], [**FZ95**], [**Fei87**], [**FG89**], [**FW01**], [**Grö91**], [**Grö93**], [**HK03**], [**OS92**], [**RS95**], [**SZ00**], [**SZ01**], [**SZ02**], [**SZ03**], [**SZ03**]), most of the methods that have been developed are small perturbations of wavelet frames on a regular grid and with a fixed dilation matrix. In contrast, our approach presented in [**ACM03**] is not a perturbation method and is very general, allowing quite general constructions. The setting includes as particular cases, wavelet frames on irregular lattices and with a set of dilations or transformations that do not have a group structure. For this paper, we will be mainly concerned with an even more particular case consisting of wavelet frames on irregular lattices and with an arbitrary but fixed expansive matrix A (A is said to be expansive if $|\lambda| > 1$ for every eigenvalue λ of A). The

2000 *Mathematics Subject Classification.* 42C40.
Key words and phrases. Frames, irregular sampling, wavelet sets, wavelets.
The research of Akram Aldroubi is supported in part by NSF grant DMS-0103104, and by DMS-0139740. The research of Carlos Cabrelli and Ursula Molter is partially supported by Grants: PICT 03134, and CONICET, PIP456/98.

case of regular lattices can also be obtained by our system, producing a substantial part of the systems recently characterized by the fundamental work of Guido Weiss and his group [**HDW02, HDW03, Lab02**] on the decomposition of $L^2(\mathbb{R}^d)$. The wavelet frames obtained by Chui, He, Stöeckler and Sun [**CHS**], [**CHSS03**], [**CS00**] are also included in our setting. Wavelet sets and wavelet frame sets studied in [**BMM99**], [**BL99**], [**BL01**], [**BS03**], [**DLS97**], [**DLS98**], [**HL00**], [**Ola03**], [**OS03**] can also be produced by our methods. Furthermore we can obtain wavelet sets with translations on irregular grids.

The method we present relies on combining ideas from four related, but different subjects: 1) Sampling theory; 2) Frame theory; 3) Wavelet theory; and 4) Geometry of $\mathbb{R}^d$. The approach can be considered in the spirit of the classic construction in 1 dimension of smooth regular tight frames done by Daubechies, Grossmann and Meyer in [**DGM86**]. (See also [**HW89**] for an expository treatment.) We will say that a set $X = \{x_k \in \mathbb{R}^d : k \in K\}$ is *separated* if

$$\inf_{k,s \in K, k \neq s} |x_k - x_s| > 0.$$

Throughout the paper J and K will denote countable index sets. One of the main ingredients in sampling theory is the notion of *lower Beurling density* $D^-(X)$ [**Beu66**] of a separated set $X = \{x_k \in \mathbb{R}^d : k \in K\}$, which is defined as:

$$D^-(X) = \lim_{r \to \infty} \frac{\nu^-(r)}{(2r)^d}$$

where $\nu^-(r) := \min_{y \in \mathbb{R}^d} \#\left(X \cap (y + [-r,r]^d)\right)$. $\#(Z)$ denotes the cardinal of the set Z.

The *upper Beurling Density* $D^+(X)$ is defined in a similar way:

$$D^+(X) = \lim_{r \to \infty} \frac{\nu^+(r)}{(2r)^d}$$

where $\nu^+(r) := \max_{y \in \mathbb{R}} \#\left(X \cap (y + [-r,r]^d)\right)$. If $D^-(X) = D^+(X) = D(X)$, then X is said to have *uniform Beurling density* $D(X)$.

Remark. Since X is separated, the limits in the definitions of $D^+(X)$ and $D^-(X)$ exist (see [**BW99**]).

Beurling [**Beu66**] introduced also the following notion of density: The *gap* ρ of the set $X = \{x_k : k \in K\}$ is defined as

$$\rho = \rho(X) = \inf \left\{ r > 0 : \bigcup_{k \in K} B_r(x_k) = \mathbb{R}^d \right\}.$$

Equivalently, the gap ρ can be defined as

$$\rho = \rho(X) = \sup_{x \in \mathbb{R}^d} \inf_{x_k \in X} |x - x_k|.$$

A family $\{Q_j : j \in J\}$ is a *covering* of $\mathbb{R}^d$ if $\mathbb{R}^d \setminus \bigcup_j Q_j$ has measure zero. A covering $\{Q_j : j \in J\}$ has *finite index* if every $x \in \mathbb{R}^d$ is at most in i sets of the covering for some fixed positive integer i. The minimum i with this property is called the *covering index.* We will denote by $e_x(w)$ the exponential of frequency x at w, that is $e_x(w) = e^{-2\pi i x w}$.

Let us now state a general theorem on wavelet frames:

THEOREM 1.1 (Wavelets). *Let A be an expansive matrix and $V \subset \mathbb{R}^d$ be any measurable bounded set containing* 0 *in its interior and such that its boundary ∂V has measure zero. Set $Q = \overline{A^T V \setminus V}$, and choose any function $h \in C^r(\mathbb{R}^d), r > 0, h \neq 0$ on Q such that* $\operatorname{Supp} h \in Q_\varepsilon$ *where $Q_\varepsilon := \{x \in \mathbb{R}^d : \operatorname{dist}(x, Q) \leq \varepsilon\}$. If the set $X = \{x_k \in \mathbb{R}^d : k \in K\}$ is separated and such that $\rho(X) < \frac{1}{4\delta}$ where* $\delta = \operatorname{Diameter}(Q)$, *then the following collection is a wavelet frame for $L^2(\mathbb{R}^d)$*

$$\{|\det A|^{j/2}\psi(A^j x - x_k) : k \in K,\ j \in \mathbb{Z})\}, \tag{1.1}$$

where ψ is the inverse Fourier transform of h.

REMARKS.

(1) The result of the theorem remains valid even if the matrix A is not expansive. For example let Q be any closed subset of $\mathbb{R}^d$, and A any invertible matrix. If $\mathbb{R}^d = \cup_j A^j Q_\varepsilon$ with finite covering index, then (1.1) is a frame for $L^2(\mathbb{R}^d)$.
(2) Instead of taking the powers A^j of a single matrix A we can choose a set of invertible matrices $\{A_j \in GL_d(\mathbb{R}) : j \in J\}$ without a particular group structure. In particular the index j can be a multi-index. For example, the set $J = \mathbb{Z} \times \{1, \cdots, N\}$, and the matrices $A_{(i,j)} = D^i R^j$ where R is a rotation and D a dilation matrix, will be used to construct directional wavelets.
(3) The wavelet can be constructed to have polynomial decay of any order by choosing r sufficiently large.
(4) The sets of translations $X_j = \{A^{-j}x_k : k \in K\}$ for each resolution level are not nested. However, the theorem can be easily modified to produce nested sets of translations $X_{j+1} \subset X_j$ for all $j \in \mathbb{Z}$ (c.f. [**ACM03**]).
(5) For the one dimensional case, ρ can be replaced by the Beurling density $D^-(X)$ which is a weaker condition and allows for arbitrary gaps between sampling points.
(6) If we choose h to be the characteristic function of the set Q_ε, then we obtain a wavelet frame set, and our construction (1.1) gives wavelet sets with translations on irregular grids.

Although the set $\{|\det A|^{j/2}\psi(A^j x - x_k) : j \in J,\ k \in K\}$ in Theorem 1.1 is a wavelet frame for $L^2(\mathbb{R}^d)$, it is not in general true that for a fixed j the set $\{\psi_{j,x_k}(x) = |\det A|^{j/2}\psi(A^j x - x_k) : k \in K\}$ is a frame. Thus, it appears at first, that the reconstruction of a function $f \in L^2(\mathbb{R}^d)$ from the wavelet coefficients $\{< f, \psi_{j,x_k} >: j \in J,\ k \in K\}$ cannot be obtained in a stable way by first reconstructing at each level j and then obtaining f by summing over all levels j. But in fact it is always possible to reconstruct each f_j in a stable way and then obtain f by summing up over all levels j, as is described in [**ACM03**].

2. Examples of wavelet frames on irregular lattices and with arbitrary set of dilation matrices and other transformations

2.1. Examples of Wavelet frames in $\mathbb{R}^d$.

(1) *Isotropic, well-localized wavelets*: Let V be the ball of radius $1/2$ centered at the origin. Let $A = 2I$, then $Q = \overline{AV \setminus V} = \{x \in \mathbb{R}^2 : 1/2 \leq \|x\| \leq 1\}$. Let $\varepsilon = 1/4$, $h(\xi_1, \xi_2) = n\beta_{n-1}\left((\xi_1^2 + \xi_2^2 - 1/4)n\right)$, where β_n is the B-spline of degree n, i.e., the $\beta_n = \chi_{[0,1]} * \cdots * \chi_{[0,1]}$ is the n-fold convolution

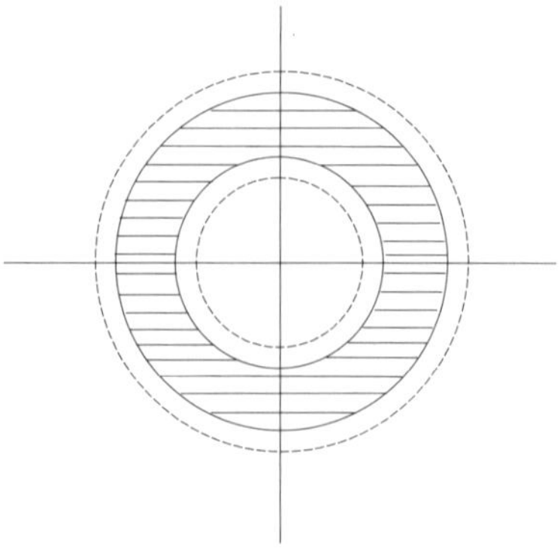

FIGURE 1. Radial wavelet frames that are well localized in space: Region between continuous lines correspond to region Q and region between dashed lines corresponds to Q_ε.

of the characteristic function on $[0,1]$. Let $X = \{x_k \in \mathbb{R}^2 : k \in K\}$ be such that its maximal gap satisfies $\rho(X) < \frac{1}{5}$. Then by Theorem 1.1 the set (1.1) is a wavelet frame for $L^2(\mathbb{R}^2)$ (see Figure 1). By construction, the wavelet ψ is isotropic and has polynomial decay of degree n in space, i.e, $|\psi(x)| \le C(1+|x|)^{-n}$. A theorem in [**ACM03**] more general than Theorem 1.1 shows that the same wavelet ψ generates a frame for $L^2(\mathbb{R}^2)$ if we replace the dilation matrix $A = 2I$ by any dilation matrix $A = bI$ where $1 < b \le 5/4$. Non-separable radial MRA frame wavelets can be found in [**PGKKH1**], [**PGKKH2**]. However, these radial MRA frame wavelets have slow spatial decay, i.e., $|\psi(x)| \le C(1+|x|)^{-1}$.

(2) *Directional wavelet frames*: To construct directional wavelet frames we use a modified version of Theorem 1.1 as suggested by remark (2). Let Q_1 be a region defined by $Q_1 = \{(x,y) \in \mathbb{R}^2 : x = r\cos(\theta),\ y = r\sin(\theta),\ 1/2 \le r \le 1,\ |\theta| \le \frac{\pi}{2N}\}$ as in Figure 2, and define $Q = (-Q_1) \cup Q_1$. Let $B = 2I$, and R be the matrix of a rotation by an angle π/N, where N is any positive integer, $\varepsilon = 0.1$, and $J = \{(j_1, j_2) : j_1 \in \mathbb{Z},\ j_2 = 0, \cdots, N-1\}$. Then $\mathbb{R}^2 = \cup_{(j_1,j_2)\in J} B^{j_1} R^{j_2} Q$. Let $X_j = \{x_{j,k} \in \mathbb{R}^2 : k \in K\}$ be a set of points such that $\rho(X_j) < \frac{5}{22}$ for each $j = (j_1, j_2) \in J$. Then the function ψ, such that $\hat{\psi} = \chi_{Q_\varepsilon}$, generates a wavelet frame for $L^2(\mathbb{R}^2)$ of the form $\{\psi_{j,k} = 2^{j_1}\psi_j(2^{j_1}R^{j_2} \cdot - x_{j,k}) : j \in J,\ k \in \mathbb{Z}\}$. The index j_1 codes for the resolution of the wavelet, while the index j_2 codes for N possible directions. Thus the wavelet frame coefficients encode time scale as well as directional information. The wavelet constructed in this example does not have good space localization, because it is the inverse Fourier transform of a characteristic function. However, we can replace the characteristic function χ_{Q_ε} by a smoother function $h \in C^r$ where $r > 0$ is any positive real number we choose, $h > 0$ on Q and $\operatorname{Supp} h \in Q_\varepsilon$. With this modification we obtain a directional wavelet ψ with spatial localization satisfying $|\psi(x)| \le C(1+|x|)^{-r}$. Moreover the wavelet can be constructed to be real and symmetric with respect to the origin, by choosing a function h who is real and symmetric with respect to the origin.

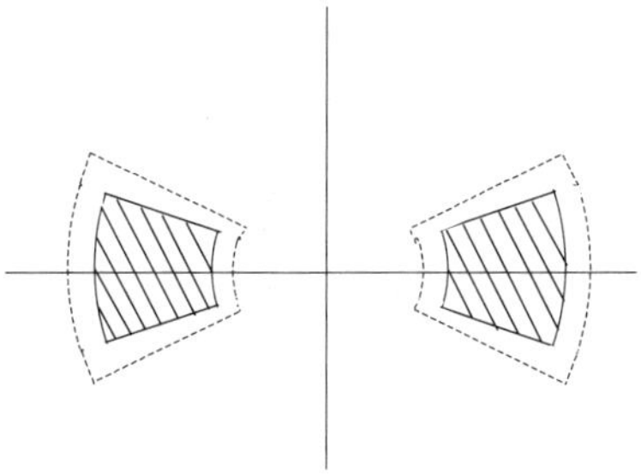

FIGURE 2. Well-localized directional wavelet: Q is the interior of the two regions delineated by continuous lines, and Q_ε is the interior of the two regions delimited by the dashed lines.

Very nice constructions of smooth directional wavelet frames on regular grids were obtained before in [**AHNV01, ADH$^+$03**].

(3) *Spiral wavelet: [ACM03]* In this example we will define a dilation A such that its powers A^jQ applied to an annulus sector Q covers $\mathbb{R}^2$ by spiral annulus sectors.

Let $a, b > 1$, and Γ the spiral curve defined by

$$\Gamma(t) = (a^t \cos(bt), a^t \sin(bt)) \quad t \in \mathbb{R}.$$

For $\alpha \in \mathbb{R}$ define R_α to be the rotation of angle α: $R_\alpha = \left[\begin{smallmatrix} \cos\alpha & -\sin\alpha \\ \sin\alpha & \cos\alpha \end{smallmatrix}\right]$. The curve Γ satisfies:

$$\Gamma(t + \alpha) = a^\alpha R_{b\alpha} \Gamma(t).$$

Note that for positive α the matrix $A = a^\alpha R_{b\alpha}$ is expansive.

Now we are ready to define the covering elements. Set $b = 2\pi$ and $\alpha = \frac{1}{m}$, for some integer $m \geq 2$ so that $A^m = aI_d$. Define the spiral annulus sector $Q = \{x \in \mathbb{R}^2 : x = \lambda\Gamma(\beta),\ 1 \leq \lambda \leq a,\ 0 \leq \beta \leq \alpha\}$ (see Figure 3). So Q is compact and $\{A^jQ : j \in \mathbb{Z}\}$ is a disjoint covering of $\mathbb{R}^d \setminus \{0\}$.

Choose $\varepsilon > 0$ sufficiently small and h a smooth function that does not vanish in Q and with support in Q_ε. Define $\hat{\psi} = h$. Select a set $X = \{x_k\}_{k\in\mathbb{Z}} \subset \mathbb{R}^2$ such that $\rho(X) < \frac{1}{2\mathrm{diam}(Q)}$.

The set $\{a^{j/m}\psi(a^{j/m}R_{-2\pi j/m}x - x_k) : k \in \mathbb{Z},\ j \in \mathbb{Z}\}$ is a wavelet frame of $L^2(\mathbb{R}^2)$ generated by a single wavelet ψ that is band-limited, with good decay and directional in frequency.

REMARKS.

(1) Obviously, all the constructions above can be generalized to $\mathbb{R}^d$ for any dimension $d > 2$.

(2) Some of the wavelet frames may be associated with MRAs. For example, the so called Shannon wavelet frame constructed above, is associated with the Shannon MRA $V_j = \{f \in L^2(\mathbb{R}) : \mathrm{Supp}\,\hat{f} \subset [-2^{-j-1}, 2^{-j-1}]\}$, $j \in \mathbb{Z}$. In general however, the precise relation needs further investigation.

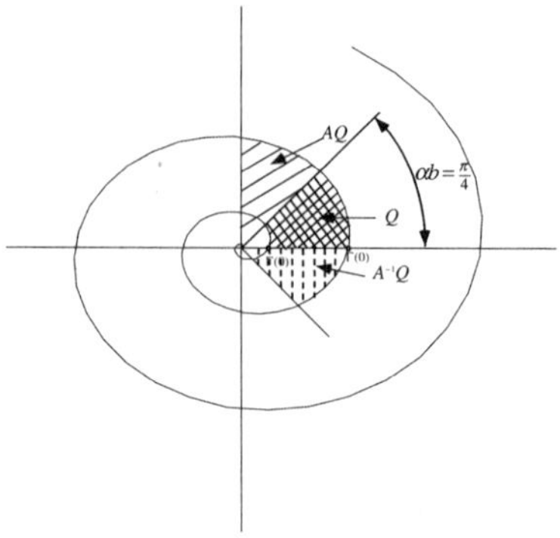

FIGURE 3. Spiral wavelet frames.

3. General Irregular Systems

From the last section we can see that there are three common ingredients in all the examples that we considered.

(1) A covering of the space $\mathbb{R}^d$, whose covering elements are obtained by the action of a countable family of matrices A_j on a basic covering element Q of $\mathbb{R}^d$.
(2) A local frame of non-harmonic exponentials controlled by the Beurling density (in dimension $d = 1$), or the Beurling gap (in dimensions $d \geq 2$) in each covering element.
(3) A smoothing gluing element: the function h with support in Q_ε.

The combination of these three elements produces a wavelet frame of $L^2(\mathbb{R}^d)$ with good decay.

A simple observation with far reaching consequences is the fact that these ideas can be further developed to obtain quite more general constructions of frames of $L^2(\mathbb{R}^d)$. These general constructions include as particular examples irregular wavelet frames and irregular Gabor frames. One nice property of these more general systems of functions is that the frame-elements can still be chosen having good decay in time, or space.

In what follows we will briefly explain the general idea of how this construction works. We will state some of the more relevant results for this exposition and refer the reader to [**ACM03**] for proofs and the more abstract theory.

In order to understand the heart of the problem, we first pose the following question: Assume that $G = \{g_j\}_{j \in J}$ is a countable family of functions in $L^2(\mathbb{R}^d)$ and let $X = \{x_{j,k} \in \mathbb{R}^d : j \in J,\ k \in K\}$ be a double indexed countable set in $\mathbb{R}^d$. Consider the system $\mathcal{G}(G, X) = \{g_j(x - x_{j,k}) : j \in J,\ k \in K\}$ associated to the pair (G, X). The fundamental question is for which pairs (G, X) the system $\mathcal{G}(G, X)$ is a frame of $L^2(\mathbb{R}^d)$. Guido Weiss and his group obtained a characterization of these systems for the case of tight frames on regular grids. The Wavelet set systems mentioned in the introduction, are also particular cases of the general systems $\mathcal{G}(G, X)$.

The general case seems to be much more difficult and probably requires other proof techniques. Our results will produce a substantial part of these systems and provide methods to specifically construct them.

Our method can be described as follows: We start by choosing a completely arbitrary covering of $\mathbb{R}^d$ by bounded sets. This covering is the first ingredient in the description above. Now we choose for the jth covering element a non-harmonic Fourier frame by choosing first a separated family $X_j = \{x_{j,k} \in \mathbb{R}^d : k \in K\}$. The gap of this family should be small enough, in order that $\{e_{x_{j,k}}\}_{k\in K}$ form a frame for the functions supported in a ball centered at the origin, whose diameter is at least equal to the diameter of the considered covering element (or twice the diameter if we are interested in real valued frames or wavelets). This guarantees that this set of exponentials form a local frame for the functions supported in the covering element.

Finally our third ingredient are functions $h_j, j \in J$ that we choose to be C^r, with support included in the ε expanded j^{th} element of the covering, and bounded away from zero on the element itself. We will also need this family to be uniformly bounded. So, if the covering has finite index, we can conclude that the system $\{h_j e_{x_{jk}}\}_{j\in J,k\in K}$ is a frame of $L^2(\mathbb{R}^d)$ provided that the frame bounds for the exponentials are uniformly bounded below and above.

With the construction above, if $\hat{g}_j = h_j$, then the system $\mathcal{G}(G, X) = \{g_j(x - x_{j,k}) : j \in J,\ k \in K\}$ is a frame of $L^2(\mathbb{R}^d)$ where $X = \bigcup_j X_j$. So we have the following theorem:

THEOREM 3.1 (General-Systems). *Let $\{Q_j : j \in J\}$ be a covering of $\mathbb{R}^d$ by bounded sets with finite covering index. Let $h_j \in L^2(\mathbb{R}^d)$ such that* $\operatorname{Supp}(h_j) \subset Q_{\varepsilon_j}$, *for some $\varepsilon_j > 0$ and $h_j \in C^r$ with $h_j \neq 0$ in Q_j and $|h_j| < c$ for all j. Let $X_j = \{x_{j,k} : k \in K\}$ be a separated set with gap $\rho(X_j) < \frac{1}{4\delta_j}$ where $\delta_j = \operatorname{diameter}(Q_j)$ and such that the frame bounds of the frame sequence $\{e_{x_{j,k}}\chi_{Q_j}\}_{k\in K}$ are uniformly bounded in J. If $X = \bigcup_j X_j$ and $G = \{g_j =^\vee h_j : j \in J\}$, then the system $\mathcal{G}(G, X)$ is a frame of $L^2(\mathbb{R}^d)$.*

4. Concluding remarks

The results of Theorem 3.1 can be extended in order to obtain atomic decompositions of the space $L^2(\mathbb{R}^d)$ with smooth well-localized frame atoms. The fact that we concentrated on local Fourier frames is to obtain frames of $L^2(\mathbb{R}^d)$ that are localized on irregular grids, e.g., wavelet frames on irregular grids. The general theory allows the construction of general atomic decompositions by pasting together arbitrary local decompositions associated to a covering of the space as is detailed in [**ACM03**].

References

[ACM03] A. Aldroubi, C. Cabrelli and U. Molter. *Wavelets on irregular grids with arbitrary dilation matrices, and frame atoms for $L^2(\mathbb{R}^d)$* preprint 2003.

[ADH+03] R. Ashino, S. J. Desjardins, C. Heil, M. Nagase, and R. Vaillancourt, *Smooth tight frame wavelets and image analysis in Fourier space*, Comp. Math. Applic. **45** (2003), 1551 – 1579.

[AHNV01] R. Ashino, C. Heil, M. Nagase, and R. Vaillancourt, *Microlocal filtering with multiwavelets*, Computers Math. Applic. **41** (2001), 111 – 133.

[BMM99] L. Baggett, H. Medina, and K. Merrill *Generalized multi-resolution analyses and a construction procedure for all wavelet sets in* $\mathbf{R}^n$, J. Fourier Anal. Appl. **5** (1999), 563–573.

[Bal97] R. Balan, *Stability theorems for Fourier frames and wavelet Riesz bases*, J. Fourier Anal. Appl. **3** (1997), no. 5, 499–504, Dedicated to the memory of Richard J. Duffin.

[BCHL03] R. Balan, P. G. Casazza, C. Heil, and Z. Landau, *Excesses of Gabor frames*, Appl. Comput. Harmon. Anal., **14** (2003), pp. 87–106.

[BL99] J. J. Benedetto and M. Leon, *The construction of multiple diadic minimally supported frequency wavelets on* R^d, AMS Contemp. Math. **247** (1999), 43 – 74.

[BL01] J. J. Benedetto and M. Leon, *The construction of single wavelets in d-dimensions*, J. Geometric Analysis (2001), 1–15.

[BS03] J. J. Benedetto and S. Sumetkijakan, *Tight frames and geometric properties of wavelet sets*, preprint, 2003.

[BW99] J. J. Benedetto and H.-C. Wu, *Non-uniform sampling and spiral MRI reconstruction*, Proceedings of SampTA 1999, Trondheim, Norway (1999).

[Beu66] A. Beurling, *Local harmonic analysis with some applications to differential operators*, Proc. Ann. Science Conf., Belfer Grad. School of Science, (1966), 109 – 125.

[CHS] C. K. Chui, W. He, and J. Stöckler, *Compactly supported tight and sibling frames with maximal vanishing moments*, Appl. Comput. Harmon. Anal., to appear.

[CHSS03] C. K. Chui, W. He, J. Stöckler and Q. Sun, *Compactly supported tight affine frames with integer dilations and maximum vanishing momemnts*, Advances Computational Math., Special Issue on frames, **18** (2003), no. 2, 159–187.

[CS00] C. K. Chui and X. Shi *Orthonormal wavelets and tight frames with arbitrary real dilations*, Appl. Comput. Harmon. Anal. **17** (2000), 243–264.

[Chr96] O. Christensen, *Moment problems and stability results for frames with applications to irregular sampling and Gabor frames*, Appl. Comput. Harmon. Anal. **3** (1996), no. 1, 82–86.

[Chr97] ———, *Perturbation of operators and applications to frame theory*, J. of Fourier Anal. Appl. **3** (1997), 543–557.

[CDH99] O. Christensen, B. Deng, and C. Heil, *Density of Gabor frames*, Appl. Comput. Harmon. Anal. **7** (1999), no. 3, 292–304.

[CH97] O. Christensen and C. Heil, *Perturbations of Banach frames and atomic decompositions*, Math. Nachr., **185** (1997), pp. 33–47.

[DLS97] Xingde Dai, David R. Larson, and Darrin M. Speegle, *Wavelet sets in* $\mathbf{R}^n$, J. Fourier Anal. Appl. **3** (1997), no. 4, 451–456.

[DLS98] ———, *Wavelet sets in* $\mathbf{R}^n$*. II*, Wavelets, multiwavelets, and their applications (San Diego, CA, 1997), Amer. Math. Soc., Providence, RI, 1998, pp. 15–40.

[DGM86] I. Daubechies, A. Grossmann, and Y. Meyer, *Painless nonorthogonal expansions*, J. Math. Phys. **27** (1986), no. 5, 1271–1283.

[FG89] H. G. Feichtinger and K. Gröchenig, *Banach spaces related to integrable group representations and their atomic decomposition*, Journal of Functional Analysis (1989), 307–340.

[FZ95] S. J. Favier and R. A. Zalik, *On the stability of frames and Riesz bases*, Appl. Comput. Harmon. Anal. **2** (1995), no. 2, 160–173.

[Fei87] H. G. Feichtinger, *Banach spaces of distributions defined by decomposition methods II*, Math. Nachr. **132** (1987), 207–237.

[FW01] H. G. Feichtinger and T. Werther, *Atomic systems for subspaces*, Proceedings SampTA 2001 (L.Zayed, ed.), Orlando, FL, 2001, pp. 163–165.

[Grö91] K. Gröchenig, *Describing functions: atomic decompositions versus frames*, Monatsh. Math. **112** (1991), no. 1, 1–42.

[Grö93] ———, *Irregular sampling of wavelet and short-time Fourier transforms*, Constr. Approx. **9** (1993), no. 2-3, 283–297.

[HL00] D. Han and D. Larson, *Frames, bases and group representations*, Memoirs of the American Mathematical Society, **147** (2000).

[HK03] C. Heil and G. Kutyniok, *Density of weighted wavelet frames*, J. Geometric Analysis, **13** (2003), pp. 479–493.

[HW89] C. Heil and D. Walnut, *Continuous and discrete wavelet transforms*, SIAM Reviews **31** (1989), 628–666.

[HDW02] E. Hernández, D. Labate, and G. Weiss, *A unified characterization of reproducing systems generated by a finite family ii*, J. Geom. Anal. **12** (2002), no. 4, 615–662.

[HDW03] E. Hernández, D. Labate, G. Weiss, and E. Wilson, *Oversampling quasi affine frames and wave packets*, to appear, 2003.

[Lab02] D. Labate, *A unified characterization of reproducing systems generated by a finite family*, J. Geom. Anal. **12** (2002), no. 3, 469 – 491.

[Ola03] G. Ólafsson, Continuous action of Lie groups on $\mathbb{R}^n$ and frames, preprint 2003.

[OS03] G. Ólafsson and D. Speegle, Groups, Wavelets and wavelet sets, preprint 2003.

[OS92] P. A. Olsen and K. Seip, *A note on irregular discrete wavelet transforms*, IEEE Trans. Inform. Theory **38** (1992), no. 2, part 2, 861–863.

[PGKKH1] M. Papadakis, G. Gogoshin, I. A. Kakadiaris, D. J. Kouri, and D. K. Hoffman, *Nonseparable radial frame multiresolution analysis in multidimensions*, Numer. Funct. Anal. Optimiz., to appear.

[PGKKH2] M. Papadakis, G. Gogoshin, I. A. Kakadiaris, D. J. Kouri, and D. K. Hoffman, *Nonseparable radial frame multiresolution analysis in multidimensions and isotropic fast wavelet algorithms*, SPIE-Wavelet X, to appear.

[RS95] J. Ramanathan and T. Steger, *Incompleteness of sparse coherent states*, Appl. Comput. Harmon. Anal. **2** (1995), no. 2, 148–153.

[SZ00] W. Sun and X. Zhou, *Irregular wavelet frames*, Sci. China Ser. A **43** (2000), no. 2, 122–127.

[SZ01] ———, *On the stability of Gabor frames*, Adv. in Appl. Math. **26** (2001), no. 3, 181–191.

[SZ02] ———, *Irregular wavelet/Gabor frames*, Appl. Comput. Harmon. Anal. **13** (2002), 63–76.

[SZ03] ———, *Irregular Gabor frames and their stability*, Proc. Amer. Math. Soc., **131** (2003), no. 9, 2883 – 2893.

[SZ03] ———, *Density of irregular wavelet frames*, Proc. Amer. Math. Soc., (2003), to appear.

[Wan02] Y. Wang, *Wavelets, tiling, and spectral sets*, Duke Math. J. **114** (2002), no. 1, 43–57.

(Akram Aldroubi) DEPARTMENT OF MATHEMATICS, VANDERBILT UNIVERSITY, 1326 STEVENSON CENTER, NASHVILLE, TN 37240

E-mail address, Akram Aldroubi: `aldroubi@math.vanderbilt.edu`

(Carlos Cabrelli and Ursula M. Molter) DEPARTAMENTO DE MATEMÁTICA, FACULTAD DE CIENCIAS EXACTAS Y NATURALES, UNIVERSIDAD DE BUENOS AIRES, CIUDAD UNIVERSITARIA, PABELLÓN I, 1428 CAPITAL FEDERAL, ARGENTINA, AND CONICET, ARGENTINA

E-mail address, Carlos Cabrelli: `cabrelli@dm.uba.ar`

E-mail address, Ursula M. Molter: `umolter@dm.uba.ar`

Contemporary Mathematics
Volume **345**, 2004

An analogue of Bratteli-Jorgensen loop group actions for GMRA's

L. W. Baggett, P. E. T. Jorgensen, K. D. Merrill, and J. A. Packer

ABSTRACT. Several years ago, O. Bratelli and P. Jorgensen developed the concept of m-systems of filters for dilation by a positive integer $N > 1$ on $L^2(\mathbb{R})$. They constructed a loop group action on m-systems. By work of Mallat and Meyer, these m-systems are important in constructing multi-resolution analyses and wavelets associated to dilation by N and translation by $\mathbb{Z}$ on $L^2(\mathbb{R})$. In this paper, we discuss an extension of this loop-group construction to generalized filter systems, which we will call "M-systems", associated with generalized multiresolution analyses. In particular, we show that every multiplicity function has an associated generalized loop group which acts freely and transitively on the set of M-systems corresponding to the multiplicity function. The results of Bratteli and Jorgensen correspond to the case where the multiplicity function is identically equal to 1.

1. Introduction

In work that first appeared in the late 1990's and has since been elaborated upon in their book [**BJ**], O. Bratteli and P. Jorgensen related filter functions corresponding to a multiresolution analysis for dilation by an integer $N > 1$ to representations of the Cuntz algebras. Brattelli and Jorgensen obtained results about frame wavelet families and orthonormal wavelet families arising from multiresolution analyses using techniques from operator theory, and in addition they were able to construct a free and transitive action of a "loop group" on "Lip_1 m-systems", consisting of families of Hölder-continuous filter functions for dilation by N.

At the same time, L. Baggett, H. Medina and K. Merrill developed a theory of generalized multiresolution analyses for dilation by N [**BMM**], which we abbreviate here using the acronym "GMRA". Somewhat later, in [**BCM**], L. Baggett, J. Courter and K. Merrill associated to these objects generalized filter functions, which in this paper we will generalize to the concept of "M- systems". Under appropriate conditions, Baggett, Courter and Merrill were able to construct a GMRA and an associated frame wavelet system by using the M-systems. In both of these

---

2000 *Mathematics Subject Classification.* Primary 54C40, 14E20; Secondary 46E25, 20C20.
Key words and phrases. Wavelet, multiresolution analysis, frame, loop group.
The first two named authors were supported by a US-NSF Focused Research Group (FRG) grant.

papers, the so-called "multiplicity function" μ associated to a GMRA and the "consistency equation" that the multiplicity function satisfies turned out to be of great importance in the proofs of theorems, and that will be the case here as well.

In this paper, we will review the loop group action for "classical m-systems." We will then define the notion of M-systems corresponding to a fixed multiplicity function μ. The M-systems turn out to be a set of Borel cross-sections of a Borel vector bundle over a certain subset of $\bigsqcup_{i=1}^{c} \mathbb{T}$, where c the essential supremum of μ. The generalized filter systems of Baggett, Courter and Merrill (see [**BCM**])turn out to be prime examples of M-systems, and of course motivate the general definition. Our main result is the construction of a generalized loop group corresponding to a multiplicity function μ. We describe this group here as the group of sections of a group bundle over $\mathbb{T}$. We will show that this generalized loop group acts freely and transitively on the set of M-systems.

For the purposes of this article, we will for the most part regard a family of M-systems corresponding to a given multiplicity function as an abstract set which is acted on by the loop group. However, our ultimate aim, which we plan to present in a later work, is to use the operator approach pioneered by Bratteli and Jorgensen in the classical case to weaken the conditions given in [**BCM**] under which an M-system gives rise to a GMRA, and hence to a normalized tight frame wavelet family in $L^2(\mathbb{R})$ for dilation by N. This would provide another approach to the construction of normalized tight frame wavelets.

2. Preliminaries: the classical case, and a loop group action on m-systems

We first review the notion of a normalized tight frame in a Hilbert space.

DEFINITION 2.1. A sequence $\{ x_n : n \in \mathbb{N} \}$ of elements in a Hilbert space $\mathcal{H}$ is said to be a *frame* if there are real constants $C, D > 0$ such that

$$C \cdot \|x\|^2 \leq \sum_{n=1}^{\infty} |\langle x, x_j \rangle|^2 \leq D \cdot \|x\|^2 \tag{2.1}$$

for every $x \in \mathcal{H}$. If $C = D = 1$, we say the set $\{ x_n : n \in \mathbb{N} \}$ is a *normalized tight frame* (abbreviated NTF) for $\mathcal{H}$.

It is possible to show that a subset $\{ x_n : n \in \mathbb{N} \}$ of a Hilbert space $\mathcal{H}$ is a NTF for $\mathcal{H}$ if and only if for every $x \in \mathcal{H}$ the following reconstruction formula is satisfied:

$$x = \sum_{n=1}^{\infty} \langle x, x_n \rangle x_n. \tag{2.2}$$

Any orthonormal basis for $\mathcal{H}$ is a normalized tight frame, but not conversely, as frames can have redundancy which does not occur in the orthonormal basis case.

We now move on to the case where $\mathcal{H} = L^2(\mathbb{R})$ and review the definition of a normalized tight frame wavelet family in this context. Fix an integer $N > 1$. In the standard way, we define dilation and translation operators on $L^2(\mathbb{R})$ by

$$D(f)(t) = \sqrt{N} f(Nt),$$

$$T(f)(t) = f(t-1), \; f \in L^2(\mathbb{R}).$$

As this paper is mainly concerned with low and high pass filters, it is natural to define orthonormal wavelets and tight frame wavelets in the frequency domain. Let $\mathcal{F}$ denote the Fourier transform on $L^2(\mathbb{R})$, defined by

$$\mathcal{F}(f)(x) = \int_{\mathbb{R}} f(t)e^{-2\pi ixt}dt.$$

Set

$$\widehat{D} = \mathcal{F}D\mathcal{F}^*, \ \widehat{T} = \mathcal{F}T\mathcal{F}^*.$$

Then

$$\widehat{D}(f)(x) = (\sqrt{N})^{-1}f(N^{-1}x),$$

and

$$\widehat{T}(f)(x) = e^{-2\pi ix}f(x), \ f \in L^2(\mathbb{R}).$$

DEFINITION 2.2. A *normalized tight frame wavelet family in the frequency domain* for dilation by $N > 1$ is a subset $\{\Psi_1, \ldots, \Psi_m\} \subseteq L^2(\mathbb{R})$ such that

$$\{\,\widehat{D}^j\widehat{T}^v(\Psi_i) : 1 \leq i \leq m,\ j \in \mathbb{Z},\ v \in \mathbb{Z}\,\}$$

is an NTF for $L^2(\mathbb{R})$. If the $\{\,\widehat{D}^j\widehat{T}^v(\Psi_i) : 1 \leq i \leq m,\ j \in \mathbb{Z},\ v \in \mathbb{Z}\}$ form an orthonormal basis for $L^2(\mathbb{R})$, the family $\{\Psi_1, \ldots, \Psi_m\}$ is called an *orthonormal wavelet family* for dilation by N.

A fundamental algorithm for constructing wavelet families is the concept of multiresolution analysis (MRA) developed by S. Mallat and Y. Meyer [**Ma**], and key tools for constructing the MRA's are filter functions for dilation by N.

DEFINITION 2.3. Let N be a positive integer greater than 1. A *low-pass filter* m_0 for dilation by N is a $\mathbb{Z}$-periodic function $m_0 \colon \mathbb{R} \to \mathbb{C}$ which satisfies the following conditions:

(i) $m_0(0) = \sqrt{N}$ ("low-pass condition")
(ii) $\sum_{l=0}^{N-1} |m_0(x + l/N)|^2 = N$ a.e.;
(iii) m_0 is Hölder continuous at 0;
(iv) m_0 is non-zero in a compact neighborhood K of 0 which satisfies $\frac{1}{N}K \subseteq K$ and $K + \frac{1}{N}\mathbb{Z} = \mathbb{R}$ ("Cohen's condition", cf. [**Ch**]).

Sometimes in the above definition, condition (iv), Cohen's condition, is dropped.

Given a low-pass filter m_0 for dilation by N, then there is a canonical way to construct a "scaling function", or "father wavelet" associated to the filter. We set

$$\Phi(x) = \prod_{i=1}^{\infty}[\frac{m_0(N^{-i}x)}{\sqrt{N}}].$$

Then Φ converges a.e. and is an element of $L^2(\mathbb{R})$. We call Φ a *scaling function in the frequency domain* for dilation by N.

To use the low-pass filter and the associated scaling function to construct a wavelet family for dilation by N, we need $N - 1$ other functions that are high pass filters in the sense of following definition.

DEFINITION 2.4. Let N be a positive integer greater than 1, and let m_0 be a low-pass filter for dilation by N. A set of essentially bounded measurable $\mathbb{Z}$ periodic functions $m_1, m_2, \ldots, m_{N-1}$ defined on $\mathbb{R}$ are called *high-pass filters* associated to m_0, if

$$\sum_{l=0}^{N-1} \overline{m_i(x+\frac{l}{N})}\, m_j(x+\frac{l}{N}) = \delta_{i,j} N,\ 0 \leq i,j \leq N-1.$$

We can express the filter conditions more concisely as follows: the $N \times N$ matrix-valued function on $x \in \mathbb{R}/\mathbb{Z}$

$$x \mapsto \left(\frac{m_j(x+\frac{l}{N})}{\sqrt{N}}\right)_{0\leq j,l\leq N-1} \tag{2.3}$$

is a unitary matrix-valued function $\forall x \in \mathbb{T}$.

DEFINITION 2.5. [**BJ**] Let m_0 be a Lip_1 (i.e., Hölder continuous) low-pass filter for dilation by N, which need not satisfy Cohen's condition. Suppose that $m_1, m_2, \ldots, m_{N-1}$ are a collection of $N-1$ Lip_1 high-pass filters associated to m_0. The N-tuple $(m_0, m_1, \ldots, m_{N-1})$ is called a *classical* Lip_1 *m-system* for dilation by N.

THEOREM 2.6. ([**Ma**], [**Me**], [**L**], [**BJ**]) *Let N be a positive integer greater than 1, let $(m_0, m_1, \ldots, m_{N-1})$ be a classical m-system for dilation by N, and let Φ be the scaling function constructed from m_0 as above. Then*

$$\{\Psi_1 = \widehat{D}(m_1\Phi),\ \Psi_2 = \widehat{D}(m_2\Phi),\ \ldots,\ \Psi_{N-1} = \widehat{D}(m_{N-1}\Phi)\}$$

is an NTF wavelet family in the frequency domain for dilation by N. If Cohen's condition is satisfied, the $\{\Psi_k\}$'s form an orthonormal wavelet family.

REMARK 2.7. In their proof of this result, Bratteli and Jorgensen used a representation of the Cuntz algebra $\mathcal{O}_N$ arising from the m-system.

So we see that classical m-systems are very useful in the construction of NTF wavelet families.

In their book [**BJ**], Bratteli and Jorgensen constructed a so-called "loop group" which acts freely and transitively on the set of m-systems. We modify the conventions slightly in the definition below, as it will be more easy to adapt to the GMRA M-system case.

DEFINITION 2.8. Fix a positive integer $N > 1$. We say a matrix valued function $K\colon \mathbb{R}/\mathbb{Z} :\to U(N,\mathbb{C})$ is $\mathrm{Lip}_1(N)$ if

(i) Each component function $k_{i,j}$ is Lip_1, $0 \leq i,j \leq N-1$,
(ii) $K(0) = \mathrm{Id}_{N\times N}$. We denote the set of all $\mathrm{Lip}_1(N)$ matrix-valued functions by $\mathcal{G}_N$; under pointwise operations $\mathcal{G}_N$ has a group structure, and we call it the *N-dimensional loop group*.

Then the main result of Bratteli and Jorgensen about the loop group is as follows:

THEOREM 2.9. [**BJ**] *The loop group $\mathcal{G}_N$ acts freely and transitively on* Lip_1 *m-systems for dilation by N via the formula*

$$[K\cdot(m_0, m_1, \ldots, m_{N-1})^T](x) = K(Nx)(m_0(x), m_1(x), \ldots, m_{N-1}(x))^T.$$

PROOF. We sketch the proof of this result for completeness. We first prove transitivity. If $(m_0, m_1, \ldots, m_{N-1})$ and $(\widetilde{m}_0, \widetilde{m}_1, \ldots, \widetilde{m}_{N-1})$ are two Lip_1 m-systems for dilation by N, then defining $K(x) = (k_{i,j}(x))$ by

$$k_{i,j}(x) = \frac{1}{N}\sum_{l=0}^{N-1} \widetilde{m}_i(\frac{x+l}{N})\,\overline{m_j(\frac{x+l}{N})},$$

one calculates that

$$\begin{aligned}
&[K(Nx)(m_0,(x)m_1(x),\ldots,m_{N-1}(x))^T]_i \\
&= \sum_{j=0}^{N-1} k_{i,j}(Nx)m_j(x) \\
&= \sum_{j=0}^{N-1}\frac{1}{N}\sum_{l=0}^{N-1}\widetilde{m}_i(\frac{Nx+l}{N})\,\overline{m_j(\frac{Nx+l}{N})}\,m_j(x) \\
&= \sum_{l=0}^{N-1}\widetilde{m}_i(x+\frac{l}{N})[\frac{1}{N}\sum_{j=0}^{N-1}\overline{m_j(x+\frac{l}{N})}\,m_j(x)] \\
&= \sum_{l=0}^{N-1}\widetilde{m}_i(x+\frac{l}{N})\delta_{l,0} = \widetilde{m}_i(x),
\end{aligned}$$

by the conditions on the filter systems coming from the fact that the matrix described in (2.3) is unitary. Thus

$$K\cdot(m_0(x), m_1(x), \ldots, m_{N-1}(x))^T = [(\widetilde{m}_0, \widetilde{m}_1, \ldots, \widetilde{m}_{N-1})(x)]^T,$$

and $\mathcal{G}_N$ acts freely on the set of m-systems.

As for freeness, let $(m_0, m_1, \ldots, m_{N-1})$ be a m-system. Recall that we can associate to this m-system the $N\times N$ matrix whose entries are in $\mathrm{Lip}_1(\mathbb{T})$ which is unitary $\forall x\in\mathbb{T}$ given by

$$\mathcal{M}(x) = (\frac{m_i(x+\frac{j}{N})}{\sqrt{N}})_{0\le i,j\le N-1}. \tag{2.4}$$

Suppose that there exists $K\in\mathcal{G}_N$ satisfying

$$K\cdot[m_0(x), m_1(x), \ldots, m_{N-1}(x)]^T = [m_0(x), m_1(x), \ldots, m_{N-1}(x)]^T,$$

for some m system $(m_0, m_1, \ldots, m_{N-1})$, for all $x\in\mathbb{T}$. But then for all $x\in\mathbb{T}$ and for $0\le j\le N-1$ we have

$$\begin{aligned}
&K\cdot[m_0(x+\frac{j}{N}), m_1(x+\frac{j}{N}), \ldots, m_{N-1}(x+\frac{j}{N})]^T \\
&\qquad = [m_0(x+\frac{j}{N}), m_1(x+\frac{j}{N}), \ldots, m_{N-1}(x+\frac{j}{N})]^T,
\end{aligned}$$

i.e.,

$$\begin{aligned}
&K(N(x+\frac{j}{N}))[m_0(x+\frac{j}{N}), m_1(x+\frac{j}{N}), \ldots, m_{N-1}(x+\frac{j}{N})]^T \\
&\qquad = [m_0(x+\frac{j}{N}), m_1(x+\frac{j}{N}), \ldots, m_{N-1}(x+\frac{j}{N})]^T,
\end{aligned}$$

for $0 \leq j \leq N-1$. But, since the entries of K are $\mathbb{Z}$-periodic, this means (in terms of matrix multiplication) exactly that

$$K(Nx)\mathcal{M}(x) = \mathcal{M}(x), \ \forall x \in \mathbb{T},$$

where $\mathcal{M}(x) = (m_i(x+j/N)/\sqrt{N})_{0\leq i,k\leq N-1}$ is the unitary-valued matrix function defined in (2.4). We thus obtain $K(Nx) = \mathcal{M}(x)\mathcal{M}(x)^* \ \forall x \in \mathbb{T}$. Thus $K(x)$ is the identity matrix for every $x \in \mathbb{T}$, as we desired to show. □

3. Generalized multiresolution analyses and generalized filter systems

Not all wavelet families come come from MRA's and m-systems. There are various ways to approach the general case, cf. [**BL**], [**HLPS**]. In 1999, L. Baggett, H. Medina and K. Merrill developed the theory of generalized multiresolution analyses (GMRA's) to deal with this more general setting [**BMM**]. Again to simplify matters we limit our definition to the frequency domain for now.

DEFINITION 3.1. [**BMM**] A *generalized multiresolution analysis* (GMRA) for dilation by N is a sequence $\{\widehat{V_i}\}_{i\in\mathbb{Z}}$ of closed subspaces of $L^2(\mathbb{R})$ satisfying the following conditions:

(i) $\cdots \subseteq \widehat{V_{-1}} \subseteq \widehat{V_0} \subseteq \widehat{V_1} \subseteq \cdots$ (the $\widehat{V_i}$ are nested)
(ii) $(\widehat{D})^i(\widehat{V_0}) = \widehat{V_i}$, $i \in \mathbb{Z}$;
(iii) $\overline{\bigcup_{i\in\mathbb{Z}} \widehat{V_i}} = L^2(\mathbb{R})$, $\bigcap_{i\in\mathbb{Z}} \widehat{V_i} = \{0\}$;
(iv) $\widehat{V_0}$ is invariant under all powers of $\widehat{T}$.

Baggett, Merrill and Medina were able to develop several characteristic invariants associated to a GMRA using spectral theory. We summarize their key results in the following theorem.

THEOREM 3.2. [**BMM**] *Given a GMRA in $L^2(\mathbb{R})$ corresponding to dilation $\widehat{D}$ by N and the transform of integer translation $\widehat{T}$, there is a unique sequence of Borel subsets $S_1 \supseteq S_2 \supseteq \cdots$ of $\mathbb{T}$ and a unitary operator $J\colon \widehat{V_0} \to \bigoplus_j L^2(S_j)$ such that*

$$[J(\widehat{T}(f)]_j(x) = e^{2\pi i x}[J(f)]_j(x).$$

The function $\mu(x) = \sum_j \chi_{S_j}(x)$ defined on $\mathbb{T}$ is called the multiplicity function *corresponding to the GMRA $\{\widehat{V_i}\}_{i\in\mathbb{Z}}$, and satisfies*

$$\mu(x) \leq \sum_{l=0}^{N-1} \mu(\frac{x+l}{N}) \ a.e..$$

If μ is essentially bounded, with $c = \operatorname{ess\,sup} \mu(x)$, it is possible to define the conjugate multiplicity function

$$\tilde{\mu}(x) = \sum_{l=0}^{N-1} \mu(\frac{x+l}{N}) - \mu(x)$$

By definition, μ and $\tilde{\mu}$ satisfy the so-called "consistency equation":

$$\mu(x) + \tilde{\mu}(x) = \sum_{l=0}^{N-1} \mu(\frac{x+l}{N}). \tag{3.1}$$

We remark that the case $\mu \equiv 1$ corresponds to the MRA case, and the case $\tilde{\mu} \equiv 1$ corresponds to the case where there is a single orthonormal wavelet. Note also by the definition of μ that the sets $\{S_i\}_{i=1}^c$ satisfy the following condition:

$$S_i = \{\, x \in \mathbb{T} \mid \mu(x) \geq i \,\}. \tag{3.2}$$

Sets $\widetilde{S}_i$ are defined analogously to the sets S_i by $\widetilde{S}_i = \{\, x \in \mathbb{T} \mid \tilde{\mu}(x) \geq i \,\}$. If $d = \text{ess sup}\, \tilde{\mu}(x)$, we have

$$\widetilde{S}_1 \supseteq \widetilde{S}_2 \supseteq \cdots \supseteq \widetilde{S}_d.$$

Baggett, Courter and Merrill in [**C2**] and [**BCM**] then generalized the Mallat and Meyer algorithm for constructing wavelets from filters to this GMRA setting. They first generalized the concept of low and high pass filters. Suppose μ is a Borel integer-valued function that is essentially bounded by c on $\mathbb{T}$, is constant in a neighborhood of the origin, and satisfies technical conditions that allow it to be a multiplicity function for a GMRA (cf. [**BM**]). They then defined "generalized conjugate mirror filters", which correspond to low pass filters in the classical case, to be functions $\{h_{i,j}\}_{1 \leq i,j \leq c}$, where each $h_{i,j}$ is supported on S_j, and the following orthogonality condition holds:

$$\sum_{j=1}^{c} \sum_{l=0}^{N-1} h_{i,j}(\frac{x+l}{N}) \overline{h_{k,j}(\frac{x+l}{N})} = N\delta_{i,k}\chi_{S_i}(x). \tag{3.3}$$

Similarly, they defined "complementary conjugate mirror filters", an analogue of classical high pass filters, to be functions $\{g_{k,j}\}_{1 \leq k \leq d,\ 1 \leq j \leq c}$, where each $g_{k,j}$ is supported on S_j, and

$$\sum_{j=1}^{c} \sum_{l=0}^{N-1} g_{k,j}(\frac{x+l}{N}) \overline{g_{k',j}(\frac{x+l}{N})} = N\delta_{k,k'}\chi_{\widetilde{S}_k}(x), \tag{3.4}$$

and

$$\sum_{j=1}^{c} \sum_{l=0}^{N-1} h_{i,j}(\frac{x+l}{N}) \overline{g_{k,j}(\frac{x+l}{N})} = 0, \ \forall i, k. \tag{3.5}$$

Examples of functions $g_{k,j}$ and $h_{i,j}$ that satisfy these conditions can be built by an explicit algorithm, and then modified to mimic examples of classical filters [**BCM**].

Under appropriate conditions on these generalized filter functions, Baggett, Courter and Merrill then used them to construct a finite tight frame wavelet family $\{\Psi_1, \ldots, \Psi_d\} \subseteq L^2(\mathbb{R})$. They first built generalized scaling functions $\{\Phi_1, \Phi_2, \ldots, \Phi_c\} \subseteq \widehat{V_0}$ using an infinite product construction involving dilates of a matrix with periodizations of $\{h_{i_j}\}$ as entries (cf. Theorem 3.4 [**BCM**]). The $\{\Phi_i\}_{i=1}^c$ appear as the first column in the infinite product matrix.

Given the above notation and construction, they then have:

THEOREM 3.3. [**BCM**] *Let* $\{(h_{i,j}\}_{1 \leq i,j \leq c}$ *and* $\{g_{k,j}\}_{1 \leq k \leq d,\ 1 \leq j \leq c}$ *be generalized filter functions associated to the multiplicity function* μ*, that satisfy appropriate conditions and let* $\{\Phi_1, \Phi_2, \ldots, \Phi_c\} \subseteq \widehat{V_0}$ *be generalized scaling functions constructed as described above. Setting*

$$\Psi_k = \widehat{D}(\sum_{j=1}^{c} g_{k,j}\Phi_j), \ 1 \leq k \leq d,$$

the $\{\, \Psi_k \mid 1 \leq k \leq d \,\}$ *form a NTF wavelet family for dilation by* N.

As mentioned above, our ultimate aim is to use the approach of Bratteli and Jorgensen to weaken the restrictive technical conditions required in this theorem (which we do not make explicit here) to hypotheses more in line with the classical result. To move toward that goal, we now use these generalized filter functions to define precise analogues of the classical high and low pass filters. In the next section, these definitions will allow us to define generalized M systems and develop the loop group action on them.

DEFINITION 3.4. Given a multiplicity function μ that is constant in a neighborhood of the origin, with its associated sequence of sets, $\{ S_i \mid 1 \leq i \leq c\}$, a *generalized low-pass filter* for dilation by N is a collection of functions $\{h_{i,j}\}_{1\leq i,j\leq c}$, that satisfy the following conditions:

(i) $h_{i,j}$ is supported on S_j;
(ii) $h_{i,j}(0) = \sqrt{N}\delta_{1,i}\delta_{1,j}$ ("low-pass condition");
(iii) $\{h_{i,j}\}$ satisfy the orthogonality condition (3.3);
(iv) $h_{i,j}$ is Lip_1 at 0.

DEFINITION 3.5. Let $\{h_{i,j}\}_{1\leq i,j\leq c}$ be a generalized low-pass filter associated with the multiplicity function μ. Suppose that both μ and the conjugate multiplicity function $\tilde{\mu}$ are constant in a neighborhood of the origin. Define the sets $\{ \widetilde{S}_k \mid 1 \leq k \leq d\}$ as above. An associated *generalized high-pass filter* is a collection of functions $\{g_{k,j}\}_{1\leq k\leq d,\ 1\leq j\leq c}$ that satisfy the following conditions:

(i) $g_{k,j}$ is supported on S_j;
(ii) $\{g_{k,j}\}$ satisfy the orthogonality conditions (3.4) and (3.5);
(iii) $g_{k,j}$ is Lip_1 at 0.

4. A generalized loop group action on the generalized filter systems

Let $\{h_{i,j}\}_{1\leq i,j\leq c}$ and $\{g_{k,j}\}_{1\leq k\leq d,\ 1\leq j\leq c}$ be generalized low-pass and high-pass filter functions defined as in the previous section. Since $\bigoplus_{i=1}^c L^2(S_i) \cong L^2(\bigsqcup_{i=1}^c S_i)$, we can suppress the second index of the filter functions and view generalized filter functions as a vector ($c+d$-tuple) of functions:

$$(h_1, h_2, \dots, h_c, g_1, g_2, \dots, g_d) \in [L^2(\bigsqcup_{i=1}^{c} S_i)]^{c+d}.$$

Further, we note that for any fixed x, the output of the vector of functions $(h_1, h_2, \dots, h_c, g_1, g_2, \dots, g_d)$ is actually in $\mathbb{C}^{\mu(Nx)+\tilde{\mu}(Nx)}$, since by (3.3), $h_i(x) = 0$ if $i > \mu(Nx)$ and by (3.4), $g_k(x) = 0$ if $k > \tilde{\mu}(Nx)$.

Before we generalize the Bratteli and Jorgensen definitions of m system, we review the notions of Borel vector bundles over a Borel space and Borel cross-sections for Borel vector bundles.

DEFINITION 4.1. [**FD**] Let X be a topological space. A (finite dimensional) vector bundle over the space X, denoted by (E, p, X), is a topological space E, together with a continuous open surjection $p\colon E \to X$, and operations and norms making each fiber $E_x = p^{-1}(X)$ into a (finite dimensional) vector space, which in addition satisfies the following conditions:

(i) $y \mapsto \|y\|$ is continous from E to $\mathbb{R}$,
(ii) The operation $+$ is continuous as a function from $\{ (y, z) \in E \times E : p(y) = p(z) \}$ to E.

(iii) For each $\lambda \in \mathbb{C}$, the map $y \mapsto \lambda \cdot y$ is continuous from E to E.
(iv) If $x \in X$ and $\{y_i\}$ is any net of elements of E such that $\|y_i\| \to 0$ and $p(y_i) \to x$ in X, then $y_i \to \vec{0} \in E_x$ in E.

A Borel map $s\colon X \to E$ is called a Borel cross-section if $p \circ s(x) = x, \forall x \in X$.

Given an essentially bounded multiplicity function μ on $\mathbb{T}$, let $c = \operatorname{ess\,sup} \mu$, $d = \operatorname{ess\,sup} \tilde{\mu}$, and let $T_j = \{\, x \in \mathbb{T} : \mu(Nx) + \tilde{\mu}(Nx) = j \,\}$, $0 \le j \le c+d$. Set $T_{i,j} = S_i \cap T_j$, $0 \le j \le c+d$; then each $T_{i,j}$ is Borel and $S_i = \bigsqcup_{j=0}^{c+d} T_{i,j}$.

DEFINITION 4.2. Let E be the Borel space given by

$$E = \bigsqcup_{i=1}^{c} \bigsqcup_{j=0}^{c+d} [T_{i,j} \times \mathbb{C}^j].$$

Let $(E, p, \bigsqcup_{i=1}^{c} S_i)$ be the Borel vector bundle where the map $p\colon E \to \bigsqcup_{i=1}^{c} S_i$ is defined by $p(x, \vec{v}) = x$, $(x, \vec{v}) \in T_{i,j} \times \mathbb{C}^j$. By definition, $(E, p, \bigsqcup_{i=1}^{c} S_i)$ is a vector bundle $\bigsqcup_{i=1}^{c} S_i$ whose fiber over $x \in S_i$ is a complex vector space of dimension $\mu(Nx) + \tilde{\mu}(Nx)$. An *$M$-system associated to the multiplicity function μ* is a Borel cross-section $M\colon \bigsqcup_{j=1}^{c} S_j \to E$ of this bundle whose values, $(M_1(x), M_2(x), \ldots, M_{\mu(Nx)+\tilde{\mu}(Nx)})$, are the output of a vector of generalized low and high pass filters.

Note all the information about the generalized filters $\{h_{i,j}\}$ and $\{g_{k,j}\}$ is encoded in the M-system. In particular, for any fixed multiplicity function μ, such that both μ and $\tilde{\mu}$ are constant in a neighborhood of the origin, we have a one-to-one correspondence between M-systems and collections of generalized filter functions as defined in the previous section.

To develop the loop group action on these M-systems, we start by defining an endomorphism $\Pi_N\colon \bigsqcup_{i=1}^{c} S_i \to \mathbb{T}$ by $\Pi_N(x) = Nx \bmod 1$. Each $x \in \mathbb{T}$ has $\sum_{l=0}^{N-1} \mu(\frac{x+l}{N}) = \mu(x) + \tilde{\mu}(x)$ preimages in $\bigsqcup_{i=1}^{c} S_i$. For convenience of notation, we label these preimage maps $r_{(l,j)}$, where $r_{(l,j)}(x) = \frac{x+l}{N} \in S_j \subseteq \bigsqcup_{i=1}^{c} S_i$ for $1 \le j \le \mu(\frac{x+l}{N})$. (Note that this range on j, as l varies from 0 to $N-1$, gives all the preimages, since if $j > \mu(\frac{x+l}{N})$, by definition $\frac{x+l}{N}$ is not an element of S_j.) For each fixed x, we give the pairs (l,j) the lexicographical order, and thus implicitly define a 1-1 map λ_x taking the pairs (l,j) onto the integers from 1 to $\mu(x) + \tilde{\mu}(x)$.

We now construct a unitary group bundle $(F, q, \mathbb{T})$ as follows. For each $j \in \{1, \ldots, c+d\}$, let $Z_j = \{\, x \in \mathbb{T} : \mu(x) + \widetilde{\mu(x)} = j \,\}$. Let

$$\widetilde{E} = \bigsqcup_{j=0}^{c+d} [Z_j \times M_j(\mathbb{C})],$$

where $M_j(\mathbb{C})$ is viewed as a j^2-dimensional normed vector space with the operator norm. Defining $q\colon \widetilde{E} \to \mathbb{T}$ by $q(x, A) = = x)$, for $(x, A) \in E$, we see that $(\widetilde{E}, q, \mathbb{T})$ is a Borel vector bundle in the sense of Definition 4.1. Now let F be the subspace of $\widetilde{E}$ defined by

$$F = \{\, (x, A) \in \widetilde{E} : A \in U(\mu(x) + \tilde{\mu}(x), \mathbb{C}) \,\}.$$

Then $q\colon F \to \mathbb{T}$ is a continuous open surjection, and the fiber $q^{-1}(x)$ of the bundle consists of the group of complex unitary matrices $U(\mu(x) + \tilde{\mu}(x), \mathbb{C})$. Borel cross sections to this group bundle consist of Borel maps $K\colon \mathbb{T} \to F$ such that $q \circ K(x) =$

x. We denote the set of sections of this bundle by $\Gamma(F,q)$. Note $\Gamma(F,q)$ is a group under pointwise operations on $\mathbb{T}$, where the identity element of the group is given by that section whose value at x is equal to $\mathrm{Id}_{\mu(x)+\tilde{\mu}(x)}$.

We are now ready to state a key theorem about M-systems that can be derived from the orthogonality relations:

THEOREM 4.3. *Let $\Gamma(F,q)$ be the group of cross sections of the group bundle associated to the multiplicity function μ defined above. Let $M\colon \bigsqcup_{j=1}^{c} S_j \to E$ be an M-system associated to μ. Then $x \mapsto (K_{i,\lambda_x(l,j)}(x))$, where*

$$K_{i,\lambda_x(l,j)}(x) = \sqrt{\frac{1}{N}}M_i(r_{(l,j)}(x))$$

is an element of $\Gamma(F,q)$.

PROOF. As noted above $1 \le \lambda_x(l,j) \le \mu(x)+\tilde{\mu}(x)$, so for each $x \in \mathbb{T}$, the matrix $(K_{i,\lambda_x(l,j)}(x))$ is a square matrix of the correct dimension. We shall show that for all $x \in \mathbb{T}$, the rows of $(K_{i,\lambda_x(l,j)}(x))$ are orthonormal. We use the orthogonality relations (3.3), (3.4), and (3.5) for generalized filter functions in this proof.

Write K_i for the ith row of $(K_{i,\lambda_x(l,j)}(x))$, and suppose first that $1 \le i \le i' \le \mu(x)$. Then

$$\begin{aligned}
\langle K_i, K_{i'}\rangle &= \sum_{\lambda_x(l,j)=1}^{\mu(x)+\tilde{\mu}(x)} K_{i,\lambda_x(l,j)}(x)\,\overline{K_{i,\lambda_x(l,j)}(x)} \\
&= \sum_{l=0}^{N-1}\sum_{j=1}^{\mu(\frac{x+l}{N})} \sqrt{\frac{1}{N}}M_i(r_{(l,j)}(x))\,\overline{\sqrt{\frac{1}{N}}M_{i'}(r_{(l,j)}(x))} \\
&= \sum_{l=0}^{N-1}\sum_{j=1}^{\mu(\frac{x+l}{N})} \frac{1}{N}h_{i,j}(\frac{x+l}{N})\,\overline{h_{i',j}(\frac{x+l}{N})} \\
&= \frac{1}{N}\sum_{j=1}^{\mu(\frac{x+l}{N})}\sum_{l=0}^{N-1} h_{i,j}(\frac{x+l}{N})\,\overline{h_{i',j}(\frac{x+l}{N})} \\
&= \frac{1}{N}\sum_{j=1}^{c}\sum_{l=0}^{N-1} h_{i,j}(\frac{x+l}{N})\,\overline{h_{i',j}(\frac{x+l}{N})}.
\end{aligned}$$

The last equality is due to the fact that $h_{i,j}(\frac{x+l}{N}) = 0$ for $j > \mu(\frac{x+l}{N})$, since $\frac{x+l}{N} \notin S_j$ in that case. Then by (3.3) we have

$$\langle K_i, K_{i'}\rangle = N\frac{1}{N}\delta_{i,i'}\chi_{S_i}(x) = \delta_{i,i'},$$

where we have noted that $\chi_{S_i}(x) = 1$, since we have $i \le \mu(x)$ and, for those values of i, $x \in S_i$ by the definition of $\mu(x)$.

The cases $\mu(x) < i \le i' \le \mu(x)+\tilde{\mu}(x)$ and $1 \le i \le \mu(x) < i' \le \mu(x)+\tilde{\mu}(x)$ follow from similar arguments using (3.4) and (3.5). Thus we have that in all cases, the rows of $(K_{i,\lambda_x(l,j)}(x))$ are orthonormal, and we have the desired unitary matrix. □

The results of Theorem 4.3 imply that the columns of $(K_{i,\lambda_x(l,j)}(x))$ are orthonormal as well, so we can deduce as a corollary:

COROLLARY 4.4. *Let μ and $\tilde{\mu}$ be multiplicity and "conjugate" multiplicity functions that are constant in a neighborhood of the origin, with related sequences of sets $\{\, S_i \mid 1 \le i \le c \}$ and $\{\, \tilde{S}_k \mid 1 \le k \le d \}$. Suppose $\{h_{i,j}\}_{1 \le i,j \le c}$ and $\{g_{k,j}\}_{1 \le k \le d,\, 1 \le j \le c}$ are generalized low-pass and high-pass filter functions with associated to the multiplicity function. Then for all $x \in \mathbb{T}$, and for all $l, l' \in \{0, 1, \dots, N-1\}$ and $j, j' \in \{1, 2, \dots, c)\}$, we have*

$$\sum_{i=1}^{c+d} \frac{1}{N} M_i(r_{(l,j)}(x))\, \overline{M_i(r_{(l',j')}(x))}$$
$$= \sum_{i=1}^{c} \frac{1}{N} h_{i,j}(\frac{x+l}{N})\, \overline{h_{i,j'}(\frac{x+l'}{N})} + \sum_{k=1}^{d} \frac{1}{N} g_{k,j}(\frac{x+l}{N})\, \overline{g_{k,j'}(\frac{x+l'}{N})} = \delta_{j,j'}\delta_{l,l'},$$

where here the M_i correspond to the generalized filter functions $h_{i,j}$ and $g_{k,j}$ as in Definition 4.2.

PROOF. The statement that orthonormality of the rows implies orthonormality of the columns for finite-dimensional unitary matrices together with Theorem 4.3 tells us that for all $x \in \mathbb{T}$, for all $l, l' \in \{0, 1, \dots, N-1\}$ and $j \in \{1, \dots, \mu(\frac{x+l}{N})\}$, $j' \in \{1, \dots, \mu(\frac{x+l'}{N})\}$ we have

$$\sum_{i=1}^{\mu(x)+\tilde{\mu}(x)} \frac{1}{N} M_i(r_{(l,j)}(x))\, \overline{M_i(r_{(l',j')}(x))}$$
$$= \sum_{i=1}^{\mu(x)} \frac{1}{N} h_{i,j}(\frac{x+l}{N})\, \overline{h_{i,j'}(\frac{x+l'}{N})} + \sum_{k=1}^{\tilde{\mu}(x)} \frac{1}{N} g_{k,j}(\frac{x+l}{N})\, \overline{g_{k,j'}(\frac{x+l'}{N})} = \delta_{j,j'}\delta_{l,l'}.$$

But now we recall (3.3), which tells us that for fixed $x \in \mathbb{T}$, if i is a fixed integer satisfying $\mu(x) < i \le c$, then

$$\sum_{j=1}^{c} \sum_{l=0}^{N-1} |h_{i,j}(\frac{x+l}{N})|^2 = 0$$

(since $x \notin S_i$ for $i > \mu(x)$ by definition of $\mu(x)$.) Thus we must have

$$h_{i,j}(\frac{x+l}{N}) = 0, \mu(x) < i \le c,\ 1 \le j \le c,\ 0 \le l \le N-1.$$

Similarly, (3.4) tells us that for fixed $x \in \mathbb{T}$, if k satisfies $\tilde{\mu(x)} < k \le d$, then

$$\sum_{j=1}^{c} \sum_{l=0}^{N-1} |g_{k,j}(\frac{x+l}{N})|^2 = 0$$

(since $x \notin \widetilde{S_k}$ for $k > \tilde{\mu}(x)$, by definition of the $\widetilde{S_k}$). Thus we must have

$$g_{k,j}(\frac{x+l}{N}) = 0,\ \tilde{\mu}(x) < k \le d,\ 1 \le j \le c,\ 0 \le l \le N-1.$$

Finally we note that whenever $j > \mu(\frac{x+l}{N})$, we will have the identities $h_{i,j}(\frac{x+l}{N}) = g_{k,j}(\frac{x+l}{N}) = 0$ for all i between 1 and c and for all k between 1 and d. The reason for this is that in this case, $\frac{x+l}{N} \notin S_j$ and for fixed j, all the functions

$\{h_{i,j}, g_{k,j}\}$ are supported on S_j. From this we see that for all $x \in \mathbb{T}$, and for all $l, l' \in \{0, 1, \ldots, N-1\}$ and $j, j' \in \{1, \ldots, \mu(\frac{x+l}{N})\}$, we have

$$\sum_{i=1}^{c} \frac{1}{N} h_{i,j}(\frac{x+l}{N}) \overline{h_{i,j'}(\frac{x+l'}{N})} + \sum_{k=1}^{d} \frac{1}{N} g_{k,j}(\frac{x+l}{N}) \overline{g_{k,j'}(\frac{x+l'}{N})} = \delta_{j,j'}\delta_{l,l'},$$

since the extra terms we are adding on to the left hand side not coming from the orthonormality of columns are all zero, by the remarks above. □

The following example explicitly shows how Theorem 4.3 works in the particular case of the Journé wavelet. The construction of the generalized filters for the Journé wavelet was first done in the thesis of J. Courter [**C**].

EXAMPLE 4.5. The Journé wavelet in the frequency domain is the characteristic function of the set

$$[-\frac{16}{7}, -2) \cup [-\frac{1}{2}, -\frac{2}{7}) \cup [\frac{2}{7}, \frac{1}{2}] \cup [2, \frac{16}{7}).$$

Here μ takes on the values 0, 1, and 2, and $\widetilde{\mu(x)} \equiv 1$, since the Journé wavelet is a single orthonormal wavelet. If we identify $\mathbb{T}$ with $[-\frac{1}{2}, \frac{1}{2})$, we can write $S_1 = [-\frac{1}{2}, -\frac{3}{7}) \cup [-\frac{2}{7}, \frac{2}{7}) \cup [\frac{3}{7}, \frac{1}{2})$, $S_2 = [-\frac{1}{7}, \frac{1}{7})$, and $\widetilde{S_1} = [-\frac{1}{2}, \frac{1}{2}]$. The generalized filter functions then are:

$$\begin{aligned} h_{1,1}(x) &= \chi_{[-\frac{2}{7}, -\frac{1}{4}) \cup (-\frac{1}{7}, \frac{1}{7}) \cup [\frac{1}{4}, \frac{2}{7})}(x), \\ h_{1,2}(x) &= 0, \\ h_{2,1} &= \chi_{[-\frac{1}{2}, -\frac{3}{7}) \cup [\frac{3}{7}, \frac{1}{2})}(x), \\ h_{2,2}(x) &= 0; \\ g_1(x) &= \chi_{[-\frac{1}{4}, -\frac{1}{7}) \cup [\frac{1}{7}, \frac{1}{4})}(x), \\ g_2(x) &= \chi_{[-\frac{1}{7}, \frac{1}{7})}(x). \end{aligned}$$

Consider the decomposition of the circle $\mathbb{T}$ (identified with $[-\frac{1}{2}, \frac{1}{2})$) given by

$$\begin{aligned} P_1 &= [-\tfrac{1}{7}, \tfrac{1}{7}) && (\text{here } \mu(x) = 2,\ \mu(\tfrac{x}{2}) = 2,\ \mu(\tfrac{x+1}{2}) = 1), \\ P_2 &= \pm[\tfrac{1}{7}, \tfrac{2}{7}) && (\text{here } \mu(x) = 1,\ \mu(\tfrac{x}{2}) = 2,\ \mu(\tfrac{x+1}{2}) = 0), \\ P_3 &= \pm[\tfrac{2}{7}, \tfrac{3}{7}) && (\text{here } \mu(x) = 0,\ \mu(\tfrac{x}{2}) = 1,\ \mu(\tfrac{x+1}{2}) = 0), \\ P_4 &= \pm[\tfrac{3}{7}, \tfrac{1}{2}) && (\text{here } \mu(x) = 1,\ \mu(\tfrac{x}{2}) = 1,\ \mu(\tfrac{x+1}{2}) = 1). \end{aligned}$$

The associated cross-section matrix bundle is:

$$K_{i,\lambda_x(l,j)}(x) = \begin{cases} \begin{pmatrix} 1 & 0 & 0 \\ 0 & 0 & 1 \\ 0 & 1 & 0 \end{pmatrix} & \text{if } x \in P_1, \\ \begin{pmatrix} 1 & 0 \\ 0 & 1 \end{pmatrix} & \text{if } x \in P_2, \\ \quad 1 & \text{if } x \in P_3, \\ \begin{pmatrix} 0 & 1 \\ 1 & 0 \end{pmatrix} & \text{if } x \in P_4. \end{cases}$$

We are ready to define the generalized loop group and its associated action on the set of M-systems associated to a multiplicity function μ.

DEFINITION 4.6. The *loop group* associated to the multiplicity function μ is defined to be the subgroup $\mathrm{Loop}(F,q)$ of the group of Borel sections $\Gamma(F,q)$ whose elements K satisfy $K(0) = \mathrm{Id}_{\mu(0)+\tilde{\mu}(0)}$, and $K_{i,\lambda_x(j,l)}$ are Lip_1 in a neighborhood of the origin.

We now come to the main theorem of the paper. Just as in the classical case, it is possible to show that the generalized loop group acts freely and transitively on the set of M-systems:

THEOREM 4.7. *There is a free and transitive action of* $\mathrm{Loop}(F,q)$ *on the set of M-systems associated to an essentially bounded multiplicity function μ such that μ is constant in neighborhoods of $\frac{l}{N}$, $0 \le l \le N-1$. This action is given by*

$$K \cdot M(x) = K(\Pi_N(x))[(M_1(x), M_2(x), \ldots, M_{\mu(Nx)+\tilde{\mu}(Nx)}(x))]^T.$$

PROOF. We prove the transitivity first. Suppose we are given two different M-systems, labeled $M = (M_i)$ and $\widetilde{M} = (\widetilde{M}_i)$. Define an element K of the group bundle, that is, an element of $\Gamma(F,q)$, where $K(x)$ has dimension $\mu(x)+\tilde{\mu}(x)$, as follows:

$$K_{i,i'}(x) = \frac{1}{N}\sum_{\lambda_x(l,j)=1}^{\mu(x)+\tilde{\mu}(x)} \overline{M_i'(r_{(l,j)}(x))}\,\widetilde{M}_i(r_{(l,j)}(x)).$$

As in the classical case, we have

$$\begin{aligned}
[K\cdot M]_i(x) &= \sum_{i'=1}^{\mu(Nx)+\tilde{\mu}(Nx)} K_{i,i'}(\Pi_N(x))M_i'(x)\\
&= \sum_{i'=1}^{\mu(Nx)+\tilde{\mu}(Nx)} \left(\frac{1}{N}\sum_{\lambda_{Nx}(l,j)=1}^{\mu(Nx)+\tilde{\mu}(Nx)} \overline{M_i'(r_{(l,j)}(Nx))}\,\widetilde{M}_i(r_{(l,j)}(Nx))\right) M_i'(x)\\
&= \sum_{\lambda_{Nx}(l,j)=1}^{\mu(Nx)+\tilde{\mu}(Nx)} \widetilde{M}_i(r_{(l,j)}(Nx)) \left(\sum_{i'=1}^{\mu(Nx)+\tilde{\mu}(Nx)} \overline{M_i'(r_{(l,j)}(Nx))}\,M_i'(x)\right)\\
&= \widetilde{M}_i(x),
\end{aligned}$$

where the last equality follows since, by the orthogonality of the columns of M as established in Corollary 4.4, we have that inside sum is 0 except for the single values of l and j where $r_{l,j}(Nx) = x$.

Now to establish that the action is free, suppose $M = (M_i)$ is a M-system associated to μ and $K \in \mathrm{Loop}(F,q)$ satisfies

$$\begin{aligned}
K(\Pi_N(x))&[(M_1(x), M_2(x), \ldots, M_{\mu(Nx)+\tilde{\mu}(Nx)}(x))]^T\\
&= [(M_1(x), M_2(x), \ldots, M_{\mu(Nx)+\tilde{\mu}(Nx)}(x))]^T.
\end{aligned}$$

Then, analogously to the classical case, for each $x \in \mathbb{T}$ we define a $(\mu(Nx)+\tilde{\mu}(Nx)) \times (\mu(Nx)+\tilde{\mu}(Nx))$ unitary matrix $\mathcal{M}$ by

$$\mathcal{M}_{i,\lambda_{Nx}(l,j)}(x) = \sqrt{\frac{1}{N}}M_i(r_{(l,j)}(Nx)),$$
$$1 \le i \le \mu(Nx)+\tilde{\mu}(Nx),\ 0 \le l \le N-1,\ 1 \le j \le \mu(x+\frac{l}{N}).$$

We then see that $K(Nx)\mathcal{M}(x) = \mathcal{M}(x)$ for all $x \in \mathbb{T}$. By unitarity of $\mathcal{M}(x)$, this shows that that $K(Nx)$ is the $(\mu(Nx) + \tilde{\mu}(Nx)) \times (\mu(Nx) + \tilde{\mu}(Nx))$ identity matrix for all $x \in \mathbb{T}$, which implies that K is the identity element of $\mathrm{Loop}(F, q)$, as desired. □

We note this result is set purely in the language of transformation groups, i.e., we have defined a set corresponding to a fixed multiplicity function (the M-systems), and we have described a group which acts freely and transitively on this set. In a later paper we intend to impose conditions on M-systems under which we can mimic the classical construction and use them to obtain a normalized tight frame wavelet family for dilation by N.

ACKNOWLEDGMENTS. The authors gratefully acknowledge helpful suggestions from Astrid An Huef and Iain Raeburn. They also thank Brian Treadway for coordinating and merging TEX files.

References

[BCM] L.W. Baggett, J.E. Courter, and K.D. Merrill, *The construction of wavelets from generalized conjugate mirror filters in $L^2(\mathbb{R}^n)$*, Appl. Comput. Harmon. Anal. **13** (2002), 201–223.

[BMM] L.W. Baggett, H.A. Medina, and K.D. Merrill, *Generalized multi-resolution analyses and a construction procedure for all wavelet sets in $\mathbb{R}^n$*, J. Fourier Anal. Appl. **5** (1999), 563–573.

[BM] L.W. Baggett and K.D. Merrill, *Abstract harmonic analysis and wavelets in $\mathbb{R}^n$*, in "The Functional and Harmonic Analysis of Wavelets and Frames (San Antonio, TX, 1999)", L.W. Baggett and D.R. Larson, eds., Contemp. Math., vol. 247, American Mathematical Society, Providence, RI, 1999, pp. 17–27.

[BL] J.J. Benedetto and S. Li, *The theory of multiresolution analysis frames and applications to filter banks*, Appl. Comput. Harmon. Anal. **5** (1998), 389–427.

[BJ1] O. Bratteli and P.E.T. Jorgensen, *Isometries, shifts, Cuntz algebras and multiresolution wavelet analysis of scale N*, Integral Equations Operator Theory **28** (1997), 382–443.

[BJ] O. Bratteli and P.E.T. Jorgensen, "Wavelets Through a Looking Glass: the World of the Spectrum", Applied and Numerical Harmonic Analysis, Birkäuser, Boston, Basel, Berlin, 2002.

[Ch] A. Cohen, "Wavelets and Multiscale Signal Processing", translated by R. Ryan, Chapman & Hall, London, 1995.

[C] J. Courter, Ph.D. Thesis, University of Colorado, 1999.

[C2] J. Courter, *Construction of dilation-d wavelets*, in "The Functional and Harmonic Analysis of Wavelets and Frames (San Antonio, TX, 1999)", L.W. Baggett and D.R. Larson, eds., Contemp. Math., vol. 247, American Mathematical Society, Providence, RI, 1999, pp. 183–205.

[D] I. Daubechies, "Ten Lectures on Wavelets", CBMS-NSF Regional Conf. Ser. in Appl. Math., vol. 61, Society for Industrial and Applied Mathematics (SIAM), Philadelphia, 1992.

[FD] J.M.G. Fell and R.S. Doran, "Representations of $*$-Algebras, Locally Compact Groups, and Banach $*$-Algebraic Bundles, Vol. 1: Basic Representation Theory of Groups and Algebras", Pure and Applied Mathematics, vol. 125, Academic Press, San Diego, CA, 1988.

[HLPS] D. Han, D.R. Larson, M. Papadakis, and T. Stavropoulos, *Multiresolution analyses of abstract Hilbert spaces and wandering subspaces*, in "The Functional and Harmonic Analysis of Wavelets and Frames (San Antonio, TX, 1999)", L.W. Baggett and D.R. Larson, eds., Contemp. Math., vol. 247, American Mathematical Society, Providence, RI, 1999, pp. 259–284.

[Jor03] P.E.T. Jorgensen, *Matrix factorizations, algorithms, wavelets*, Notices Amer. Math. Soc. **50** (2003), 880–894.

[L] W.M. Lawton, *Tight frames of compactly supported affine wavelets*, J. Math. Phys. **31** (1990), 1898–1901.

[L2] W.M. Lawton, *Multiresolution properties of the wavelet Galerkin operator*, J. Math. Phys. **32** (1991), 1440–1443.

[Ma] S.G. Mallat, *Multiresolution approximations and wavelet orthonormal bases of* $L^2(\mathbb{R})$, Trans. Amer. Math. Soc. **315** (1989), 69–87.

[Mal99] S.G. Mallat, "A Wavelet Tour of Signal Processing", 2nd ed., Academic Press, Orlando, San Diego, 1999.

[Me] Y. Meyer, "Wavelets and Operators", Cambridge Studies in Advanced Mathematics vol. 37, Cambridge University Press, Cambridge, England, 1992.

[Mey93] Y. Meyer, *Wavelets and operators*, in "Different Perspectives on Wavelets (San Antonio, TX, 1993)", Ingrid Daubechies, ed., Proc. Sympos. Appl. Math., vol. 47, American Mathematical Society, Providence, RI, 1993, pp. 35–58.

Department of Mathematics, Campus Box 395, University of Colorado, Boulder, CO, 80309-0395

E-mail address: baggett@euclid.colorado.edu

Department of Mathematics, University of Iowa, 14 MacLean Hall, Iowa City, IA, 52242-1419

E-mail address: jorgen@math.uiowa.edu

Department of Mathematics, Colorado College, Colorado Springs, CO 80903-3294

E-mail address: kmerrill@coloradocollege.edu

Department of Mathematics, Campus Box 395, University of Colorado, Boulder, CO, 80309-0395

E-mail address: packer@euclid.colorado.edu

Contemporary Mathematics
Volume **345**, 2004

Examples of wavelets for local fields

Robert L. Benedetto

ABSTRACT. It is well known that the Haar and Shannon wavelets in $L^2(\mathbb{R})$ are at opposite extremes, in the sense that the Haar wavelet is localized in time but not in frequency, whereas the Shannon wavelet is localized in freqency but not in time. We present a rich setting where the Haar and Shannon wavelets coincide and are localized both in time and in frequency.

More generally, if $\mathbb{R}$ is replaced by a group G with certain properties, J. Benedetto and the author have proposed a theory of wavelets on G, including the construction of wavelet sets [**2**]. Examples of such groups G include the p-adic rational group $G = \mathbb{Q}_p$, which is simply the completion of $\mathbb{Q}$ with respect to a certain natural metric topology, and the Cantor dyadic group $\mathbb{F}_2((t))$ of formal Laurent series with coefficients 0 or 1.

In this expository paper, we consider some specific examples of the wavelet theory on such groups G. In particular, we show that Shannon wavelets on G are the same as Haar wavelets on G. We also give several examples of specific groups (such as $\mathbb{Q}_p$ and $\mathbb{F}_p((t))$, for any prime number p) and of various wavelets on those groups. All of our wavelets are localized in frequency; the Haar/Shannon wavelets are localized both in time and in frequency.

One of the principal goals of wavelet theory has been the construction of useful orthonormal bases for $L^2(\mathbb{R}^d)$. The group $\mathbb{R}^d$ has been an appropriate setting both because of its use in applications and because of its special property of containing lattices such as $\mathbb{Z}^d$ which induce discrete groups of translation operators on $L^2(\mathbb{R}^d)$. Of such wavelets, the easiest to describe are the Haar wavelet and the Shannon wavelet in $L^2(\mathbb{R})$. The Haar is a compactly supported step function in time but has noncompact support and slow decay in frequency. On the other hand, the Shannon is considered to be at the opposite extreme, being a compactly supported step function in frequency but with noncompact support and slow decay in time.

This distinction between Haar and Shannon wavelets appears to caused by certain properties of the topological group $\mathbb{R}^d$. J. Benedetto and the author [**2**] have presented a theory of wavelets on $L^2(G)$, for groups G with different properties to be described in Section 2 below. In this paper, we shall present some of that theory and some examples. Our main result, Theorem 4.3, is the surprising fact that Haar and Shannon wavelets not only exist in $L^2(G)$, but the two are the same. Moreover,

2000 *Mathematics Subject Classification.* 11S85, 42C40.
Key words and phrases. Wavelet, wavelet set, p-adic field.

these Haar/Shannon wavelets are localized both in time and in frequency; in fact, both the wavelets and their transforms are compactly supported step functions.

To reach that goal, we must first generalize wavelet theory for $\mathbb{R}^d$ to a more abstract setting. A number of authors have extended wavelet theories to locally compact groups G other than $\mathbb{R}^d$; see, for example, [**7, 10, 12, 15, 19, 21**]. However, those generalizations required the group G to contain a discrete subgroup, just as $\mathbb{R}^d$ contains the discrete lattice $\mathbb{Z}^d$. The more abstract generalized multiresolution analyses in [**1, 22**] are even broader, though the discrete group of translation operators is, naturally, a requirement and indeed one of the defining characteristics of a wavelet theory. Indeed, the standard constructions of multiresolution analysis and of wavelet sets in $L^2(\mathbb{R}^d)$ rely crucially on a discrete group of translation operators (as can be seen in expositions and investigations such as [**3, 4, 9, 8, 20, 26**]).

Beyond the identification of Haar and Shannon wavelets, we are also motivated by the study of analysis on local fields, which are of great importance in number theory; see [**11, 16, 24, 25**] for background. The study of Fourier analysis on local fields and on adèle groups arises in class field theory and in the study of certain zeta functions; see [**5, 6, 23, 27**], for example. However, no prior knowledge of such topics are needed to read this paper.

One simple example of a local field we wish to consider is the field $\mathbb{F}_p((t))$ of formal Laurent series over the field $\mathbb{F}_p$ of p elements, where p is a prime; see Examples 2.5, 3.4, and 5.4. As an additive group, $\mathbb{F}_p((t))$ contains a compact open subgroup $\mathbb{F}_p[[t]]$ (consisting of formal power series) as well as a discrete subgroup Γ, with $\mathbb{F}_p((t)) = \mathbb{F}_p[[t]] + \Gamma$. Using the elements of Γ as translations, one can study wavelets in $L^2(\mathbb{F}_p((t)))$. For example, Lang [**18**] examined the MRA theory in the case $p = 2$; we shall consider this case briefly in Section 1 to motivate the generalizations to follow.

However, while some of the groups G we wish to consider contain a discrete lattice, others do not. For example, the p-adic field $\mathbb{Q}_p$, which we shall study in Examples 2.4, 3.3, 5.2, and 5.5, is a completion of $\mathbb{Q}$ with respect to a certain metric and is a natural analogue of the real line $\mathbb{R}$. Unfortunately, as an additive group, $\mathbb{Q}_p$ contains no nontrivial discrete subgroup, thus presenting a serious obstacle to the development of a theory of wavelets in $L^2(\mathbb{Q}_p)$. Kozyrev [**17**] produced one specific set of wavelet generators, analogous to Haar wavelets on $\mathbb{R}^d$, using a discrete set of translation operators which do not form a group. However, as we shall see, those operators allow the Haar wavelets, but they preclude the possibility of a general theory of wavelet sets or multiresolution analysis in $L^2(\mathbb{Q}_p)$.

Fortunately, although it lacks a lattice, $\mathbb{Q}_p$ does contain the compact open subgroup $\mathbb{Z}_p$. Whereas $\mathbb{R}^d$ contains a discrete lattice $\mathbb{Z}^d$ with compact quotient $\mathbb{R}^d/\mathbb{Z}^d$, then, the group $\mathbb{Q}_p$ contains a *compact* subgroup $\mathbb{Z}_p$ with *discrete* quotient $\mathbb{Q}_p/\mathbb{Z}_p$. Thus, the situation for $\mathbb{Q}_p$ is the reverse of that for $\mathbb{R}^d$, but it still permits the construction of a discrete *group* of operators on $L^2(\mathbb{Q}_p)$. The resulting operators $\tau_{[s]}$ are not simple translation operators, but one may compute with them using the results of Section 3 below.

Of course, one also needs a dilation, which we shall call an expansive automorphism, to be defined in Section 3. J. Benedetto and the author [**2**] therefore considered wavelets on any locally compact abelian group G with a compact open subgroup H and an expansive automorphism A. The resulting wavelets, developed independently of Kozyrev's work, form a large class of orthonormal wavelet bases over $\mathbb{Q}_p$, $\mathbb{F}_p((t))$, and any other such group G. The theory also unifies both Lang's and Kozyrev's wavelets within a much broader setting.

Besides identifying Haar and Shannon wavelets in this context, the main purpose of this paper is to give examples of some of the groups and constructions considered in [**2**] in an expository setting. In Section 1, we present a concrete example of a Haar/Shannon wavelet, in the Cantor dyadic group $\mathbb{F}_2((t))$. In Section 2, we list some basic properties of a group G with compact open subgroup H and expansive automorphism A; we also present several examples of such groups. In Section 3 we describe the discrete group of translation-like operators on $L^2(G)$ defined in [**2**], and we consider specific examples by revisiting several of the groups from Section 2. In Section 4 we define wavelets using those operators as in [**2**]. We also present an explicit analogue of Shannon wavelets on G in Theorem 4.2. Then, in Theorem 4.3, we prove the promised result that our Shannon wavelets *are* Haar wavelets. Finally, in Section 5, we present several examples of wavelet sets produced by the results of [**2**] over certain groups G.

1. A motivating example

We begin by observing the Haar/Shannon phenomenon in a specific and relatively simple context. Let $\mathbb{F}_2 = \{0, 1\}$ be the field of two elements, with operations $0+0 = 1+1 = 0$, $1+0 = 0+1 = 1$, $0\cdot 0 = 0\cdot 1 = 1\cdot 0 = 0$, and $1\cdot 1 = 1$. The field $G = \mathbb{F}_2((t))$ of formal Laurent series consists of all infinite formal sums

$$c_{n_0}t^{n_0} + c_{n_0+1}t^{n_0+1} + c_{n_0+2}t^{n_0+2} + \cdots,$$

where $n_0 \in \mathbb{Z}$, each $c_n \in \mathbb{F}_2$, and no consideration is given to convergence or divergence. One can add or multiply elements of G as formal power series: addition follows the rule $b_n t^n + c_n t^n = (b_n + c_n)t^n$, and multiplication follows the rule $(c_n t^n)\cdot(c_m t^m) = (c_n \cdot c_m)t^{n+m}$. One can also put a metric topology on G by declaring the absolute value of $g \in G$ to be $|g| = 2^{-n_0}$, where n_0 is the smallest integer such that $c_{n_0} \neq 0$. (We will see the more general fields $\mathbb{F}_p((t))$ later in Examples 2.5, 3.4, and 5.4.)

Lang [**18**] considers the "Cantor dyadic group", which is the group $G = \mathbb{F}_2((t))$ under addition. He views G as the subgroup of $\prod_{n\in\mathbb{Z}}\mathbb{F}_2$ consisting of elements $(\cdots, c_{-2}, c_{-1}, c_0, c_1, c_2, \ldots)$ for which there exists an $n_0 \in \mathbb{Z}$ such that $c_n = 0$ for all $n < n_0$. The metric topology we have described on G coincides with the product topology G inherits from $\prod_{n\in\mathbb{Z}}\mathbb{F}_2$.

One can show that G is self-dual. Whereas the duality pairing on $\mathbb{R}$ takes $x, y \in \mathbb{R}$ to $e^{2\pi i x y}$, the pairing on G takes $g_1, g_2 \in G$ to $e^{\pi i c}$, where $c = c_{-1}$ is the t^{-1}-coefficient of $g_1 \cdot g_2$. Thus, the pairing of g_1 and g_2 is 1 if $c = 0$, and -1 if $c = 1$.

G contains a discrete subgroup Γ consisting of elements for which $c_0 = c_1 = c_2 = \cdots = 0$. (That is, Γ consists of all the formal sums which have only finitely many nonzero coefficients, and all nonnegative power terms $c_n t^n$ are zero.) The compact subgroup $H = \mathbb{F}_2[[t]]$ of G, consisting of all power series $c_0 + c_1 t + c_2 t^2 + \cdots$, is a fundamental domain for Γ, in the sense that $H \cap \Gamma = \{0\}$ and $H + \Gamma = G$. It may be helpful to think of Γ as analogous to the discrete subgroup $\mathbb{Z}$ of $\mathbb{R}$, while H is analogous to the fundamental domain $[-1/2, 1/2)$; however, in our case, H is itself a subgroup of G.

We may normalize a Haar measure μ on G with $\mu(H) = 1$. Note that $tH = \{c_1 t + c_2 t^2 + \cdots\}$ is a subgroup of H of measure $1/2$, because H is the disjoint union $(1+tH) \cup tH$, and $\mu(tH) = \mu(1+tH)$ by translation invariance. Note also that H is homeomorphic to the Cantor set; it consists of two disjoint pieces, tH and $1+tH$, each of which is homeomorphic (via multiplication-by-t^{-1}) to H.

Similarly, $t^{-1}H = \{c_{-1}t^{-1} + c_0 + c_1 t + \cdots\}$ is a subgroup of G which contains H and which has measure 2. It is easy to see that

$$\cdots \supset t^{-2}H \supset t^{-1}H \supset H \supset tH \supset t^2 H \supset \cdots,$$

and

$$\bigcap_{n\geq 0} t^n H = \emptyset, \quad \text{and} \quad \bigcup_{n\geq 0} t^{-n} H = G.$$

Thus, multiplication-by-t^{-1} is a dilation which expands areas by a factor of 2, just as multiplication-by-2 is a dilation of $\mathbb{R}$ which expands areas by a factor of 2.

Lang [**18**, Example 5.2] observed that the function

$$f(g) = \begin{cases} 1, & \text{if } g \in tH; \\ -1, & \text{if } g \in 1+tH; \\ 0, & \text{otherwise} \end{cases}$$

is a wavelet, in the sense that the set of dilated translates $\{f(t^n g - s) : n \in \mathbb{Z}, s \in \Gamma\}$ is an orthonormal basis for $L^2(G)$. Indeed, this function f is an obvious analogue of the usual Haar wavelet on $\mathbb{R}$. The surprising fact that motivates much of the rest of this paper is that using the duality pairing on G, one can easily compute that the Fourier transform of f is

$$\widehat{f}(\omega) = \begin{cases} 1, & \text{if } \omega \in t^{-1} + H; \\ 0, & \text{otherwise,} \end{cases}$$

which shows that $\widehat{f}$ is a step function with compact support in the frequency domain, and that it is therefore natural to call f a Shannon wavelet. (The skeptical reader may observe that $t^{-1}+H$ may be written as the disjoint union of $t^{-1}+(tH)$ and $-t^{-1}+(1+tH)$, just as the Shannon wavelet in $L^2(\mathbb{R})$ has Fourier transform equal to the characteristic function of $[-1,-1/2) \cup [1/2,1)$.)

It turns out that the key property of G that allows the Haar/Shannon identification is the fact that the fundamental domain H is itself a subgroup of G. Moreover, the lattice Γ turns out to be unnecessary, provided we define operators to act as translations in some appropriate manner. Thus, we will be able to work with groups such as $\mathbb{Q}_p$, which have no such lattice. With these ideas in mind, we are now prepared to generalize.

2. The group, subgroup, and automorphism

We fix the following notation.

G	a locally compact abelian group
$\widehat{G}$	the dual group of G
H	a compact open subgroup of G
$H^\perp$	the annihilator of H in $\widehat{G}$
μ	Haar measure on G, normalized so that $\mu(H) = 1$
ν	Haar measure on $\widehat{G}$, normalized so that $\nu(H^\perp) = 1$

Given $x \in G$ and $\gamma \in \widehat{G}$, the pairing $(x, \gamma) \in \mathbb{T} = \{z \in \mathbb{C} : |z| = 1\}$ shall denote the action of γ on x. $H^\perp$, which by definition consists of all $\gamma \in \widehat{G}$ such that $(x, \gamma) = 1$ for each $x \in H$, is a compact open subgroup of $\widehat{G}$. The quotient groups G/H and $\widehat{G}/H^\perp$ are discrete, with counting measure induced by μ and ν, respectively. Furthermore, the discrete group $\widehat{H}$ is naturally isomorphic to $\widehat{G}/H^\perp$, and the compact group $\widehat{G/H}$ is naturally isomorphic to $H^\perp$. The normalizations of μ and ν are compatible in the sense that the Fourier inversion formula holds; see [**14**, §31.1(c)].

The study of wavelets in $L^2(\mathbb{R}^d)$ requires a dilation operator and a discrete group of translation operators. For the translations, one usually considers a discrete lattice $\Lambda \subset \mathbb{R}^d$ with a fundamental domain F which is also a neighborhood of the origin. However, our group G may not actually contain a discrete subgroup. Fortunately, the compact subgroup H can play the role of the fundamental domain, even without an actual lattice. Meanwhile, the discrete group G/H shall act as a group of translation operators, in a manner to be described in Section 3.

The dilation over $\mathbb{R}^d$ is usually an expansive matrix B; as a linear map, B satisfies $d(\lambda \circ B) = |\det B| d\lambda$, where λ denotes Lebesgue measure on $\mathbb{R}^d$. In our setting, let A be an automorphism of G; that is, A is an algebraic isomorphism from G to G such that A is also a topological homeomorphism. Then $\mu \circ A$ is a nontrivial Haar measure on G. Define the *modulus* of A to be $|A| = \mu(AH) > 0$, so that $d\mu(Ax) = |A| d\mu(x)$. It is easy to verify that $|A^{-1}| = |A|^{-1}$. Furthermore, the adjoint A^* of A is an automorphism of $\widehat{G}$, and $|A^*| = |A|$ in the sense that $\nu(A^* H^\perp) = |A|$ and $d\nu(A^*\gamma) = |A| d\nu(\gamma)$.

If the Fourier transform $\widehat{f}$ of a function $f \in L^2(G)$ has compact support, then as usual, f is smooth. Proposition 2.1 below shall make that notion of smoothness more precise. As a preface to that result, note that for any fixed $r \in \mathbb{Z}$, $A^{-r}H$ is an open subgroup of G, and therefore the sets of the form $c + A^{-r}H$ form a partition of G into open sets. The proposition shall show that if the support of $\widehat{f}$ is contained in the compact set $(A^*)^r H^\perp$, then f is constant on all the sets $c + A^{-r}H$. Thus, f is locally constant.

PROPOSITION 2.1. *Let G be a locally compact abelian group with compact open subgroup H, let A be an automorphism of G, let $r \in \mathbb{Z}$, and let $f \in L^2(G)$.*

Then $\operatorname{Supp} \widehat{f} \subset (A^*)^r H^\perp$ *if and only if f is constant on every (open) set of the form $c + A^{-r}H$.*

PROOF. To prove the forward implication, pick any $c \in G$ and $x \in c + A^{-r}H$. Note that $x - c \in A^{-r}H$, and therefore $(x - c, \gamma) = 1$ for every $\gamma \in (A^*)^r H^\perp$. Thus,

$$f(x) = \int_{(A^*)^r H^\perp} (x, \gamma) \widehat{f}(\gamma) d\nu(\gamma) = \int_{(A^*)^r H^\perp} (c, \gamma) \widehat{f}(\gamma) d\nu(\gamma) = f(c).$$

For the converse, pick $c \in G$ and $r \in \mathbb{Z}$. We shall show that the desired implication holds for the characteristic function $f = \mathbf{1}_{c + A^{-r}H}$; the general result shall then follow by linearity. We compute

$$\begin{aligned}\widehat{f}(\gamma) &= \int_{c+A^{-r}H} \overline{(x,\gamma)}d\mu(x) = \overline{(c,\gamma)} \int_{A^{-r}H} \overline{(x,\gamma)}d\mu(x)\\ &= \overline{(c,\gamma)}|A|^{-r} \int_H \overline{(A^{-r}x,\gamma)}d\mu(x)\\ &= \overline{(c,\gamma)}|A|^{-r} \int_H \overline{(x,(A^*)^{-r}\gamma)}d\mu(x)\\ &= \overline{(c,\gamma)}|A|^{-r}\mathbf{1}_{H^\perp}((A^*)^{-r}\gamma) = \overline{(c,\gamma)}|A|^{-r}\mathbf{1}_{(A^*)^r H^\perp}(\gamma), \end{aligned} \tag{2.1}$$

which has support contained in the required domain. □

DEFINITION 2.2. Let G be a locally compact abelian group with compact open subgroup H. Let A be an automorphism of G. We say that A is *expansive* with respect to H if both of the following conditions hold:

(1) $AH \supsetneq H$, and
(2) $\bigcap_{n\leq 0} A^n H = \{0\}$.

The idea of Definition 2.2 is that an expansive automorphism acts on G like an expansive integer matrix B acts on $\mathbb{R}^d$; the fundamental domain F for the lattice $\mathbb{Z}^d$ is contained in its image BF, and the intersection of all inverse images $B^{-k}F$ is just the origin. Moreover, if A is an expansive automorphism of G, then $|A|$ is an integer greater than 1, just as is true of an expansive integer matrix. Indeed, H is a proper subgroup of AH, so that AH may be covered by (disjoint) cosets $s+H$. Each coset has measure 1, so that AH must have measure equal to the number of cosets. Thus, $|A| = \mu(AH)$ is just the number of elements in the finite quotient group $(AH)/H$.

The following alternate characterization of expansiveness was proven in [**2**].

PROPOSITION 2.3 (From [**2**]). *Let G be a locally compact abelian group with compact open subgroup H. Let A be an automorphism of G. Then A is expansive with respect to H if and only if both of the following conditions hold:*

(1) $A^*H^\perp \supsetneq H^\perp$, *and*
(2) $\bigcup_{n\geq 0}(A^*)^n H^\perp = \widehat{G}$.

EXAMPLE 2.4. Let $p\geq 2$ be a prime number. Let $G=\mathbb{Q}_p$ be the set of p-adic rational numbers, and $H=\mathbb{Z}_p$ the set of p-adic integers. That is,

$$G = \mathbb{Q}_p = \left\{ \sum_{n\geq n_0} c_n p^n : n_0\in\mathbb{Z}, c_n = 0,1,\ldots,p-1 \right\}, \quad \text{and}$$

$$H = \mathbb{Z}_p = \left\{ \sum_{n\geq 0} c_n p^n : c_n = 0,1,\ldots,p-1 \right\}.$$

Here, the sums are formal sums, but they indicate that the group law is addition with carrying of digits, so that in $\mathbb{Q}_3$, for example, we have

$$\begin{aligned}(2+2\cdot 3+2\cdot 3^2+2\cdot 3^3+2\cdot 3^4+\ldots)+(1+2\cdot 3+1\cdot 3^2+2\cdot 3^3+1\cdot 3^4+\ldots)\\ = (0+2\cdot 3+1\cdot 3^2+2\cdot 3^3+1\cdot 3^4+\ldots).\end{aligned}$$

The discrete quotient $G/H = \mathbb{Q}_p/\mathbb{Z}_p$ is naturally isomorphic to μ_{p^∞}, the multiplicative group of all p^n-th roots of unity (as n ranges through all nonnegative integers) in $\mathbb{C}$. However, $\mathbb{Q}_p$ itself contains no nontrivial discrete subgroups; if a closed subgroup E contains some nonzero element $x \in \mathbb{Q}_p$, then E contains the open subgroup $x\mathbb{Z}_p$.

$\mathbb{Q}_p$ is in fact a field, with multiplication given by multiplication of formal Laurent series together with carrying of digits, though we are considering it only as an additive group. For more of the standard background on $\mathbb{Q}_p$, see [**11, 16, 23**]; see [**24, 25**] for more advanced expositions.

Define a character χ on G by

$$\chi\left(\sum_{n\geq n_0} c_n p^n\right) = \exp\left(2\pi i \sum_{n=n_0}^{-1} c_n p^n\right).$$

Note that $H = \{x \in G : \chi(x) = 1\}$. G is self-dual, with pairing given by $(x, \gamma) = \chi(x\gamma)$, where $x\gamma$ means the product in the field $\mathbb{Q}_p$. Under this pairing, $H^\perp \subset \mathbb{Q}_p$ is again just $\mathbb{Z}_p$.

For any nonzero $a \in \mathbb{Q}_p$, the multiplication-by-a map $A : x \mapsto ax$ is an automorphism of G. (In fact, any automorphism must be of this form.) If $a \notin \mathbb{Z}_p$, then A is expansive; in that case, if $a = c_{-m}p^{-m} + c_{-m+1}p^{-m+1} + \dots$ with $c_{-m} \neq 0$, then $|A| = p^m$. (For example, multiplication-by-$1/p$ is an expansive map with modulus p.) The adjoint A^* on $\widehat{G} = \mathbb{Q}_p$ is again multiplication-by-a.

EXAMPLE 2.5. Let $p \geq 2$ be a prime number. Let $\mathbb{F}_p$ denote the field of order p, with elements $\{0, 1, \dots, p-1\}$. Let $G = \mathbb{F}_p((t))$, the set of formal Laurent series in the variable t with coefficients in $\mathbb{F}_p$, and let $H = \mathbb{F}_p[[t]]$ be the subset consisting of power series (i.e., nonnegative powers of t only). G is a field under addition and multiplication of formal power series. Although G is topologically homeomorphic to $\mathbb{Q}_p$, the algebraic structure is very different because there is no carrying of digits. In $\mathbb{F}_3((t))$, for example, we have

$$\begin{aligned}(2 + 2\cdot t + 2\cdot t^2 + 2\cdot t^3 + 2\cdot t^4 + \dots) + (1 + 2\cdot t + 1\cdot t^2 + 2\cdot t^3 + 1\cdot t^4 + \dots)\\ = (0 + 1\cdot t + 0\cdot t^2 + 1\cdot t^3 + 0\cdot t^4 + \dots).\end{aligned}$$

Again, we consider G and H as groups under addition; note that for any $x \in G$,

$$\underbrace{x + x + \dots + x}_{p} = 0.$$

(That is, $\mathbb{F}_p((t))$ is a field of characteristic p.) The discrete quotient G/H is naturally isomorphic to a direct sum of a countable number of copies of $\mathbb{F}_p$. In fact, G/H is isomorphic to the subgroup $\Gamma \subset G$ consisting of all elements of the form $c_{-m}t^{-m} + \dots + c_{-1}t^{-1}$. See [**23**] or [**25**] for more on such fields.

Define a character χ on G by

$$\chi\left(\sum_{n\geq n_0} c_n t^n\right) = \exp\left(2\pi i c_{-1}/p\right).$$

Note that the image of χ consists only of the p-th roots of unity, whereas the corresponding character on $\mathbb{Q}_p$ included all p^n-th roots of unity in its image. Another contrast with $\mathbb{Q}_p$ is that this time, $H \subsetneq \{x \in G : \chi(x) = 1\}$. Still, as before, G is self-dual, with pairing given by $(x, \gamma) = \chi(x\gamma)$, where $x\gamma$ means the product in the field $\mathbb{F}_p((t))$. Under this pairing, $H^\perp \subset \mathbb{F}_p((t))$ is again just $\mathbb{F}_p[[t]]$.

For any nonzero $a \in \mathbb{F}_p((t))$, the multiplication-by-a map $A : x \mapsto ax$ is an automorphism of G. (This time, however, many other automorphisms are possible.) If $a \notin \mathbb{F}_p[[t]]$, then A is expansive; in that case, if $a = c_{-m}t^{-m} + c_{-m+1}t^{-m+1} + \ldots$ with $c_{-m} \neq 0$, then $|A| = p^m$. (For example, multiplication-by-$1/t$ is an expansive map with modulus p.) The adjoint A^* on $\widehat{G} = \mathbb{F}_p((t))$ is again multiplication-by-a.

EXAMPLE 2.6. Let K be a field which is a finite extension of $\mathbb{Q}_p$. In general, $\mathbb{Q}_p$ is equipped with an absolute value function $|\cdot| : \mathbb{Q}_p \to \mathbb{R}_{\geq 0}$ with $|p| = 1/p$ and which extends uniquely to the algebraic closure of $\mathbb{Q}_p$, and hence to K. The ring of integers $\mathcal{O}_K$ is precisely the set of elements of K with absolute value at most 1, just as $\mathbb{Z}_p$ is the set of elements in $\mathbb{Q}_p$ with absolute value at most 1. Equivalently, $\mathcal{O}_K$ consists of all elements of K which are roots of monic polynomials with coefficients in $\mathbb{Z}_p$.

For example, if $K = \mathbb{Q}_3(\sqrt{-1})$, then elements of K may be represented as formal Laurent series $\sum c_n 3^n$, where each c_n is one of the nine numbers $a+b\sqrt{-1}$, with $a, b = 0, 1, 2$. (The addition law for carrying digits requires some extra computation in this case.) Similarly, if $K = \mathbb{Q}_3(\sqrt{3})$, then elements of K may be represented as formal Laurent series $\sum c_n 3^{n/2}$, where each c_n is one of the three numbers $0, 1, 2$, and with carrying from the $3^{n/2}$ term to the $3^{(n+2)/2}$ term. In both of these cases, $\mathcal{O}_K$ consists of those elements of K with no negative power terms.

Let G be the additive group of K, and $H \subset G$ be the additive group of $\mathcal{O}_K$. Define a character χ_K on G by

$$\chi_K(x) = \chi(\operatorname{Tr} x),$$

where χ was the character defined in Example 2.4 for $\mathbb{Q}_p$, and Tr is the trace map from K down to $\mathbb{Q}_p$. G is self-dual, with pairing given by $(x, \gamma) = \chi_K(x\gamma)$, where $x\gamma$ means the product in K. This time, $H^\perp \subset K$ need not be the same as H. Instead, $H^\perp$ is the inverse different of K, which is a (possibly larger) subgroup of K of the form $b\mathcal{O}_K$, for some $b \in K$. (See [**23**, §B.2] or [**25**, III] for more on the different and inverse different.) For example, if $K = \mathbb{Q}_3(\sqrt{-1})$, then the inverse different is just $\mathcal{O}_K = \mathbb{Z}_3[\sqrt{-1}]$; but if $K = \mathbb{Q}_3(\sqrt{3})$, then the inverse different is $3^{-1/2}\mathcal{O}_K = 3^{-1/2}\mathbb{Z}_3[\sqrt{3}]$. (The inverse different roughly measures the degree of *ramification* in K, which, speaking informally, means the largest denominator that appears in the powers of p in expansions of elements of K. So $\mathbb{Q}_3(\sqrt{3})$ has ramification degree 2, since its elements involve half-integral powers of 3.)

As before, for any nonzero $a \in K$, the multiplication $A : x \mapsto ax$ is an automorphism of G. (Other automorphisms are possible, incorporating Galois automorphisms of K over $\mathbb{Q}_p$.) If $a \notin \mathcal{O}_K$, then A is expansive; the modulus $|A|$ is, as usual, the order of the group $(AH)/H$, which is always be a power of p.

One could also choose G to be the additive group of a field which is a finite extension of $\mathbb{F}_p((t))$. Again, G would be self-dual, with pairing given by composing the original χ for $\mathbb{F}_p((t))$ with the trace map, and with $H^\perp$ given by the inverse different. We omit the details here.

EXAMPLE 2.7. Let $G_1, \ldots, G_N$ be locally compact abelian groups with compact open subgroups $H_1, \ldots, H_N$, and expansive automorphisms $A_1, \ldots, A_N$, respectively. Then $G = G_1 \times \cdots \times G_N$ has open compact subgroup $H = H_1 \times \cdots \times H_N$ with expansive automorphism $A = A_1 \times \cdots \times A_N$.

As a special case, if $G_1 = \cdots = G_N = K$ is the additive group of one of the fields in the previous three examples, then $G = K^{\times N}$ is a locally compact abelian group with compact open subgroup $H = \mathcal{O}_K^{\times N}$. Moreover, if $A \in GL(N, K)$ is a matrix with all eigenvalues of absolute value greater than 1 and such that $AH \subsetneq H$, then A is expansive. The modulus of A is $|\det A|$, where $|\cdot|$ is a certain appropriately chosen absolute value on K.

EXAMPLE 2.8. The preceding examples are of self-dual σ-compact groups; the following example is of a group which is not self-dual and may not be σ-compact.

Let G_0 be a locally compact abelian group with compact open subgroup H_0 and expansive automorphism A_0, and let W be any discrete group (possibly uncountable). Let $G = G_0 \times W$, with compact open subgroup $H = H_0 \times \{0\}$. Then $\widehat{G}$ is isomorphic to $\widehat{G}_0 \times \widehat{W}$, and the image of $H^{\perp}$ under that isomorphism is $H_0^{\perp} \times \widehat{W}$. If W is uncountable, note that G is not σ-compact.

It is easy to check that the automorphism A of G given by $A = A_0 \times \mathrm{id}_W$ (where id_W is the identity automorphism of W) is expansive, with adjoint $A^* = A_0^* \times \mathrm{id}_{\widehat{W}}$. However, it should be noted that while $\bigcap_{n \leq 0} A^n H = \{0\}$, as required for expansiveness, the union $\bigcup_{n \geq 0} A^n H$ is *not* all of G, but only $G_0 \times \{0\}$. Similarly, on the dual side, the union $\bigcup_{n \geq 0} (A^*)^n H^{\perp} = \widehat{G}$ is the whole group, but the intersection $\bigcap_{n \leq 0} (A^*)^n H^{\perp} = \{0\} \times \widehat{W} \neq \{0\}$ includes more than just the identity.

3. Dilation and translation operators

We are now prepared to present the operators on $L^2(G)$ introduced in [**2**]. Given our group G and automorphism A, we can define a dilation operator δ_A on $L^2(G)$ by

$$\delta_A(f)(x) = |A|^{1/2} f(Ax).$$

It is easy to verify that δ_A is a unitary operator. If A is expansive, then certainly A does not have finite order, and therefore $\{\delta_A^n : n \in \mathbb{Z}\}$ forms a group isomorphic to $\mathbb{Z}$.

The construction of translation operators requires a little more work. If G contains a discrete subgroup Γ such that $G \cong \Gamma \times H$, then Γ plays the same role that the lattice $\mathbb{Z}^d$ plays in $\mathbb{R}^d$. That is, for any $s \in \Gamma$, we define τ_s on $L^2(G)$ by $\tau_s(f)(x) = f(x - s)$. For example, if $G = \mathbb{F}_p((t))$, then the choice of Γ described in Example 2.5 is such a lattice. Indeed, Lang [**18**] used precisely this lattice to define wavelets on $\mathbb{F}_2((t))$, the Cantor dyadic group.

On the other hand, if G contains no such lattice, as is the case for $\mathbb{Q}_p$, the situation is a little more complicated. Kozyrev's [**17**] solution for the special case of $\mathbb{Q}_p$ was to let $\mathcal{C} \subset \mathbb{Q}_p$ be the discrete subset (but not subgroup) of elements of the form $c_{-m}p^{-m} + \cdots + c_{-1}p^{-1}$. Then, for each $s \in \mathcal{C}$, the translation operator τ_s is defined in the usual way. He was then able to construct Haar wavelets on $\mathbb{Q}_p$ using dilation-by-$1/p$ and translation by elements of $\mathcal{C}$.

More generally, one could choose a set $\mathcal{C}$ of coset representatives for the quotient G/H. That is, choose a subset $\mathcal{C} \subset G$ such that the translations $s + H$ of H by elements $s \in \mathcal{C}$ are mutually disjoint, and such that $\bigcup_{s \in \mathcal{C}} (s + H) = G$. (Such a $\mathcal{C}$ always exists if one accepts the Axiom of Choice; but in many cases, it is possible to construct such a set $\mathcal{C}$ without invoking the Axiom of Choice. See, for example, Proposition 3.2.)

Unfortunately, Kozyrev's method has its limitations. For one thing, the resulting group of translation operators does not form a group unless $\mathcal{C}$ itself is a group, which would be impossible for $\mathbb{Q}_p$. Furthermore, it appears to be difficult or impossible to construct wavelets other than Haar wavelets using such translation operators. More precisely, it appears that any sort of analogue of a multiresolution analysis or of minimally supported frequency wavelets requires the translation operators to behave well on the dual group $\widehat{G}$, as follows. If $\mathcal{C} = \Gamma$ is indeed a group, then the dual lattice $\Gamma^\perp \subset \widehat{G}$ plays a crucial role in wavelet analysis; but if $\mathcal{C}$ is not a group, then there does not seem to be any reasonable object which can assume the role of the dual lattice.

In [**2**], the authors solved this problem by choosing a set $\mathcal{D} \subset \widehat{G}$ to play the role of the dual lattice first, and then building the operators to respect $\mathcal{D}$. Specifically, let $\mathcal{D}$ be a choice of coset representatives for the quotient $\widehat{G}/H^\perp$. Define functions $\theta = \theta_{\mathcal{D}} : \widehat{G} \to \mathcal{D}$ and $\eta = \eta_{\mathcal{D}} : \widehat{G} \to H^\perp$ given by

$$\begin{aligned} \theta(\gamma) &= \text{ the unique } \sigma \in \mathcal{D} \text{ such that } \gamma - \sigma \in H^\perp, \\ \eta(\gamma) &= \gamma - \theta(\gamma). \end{aligned} \tag{3.1}$$

Speaking informally, $\eta(\gamma)$ is the unique point in the fundamental domain $H^\perp$ which is congruent to γ modulo the "lattice" $\mathcal{D}$, and $\theta(\gamma)$ is the unique "lattice" point such that $\gamma = \eta(\gamma) + \theta(\gamma)$.

Any element $[s] \in G/H$ of the discrete quotient is, by definition, a coset of the form $s + H$, with $s \in G$. We wish to define a translation operator associated with $[s]$ which depends only on the coset, not the particular representative s. Note that on the transform side, the translation-by-s operator sends $\widehat{f}(\gamma)$ to $\overline{(s,\gamma)}\widehat{f}(\gamma)$. Thus, we define the translation-by-the-coset operator $\tau_{[s]}$ by defining its action $\tilde{\tau}_{[s]}$ on $L^2(\widehat{G})$ as follows:

$$\tilde{\tau}_{[s]}(\widehat{f})(\gamma) = \overline{(s, \eta(\gamma))}\widehat{f}(\gamma). \tag{3.2}$$

It is easy to check that $\tilde{\tau}_{[s]}$, and hence $\tau_{[s]}$, are unitary operators.

To simplify notation, define

$$w_{[s]}(\gamma) = \overline{(s, \eta(\gamma))},$$

which is a function in $L^\infty(\widehat{G})$, since $|w_{[s]}(\gamma)| = 1$ for all $\gamma \in \widehat{G}$. We compute

$$w_{[s+t]}(\gamma) = \overline{(s+t, \eta(\gamma))} = \overline{(s, \eta(\gamma))(t, \eta(\gamma))} = w_{[s]}(\gamma) w_{[t]}(\gamma). \tag{3.3}$$

If we consider $t \in H$, so that $w_{[t]}$ is identically 1, then equation (3.3) shows that $w_{[s+t]} = w_{[s]}$. Thus, as promised, $w_{[s]}$ and hence $\tau_{[s]}$ depend only on $[s]$, not on the particular representative s. Equation (3.3) also shows that the set of operators

$$\Gamma_{G/H} = \{\tau_{[s]} : [s] \in G/H\}$$

forms a group isomorphic to G/H.

On $L^2(G)$, the action of $\tau_{[s]}$ is convolution by the pseudo-measure $\check{w}_{[s]}$, defined to be the inverse transform of $w_{[s]}$. By way of comparison, the usual translation-by-s operator is convolution by a delta measure centered at s. Proposition 3.1 below computes the action of our operator $\tau_{[s]}$, and the result shows that our pseudo-measure $\check{w}_{[s]}$ has support contained in $s + H$, in the sense that it is trivial on any measurable set $U \subset G$ disjoint from $s + H$. Thus, $\check{w}_{[s]}$ may be thought of as a perturbation of the original delta measure. To see this, we consider the action of

$\tau_{[s]}$ on the characteristic function of a set of the form $c + A^{-r}H$ in the following proposition.

PROPOSITION 3.1. *Let G be a locally compact abelian group with compact open subgroup H, and let $\mathcal{D}$ be a set of coset representatives for the quotient $\widehat{G}/H^{\perp}$. Let A be an expansive automorphism of G, and let $M = |A| \geq 2$ be the modulus of A. For any $[s] \in G/H$, define the operator $\tau_{[s]}$ as in equation (3.2) above. For any $c \in G$ and any $r \in \mathbb{Z}$, let $\mathbf{1}_{c+A^{-r}H}$ be the characteristic function of the set $c+A^{-r}H$. If $r \leq 0$, let σ_0 be the unique element of $\mathcal{D}$ in $H^{\perp}$. If $r > 0$, let $\{\sigma_0, \sigma_1, \ldots, \sigma_{M^r-1}\}$ be all the elements of $\mathcal{D} \cap (A^*)^r H^{\perp}$.*

(1) *If $r \leq 0$, then*

$$\tau_{[s]}\mathbf{1}_{c+A^{-r}H}(x) = (s, \sigma_0)\mathbf{1}_{s+c+A^{-r}H}(x).$$

(2) *If $r > 0$, then*

$$\tau_{[s]}\mathbf{1}_{c+A^{-r}H}(x) = \frac{1}{M^r}\left(\sum_{i=0}^{M^r-1}(x - c, \sigma_i)\right)\mathbf{1}_{s+c+H}(x).$$

Note that in the case $r \leq 0$ (in which case $A^{-r}H$ is large), $\tau_{[s]}$ acts as translation-by-s and multiplication by the constant (s, σ_0) on $\mathbf{1}_{c+A^{-r}H}$. For $r > 0$ (in which case $A^{-r}H$ is small), $\tau_{[s]}$ is similar to translation by s, but it spreads the function out over the full coset $s + c + H$, not the smaller $s + c + A^{-r}H$. Still, the translated function is constant on any region of the form $b+A^{-r}H$. Moreover, when that constant is nonzero (i.e., when $b \in s + c + H$), its value $M^{-r}\sum_i(b, \sigma_i)$ usually is small, due to a lot of cancellation (but usually not complete cancellation) within the sum; see the examples later in this section. Note also that we can easily verify directly from the proposition that $\tau_{[s]}$ is independent of the coset representative s, and that it is the identity operator if $s \in H$.

If a function $f \in L^2(G)$ is constant on all such sets $c+A^{-r}H$, then f is a linear combination of characteristic functions $\mathbf{1}_{c+A^{-r}H}$. Thus, $\tau_{[s]}f$ may be computed using Proposition 3.1 and linearity. In particular, the action of $\tau_{[s]}$ on such a function f is determined not by all of $\mathcal{D}$ but only by a subset of size M^r. Furthermore, the computation of $\tau_{[s]}f$ involves only a finite sum and is therefore easily computable in practice.

PROOF OF PROPOSITION 3.1. Let $f = \mathbf{1}_{c+A^{-r}H}$. By Equation (2.1),

$$\widehat{f}(\gamma) = M^{-r}\overline{(c, \gamma)}\mathbf{1}_{(A^*)^r H^{\perp}}(\gamma),$$

and therefore

$$\tilde{\tau}_{[s]}\widehat{f}(\gamma) = M^{-r}\overline{(s, \eta(\gamma))(c, \gamma)}\mathbf{1}_{(A^*)^r H^{\perp}}(\gamma).$$

If $r \leq 0$, then $(A^*)^r H^{\perp} \subset H^{\perp}$, so that $\eta(\gamma) = \gamma - \sigma_0$ for $\gamma \in (A^*)^r H^{\perp}$. In that case,

$$\tilde{\tau}_{[s]}\widehat{f}(\gamma) = M^{-r}(s, \sigma_0)\overline{(s + c, \gamma)}\mathbf{1}_{(A^*)^r H^{\perp}}(\gamma).$$

Taking the inverse transform gives

$$\tau_{[s]}f(x) = M^{-r}(s, \sigma_0)\int_{(A^*)^r}(x - s - c, \gamma)d\nu(\gamma) = (s, \sigma_0)\mathbf{1}_{s+c+A^{-r}H}(x).$$

On the other hand, if $r > 0$, then $(A^*)^r H^\perp \supsetneq H^\perp$. To compute $\eta(\gamma)$, then, we partition $(A^*)^r H^\perp$ into the M^r regions $\{\sigma_i + H^\perp : i = 0, 1, \ldots, M^r - 1\}$. We have $\eta(\gamma) = \gamma - \sigma_i$ for $\gamma \in \sigma_i + H^\perp$. Thus,

$$\tilde{\tau}_{[s]}\widehat{f}(\gamma) = M^{-r} \sum_{i=0}^{M^r-1} (s, \sigma_i)\overline{(s + c, \gamma)}\mathbf{1}_{\sigma_i + H^\perp}(\gamma).$$

Taking the inverse transform,

$$\begin{aligned} \tau_{[s]}f(x) &= M^{-r} \sum_{i=0}^{M^r-1} (s, \sigma_i) \int_{\sigma_i + H^\perp} (x - s - c, \gamma) d\nu(\gamma) \\ &= M^{-r} \sum_{i=0}^{M^r-1} (x - c, \sigma_i)\mathbf{1}_{s+c+H}(x). \end{aligned}$$ □

Of course, the operator $\tau_{[s]}$ depends on the choice $\mathcal{D}$ of coset representatives. In [**2**] it was shown that given an expansive automorphism A of G, it is possible to construct such a $\mathcal{D}$ while making only finitely many choices. We restate that result here as a proposition.

PROPOSITION 3.2 (From [**2**]). *Let G be a locally compact abelian group with compact open subgroup H. Let A be an expansive automorphism of G with modulus $|A| = M \geq 2$, and let $\mathcal{D}_1 = \{\rho_0, \ldots, \rho_{M-1}\}$ be a set of coset representatives for the (finite) quotient $H^\perp/((A^*)^{-1}H^\perp)$, with $\rho_0 = 0$. Define $\mathcal{D} \subset \widehat{G}$ to be the (infinite) set of all elements $\sigma \in \widehat{G}$ of the form*

$$\sigma = \sum_{j=1}^{n} (A^*)^j \rho_{i_j}, \quad \text{where } n \geq 1 \text{ and } i_j \in \{0, 1, \ldots, M - 1\}. \tag{3.4}$$

Then $\mathcal{D}$ is a set of coset representatives for the quotient $\widehat{G}/H^\perp$. Moreover, $A^*\mathcal{D} \subsetneq \mathcal{D}$.

EXAMPLE 3.3. Let $p \geq 2$ be a prime number, $G = \mathbb{Q}_p$, and $H = \mathbb{Z}_p$; identify $\widehat{G}$ as $\mathbb{Q}_p$ and $H^\perp$ as $\mathbb{Z}_p$, as in Example 2.4. Let the automorphism A be multiplication-by-$1/p$. As mentioned previously, Kozyrev [**17**] chose a certain standard set of coset representatives in $\mathbb{Q}_p$ for the quotient G/H; the same set in $\mathbb{Q}_p$ can be used as our choice of coset representatives for $\widehat{G}/H^\perp$. Specifically, let $\mathcal{D}$ consist of all elements of $\widehat{G} = \mathbb{Q}_p$ of the form

$$\sigma = \sum_{i=1}^{n} c_i p^{-i}, \quad \text{with } c_i \in \{0, 1, \ldots, p - 1\}.$$

This is the same set $\mathcal{D}$ that would come from $\mathcal{D}_1 = \{0, 1, 2 \ldots, (p - 1)\}$ using Proposition 3.2.

The resulting operator $\tau_{[s]}$ takes $f = \mathbf{1}_{c+p^n\mathbb{Z}_p}$ to

$$\tau_{[s]}f(x) = p^{-n} \left[\sum_{j=0}^{p^n-1} (x - c, \sigma_j) \right] \mathbf{1}_{s+c+\mathbb{Z}_p}(x)$$

by Proposition 3.1, with $r = n > 0$. If $s \in \mathbb{Z}_p$ (that is, if $[s] = [0]$), then it is easy to verify from the above formula that $\tau_{[s]}f = f$. On the other hand, if $s \notin \mathbb{Z}_p$, write $x \in s + c + \mathbb{Z}_p$ as $x \in c + m/p^\ell + p^n\mathbb{Z}_p$, for some integer m prime to p and some

$\ell > 1$. (Note that m is only determined modulo $p^{n+\ell}$.) The sum in the formula for $\tau_{[s]}f$ becomes

$$p^{-n}\left[\sum_{j=0}^{p^n-1}(x-c,\sigma_j)\right] = p^{-n}\sum_{j=0}^{p^n-1}(m/p^\ell, j/p^n) = p^{-n}\sum_{j=0}^{p^n-1}e^{2\pi imj/p^{\ell+n}}.$$

We may assume that $1 \leq m \leq p^{\ell+n}-1$. For values of m larger than p^ℓ, we should expect a great deal of cancellation in the sum, because the terms must be distributed fairly evenly around the unit circle. Thus, most of the weight of $\tau_{[s]}\mathbf{1}_{c+p^n\mathbb{Z}_p}$ should be concentrated in the (relatively few) regions $b + p^n\mathbb{Z}_p$ corresponding to small values of m.

We note that another standard choice for $\mathcal{D}$ would be to use $\mathcal{D}_1 = \{0\} \cup \{\zeta_{p-1}^i : i = 1, \ldots, (p-1)\}$, where $\zeta_{p-1} \in \mathbb{Q}_p$ is a primitive $(p-1)$st root of unity.

EXAMPLE 3.4. Let $p \geq 2$ be a prime number, $G = \mathbb{F}_p((t))$, and $H = \mathbb{F}_p[[t]]$; identify $\widehat{G}$ as $\mathbb{F}_p((t))$ and $H^\perp$ as $\mathbb{F}_p[[t]]$, as in Example 2.5. Let the automorphism A be multiplication-by-$1/t$. We may choose our set $\mathcal{D}$ of coset representatives for $\widehat{G}/H^\perp$ to be all elements of $\widehat{G}$ of the form

$$\sigma = \sum_{i=1}^{n} c_i t^{-i}, \quad \text{with } c_i \in \mathbb{F}_p.$$

This is the same set $\mathcal{D}$ that would come from Proposition 3.2 using $\mathcal{D}_1 = \mathbb{F}_p$ (the constants in $\mathbb{F}_p[[t]]$). As a subset of $\mathbb{F}_p((t))$, it is the same as the group Γ mentioned in Example 2.5. This time we are considering it as a subgroup of $\widehat{G}$, not of G, and so we call the group $\Gamma' = \mathcal{D}$.

In particular, $\mathcal{D}$ is actually a subgroup of $\widehat{G}$. Thus, the sum

$$p^{-n}\sum_{j=0}^{p^n-1}(x-c,\sigma_j)$$

(where σ_j ranges over the p^n elements of the group $\Gamma' \cap t^{-n}\mathbb{F}_p[[t]]$) from Proposition 3.1 simplifies to zero unless $(x-c,\sigma_j)$ is trivial for all $j = 0, 1, \ldots, p^n-1$. It follows that the operator $\tau_{[s]}$ is the usual translation-by-u operator, where $u \in G$ is the unique element of $(s+H) \cap \Gamma$. In particular, for $p = 2$, in which case $G = \mathbb{F}_2((t))$ is sometimes known as the "Cantor dyadic group", the translation operators induced by $\mathcal{D}$ are the same as the translation operators used by Lang in his study of wavelets on that group [**18**]. Indeed, for any p, $\mathbb{F}_p((t))$ is a locally compact abelian group with a discrete lattice Γ and dual lattice Γ', which is the situation studied by Dahlke in [**7**].

We close this example by noting that one could choose $\mathcal{D}$ in other ways so as not to form a subgroup. In that case, the operators $\tau_{[s]}$ would not be true translation operators, but would behave more like the operators of Example 3.3 above.

EXAMPLE 3.5. Let $G = \mathbb{Q}_3(\sqrt{-1})$ and $H = \mathbb{Z}_3[\sqrt{-1}]$; identify $\widehat{G}$ as $\mathbb{Q}_3(\sqrt{-1})$ and $H^\perp$ as $\mathbb{Z}_3[\sqrt{-1}]$. Let A be multiplication-by-$1/3$, so that $|A| = 9$. We may choose $\mathcal{D}_1$ to be the set of nine elements

$$\mathcal{D}_1 = \{a + b\sqrt{-1} : a, b = 0, 1, 2\},$$

so that the resulting full set $\mathcal{D}$ of coset representatives for $\widehat{G}/H^{\perp}$ consists of all elements of the form

$$\sigma = \sum_{i=1}^{n} (a_i + b_i\sqrt{-1})3^{-i}, \quad \text{with } a_i, b_i \in \{0,1,2\}.$$

Thus, $\mathcal{D}$ consists of elements of the form $q_1 + q_2\sqrt{-1}$, with $q_j = m_j/3^{n_j}$ and $0 \le m_j \le 3^{n_j} - 1$.

The resulting operator $\tau_{[s]}$ takes $f = \mathbf{1}_{c+3^n H}$ to

$$\tau_{[s]}f(x) = 9^{-n}\left[\sum_{m_1=0}^{3^n-1}\sum_{m_2=0}^{3^n-1}(x-c,(m_1+m_2\sqrt{-1})/3^n)\right]\mathbf{1}_{s+c+H}(x)$$

by Proposition 3.1, with $r = n > 0$. If $s \in H$ (that is, if $[s] = [0]$), then it is easy to verify from the above formula that $\tau_{[s]}f = f$. On the other hand, if $s \notin H$, write $x \in s+c+\mathbb{Z}_p$ as $x \in c+(m_3+m_4\sqrt{-1})/3^{\ell}+3^n H$, for some $\ell > 1$ and some integers m_3, m_4 not both divisible by 3. (Note that m_3 and m_4 are only determined modulo $p^{n+\ell}$.) The trace map from $\mathbb{Q}_3(\sqrt{-1})$ to $\mathbb{Q}_3$ takes $r_1 + r_2\sqrt{-1}$ to $2r_1$, and so the sum in the formula for $\tau_{[s]}f$ becomes

$$\begin{aligned}\tau_{[s]}f(x) &= 9^{-n}\left[\sum_{m_1=0}^{3^n-1}\sum_{m_2=0}^{3^n-1}(x-c,(m_1+m_2\sqrt{-1})/3^n)\right]\\ &= 9^{-n}\sum_{m_1=0}^{3^n-1}\sum_{m_2=0}^{3^n-1} e^{4\pi i(m_1m_3-m_2m_4)/3^{\ell+n}}.\end{aligned}$$

We may assume that $1 \le m_3, m_4 \le p^{\ell+n} - 1$. As in Example 3.3, we should expect a great deal of cancellation in the sum if either or both of m_3 and m_4 are larger than p^{ℓ}. Thus, most of the weight of $\tau_{[s]}\mathbf{1}_{c+3^n H}$ should be concentrated in the (relatively few) regions $b + 3^n H$ corresponding to small values of m_3 and m_4.

4. Haar and Shannon wavelets

Given our dilation operator δ_A and our group of translation operators $\Gamma_{G/H} = \{\tau_{[s]}\}$, we may now define wavelets on G as in [**2**].

DEFINITION 4.1. Let G be a locally compact abelian group with compact open subgroup $H \subseteq G$. Let A be an automorphism of G, expansive with respect to H. Let $\mathcal{D}$ be a choice of coset representatives in $\widehat{G}$ for $\widehat{H} = \widehat{G}/H^{\perp}$, and let $\Gamma_{G/H} = \{\tau_{[s]} : [s] \in G/H\}$ be the group of translation operators on $L^2(G)$ determined by $\mathcal{D}$ via equation (3.2).

Let $\Psi = \{\psi_1, \ldots, \psi_N\} \subseteq L^2(G)$ be a finite subset of $L^2(G)$. Ψ is a *set of wavelet generators* for $L^2(G)$ with respect to $\mathcal{D}$ and A if

$$\{\delta_A^n \tau_{[s]}\psi_i : 1 \le i \le N, n \in \mathbb{Z}, [s] \in G/H\}$$

forms an orthonormal basis for $L^2(G)$. In that case, the resulting basis is called a *wavelet basis* for $L^2(G)$.

If $N = 1$ and $\Psi = \{\psi\}$, then ψ is a *wavelet* for $L^2(G)$.

The following theorems show that any group G of the type we have considered has both Haar and Shannon wavelets, and, perhaps surprisingly, that the Haar wavelets are precisely the same as the Shannon wavelets.

THEOREM 4.2. *Let G be a locally compact abelian group with compact open subgroup H, and let $\mathcal{D}$ be a set of coset representatives for the quotient $\widehat{G}/H^{\perp}$. Let A be an expansive automorphism of G, and let $M = |A| \geq 2$ be the modulus of A. Let $\{\sigma_0, \sigma_1, \ldots, \sigma_{M-1}\}$ be the M elements of $(A^* H^{\perp})/H^{\perp}$, with $\sigma_0 \in H^{\perp}$.*

Define

$$\Omega_i = H^{\perp} + \sigma_i, \quad \text{and} \quad \psi_i = \check{\mathbf{1}}_{\Omega_i}, \quad \text{for all } i = 1, \ldots, M-1.$$

That is, ψ_i is the inverse transform of the characteristic function of $H^{\perp} + \sigma_i$. Then $\{\psi_1, \ldots, \psi_{M-1}\}$ is a set of wavelet generators for $L^2(G)$ with respect to $\mathcal{D}$ and A.

THEOREM 4.3. *Let G, H, $\mathcal{D}$, A, M, $\{\sigma_i : i = 0, \ldots, M-1\}$, and $\{\psi_i : i = 1, \ldots, M-1\}$ be as in Theorem 4.2. Then for any $i = 1, \ldots, M-1$ and any $[s] \in G/H$, the translated function $\tau_{[s]}\psi_i$ is*

$$\tau_{[s]}\psi_i(x) = (x, \sigma_i)\mathbf{1}_{s+H}(x).$$

Moreover, $\tau_{[s]}\psi_i$ is locally constant; specifically, it is constant on every set of the form $c + A^{-1}H$.

PROOF OF THEOREM 4.2. The sets $(A^*)^n \left(\sigma_i + H^{\perp}\right)$ are pairwise disjoint as n ranges over $\mathbb{Z}$ and i ranges over $\{1, \ldots, M-1\}$. Moreover, the union of all those sets is $\widehat{G} \setminus \{0\}$. Thus, up to sets of measure zero,

$$\{(A^*)^n \Omega_i : n \in \mathbb{Z}, i = 1, 2, \ldots, M-1\}$$

tiles $\widehat{G}$. At the same time, each Ω_i is the translate of $H^{\perp}$ by $\sigma_i - \sigma_0$. By the results of [**2**], then, $\{\psi_1, \ldots, \psi_{M-1}\}$ is a set of wavelet generators. □

PROOF OF THEOREM 4.3. By definition, $\tilde{\tau}_{[s]}\widehat{\psi_i}$ is the function

$$\tilde{\tau}_{[s]}\widehat{\psi_i}(\gamma) = \overline{(s, \eta(\gamma))}\widehat{\psi_i}(\gamma) = \overline{(s, \gamma - \sigma_i)}\mathbf{1}_{\sigma_i + H^{\perp}}(\gamma).$$

We compute $\tau_{[s]}\psi_i$ by taking the inverse transform:

$$\begin{aligned}
\tau_{[s]}\psi_i(x) &= \int_{\sigma_i + H^{\perp}} \overline{(s, \gamma - \sigma_i)}\,(x, \gamma)\, d\nu(\gamma) = \int_{H^{\perp}} \overline{(s, \gamma)}\,(x, \gamma)\,(x, \sigma_i)\, d\nu(\gamma) \\
&= (x, \sigma_i) \int_{H^{\perp}} \overline{(x - s, \gamma)} d\nu(\gamma) = (x, \sigma_i)\, \mathbf{1}_{s+H}(x)
\end{aligned}$$

as claimed.

Finally, $\tau_{[s]}\psi_i$ is constant on any set of the form $c + A^{-1}H$, by Propostion 2.1. This local constancy condition may also be proven directly from the above formula for $\tau_{[s]}\psi_i$; we leave the details to the reader. □

By Theorem 4.2, the transform $\widehat{\psi_i}$ of each ψ_i is the characteristic function of an uncomplicated set $\Omega_i = \sigma_i + H^{\perp}$ formed by translating the fundamental domain $H^{\perp}$. Moreover, these $M-1$ sets, together with $\Omega_0 = H^{\perp}$ itself, tile $A^* H^{\perp}$. Thus, the functions $\{\psi_i\}$ may be considered analogues for G of the standard Shannon wavelets from $\mathbb{R}^d$.

On the other hand, by Theorem 4.3, each ψ_i is a function supported in H and which is constant on each of the M subsets of H of the form $c + A^{-1}H$. Furthermore, each of those constant values is an M-th root of unity in $\mathbb{C}$. (After all, $M\sigma_i \in H$, since AH/H is a group of order M.) Therefore, the functions $\{\psi_i\}$ are analogues for G of the standard Haar wavelets for $\mathbb{R}^d$.

As claimed, then, Shannon wavelets are the same as Haar wavelets for our group G. Moreover, these wavelets ψ_i are locally constant and have compact support; the same is true of their transforms $\widehat{\psi}_i$. Thus, our Haar/Shannon wavelets are simultaneously smooth and of compact support in both the space and transform domains.

According to Theorem 4.3, the Haar wavelet ψ_i satisfies

$$\psi_i(x) = (x, \sigma_i)\mathbf{1}_H(x), \quad \text{and} \quad \tau_{[s]}\psi_i(x) = (x, \sigma_i)\mathbf{1}_{s+H}(x).$$

On the other hand, the usual translation-by-s operator sends ψ_i to

$$\psi_i(x - s) = \overline{(s, \sigma_i)}(x, \sigma_i)\mathbf{1}_{s+H}(x),$$

which is simply the constant $\overline{(s, \sigma_i)}$ multiplied by $\tau_{[s]}\psi_i(x)$. In particular, as Kozyrev [**17**] showed for the special case of $\mathbb{Q}_p$, $\{\psi_i : 1 \leq i \leq M - 1\}$ is a set of wavelet generators even if the usual translation-by-s operators are used, as s ranges over a set $\mathcal{C}$ of coset representatives for G/H.

Similarly, if we replace our "dual lattice" $\mathcal{D}$ by another choice $\mathcal{D}'$ of coset representatives for $\widehat{G}/H^\perp$, then the functions $\tau'_{[s]}\psi_i$ are different from $\tau_{[s]}\psi_i$, but only by a constant multiple $(s, \sigma_i - \sigma'_i)$. Thus, $\{\psi_i : 1 \leq i \leq M - 1\}$ is still a set of wavelet generators for the translation operators $\tau'_{[s]}$ induced by $\mathcal{D}'$.

The fact that waveletness is preserved even when the translation operators are changed so arbitrarily is unique to these particular wavelets. It happens mainly because the support of $\widehat{\psi}_i$ is contained in a single coset $\sigma + H^\perp$, and therefore ψ_i is the exponential (x, σ) multiplied by the characteristic function of H. There are simply not very many candidates for a new function $\tau_{[s]}\psi_i$ which could reasonably be considered some kind of translation of ψ_i by the coset $s + H$.

Other wavelets have transforms $\widehat{\psi}$ with support in at least two cosets $\sigma + H^\perp$, and therefore the usual translation operators do not work as well. In particular, the fact that there is no lattice in $G = \mathbb{Q}_p$ (or many other such groups) means that no discrete set of usual translation operators can form a group, making it very difficult for all the translations of a non-Haar ψ to be orthonormal. In the next section we shall consider a few examples of such non-Haar wavelets.

5. Other examples

If our expansive map A has modulus $|A| \geq 3$, then the sets of wavelet generators described in Section 4 have at least two elements. Using wavelet sets, J. Benedetto and the author produced (single) wavelets in the same context [**2**]. We state a special case of that result here without proof:

THEOREM 5.1 (From [**2**]). *Let G be a locally compact abelian group with compact open subgroup H, let A be an expansive automorphism of G, and let $M = |A| \geq 2$. Let $\mathcal{D}$ be a set of coset representatives for the quotient $\widehat{G}/H^\perp$, and let σ_0 be the unique element of $\mathcal{D} \cap H^\perp$. Define*

$$\mathcal{E} = \left(\mathcal{D} \cap A^* H^\perp\right) \setminus \{\sigma_0\},$$

and let $\{V_\sigma : \sigma \in \mathcal{E}\}$ be a partition of $H^\perp$ into $M - 1$ measurable subsets.

Define a function $T : H^\perp \to \widehat{G}$ by

$$T(\gamma) = \gamma + \sigma - \sigma_0 \quad \text{for } \gamma \in V_\sigma.$$

Then there is a measurable set $\Omega \subset A^* H^\perp$ such that, up to measure zero subsets,

$$\Omega = \left(H^\perp \setminus \bigcup_{n \geq 1} (A^*)^{-n} \Omega \right) \cup T \left(\bigcup_{n \geq 1} (A^*)^{-n} \Omega \right). \tag{5.1}$$

Let $\psi = \check{\mathbf{1}}_\Omega$ be the inverse transform of the characteristic function of Ω. Then ψ is a wavelet for $L^2(G)$.

The set Ω may be constructed from T inductively as follows. Define $\Omega_0 = H^\perp$; for every $n \geq 1$, define

$$\Lambda_n = \Omega_{n-1} \cap \left(\bigcup_{i \geq 1} (A^*)^{-i} \Omega_{n-1} \right) \quad \text{and} \quad \Omega_n = (\Omega_{n-1} \setminus \Lambda_n) \cup T\Lambda_n.$$

Then let

$$\Lambda = \bigcup_{n \geq 1} \Lambda_n, \qquad \text{and} \qquad \Omega = \left(H^\perp \setminus \Lambda \right) \cup T\Lambda. \tag{5.2}$$

We may compute the resulting wavelet ψ, and any translate $\tau_{[s]}\psi$, as follows. With notation as in Theorem 5.1, we have

$$\begin{aligned}
\tau_{[s]}\psi(x) &= \int_\Omega (x, \gamma)\overline{(s, \eta(\gamma))} d\nu(\gamma) \\
&= \int_{H^\perp \setminus \Lambda} (x, \gamma)\overline{(s, \gamma - \sigma_0)} d\nu(\gamma) + \sum_{\sigma \in \mathcal{E}} \int_{T(\Lambda \cap V_\sigma)} (x, \gamma)\overline{(s, \gamma - \sigma)} d\nu(\gamma) \\
&= (s, \sigma_0) \left[\int_{H^\perp \setminus \Lambda} (x - s, \gamma) d\nu(\gamma) + \sum_{\sigma \in \mathcal{E}} \int_{(\Lambda \cap V_\sigma)} (x, \gamma + \sigma - \sigma_0)\overline{(s, \gamma)} d\nu(\gamma) \right].
\end{aligned}$$

If we add and subtract $(s, \sigma_0) \int_\Lambda (x - s, \gamma) d\nu(\gamma)$, we see that the ψ produced by Theorem 5.1 is given by

(5.3)

$$\tau_{[s]}\psi(x) = (s, \sigma_0) \left[\mathbf{1}_H(x - s) + \sum_{\sigma \in \mathcal{E}} ((x, \sigma - \sigma_0) - 1) \int_{(\Lambda \cap V_\sigma)} (x - s, \gamma) d\nu(\gamma) \right].$$

Note that ψ has transform $\widehat{\psi} = \mathbf{1}_\Omega$ with support contained in $A^* H^\perp$. Thus, for any $[s] \in G/H$, $\tau_{[s]}\psi$ is constant on any set of the form $c + A^{-1}H$, by Proposition 2.1.

In the special case that $|A| = 2$, then $\mathcal{E}$ has only one element σ_1. Therefore there is only one choice for T (namely $T(\gamma) = \gamma + \sigma_1 - \sigma_0$ for all $\gamma \in H^\perp$), which produces $\Omega = \sigma_1 + H^\perp$. The resulting ψ is simply the Haar/Shannon wavelet from Section 4.

On the other hand, if $|A| \geq 3$, then although ψ is locally constant, it does not have compact support. We now consider examples of such wavelets ψ.

EXAMPLE 5.2. Let $p \geq 2$ be a prime number, let $G = \mathbb{Q}_p$, let $H = \mathbb{Z}_p$, and let A be multiplication-by-$1/p^r$, for some integer $r \geq 1$; note that $|A| = p^r$. Identify $\widehat{G}$ as $\mathbb{Q}_p$ and $H^\perp$ as $\mathbb{Z}_p$, as in Example 2.4. Let $\mathcal{D}$ be the set of coset representatives for $\widehat{G}/H^\perp$ defined in Example 3.3. Define $V_{1/p^r} = H^\perp$, and $V_\sigma = \emptyset$ for all $\sigma \in \mathcal{D} \setminus \{1/p^r\}$; then define T as in Theorem 5.1. That is, for all $\gamma \in H^\perp$, $T(\gamma) = \gamma + 1/p^r$. For every $n \geq 1$, the subset Λ_n of $\widehat{G}$ is

$$\Lambda_n = 1 + p^r + p^{2r} + \cdots + p^{(n-2)r} + p^{nr}\mathbb{Z}_p,$$

which is the closed ball of radius $1/p^{nr}$ about $1+p^r+\cdots+p^{(n-2)r} \in \widehat{G}$, and which has measure $1/p^{nr}$. (Note that $\Lambda_1 = p^r\mathbb{Z}_p$ is the closed ball of radius $1/p^r$ about 0.) Define Λ and Ω as in equation (5.2). Note that $\Lambda = \bigcup \Lambda_n \subset \mathbb{Z}_p$ is a disjoint union of countably many balls, and that $\nu(\Lambda) = 1/(p^r-1)$. Define $\psi = \check{\mathbf{1}}_\Omega$; then by Theorem 5.1, ψ is a (single) wavelet.

We can describe $\tau_{[s]}\psi(x)$ explicitly using equation (5.3), as follows. Since there is only one nonempty V_σ, namely $V_{1/p^r} = \mathbb{Z}_p$, we have $\Lambda \cap V_{1/p^r} = \Lambda$, and therefore

$$\tau_{[s]}\psi(x) = \mathbf{1}_{\mathbb{Z}_p}(x-s) + ((x,1/p^r)-1)\sum_{n\geq 1}\int_{\Lambda_n}(x-s,\gamma)d\gamma.$$

Given $x, s \in G$ and $n \geq 1$, we compute

$$\int_{\Lambda_n}(x-s,\gamma)d\gamma = \frac{1}{p^{nr}}(x-s, 1+p^r+\cdots+p^{(n-2)r})\mathbf{1}_{p^{-nr}\mathbb{Z}_p}(x-s).$$

Thus, for $x, s \in G$, if we let N be the smallest positive integer such that $x-s \in p^{-Nr}\mathbb{Z}_p$, then

$$\begin{aligned}\tau_{[s]}\psi(x) = \mathbf{1}_{\mathbb{Z}_p}(x-s) &+ \frac{[(x,1/p^r)-1]}{p^{Nr}}(x-s, 1+p^r+\cdots+p^{(N-2)r}) \\ &+ \frac{[(x,1/p^r)-1]}{p^{Nr}(p^r-1)}(x-s, 1+p^r+\cdots+p^{(N-1)r}),\end{aligned}$$

because $(x-s,\gamma) = 1$ for all $\gamma \in p^{Nr}\mathbb{Z}_p$. Note that the p^{Nr} in the denominator forces $\tau_{[s]}\psi$ to decay as x goes to ∞ in $\mathbb{Q}_p$.

EXAMPLE 5.3. Let $G = \mathbb{Q}_3(\sqrt{-1})$, let $H = \mathbb{Z}_3[\sqrt{-1}]$, and let A be multiplication-by-$1/3^r$, for some integer $r \geq 1$; note that $|A| = 9^r$. Identify $\widehat{G}$ as $\mathbb{Q}_3(\sqrt{-1})$ and $H^\perp$ as $\mathbb{Z}_3[\sqrt{-1}]$, as in Example 2.6. Let $\mathcal{D}$ be the set of coset representatives for $\widehat{G}/H^\perp$ defined in Example 3.5. Define $V_{1/3^r} = H^\perp$, and $V_\sigma = \emptyset$ for all $\sigma \in \mathcal{D} \setminus \{1/3^r\}$; then define T as in Theorem 5.1. That is, for all $\gamma \in H^\perp$, $T(\gamma) = \gamma + 1/3^r$. For every $n \geq 1$, the subset Λ_n of $\widehat{G}$ is

$$\Lambda_n = 1+3^r+3^{2r}+\cdots+3^{(n-2)r}+3^{nr}\mathbb{Z}_3[\sqrt{-1}],$$

which is the closed ball of radius $1/3^{nr}$ about $1+3^r+\cdots+3^{(n-2)r} \in \widehat{G}$, and which has measure $1/9^{nr}$.

As before, the set Λ from equation (5.2) is a disjoint union of countably many balls, and that $\nu(\Lambda) = 1/(9^r-1)$. Define $\psi = \check{\mathbf{1}}_\Omega$; then by Theorem 5.1, ψ is a (single) wavelet. By a computation similar to that in Example 5.2, we find that

$$\begin{aligned}\tau_{[s]}\psi(x) = \mathbf{1}_H(x-s) &+ \frac{[(x,1/3^r)-1]}{9^{Nr}}(x-s, 1+3^r+\cdots+3^{(N-2)r}) \\ &+ \frac{[(x,1/3^r)-1]}{9^{Nr}(9^r-1)}(x-s, 1+3^r+\cdots+3^{(N-1)r}),\end{aligned}$$

where N be the smallest positive integer such that $x-s \in 3^{-Nr}H$.

EXAMPLE 5.4. Let $p \geq 2$ be a prime number, let $G = \mathbb{F}_p((t))$, let $H = \mathbb{F}_p[[t]]$, let A be multiplication-by-$1/t$, and identify $\widehat{G}$ as $\mathbb{F}_p((t))$ and $H^\perp$ as $\mathbb{F}_p[[t]]$, as in Example 2.5. Let $\mathcal{D}$ be the set of coset representatives for $\widehat{G}/H^\perp$ defined in Example 3.4. Define $V_{1/t} = H^\perp$, and $V_\sigma = \emptyset$ for all $\sigma \in \mathcal{D} \setminus \{1/t\}$, and define T as in Theorem 5.1. That is, for all $\gamma \in H^\perp$, $T(\gamma) = \gamma + 1/t$.

The resulting Λ_n, Λ, and Ω are exactly analogous to those in Example 5.2, but with t in place of p. As in that case, we get

$$\tau_{[s]}\psi(x) = \mathbf{1}_{\mathbb{F}_p[[t]]}(x-s) + \left[\frac{(x,1/t)-1}{p^N(p-1)}\right](x-s, 1+t+\cdots+t^{N-1}).$$

(We still have p appearing in the denominator because the measure of Λ_n is the *real number* p^n, not an element of G.)

This time, we can simplify the expression even further. Write

$$x = a_{-N}t^{-N} + a_{-N+1}t^{-N+1} + \cdots \quad \text{and}$$
$$x - s = b_{-N}t^{-N} + b_{-N+1}t^{-N+1} + \cdots,$$

where N is the smallest positive integer for which $b_{-N} \neq 0$ (or $N = 1$ if no such integer exists). Then

$$\tau_{[s]}\psi(x) = \mathbf{1}_{\mathbb{F}_p[[t]]}(x-s) + \left[\frac{(e^{2\pi i a_0/p}-1)}{p^N(p-1)}\right]\left[e^{2\pi i c/p} + (p-1)e^{2\pi i d/p}\right],$$

where

$$c = b_{-N+1} + b_{-N+2} + \cdots + b_{-1} \in \mathbb{F}_p, \quad \text{and}$$
$$d = b_{-N} + b_{-N+1} + \cdots + b_{-1} \in \mathbb{F}_p.$$

EXAMPLE 5.5. Let $G = \mathbb{Q}_3$, let $H = \mathbb{Z}_3$, and let A be multiplication-by-1/3, so that $|A| = 3$. Identify $\widehat{G}$ as $\mathbb{Q}_3$ and $H^\perp$ as $\mathbb{Z}_3$, as in Example 2.4, for $p = 3$. Let $\mathcal{D}$ be the set of coset representatives for $\widehat{G}/H^\perp$ defined in Example 3.3. Define $V_{1/3} = (3\mathbb{Z}_3) \cup (2 + 3\mathbb{Z}_3)$, $V_{2/3} = 1 + 3\mathbb{Z}_3$, and $V_\sigma = \emptyset$ for all $\sigma \in \mathcal{D} \setminus \{1/3, 2/3\}$, and define T as in Theorem 5.1. That is, for all $\gamma \in 1 + 3\mathbb{Z}_3$, $T(\gamma) = \gamma + 2/3$; and for all $\gamma \in \mathbb{Z}_3 \setminus (1 + 3\mathbb{Z}_3)$, $T(\gamma) = \gamma + 1/3$. Then for $n \geq 1$,

$$\Lambda_n = \begin{cases} 1 + 2\cdot 3 + 1\cdot 3^2 + 2\cdot 3^3 + \cdots + 1\cdot 3^{n-2} + 3^n\mathbb{Z}_3, & \text{if } n \text{ is even;} \\ 2 + 1\cdot 3 + 2\cdot 3^2 + 1\cdot 3^3 + \cdots + 1\cdot 3^{n-2} + 3^n\mathbb{Z}_3, & \text{if } n \text{ is odd} \end{cases}$$

Thus, Λ_n is a closed ball of radius $1/3^n$ (and hence measure $1/3^n$) about a point which is in $1 + 3\mathbb{Z}_3$ if n is even, or in $2 + 3\mathbb{Z}_3$ if $n \geq 3$ is odd, or at 0 if $n = 1$. For Λ and Ω as in equation (5.2), we see as before that $\Lambda = \bigcup \Lambda_n \subset \mathbb{Z}_3$ is a disjoint union of countably many balls, and that $\nu(\Lambda) = 1/(3-1) = 1/2$.

Define $\psi = \check{\mathbf{1}}_\Omega$; then by Theorem 5.1, ψ is a (single) wavelet. And again, $\tau_{[s]}\psi$ must be constant on every set of the form $c + 3\mathbb{Z}_3 \subset G$, for any $[s] \in G/H$; but ψ does not have compact support. This time, when we compute $\tau_{[s]}\psi(x)$ by equation (5.3), we get

$$\begin{aligned}\tau_{[s]}\psi(x) &= \mathbf{1}_{\mathbb{Z}_3}(x-s) \\ &+ \frac{[(x,2/3)-1]}{3^N}(x-s, 1+2\cdot 3+1\cdot 3^2+\cdots+1\cdot 3^{(N-2)}) \\ &+ \frac{[(x,2/3)-1]}{8\cdot 3^N}(x-s, 1+2\cdot 3+1\cdot 3^2+\cdots+2\cdot 3^{(N-1)}) \\ &+ \frac{[(x,1/3)-1]}{8\cdot 3^{N-1}}(x-s, 2+1\cdot 3+2\cdot 3^2+\cdots+1\cdot 3^{(N-1)}),\end{aligned}$$

if N is even, and

$$\begin{aligned}\tau_{[s]}\psi(x) &= \mathbf{1}_{\mathbb{Z}_3}(x-s)\\ &+ \frac{[(x,1/3)-1]}{3^N}(x-s, 2+1\cdot 3+2\cdot 3^2+\cdots+1\cdot 3^{(N-2)})\\ &+ \frac{[(x,1/3)-1]}{8\cdot 3^N}(x-s, 2+1\cdot 3+2\cdot 3^2+\cdots+2\cdot 3^{(N-1)}),\\ &+ \frac{[(x,2/3)-1]}{8\cdot 3^{N-1}}(x-s, 1+2\cdot 3+1\cdot 3^2+\cdots+1\cdot 3^{(N-1)})\end{aligned}$$

if N is odd, where N is the smallest positive integer such that $x-s \in 3^{-N}\mathbb{Z}_3$.

The author would like to thank John Benedetto and the referee for their helpful comments on the exposition of this paper.

References

[1] L. Baggett, A. Medina, and K. Merrill, Generalized multi-resolution analyses and a construction procedure for all wavelet sets in $\mathbf{R}^n$, *J. Fourier Anal. Appl.* **5** (1999), 563–573.

[2] J. Benedetto and R. Benedetto, A wavelet theory for local fields and related groups, submitted.

[3] J. Benedetto and M. Leon, The construction of multiple dyadic minimally supported frequency wavelets on $\mathbb{R}^d$, *The Functional and Harmonic Analysis of Wavelets and Frames (San Antonio, TX, 1999)* (L.W. Baggett and D.R. Larson, eds.), Contemporary Mathematics, vol. 247, American Mathematical Society, Providence, RI, 1999, pp. 43–74.

[4] J. Benedetto and M. Leon, The construction of single wavelets in d-dimensions, *J. Geom. Anal.* **11** (2001), 1–15.

[5] A. Connes, Trace formula on the adèle class space and Weil positivity, *Current Developments in Mathematics, 1997*, International Press, Boston, 1999, pp. 5–64.

[6] A. Connes, Trace formula in noncommutative geometry and the zeros of the Riemann zeta function, *Selecta Math. (N.S.)* **5** (1999), 29–106.

[7] S. Dahlke, Multiresolution analysis and wavelets on locally compact abelian groups, *Wavelets, Images, and Surface Fitting (Chamonix-Mont-Blanc, 1993)*, A K Peters, Wellesley, 1994, pp. 141–156.

[8] I. Daubechies, *Ten Lectures on Wavelets*, CBMS-NSF Regional Conference Series in Applied Mathematics, vol. 61, SIAM, Philadelphia, 1992.

[9] X. Dai, D. Larson, and D. Speegle, Wavelet sets in $\mathbf{R}^n$, *J. Fourier Anal. Appl.* **3** (1997), 451–456.

[10] Yu. Farkov, Orthogonal wavelets on locally compact abelian groups, *Funktsional. Anal. i Prilozhen.* **31** (1997), 86–88.

[11] F. Gouvêa, *p-adic Numbers: An Introduction,* 2nd ed., Universitext, Springer-Verlag, Berlin, 1997.

[12] D. Han, D. Larson, M. Papadakis, and Th. Stavropoulos, Multiresolution analyses of abstract Hilbert spaces and wandering subspaces, *The Functional and Harmonic Analysis of Wavelets and Frames (San Antonio, TX, 1999)* (L.W. Baggett and D.R. Larson, eds.), Contemporary Mathematics, vol. 247, American Mathematical Society, Providence, RI, 1999, pp. 259–284.

[13] E. Hewitt and K. Ross, *Abstract Harmonic Analysis, Volume I,* Springer-Verlag, New York, 1963.

[14] E. Hewitt and K. Ross, *Abstract Harmonic Analysis, Volume II*, Springer-Verlag, New York, 1970.

[15] M. Holschneider, Wavelet analysis over abelian groups, *Appl. Comput. Harmon. Anal.* **2** (1995), 52–60.

[16] N. Koblitz, *p-adic Numbers, p-adic Analysis, and Zeta-Functions*, 2nd ed., Springer-Verlag, New York-Berlin, 1984.

[17] S. Kozyrev, Wavelet theory as p-adic spectral analysis, *Izv. Ross. Akad. Nauk Ser. Mat.* **66** (2002), 149–158.

[18] W.C. Lang, Wavelet analysis on the Cantor dyadic group, *Houston J. Math.* **24** (1998), 533–544.

[19] T. Lukashenko, Wavelets on topological groups, *Izv. Ross. Akad. Nauk Ser. Mat.* **58** (1994) 88–102.

[20] Y. Meyer, *Ondelettes et opérateurs, I*, Actualités Mathématiques, Hermann, Paris, 1990.

[21] M. Papadakis and Th. Stavropoulos, On the multiresolution analyses of abstract Hilbert spaces, *Bull. Greek Math. Soc.* **40** (1998), 79–92.

[22] M. Papadakis, Generalized frame multiresolution analysis of abstract Hilbert spaces, *Sampling Theory in Signal and Image Processing*, to appear.

[23] D. Ramakrishnan and R. Valenza, *Fourier Analysis on Number Fields*, Springer-Verlag, New York, 1999.

[24] A. Robert, *A Course in p-adic Analysis*, Springer-Verlag, New York, 2000.

[25] J.-P. Serre, *Local Fields*, translated by M. Greenberg, Springer-Verlag, New York-Berlin, 1979.

[26] P. Soardi and D. Weiland, Single wavelets in n-dimensions, *J. Fourier Anal. Appl.* **4** (1998), 299–315.

[27] J. Tate, Fourier analysis in number fields and Hecke's zeta-functions, Princeton University Ph.D. thesis, 1950; in J. Cassels, A. Frölich, eds., *Algebraic Number Theory*, Thompson, Washington, 1967.

Department of Mathematics and Computer Science, Amherst College, Amherst, MA 01002, USA

E-mail address: `rlb@cs.amherst.edu`

URL: `http://www.cs.amherst.edu/~rlb`

Contemporary Mathematics
Volume **345**, 2004

The spectral function of shift-invariant spaces on general lattices

Marcin Bownik and Ziemowit Rzeszotnik

Abstract. We extend the notion of the spectral function of shift-invariant spaces introduced by the authors in [**BRz**] to the case of general lattices. The main feature is that the spectral function is not dependent on the choice of the underlying lattice with respect to which a space is shift-invariant. We also show that in general the spectral function is not additive on the orthogonal infinite sums of SI spaces with varying lattices.

1. Introduction

The shift-invariant (SI) spaces are closed subspaces of $L^2(\mathbb{R}^n)$ that are invariant under all shifts, i.e., integer translations. The theory of shift-invariant subspaces of $L^2(\mathbb{R}^n)$ plays an important role in many areas, most notably in the theory of wavelets, spline systems, Gabor systems, and approximation theory [**BMM, BDR1, BDR2, BL, Bo1, BRz, Ji, RS1, RS2**].

Given a SI space $V \subset L^2(\mathbb{R}^n)$, we would like to associate to V some kind of a function on $\mathbb{R}^n$ which encapsulates the most important properties of V, such as the "size" of V. One of such possible functions is the dimension function (or multiplicity function) of V, which measures the size of V by counting the dimensions of "fibers" of V. Another such possible function, which contains much more information than the dimension function, is the spectral function measuring "localized size" of V.

The goal of this work is to show the existence and the fundamental properties of the spectral function associated to SI spaces on general lattices. This extends the spectral function for SI spaces with respect to the standard lattice $\mathbb{Z}^n$ introduced by the authors in [**BRz**]. Quite likely the most striking property of the spectral function shown in this work is independence of its definition with respect to the choice of the underlying lattice. This property is a unique feature of the spectral function, which is not shared by the dimension function.

The spectral function is not only interesting in itself, but it is also useful in studying wavelet and Gabor systems. Several applications of the spectral function were presented in [**BRz**]. For example, the spectral function can be employed to

---

2000 *Mathematics Subject Classification.* 42C40.

Key words and phrases. Shift-invariant spaces, spectral function, dimension function, orthogonal wavelets.

The first author was partially supported by the NSF grant DMS-0200080.

characterize approximation order of shift-invariant spaces [**BRz**, Section 2], dimension functions of refinable spaces and generalized multiresolution analyses [**BRz**, Section 3 and 4], and to provide a proof of Rieffel's incompleteness theorem for Gabor systems [**BRz**, Section 5]. In this work we concentrate on the relevance of the spectral function to the study of general shift-invariant systems with varying lattice of translations. Such systems were recently studied by Hernández, Labate, and Weiss [**HLW**].

In order to define the spectral function we need to recall a few basic facts about shift-invariant spaces.

Suppose $\Gamma \subset \mathbb{R}^n$ is a lattice of full rank, i.e., $\Gamma = P\mathbb{Z}^n$ for some $n \times n$ non-singular real matrix P. We say that a closed subspace $V \subset L^2(\mathbb{R}^n)$ is Γ-*shift-invariant* (Γ-SI) if for every function $f \in V$ we also have $T_\gamma f \in V$ when $\gamma \in \Gamma$, where $T_y f(x) = f(x-y)$ is the translation by a vector $y \in \mathbb{R}^n$. For any subset $\Phi \subset L^2(\mathbb{R}^n)$ let

$$\mathcal{S}^\Gamma(\Phi) = \overline{\operatorname{span}}\{T_\gamma \varphi : \varphi \in \Phi, \gamma \in \Gamma\}$$

be the Γ-SI space generated by Φ. A Γ-*principal shift-invariant* (Γ-PSI) space is a Γ-SI space V generated by a single function $\varphi \in L^2(\mathbb{R}^n)$, i.e., $V = \mathcal{S}^\Gamma(\{\varphi\}) = \mathcal{S}^\Gamma(\varphi)$.

A *range function* is any mapping

$$J : \mathbb{R}^n \to \{\text{closed subspaces of } \ell^2(\Gamma^*)\},$$

satisfying a consistency formula with respect to the lattice Γ^*,

$$J(\xi + \gamma) = S_\gamma(J(\xi)) \qquad \text{for all } \gamma \in \Gamma^*.$$

Here,

$$\Gamma^* = \{x \in \mathbb{R}^n : \langle x, k\rangle \in \mathbb{Z} \quad \text{for all } k \in \Gamma\}$$

is the dual lattice and $S_\gamma : \ell^2(\Gamma^*) \to \ell^2(\Gamma^*)$ is the shift operator given by

$$S_\gamma((a_k)_{k\in\Gamma^*}) = (a_{k-\gamma})_{k\in\Gamma^*}.$$

We say that J is *measurable* if the associated orthogonal projections $P_J(\xi)$ of $\ell^2(\Gamma^*)$ onto $J(\xi)$ are operator measurable, i.e., $\xi \mapsto P_J(\xi)v$ is measurable for any $v \in \ell^2(\Gamma^*)$. We note that the range function is uniquely determined by its values on the representatives of the cosets of $\mathbb{R}^n/\Gamma^*$. Therefore, it suffices to define the range function only on some fundamental domain of $\mathbb{R}^n/\Gamma^*$ and then extend it using the above consistency formula.

Define the Hilbert space

$$L^2(\mathbb{R}^n/\Gamma^*, \ell^2(\Gamma^*)) = \{\Phi : \mathbb{R}^n \to \ell^2(\Gamma^*) : \quad \Phi(\xi+\gamma) = S_\gamma \Phi(\xi) \text{ for all } \gamma \in \Gamma^*,$$
$$\text{and } ||\Phi||^2 := \int_{\mathbb{R}^n/\Gamma^*} ||\Phi(\xi)||^2_{\ell^2(\Gamma^*)} d\xi < \infty\}.$$

Let $\mathcal{T} : L^2(\mathbb{R}^n) \to L^2(\mathbb{R}^n, \ell^2(\Gamma^*))$ be an isometric isomorphism defined for $f \in L^2(\mathbb{R}^n)$ by

$$\mathcal{T}f : \mathbb{R}^n \to \ell^2(\Gamma^*), \qquad \mathcal{T}f(\xi) = (\hat{f}(\xi+\gamma))_{\gamma\in\Gamma^*},$$

where $\hat{f}(\xi) = \int f(x) e^{-2\pi i\langle x,\xi\rangle}\, dx$.

The following proposition, due to Helson [**He**, Theorem 8, p. 59], plays an important role in the theory of general SI spaces in $L^2(\mathbb{R}^n)$. A proof of Proposition 1.1 for $\mathbb{Z}^n$-SI spaces can be also found in [**Bo1**, Proposition 1.5]. A general case of Γ-SI is an immediate consequence of $\mathbb{Z}^n$-SI case.

PROPOSITION 1.1. *A closed subspace $V \subset L^2(\mathbb{R}^n)$ is Γ-SI if and only if*

$$V = \{f \in L^2(\mathbb{R}^n) : \mathcal{T}f(\xi) \in J(\xi) \text{ for a.e. } \xi \in \mathbb{R}^n\}, \tag{1.1}$$

where J is a measurable range function. The correspondence between V and J is one-to-one under the convention that the range functions are identified if they are equal a.e. Furthermore, if $V = \mathcal{S}^\Gamma(\Phi)$ for some countable $\Phi \subset L^2(\mathbb{R}^n)$, then

$$J(\xi) = \overline{\operatorname{span}}\{\mathcal{T}\varphi(\xi) : \varphi \in \Phi\}. \tag{1.2}$$

The *dimension function* of a Γ-SI space $V \subset L^2(\mathbb{R}^n)$ is a mapping $\dim_V^\Gamma : \mathbb{R}^n \to \mathbb{N} \cup \{0, \infty\}$ given by $\dim_V^\Gamma(\xi) = \dim J(\xi)$, where J is the range function corresponding to V. Alternatively, the dimension function of V can be introduced as the *multiplicity function* of the projection-valued measure coming from the representation of the lattice Γ on V via translations by Stone's Theorem, see [**Ba, BMM**].

Note also that $V = \check{L}^2(E)$ is Γ-SI for any lattice Γ, where E is a measurable subset of $\mathbb{R}^n$ and

$$\check{L}^2(E) = \{f \in L^2(\mathbb{R}^n) : \operatorname{supp} \hat{f} \subset E\}, \tag{1.3}$$

moreover its dimension function is given by

$$\dim_V^\Gamma(\xi) = \sum_{k \in \Gamma^*} \mathbf{1}_E(\xi + k). \tag{1.4}$$

2. The spectral function

In this section we show the basic properties of the spectral function associated to general shift-invariant spaces with respect to an arbitrary lattice $\Gamma \subset \mathbb{R}^n$. The spectral function for shift-invariant spaces with respect to the standard lattice $\mathbb{Z}^n$ was introduced and investigated by the authors in [**BRz**].

The spectral function for $\mathbb{Z}^n$-SI spaces.

An interesting feature of the spectral function for $\mathbb{Z}^n$-SI spaces is that it can be defined in several equivalent ways. Indeed, it was shown in [**BRz**] that the spectral function of $\mathbb{Z}^n$-SI space V can be defined using either of the following methods:

(a) orthogonal projections of the unit standard vectors of $\ell^2(\mathbb{Z}^n)$ onto the range function [**BRz**, Definition 2.1],
(b) decompositions of V into the orthogonal sum of PSI spaces [**BRz**, Proposition 2.2],
(c) generators of SI systems forming tight frame (with constant 1) for V [**BRz**, Lemma 2.3],
(d) diagonal terms of dual Gramians for SI systems as in (c) [**BRz**, Remark after Lemma 2.3],
(e) density formula based on the Lebesgue differentiation theorem [**BRz**, formula (2.15)].

More precisely, the following proposition can serve as the definition of the spectral function.

PROPOSITION 2.1. *Suppose $V \subset L^2(\mathbb{R}^n)$ is $\mathbb{Z}^n$-SI. Then for a.e. $\xi \in \mathbb{R}^n$ the following expressions are equal:*

(i)
$$||P_J(\xi)e_0||^2,$$
where $P_J(\xi)$ is the orthogonal projection onto the range function $J(\xi)$ corresponding to V and given by Proposition 1.1,

(ii)
$$\sum_{\varphi\in\Phi} |\hat{\varphi}(\xi)|^2,$$
where $\Phi \subset V$ is such that its translates by $\mathbb{Z}^n$,
$$E^{\mathbb{Z}^n}(\Phi) = \{T_k\varphi : k \in \mathbb{Z}^n, \varphi \in \Phi\},$$
form a tight frame with constant 1 for V; moreover, this sum is independent of the choice of such Φ,

(iii)
$$\lim_{r\to 0^+} \frac{||P_V(\check{\mathbf{1}}_{B(\xi,r)})||^2}{|B(\xi,r)|},$$
where P_V is the orthogonal projection of $L^2(\mathbb{R}^n)$ onto V and $B(\xi, r)$ denotes the ball with center ξ and radius r.

The spectral function of V is defined as mapping $\sigma_V : \mathbb{R}^n \to \mathbb{R}$, where $\sigma_V(\xi)$ is given by either one of (i)–(iii).

The next proposition lists the most important properties of the spectral function, see [**BRz**, Proposition 2.6]. In particular, Proposition 2.2 shows that the spectral function behaves nicely with respect to the action of modulations and dilations. This is relevant in the study of Gabor systems and wavelets. Recall, that the *modulation* by a vector $a \in \mathbb{R}^n$ of $f \in L^2(\mathbb{R}^n)$ is given by
$$M_a(f)(x) = e^{2\pi i\langle a,x\rangle} f(x).$$
The *dilation* by $n \times n$ non-singular matrix A of $f \in L^2(\mathbb{R}^n)$ is given by
$$D_A f(x) = |\det A|^{\frac{1}{2}} f(Ax).$$
In Proposition 2.2(g), we restrict our attention to dilations A preserving the lattice $\mathbb{Z}^n$ because this is exactly when in general we can expect that D_AV is $\mathbb{Z}^n$-SI if V is $\mathbb{Z}^n$-SI. Later we will see that this property holds for a general non-singular matrix A.

PROPOSITION 2.2. *Let $\mathfrak{S}$ be the collection of all $\mathbb{Z}^n$-SI subspaces of $L^2(\mathbb{R}^n)$. The spectral function satisfies the following properties: ($V, W \in \mathfrak{S}$)*
(a) $\sigma_V : \mathbb{R}^n \to [0, 1]$ is a measurable function,
(b) $V = \bigoplus_{i\in\mathbb{N}} V_i$, where $V_i \in \mathfrak{S} \implies \sigma_V(\xi) = \sum_{i\in\mathbb{N}} \sigma_{V_i}(\xi)$,
(c) $V \subset W \implies \sigma_V(\xi) \le \sigma_W(\xi)$,
(d) $V \subset W \implies (V = W \iff \sigma_V(\xi) = \sigma_W(\xi))$,
(e) $\sigma_V(\xi) = \mathbf{1}_E(\xi) \iff V = \check{L}^2(E)$,
(f) $\sigma_{M_a(V)}(\xi) = \sigma(\xi - a)$, where M_a is a modulation by $a \in \mathbb{R}^n$,
(g) $\sigma_{D_AV}(\xi) = \sigma_V((A^*)^{-1}\xi)$, where D_A is a dilation by non-singular integer matrix A,

(h) $\dim_V^{\mathbb{Z}^n}(\xi) = \sum_{k\in\mathbb{Z}^n} \sigma_V(\xi + k)$.

Finally, the following lemma will be very useful in our considerations, see [**BRz**, Lemma 2.8].

LEMMA 2.3. *Suppose V is $\mathbb{Z}^n$-SI and $K \subset \mathbb{R}^n$ is a measurable set such that $|K \cap (l + K)| = 0$ for all $l \in \mathbb{Z}^n \setminus \{0\}$. Then*

$$||P_V(\check{\mathbf{1}}_K)||^2 = \int_K \sigma_V(\xi)d\xi, \tag{2.1}$$

where P_V is an orthogonal projection of $L^2(\mathbb{R}^n)$ onto V.

The spectral function for general SI spaces.

Using Propositions 2.1 and 2.2 it is not hard to give a definition of the spectral function σ_V^Γ for a general lattice Γ. Possibly the slickest way of defining the spectral function of a Γ-SI space V, where $\Gamma = P\mathbb{Z}^n$ for some $n \times n$ non-singular matrix P is suggested by Proposition 2.2(g). In fact, we could set

$$\sigma_V^\Gamma(\xi) = \sigma_{D_P V}(P^*\xi) \qquad \text{for } \xi \in \mathbb{R}^n,$$

because $D_P V$ is $\mathbb{Z}^n$-SI and therefore has the usual spectral function. One can show that this definition does not depend on the choice of a matrix P such that $\Gamma = P\mathbb{Z}^n$. Indeed, given two matrices P_1 and P_2 with $\Gamma = P_1\mathbb{Z}^n = P_2\mathbb{Z}^n$ we can verify using Proposition 2.2(g) that

$$\sigma_{D_{P_2}V}(\xi) = \sigma_{D_{P_1^{-1}P_2}D_{P_1}V}(\xi) = \sigma_{D_{P_1}V}(((P_1^{-1}P_2)^*)^{-1}\xi) = \sigma_{D_{P_1}V}(P_1^*(P_2^*)^{-1}\xi),$$

since $P_1^{-1}P_2$ has integer entries and $D_{P_1}V$ is $\mathbb{Z}^n$-SI.

However, for the purposes of a systematic development of the spectral function it is better to use the following formal definition (we will see later in (2.5) that it coincides with the above formulation).

DEFINITION 2.4. Suppose $V \subset L^2(\mathbb{R}^n)$ is Γ-SI with the range function $J(\xi)$ and the corresponding projection $P_J(\xi)$. The *spectral function* of V is a measurable mapping $\sigma_V^\Gamma : \mathbb{R}^n \to [0,1]$ given by

$$\sigma_V^\Gamma(\xi) = ||P_J(\xi)e_0||^2 \qquad \text{for } \xi \in \mathbb{R}^n, \tag{2.2}$$

where $\{e_k\}_{k\in\Gamma^*}$ denotes the standard basis of $\ell^2(\Gamma^*)$.

Note that there is a very simple relationship between the spectral and the dimension function

$$\dim_V^\Gamma(\xi) = \sum_{k\in\Gamma^*} \sigma_V^\Gamma(\xi + k). \tag{2.3}$$

Indeed,

$$\sum_{k\in\Gamma^*} \sigma_V^\Gamma(\xi + k) = \sum_{k\in\Gamma^*} ||P_J(\xi + k)e_0||^2 = \sum_{k\in\Gamma^*} ||P_J(\xi)e_k||^2 = \dim \operatorname{Ran}(P_J(\xi))$$
$$= \dim_V^\Gamma(\xi).$$

Since the above definition is rather abstract we can give an alternative description of the spectral function of V in terms of a family generating a tight frame for V.

LEMMA 2.5. *Suppose $V \subset L^2(\mathbb{R}^n)$ is Γ-SI space and $\Phi \subset V$ is a countable family such that $E^\Gamma(\Phi) = \{T_k\varphi : k \in \Gamma, \varphi \in \Phi\}$ forms a tight frame with constant 1 for the space V. Then*

$$\sigma_V^\Gamma(\xi) = \frac{1}{|\mathbb{R}^n/\Gamma|} \sum_{\varphi\in\Phi} |\hat{\varphi}(\xi)|^2, \tag{2.4}$$

where $|\mathbb{R}^n/\Gamma|$ is the Lebesgue measure of the fundamental domain of $\mathbb{R}^n/\Gamma$, i.e., $|\mathbb{R}^n/\Gamma| = |\det P|$, if $\Gamma = P\mathbb{Z}^n$ for some $n \times n$ nonsingular matrix P.

In particular, (2.4) does not depend on the choice of Φ as long as $E^\Gamma(\Phi)$ is a tight frame with constant 1 for V.

PROOF. Let $\Gamma = P\mathbb{Z}^n$ for some $n \times n$ nonsingular matrix P. As we noticed before, D_PV is $\mathbb{Z}^n$-SI. We claim that

$$\sigma_V^\Gamma(\xi) = \sigma_{D_PV}^{\mathbb{Z}^n}(P^*\xi) \qquad \text{a.e. } \xi \in \mathbb{R}^n. \tag{2.5}$$

Indeed, let $I : \ell^2(\Gamma^*) \to \ell^2(\mathbb{Z}^n)$ be an isometric isomorphism given by $I((a_k)_{k\in\Gamma^*}) = (a_{P^*k})_{k\in\mathbb{Z}^n}$. Let $J(\xi)$ be the range function of Γ-SI space V. Then using Proposition 1.1, it is easy to verify that $\tilde{J}(\xi) = I(J((P^*)^{-1}\xi))$ is the range function of $\mathbb{Z}^n$-SI space D_PV, which shows the claim.

It is also clear that $E^\Gamma(\Phi)$ is a tight frame with constant 1 for V if and only if $E^{\mathbb{Z}^n}(D_P\Phi)$ is a tight frame with constant 1 for D_PV. Since Lemma 2.5 is valid for the standard lattice $\mathbb{Z}^n$, hence

$$\sigma_V^\Gamma(\xi) = \sigma_{D_PV}^{\mathbb{Z}^n}(P^*\xi) = \sum_{\phi\in D_P\Phi} |\hat{\phi}(P^*\xi)|^2 = \frac{1}{|\det P|} \sum_{\varphi\in\Phi} |\hat{\varphi}(\xi)|^2,$$

which completes the proof of Lemma 2.5.

THEOREM 2.6. *Suppose $V \subset L^2(\mathbb{R}^n)$ is a Γ-SI space for some lattice Γ. Then*

$$\sigma_V^\Gamma(\xi) = \lim_{r\to 0^+} \frac{||P_V(\check{\mathbf{1}}_{B(\xi,r)})||^2}{|B(\xi,r)|} \qquad \text{for a.e. } \xi \in \mathbb{R}^n. \tag{2.6}$$

PROOF. Suppose that $K \subset \mathbb{R}^n$ is a measurable set such that $|K \cap (l + K)| = 0$ for all $l \in \Gamma^* \setminus \{0\}$. Then combining Lemma 2.3 and (2.5),

$$\begin{aligned} ||P_V(\check{\mathbf{1}}_K)||^2 &= ||P_{D_PV}(D_P(\check{\mathbf{1}}_K))||^2 = \frac{1}{|\det P|} ||P_{D_PV}(\check{\mathbf{1}}_{P^*K})||^2 \\ &= \frac{1}{|\det P|} \int_{P^*K} \sigma_{D_PV}^{\mathbb{Z}^n}(\xi) d\xi = \int_K \sigma_V^\Gamma(\xi) d\xi. \end{aligned}$$

Applying the Lebesgue Differentiation Theorem yields (2.6).

REMARK. Theorem 2.6 is the key result of the paper and has interesting implications. It suggests that for any closed subspace $V \subset L^2(\mathbb{R}^n)$ we can consider

$$\sigma_V^+(\xi) = \limsup_{r\to 0^+} \frac{||P_V(\check{\mathbf{1}}_{B(\xi,r)})||^2}{|B(\xi,r)|}, \qquad \text{and} \qquad \sigma_V^-(\xi) = \liminf_{r\to 0^+} \frac{||P_V(\check{\mathbf{1}}_{B(\xi,r)})||^2}{|B(\xi,r)|}$$

Of course both these functions are zero if V is "small", e.g., V has a finite dimension. Therefore, σ_V^+ and σ_V^- are going to be interesting only if V is "big" in some sense, for example in the case when V is a Γ-SI space. In this case we can also see that $\sigma_V^+ = \sigma_V^-$.

A surprising consequence of Theorem 2.6 is the following

COROLLARY 2.7. *Suppose $V \subset L^2(\mathbb{R}^n)$ is both Γ_1-SI and Γ_2-SI space for two lattices Γ_1 and Γ_2. Then,*

$$\sigma_V^{\Gamma_1}(\xi) = \sigma_V^{\Gamma_2}(\xi) \qquad \textit{for a.e. } \xi \in \mathbb{R}^n.$$

Consequently, the spectral function is independent of the underlying lattice.

Corollary 2.7 shows a huge advantage of the spectral function over the dimension function. Namely, in order to talk meaningfully about the dimension function of a certain SI subspace V of $L^2(\mathbb{R}^n)$, one needs to specify a lattice Γ with respect to which V is SI. However, to talk about the spectral function of V, one doesn't need to specify the lattice with respect to which V is SI, since the resulting $\sigma_V(\xi)$ will be the same in each case. This suggests that $\sigma_V(\xi)$ is even more inherent notion of "size" of a SI space than the dimension function. It is also a more subtle notion, since one can always easily recover the dimension function from the spectral function using

$$\dim_V^\Gamma(\xi) = \sum_{k \in \Gamma^*} \sigma_V(\xi + k).$$

The following example illustrates the basic difference between dimension and spectral functions.

EXAMPLE. Let $V = \mathcal{S}^{\mathbb{Z}}(\varphi)$, where φ is a function in $L^2(\mathbb{R})$ whose integer shifts are orthonormal. Then $\sigma_V(\xi) = |\hat{\varphi}(\xi)|^2$ and $\dim_V(\xi) = 1$. If we consider the standard "dilation by 2" operator $Df(x) = \sqrt{2}f(2x)$, then the space DV is $\mathbb{Z}$-SI and also $\frac{1}{2}\mathbb{Z}$-SI. As we can see, $\sigma_{DV}^\Gamma(\xi) = |\hat{\varphi}(\frac{\xi}{2})|^2$ if Γ is $\mathbb{Z}$ or $\frac{1}{2}\mathbb{Z}$, while $\dim_{DV}^\Gamma(\xi) = 2$ if $\Gamma = \mathbb{Z}$ and $\dim_{DV}^\Gamma(\xi) = 1$ if $\Gamma = \frac{1}{2}\mathbb{Z}$.

If for a given closed subspace $V \subset L^2(\mathbb{R}^n)$ the functions σ_V^+ and σ_V^- coincide then we can define $\sigma_V = \sigma_V^+ = \sigma_V^-$. This function satisfies some obvious properties like additivity on orthogonal sums and $0 \le \sigma_V \le 1$. If we restrict our attention to general SI spaces we obtain the following analogue of Proposition 2.2

PROPOSITION 2.8. *Let $\mathfrak{S} = \mathfrak{S}_{\text{all}}$ be the collection of all possible SI subspaces of $L^2(\mathbb{R}^n)$, i.e., $V \in \mathfrak{S}$ if and only if there exists a lattice Γ such that V is Γ-SI. The spectral function satisfies the following properties: ($V, W \in \mathfrak{S}$)*

(a) $\sigma_V : \mathbb{R}^n \to [0,1]$ is a measurable function,

(b) $V = \bigoplus_{i=1}^N V_i$, where $V_i \in \mathfrak{S} \implies \sigma_V(\xi) = \sum_{i=1}^N \sigma_{V_i}(\xi)$,

(c) $V = \bigoplus_{i \in \mathbb{N}} V_i$, where V_i is Γ-SI for a fixed lattice $\Gamma \implies \sigma_V(\xi) = \sum_{i \in \mathbb{N}} \sigma_{V_i}(\xi)$,

(d) $V \subset W \implies \sigma_V(\xi) \le \sigma_W(\xi)$,

(e) $V \subset W \implies (V = W \iff \sigma_V(\xi) = \sigma_W(\xi))$,

(f) $\sigma_V(\xi) = \mathbf{1}_E(\xi) \iff V = \check{L}^2(E)$,

(g) $\sigma_{M_a(V)}(\xi) = \sigma(\xi - a)$, where M_a is a modulation by $a \in \mathbb{R}^n$,

(h) $\sigma_{D_A V}(\xi) = \sigma_V((A^*)^{-1}\xi)$, where D_A is a dilation by a non-singular **real** *matrix A,*

(i) if V is Γ-SI then $\dim_V^\Gamma(\xi) = \sum_{k \in \Gamma^*} \sigma_V(\xi + k)$.

PROOF. The proof of Proposition 2.8 is a routine. Part (a) is immediate by Definition 2.4. Parts (b), (d) are a consequence of Theorem 2.6. Parts (c), (f), (g), and (i) follow from the corresponding properties in Proposition 2.2 and (2.5) by considering $\mathbb{Z}^n$-SI space $D_P V$.

It is not hard to see that (2.5) implies (h) for V being $\mathbb{Z}^n$-SI. If V is Γ-SI, where $\Gamma = P\mathbb{Z}^n$, it suffices to use this observation together with

$$\sigma_{D_A V}(\xi) = \sigma_{D_{P^{-1}A} D_P V}(\xi) = \sigma_{D_P V}(P^*(A^*)^{-1}\xi) = \sigma_V((A^*)^{-1}\xi).$$

Finally, to show the least trivial property (e), assume that V is Γ_1-SI, W is Γ_2-SI, and $V \subset W$. If $V = W$ then $\sigma_V = \sigma_W$ by Theorem 2.6. To show the converse, suppose that $\sigma_V = \sigma_W$. Let V' be the smallest Γ_2-SI space containing V, i.e., $V' = \mathcal{S}^{\Gamma_2}(V)$. Clearly $V \subset V' \subset W$ thus by (d) we must have $\sigma_{V'} = \sigma_W$. Since V' and W are both Γ_2-SI, we have $V' = W$ by Proposition 2.2 (d). On the other hand, V' is also Γ_1-SI (the space is spanned by Γ_2 translations of Γ_1-SI space and translations commute) and $\sigma_V = \sigma_{V'}$. Hence, $V = V'$ again by Proposition 2.2 (d). This shows $V = W$ and completes the proof of Proposition 2.8.

We remark that unlike Proposition 2.2, Proposition 2.8(b) does not hold in general for **countable** collection of V_i's such that all V_i's are not necessarily SI with respect to some common lattice Γ. This will be illustrated by examples in Section 3. In general, we can only expect that if $V = \bigoplus_{i=1}^{\infty} V_i$, where $V_i \in \mathfrak{S}$ and $\mathfrak{S}$ is the same as in Proposition 2.8, then we always have

$$\sigma_V(\xi) \geq \sum_{i=1}^{\infty} \sigma_{V_i}(\xi) \qquad \text{a.e. } \xi \in \mathbb{R}^n.$$

Indeed, this follows easily from Proposition 2.8 (b) and (d).

We would like to point out that in some interesting cases Proposition 2.8(b) does hold for infinite sums. For example, if $\Psi = \{\psi^1, \dots, \psi^L\} \subset L^2(\mathbb{R}^n)$ is an orthogonal wavelet with respect to an expansive dilation A then the Calderón condition

$$\sum_{j\in\mathbb{Z}} \sum_{\psi\in\Psi} |\hat{\psi}((A^*)^j\xi)|^2 = 1 \qquad \text{for a.e. } \xi \in \mathbb{R}^n \tag{2.7}$$

holds. Recall that $n \times n$ matrix A is expansive if all eigenvalues λ of A satisfy $|\lambda| > 1$. Condition (2.7), originally discovered by Meyer, was investigated by many authors as one of the equations characterizing wavelets. That it is equivalent to the completeness of the corresponding orthogonal affine system was proven in [**Bo2, Bo3, Rz**]. Recently it was shown that (2.7) holds not only for expansive dilations (with real entries), but also for wider classes of dilations expanding on subspaces, see [**HLW, La**]. Since the Calderón condition (2.7) can be written as

$$\sum_{j\in\mathbb{Z}} \sigma_{D_{A^j}(\mathcal{S}(\Psi))} = \sigma_{L^2(\mathbb{R}^n)}$$

and $\bigoplus_{j\in\mathbb{Z}} D_{A^j}(\mathcal{S}(\Psi)) = L^2(\mathbb{R}^n)$ we recognize it as Proposition 2.8(b) holding for this infinite sum.

3. Examples

In this section we show that the spectral function is not additive on infinite orthogonal sums of SI spaces with respect to varying lattices.

EXAMPLE 3.1. Suppose that $\varphi \in L^2(\mathbb{R})$ and $\{\varphi(x-k)\}_{k\in\mathbb{Z}}$ is an orthonormal sequence.

Let $N \geq 3$ be a fixed integer. Define the sequence of nested lattices $\Gamma_0 \supset \Gamma_1 \supset \Gamma_2 \ldots$, where

$$\Gamma_i = N^i\mathbb{Z} \qquad \text{for } i = 0, 1, 2, \ldots \tag{3.1}$$

It is not hard to see that there exists a sequence $\{t_i\}_{i=1}^{\infty} \subset \mathbb{Z}$ such that

$$\begin{gathered}\bigcup_{i=1}^{\infty}(t_i + \Gamma_i) = \mathbb{Z}\\ (t_i+\Gamma_i)\cap(t_j+\Gamma_j) = \emptyset \qquad \text{for } i \neq j.\end{gathered} \tag{3.2}$$

Indeed, one can define $\{t_i\}$ by induction as follows. Let $t_1 = 0$. Once $t_1, \ldots, t_i$ are defined, let t_{i+1} be an integer t with the minimal absolute value $|t|$ such that

$$(t+\Gamma_{i+1}) \cap \left(\bigcup_{j=1}^{i}(t_j+\Gamma_j)\right) = \emptyset.$$

Define the sequence of spaces $\{V_i\}$ by

$$V_i = \mathcal{S}^{\Gamma_i}(T_{t_i}\varphi) = \overline{\operatorname{span}}\{\varphi(x-t_i-k) : k \in \Gamma_i\}.$$

Obviously, V_i is Γ_i-SI for $i = 1, 2, \ldots$. Moreover, by (3.2),

$$\bigoplus_{i=1}^{\infty} V_i = \mathcal{S}^{\mathbb{Z}}(\varphi). \tag{3.3}$$

Since

$$\sigma_{V_i}(\xi) = \frac{1}{N^i}|\hat{\varphi}(\xi)|^2 = \frac{1}{N^i}\sigma_{\mathcal{S}^{\mathbb{Z}}(\varphi)}(\xi),$$

hence

$$\sum_{i=1}^{\infty}\sigma_{V_i}(\xi) = \frac{1}{N-1}\sigma_{\mathcal{S}^{\mathbb{Z}}(\varphi)}(\xi). \tag{3.4}$$

Therefore, for $N \geq 3$,

$$\sum_{i=1}^{\infty}\sigma_{V_i}(\xi) \neq \sigma_{\mathcal{S}^{\mathbb{Z}}(\varphi)}(\xi),$$

despite (3.3) and the fact that V_i is Γ_i-SI for $i = 1, 2, \ldots$.

The next example is a refinement of Example 3.1.

EXAMPLE 3.2. Define the sequence of functions $\{\varphi_k\}_{k\in\mathbb{Z}}$ by $\hat{\varphi}_k = \mathbf{1}_{[k,k+1]}$. Let

$$V_i^k = \mathcal{S}^{\Gamma_i}(T_{t_i}\varphi_k) \qquad \text{for } k \in \mathbb{Z},\ i = 1, 2, \ldots,$$

where $\{\Gamma_i\}$ and $\{t_i\}$ are the same as in Example 3.1. Clearly,

$$\bigoplus_{k\in\mathbb{Z}}\bigoplus_{i=1}^{\infty} V_i^k = \bigoplus_{k\in\mathbb{Z}}\check{L}^2(k,k+1) = L^2(\mathbb{R}),$$

but

$$\sum_{k\in\mathbb{Z}}\sum_{i=1}^{\infty}\sigma_{V_i^k}(\xi)=\sum_{k\in\mathbb{Z}}\frac{1}{N-1}\mathbf{1}_{[k,k+1]}(\xi)=\frac{1}{N-1}\qquad\text{for a.e. } \xi\in\mathbb{R}. \tag{3.5}$$

Shift-invariant systems with varying lattices.

Example 3.2 shows difficulties one may encounter when trying to characterize shift-invariant systems with varying lattices which form orthonormal bases (or more generally Riesz bases or frames) for $L^2(\mathbb{R}^n)$. In one such study, Hernández, Labate, and Weiss, consider general shift-invariant systems of the form

$$\{T_{C_p k} g_p : k\in\mathbb{Z}^n,\ p\in P\}, \tag{3.6}$$

where $\{C_p\}_{p\in P}$ is a collection of non-singular $n\times n$ matrices and $\{g_p\}_{p\in P}$ is a collection of functions in $L^2(\mathbb{R}^n)$. In their main result [**HLW**, Theorem 2.1], the authors characterize when the above system (3.6) forms a tight frame for $L^2(\mathbb{R}^n)$ under a delicate technical assumption of the local integrability, see [**HLW**, (2.7)]. In particular, they show that if (3.6) forms a tight frame with constant 1 for $L^2(\mathbb{R}^n)$ then necessarily

$$\sum_{p\in P}\frac{1}{|\det C_p|}|\hat{g}_p(\xi)|^2=1\qquad\text{a.e. } \xi\in\mathbb{R}^n. \tag{3.7}$$

Example 3.2 shows that the local integrability condition of [**HLW**] can not be removed. Indeed, by a simple change of indexing set, the system in Example 3.2 can be easily written in the form (3.6). Since $\sigma_{V_i^k}(\xi)=\frac{1}{N^i}|\hat{\varphi}_k(\xi)|^2$, (3.5) implies that

$$\sum_{p\in P}\frac{1}{|\det C_p|}|\hat{g}_p(\xi)|^2=\sum_{k\in\mathbb{Z}}\sum_{i=1}^{\infty}\frac{1}{N^i}|\hat{\varphi}_k(\xi)|^2=\frac{1}{N-1}\qquad\text{for a.e. } \xi\in\mathbb{R}.$$

This obviously contradicts (3.7) and shows that the local integrability condition is essential.

Orthogonal wavelets for non-expansive dilations.

One of the longer standing problems in the theory of wavelets is the following question. For which non-singular $n\times n$ matrices A, does there exist a wavelet $\Psi=\{\psi^1,\dots,\psi^L\}\subset L^2(\mathbb{R}^n)$ associated with A such that

$$\{D_{A^j}T_k\psi : j\in\mathbb{Z},\ k\in\mathbb{Z}^n,\ \psi\in\Psi\}$$

is an orthonormal basis of $L^2(\mathbb{R}^n)$?

Only a few necessary and sufficient conditions for such A are known despite a very interesting progress obtained recently by Speegle [**Sp**]. Speegle's work suggests that an initial step toward answering this question might involve showing that the Calderón condition (2.7) must hold. At the present time, this is known to be true only for the class of dilations expanding on subspaces introduced by Hernández, Labate, and Weiss [**HLW**]. Whether (2.7) must hold for all other dilations is an open problem.

This in turn may be considered as a part of a bigger problem of understanding for which families of orthogonal SI spaces $\{W_i\}$ (with varying lattices) the spectral function is additive, i.e., $\sum \sigma_{W_i} = \sigma_{\oplus W_i}$. Indeed, in the special case of a wavelet Ψ and

$$W_i = \overline{\text{span}}\{D_{A^i} T_k \psi : k \in \mathbb{Z}^n, \ \psi \in \Psi\},$$

the additivity of the spectral function is equivalent to (2.7). Example 3.2 merely suggests that the answer to the last problem must necessarily take into account the dilation structure of spaces $\{W_i\}$.

References

[Ba] L. W. Baggett, *An abstract interpretation of the wavelet dimension function using group representations*, J. Funct. Anal. **173** (2000), 1–20.

[BMM] L. W. Baggett, H. A. Medina, and K. D. Merrill, *Generalized multi-resolution analyses and a construction procedure for all wavelet sets in* $\mathbb{R}^n$, J. Fourier Anal. Appl. **5** (1999), 563–573.

[BDR1] C. de Boor, R. A. DeVore, and A. Ron, *The structure of finitely generated shift-invariant spaces in* $L_2(\mathbb{R}^d)$, J. Funct. Anal. **119** (1994), 37–78.

[BDR2] C. de Boor, R. A. DeVore, and A. Ron, *Approximation from shift-invariant subspaces of* $L_2(\mathbb{R}^d)$, Trans. Amer. Math. Soc. **341** (1994), 787–806.

[BL] J. J. Benedetto and S. Li, *The theory of multiresolution analysis frames and applications to filter banks*, Appl. Comput. Harmon. Anal. **5** (1998), 389–427.

[Bo1] M. Bownik, *The structure of shift-invariant subspaces of* $L^2(\mathbb{R}^n)$, J. Funct. Anal. **177** (2000), 282–309.

[Bo2] M. Bownik, *On characterizations of multiwavelets in* $L^2(\mathbb{R}^n)$, Proc. Amer. Math. Soc. **129** (2001), 3265–3274.

[Bo3] M. Bownik, *Quasi-affine systems and the Calderòn condition*, Harmonic Analysis at Mount Holyoke, Contemp. Math. **320** (2003), 29–43.

[BRz] M. Bownik and Z. Rzeszotnik, *The spectral function of shift-invariant systems*, Michigan Math. J. **51** (2003), 387–414.

[BRS] M. Bownik, Z. Rzeszotnik, and D. Speegle, *A characterization of dimension functions of wavelets*, Appl. Comput. Harmon. Anal. **10** (2001), 71–92.

[Da] I. Daubechies, *Ten Lectures on Wavelets*, Society for Industrial and Applied Mathematics (SIAM), Philadelphia, PA, 1992.

[He] H. Helson, *Lectures on Invariant Subspaces*, Academic Press, New York-London, 1964.

[HW] E. Hernández and G. Weiss, *A First Course on Wavelets*, Studies in Advanced Mathematics, CRC Press, Boca Raton, FL, 1996.

[HWL] E. Hernández, D. Labate, and G. Weiss, *A unified characterization of reproducing systems generated by a finite family, II*, J. Geom. Anal. **12** (2002), 615–662.

[Ji] R.-Q. Jia, *Shift-invariant spaces and linear operator equations*, Israel J. Math. **103** (1998), 259–288.

[La] R. Laugesen, *Translational averaging for completeness, characterization and oversampling of wavelets*, Collect. Math. **53** (2002), 211–249.

[RS1] A. Ron and Z. Shen, *Frames and stable bases for shift-invariant subspaces of* $L_2(\mathbb{R}^d)$, Canad. J. Math. **47** (1995), 1051–1094.

[RS2] A. Ron and Z. Shen, *Affine systems in* $L_2(\mathbb{R}^d)$*: the analysis of the analysis operator*, J. Funct. Anal. **148** (1997), 408–447.

[Rz] Z. Rzeszotnik, *Calderón's condition and wavelets*, Collect. Math. **52** (2001), 181–191.

[Sp] D. Speegle, *On the existence of wavelets for non-expansive dilation matrices*, Collect. Math. **54** (2003), 163–179.

DEPARTMENT OF MATHEMATICS, UNIVERSITY OF MICHIGAN, ANN ARBOR, MI 48109–1109
Current address: Department of Mathematics, University of Oregon, Eugene, OR 97403-1222
E-mail address: mbownik@uoregon.edu

DEPARTMENT OF MATHEMATICS, UNIVERSITY OF TEXAS, AUSTIN, TX 78712–1082
E-mail address: zioma@mail.ma.utexas.edu

Contemporary Mathematics
Volume **345**, 2004

Custom building finite frames

Peter G. Casazza

ABSTRACT. After the introduction of *frame potentials* by Benedetto and Fickus, there was an explosion of new results in frame theory. The generalization of this notation to *weighted frame potentials* yields a physical interpretation for finite frames along the lines of Coulomb's law in physics. This allows us to use results from classical mechanics to anticipate new results in frame theory. All of this resulted in the resolution of a large number of important open questions concerning the existence and construction of finite frames.

We will put all of this into the framework of a theory as well as adding many new results along the way. As part of the theory, we include the recent "algorithmic constructions" for finite frames. We will also discuss the related important questions which are still open at this time.

Introduction

Hilbert space frames were introduced by Duffin and Schaeffer in 1952 [**DS**] while working on some deep questions in non-harmonic Fourier series (see [**Y**]). These problems grew out of G.D. Birkhoff's work on Sturm-Liouville functions in 1917; the trigonometric results of Welch in 1921; and results of Paley-Wiener in 1934 (see [**Y**]). For some reason the results of Duffin and Schaeffer were not continued (except in signal/image processing [**G**]) until in 1986 Daubechies, Grossman and Meyer [**DGS**] brought this back to life right at the dawn of the wavelet era.

Recently, many new applications of frames have arisen to internet coding (see [**GKK**] and its references and [**CK**]); multiple antenna coding (see [**HH**] and [**HM**]); Communication Theory; Sampling Theory and more.

Benedetto and Fickus [**BF**] introduced an important tool into frame theory called *frame potentials*. This gave a geometric interpretation for tight frames which resulted in an explosion of important new results in the field including a physical interpretation for tight frames along the lines of Coulomb's law in Physics [**BF, CF**]. As a result, we can now use results from quantum mechanics to anticipate new results on frames. This also led to important advances in the *algorithmic construction* of frames which is critically important for applications of frames.

We will put all of these new advances into the framework of a theory as well as singling out the important open questions at this time. Also, we include may new

2000 *Mathematics Subject Classification.* Primary 46C05, 46N99; Secondary 42C40.
The author was supported by NSF DMS 0102686.

results which did not make it into these papers. For general information concerning frames, we refer the reader to [**C, C1, Ch**].

Acknowledgement. The author thanks the referee for an exceptional job resulting in a much better paper.

1. Preliminaries

A *frame* is a redundant set of vectors in a Hilbert space that allows one "natural" representation for each vector. Formally, a set of vectors $\{\phi_n\}_{n\in I}$ in a Hilbert space H is a *frame* for H if there are numbers $A, B > 0$ so that for all $\phi \in H$,

$$A\|\phi\|^2 \leq \sum_{n\in I} |\langle \phi, \phi_n\rangle|^2 \leq B\|\phi\|^2. \tag{1.1}$$

We call A (resp. B) a *lower frame bound* (resp. *upper frame bound*) for the frame. If $A = B$ this is an A-*tight frame* and if $A = B = 1$ it is called a *Parseval frame.* If all the frame vectors have the same norm, we call this an *equal-norm frame* and if the frame vectors are all norm one, we call $\{\phi_n\}_{n\in I}$ a *unit norm frame.*

We will work with finite frames on N-dimensional (real $\mathbb{R}^N$ or complex $\mathbb{C}^N$) Hilbert spaces H_N. If $\{\phi_m\}_{m=1}^M$ is a sequence of vectors in H_N, the *analysis operator* is $F : H \to \ell_2^M$ given by

$$(F\phi)_i = \langle \phi, \phi_i\rangle$$

and the *synthesis operator* is F^*. It follows from (1.1) that

$$AI \leq F^*F \leq BI.$$

We call $S = F^*F$ the *frame operator.* So $\{\phi_m\}_{m=1}^M$ is a frame for H_N if and only if S is a positive, self-adjoint invertible operator on H_N. And it is a A-tight frame if and only if $F^*F = AI$. Also, for every $\phi \in H_N$, we have the *reconstruction formula*:

$$\phi = \sum_{m=1}^{M} \langle S^{-1}\phi, \phi_m\rangle\phi_m = \sum_{m=1}^{M} \langle \phi, S^{-\frac{1}{2}}\phi_m\rangle S^{-\frac{1}{2}}\phi_m.$$

That is, $\{S^{-\frac{1}{2}}\phi_m\}_{m=1}^M$ is a Parseval frame.

If $\{\phi_m\}_{m=1}^M$ is a frame for H_N with frame operator S having eigenvalues $\lambda_1 \geq \lambda_2 \geq \cdots \geq \lambda_N$ then

$$\sum_{k=1}^{M} \lambda_k = \sum_{m=1}^{M} \|\phi_m\|^2.$$

The frame $\{S^{-1}\phi_m\}_{m=1}^M$ is called the *canonical dual frame* of $\{\phi_m\}_{m=1}^M$. This frame carries out reconstruction. But there are other frames giving reconstruction formulas. A sequence $\{\psi_m\}_{m=1}^M$ is called a *dual frame* for $\{\phi_m\}_{m=1}^M$ if for all $\phi \in H_N$ we have

$$\phi = \sum_{m=1}^{M} \langle \phi, \phi_m\rangle\psi_m = \sum_{m=1}^{M} \langle \phi, \psi_m\rangle\phi_m.$$

2. Frame potentials

Our goal in this section is to introduce the notion of *frame potential* of Benedetto and Fickus [**BF**] and develop their basic properties and applications. The importance of frame potentials is the fact that for many families of frames, the frames of minimal frame potential are precisely the tight frames in the family. This fact has proven to be very useful in the resolution of several important open questions in applications of frames. For most applications of frames, we prefer (or need) tight frames. Now, we can consider the family of frames which are allowed for our application and search for an element of minimal frame potential. Even more important, if our application does not allow tight frames, by minimizing frame potential we obtain the allowable frames which are the closest to being tight.

DEFINITION 2.1. If $\{\varphi_m\}_{m=1}^M$ is a frame for $\mathbb{H}_N$, the *frame potential* of $\{\varphi_m\}_{m=1}^M$ is given by:

$$FP(\{\varphi_m\}_{m=1}^M) = \sum_{n,m=1}^{M} |\langle \varphi_n, \varphi_m\rangle|^2.$$

The frame potential is measuring how close a frame is to being orthogonal. In particular, we will see that if $\mathcal{F}$ is the family of frames with lower frame bound λ then the λ-tight frames are the minimizers of the frame potential over $\mathcal{F}$. This theorem gives us a way to identify tight frames. That is, they are the minimizers of the frame potential on certain families of frames. Our goal is to identify those families of frames which have minimizers of the frame potential and for which these minimizers must be tight.

If $\{\phi_m\}_{m=1}^M$ is a frame for H_N with frame operator S having eigenvalues $\lambda_1 \geq \lambda_2 \geq \cdots \geq \lambda_N > 0$, then

$$\begin{aligned} FP(\{\phi_m\}_{m=1}^M) &= \sum_{m,n=1}^{M} |\langle \phi_n, \phi_m\rangle|^2 = Tr(S^2) \\ &= \sum_{n=1}^{M} \lambda_n^2 = \sum_{m=1}^{M} \|S^{\frac{1}{2}}\phi_m\|^2. \end{aligned} \tag{2.1}$$

The important point here is the fact that the minimizers of the frame potential are tight frames [**BF**] (the proof below comes from [**CF**]).

PROPOSTION 2.2. *Let $\mathbb{H}_N$ be an N-dimensional Hilbert space, $0 < \lambda$ and let*

$$W = \left\{ \{\phi_m\}_{m=1}^M | \sum_{m=1}^{M} \|\phi_m\|^2 = \lambda \right\}.$$

(1) *If $M \leq N$, the minimum value of the frame potential on W is λ^2/M and the minimizers are orthogonal sequences of vectors all with the same norm $\sqrt{\lambda/M}$.*
(2) *If $M \geq N$, the minimum value of the frame potential on W is λ^2/N and the minimizers are the tight frames with tight frame bound λ/N.*

PROOF. (1) We compute:

$$
\begin{aligned}
FP(\{\phi_m\}_{m=1}^M) &= \sum_{n,m=1}^{M} |\langle \phi_n, \phi_m\rangle|^2 \\
&= \sum_{m=1}^{M} \|\phi_m\|^4 + \sum_{n\neq m} |\langle \phi_n, \phi_m\rangle|^2 \qquad (2.2)\\
&\geq \sum_{m=1}^{M} \|\phi_m\|^4.
\end{aligned}
$$

Since we assumed that

$$\sum_{m=1}^{M} \|\phi_m\|^2 = \lambda,$$

by a standard application of Lagrange multipliers, the right-hand side of (2.2) is minimized when

$$\|\phi_m\|^2 = \|\phi_n\|^2, \quad \text{for all } 1 \leq m, n \leq M.$$

In this case $\|\phi_m\|^2 = \lambda/M$, for all $1 \leq m \leq M$ and so

$$\sum_{m=1}^{M} \|\phi_m\|^4 = \frac{\lambda^2}{M}.$$

This minimum is achieved when we have equality in (2.2), and hence

$$\sum_{n\neq m} |\langle \phi_n, \phi_m\rangle|^2 = 0,$$

showing that $\{\phi_m\}_{m=1}^M$ is an orthogonal sequence.

(2) When $M > N$, we cannot use the same approach, since we cannot find M mutually orthogonal vectors ϕ_m. By (2.1), minimizing the frame potential under our constraint means minimizing $\sum_{n=1}^{N} \lambda_n^2$ under the constraint $\sum_{n=1}^{N} \lambda_n = \sum_{m=1}^{M} \|\phi_m\|^2 = \lambda$. Again, a standard application of Lagrange multipliers yields that the minimizers satisfy $\lambda_n = \lambda/N$, for all $1 \leq n \leq N$. Hence, the frame operator for $\{\phi_m\}_{m=1}^M$ is $(\lambda/N)I$. That is, a minimizer of the frame potential is a tight frame with the tight frame bound λ/N. Since there always exist such tight frames and these are clearly minimizers of the frame potential, we have the proof.

□

We have immediately:

COROLLARY 2.3. *Given* $\{\phi_m\}_{m=1}^M \subseteq H_N$,

$$FP(\{\phi_m\}) = \sum_{m=1}^{M}\sum_{n=1}^{M} |\langle \phi_n, \phi_m\rangle|^2 \geq \frac{\left(\sum_{m=1}^{M} \|\phi_m\|^2\right)^2}{N}$$

with equality if and only if $\{\phi_m\}_{m=1}^M$ *is a tight frame.*

These results give a method for finding and even constructing tight frames since the tight frames are minimizers of the frame potential. Suppose in a given application we need tight frames with"extra properties". We let $\mathcal{F}$ be the family of frames having these properties. Generally, a simple compactness arguement shows that $\mathcal{F}$ is "closed" in almost any choice of topology. This means that $\mathcal{F}$ has elements of minimal frame potential. Now we use our theory of frame potentials to show that the "minimizers" must be tight frames. We will see in several places later in these notes where this strategy works quite well.

Minimizing the frame potential over certain classes of frames classifies tight frames. The following result from [**CF**] shows how frame potentials also identify Parseval frames.

THEOREM 2.4. *If $\{\phi_m\}_{m=1}^M$ is a frame with frame operator S and canonical dual frame $\{S^{-1}\phi_m\}_{m=1}^M$ then*

$$FP(\{\phi_m\}_{m=1}^M) + FP(\{S^{-1}\phi_m\}_{m=1}^M)$$

is minimum if and only if $\{\phi_m\}_{m=1}^M$ is a Parseval frame.

PROOF. If $\{\lambda_n\}_{n=1}^N$ are the eigenvalues of the frame operator S, then we are just minimizing

$$\sum_{n=1}^N \lambda_n^2 + \sum_{n=1}^N \frac{1}{\lambda_n^2}.$$

□

The following result from [**CF**] shows that for any frame, the canonical dual frame is the closest to being tight of all the dual frames.

This result allows us to give quantative estimates when working with dual frames.

THEOREM 2.5. *If $\{\phi_m\}_{m=1}^M$ is a frame for H_N, then the canonical dual frame $\{S^{-1}\phi_m\}_{m=1}^M$ is the unique dual frame of minimal frame potential.*

Important applications of the frame potential come from studying its *local minimizers* (see Proposition 2.8 for the definition). Now we will examine this local behavior of the frame potential.

An important result of Benedetto and Fickus [**BF**] is that local minimizers of the frame potential for unit norm frames are also global minimizers. We present here a generalization of this which is a previously unpublished result of Casazza, Fickus, Kovačević, Leon, and Tremain.

PROPOSTION 2.6. *Let $\{\varphi_m\}_{m=1}^M$ be a frame for $\mathbb{H}_N$ with frame operator S having distinct eigenvalues $\lambda_1 > \lambda_2 > \cdots > \lambda_N$ with respective eigenspaces $\{E_{\lambda_i}\}_{i=1}^L$. Define*

$$F : S^{\mathbb{K}^N} \to \mathbb{R},$$

(where $\mathbb{K}$ is either $\mathbb{R}$ or $\mathbb{C}$) by

$$F(\Phi) = \sum_{m=1}^M |\langle \Phi, \varphi_m \rangle|^2.$$

Then the sphere of E_{λ_L} is the set of local minimizers of F. Hence, the local minimizers of F are global minimizers.

PROOF. We will do the real case since it is quite illuminating. The complex case follows from the same proof with notational changes. Since F is continuous and nonzero and $\{\varphi_m\}_{m=1}^M$ spans $\mathbb{R}^N$, by Lagrange Multipliers, there is a $\mu \neq 0$ so that the minimizers of F satisfy:

$$\nabla F = \mu \nabla G,$$

where for $\Phi = (\Phi_1, \Phi_2 \ldots, \ldots, \Phi_N)$ in $\mathbb{R}^N$ we have

$$G(\Phi) = \sum_{i=1}^{N} \Phi_i^2 - 1.$$

Hence,

$$\partial_i \lambda G = 2\mu \Phi_i, \quad \text{for all } 1 \leq i \leq N.$$

Also,

$$F(\Phi) = \sum_{m=1}^{M} \left[\sum_{k=1}^{N} \Phi_k \langle \varphi_m, e_k \rangle \right]^2 .$$

Hence,

$$\partial_i F = \sum_{m=1}^{M} 2 \left[\sum_{k=1}^{N} \Phi_k \langle \varphi_m, e_k \rangle \right] \langle \varphi_m, e_i \rangle = 2 \sum_{m=1}^{M} \langle \Phi, \varphi_m \rangle \langle \varphi_m, e_i \rangle.$$

Finally,

$$\Phi_i = (2\mu)^{-1} \sum_{m=1}^{M} \langle \Phi, \varphi_m \rangle \langle \varphi_m, e_i \rangle = (2\mu)^{-1} \left\langle \left(\sum_{m=1}^{M} \langle \Phi, \varphi_m \rangle \varphi_m \right), e_i \right\rangle .$$

Hence,

$$\Phi = (\mu)^{-1} \sum_{m=1}^{M} \langle \Phi, \varphi_m \rangle \varphi_m.$$

We now have that a minimizer of F is an eigenvector of S. Next we show that these eigenvectors are all in E_{λ_L}. We proceed by way of contradiction. If Φ is a local minimizer of F and $\Phi \in E_{\lambda_i}$ for some $1 \leq i < L$, choose $e_1 \in E_{\lambda_L}$ with $\|e_1\| = 1$. Fix $0 < \epsilon < 1$ and let

$$\Psi = \epsilon \Phi + \sqrt{(1 - \epsilon^2)} e_1.$$

So $\|\Psi\| = 1$ and

$$F(\psi) = \sum_{m=1}^{M} |\langle \Psi, \varphi_m \rangle|^2 = \epsilon^2 \lambda_i + (1 - \epsilon^2) \lambda_L < \lambda_i = F(\Phi).$$

This contradiction completes the proof. □

DEFINITION 2.7. A frame $\{\varphi_m\}_{m=1}^M$ for $\mathbb{H}_N$ with frame operator S is called a *FF-critical sequence* if each ϕ_m is an eigenvector for S.

The next two propositions and their corollaries are also previously unpublished results of Casazza, Fickus, Kovačević, Leon, and Tremain.

The main point is to identify the exact structure of local minimizers of the frame potential over quite general familes of frames.

PROPOSTION 2.8. *If $\{\varphi_m\}_{m=1}^M$, $M \geq N$ is a sequence of vectors in $\mathbb{H}_N$ with frame operator S over its span, and for a fixed i, ϕ_i is a local minimizer for the frame potential over the set:*

$$\Omega_i = \{\{\psi_m\}_{m=1}^M \mid \psi_m = \phi_m \text{ for all } m \neq i \text{ and } \|\psi_i\| = \|\phi_i\|\}$$

then ϕ_i is an eigenvector for S. Hence, any locally minimal sequence $\{\phi_m\}_{m=1}^M$ (i.e., ϕ_i is a local minimizer of the frame potential over Ω_i for all $1 \leq i \leq M$) is an FF-critical sequence. Moreover, any minimizer for the frame potential must span H_N.

PROOF. An obvious compactness argument guarantees that there is some sequence $\{\varphi_m\}_{m=1}^M \in \Omega_i$ which minimizes the frame potential. It is not clear, and will be addressed at the end of the proof, that the minimizers span the space $\mathbb{H}_N$. We note that:

$$\begin{aligned} FP(\{\{\varphi_m\}_{m\neq i} \cup \{\Phi\}) &= \sum_{n,m\neq i} |\langle \varphi_n, \varphi_m\rangle|^2 + |\langle \Phi, \Phi\rangle|^2 + 2\sum_{m\neq i} |\langle \Phi, \varphi_n\rangle|^2 \\ &= FP(\{\varphi_m\}_{m\neq i}) + 2F(\{\Phi\}) - \|\varphi_i\|^4, \end{aligned}$$

where

$$F(\{\Phi\}) = \sum_{m\neq i} |\langle \Phi, \varphi_m\rangle|^2.$$

It follows that the minimizers of FP over Ω_i are the minimizers of F. By Proposition 2.6, φ_i is an eigenvector for $\{\varphi_m\}_{m\neq i}$ with eigenvalue say λ. Now,

$$\begin{aligned} \sum_{m=1}^M \langle \varphi_i, \varphi_m\rangle \varphi_m &= \sum_{m\neq i} \langle \varphi_i, \varphi_m\rangle \varphi_m + \langle \varphi_i, \varphi_i\rangle \varphi_i \\ &= \lambda\varphi_i + \|\varphi_i\|^2\varphi_i = (\lambda + \|\varphi_i\|^2)\varphi_i. \end{aligned}$$

Suppose $\{\varphi_m\}_{m=1}^M \in \Omega_i$ is a minimizer for the frame potential over Ω_i. We proceed by way of contradiction. Suppose span $\{\varphi_m\}_{m=1}^M \neq \mathbb{H}_N$. Let S be the frame operator for $\{\varphi_m\}_{m=1}^M$ over its span with eigenspaces $\{E_{\lambda_i}\}_{i=1}^L$. By the first part of the proof, the vectors ϕ_m sit in the E_{λ_i}. For every $1 \leq j \leq L$ let $I_j = \{m \mid \phi_m \in E_{\lambda_j}\}$. If $\{\psi_m\}_{m=1}^M$ is a λ-tight frame, it is known (and a simple calculation to verify) that $\|\psi_i\|^2 \leq \lambda$ and $\|\psi_i\|^2 = \lambda$ if and only if $\psi_i \perp$ span $\{\psi_m\}_{m\neq i}$. Now, since $M \geq N$ and span $\{\varphi_m\}_{m=1}^M \neq H_N$, it follows that there is a $1 \leq j \leq L$ so that $|I_j| > \dim E_{\lambda_j}$. Hence, there is some $i \in I_j$ with $\|\varphi_i\|^2 < \lambda_j$. Now choose $e_1 \in \mathbb{H}_N$ with $\|e_1\| = 1$ and $e_1 \perp \{\varphi_m\}_{m=1}^M$. Define $\{\psi_m\}_{m=1}^M \in \Omega$ by: $\psi_m = \phi_m$ for all $1 \leq m \neq i \leq M$ and let $\psi_i = \|\varphi_i\|e_1$. We will obtain a contradiction by showing that $FP(\{\psi_m\}_{m=1}^M) < FP(\{\varphi_m\}_{m=1}^M)$.

$$\begin{aligned}FP(\{\psi_m\}_{m=1}^M) &= FP(\{\psi_m\}_{m\neq i} + \|\psi_i\|^4\\ &= FP(\{\varphi_m\}_{m\neq i}) + \|\varphi_i\|^4\\ &= \sum_{n,m\neq i} |\langle \varphi_n, \varphi_m\rangle|^2 + \|\varphi_i\|^4\\ &= \sum_{n,m=1}^M |\langle \varphi_n, \varphi_m\rangle|^2 + \|\varphi_i\|^4 - 2\sum_{n=1}^M |\langle \varphi_i, \varphi_n\rangle|^2 + \|\varphi_i\|^4\\ &= FP(\{\varphi_m\}_{m=1}^M) + 2\|\varphi_i\|^4 - 2\lambda_j\|\varphi_i\|^2\\ &= FP(\{\varphi_m\}_{m=1}^M) + 2\|\varphi_i\|^2(\|\varphi_i\|^2 - \lambda_j) < FP(\{\varphi_m\}_{m=1}^M).\end{aligned}$$

This contradiction completes the proof of the Theorem. □

In some applications of frame potentials we must switch to subsets of our frame and then see how the frame potential changes. The next two results give the basic calculations for the frame potential restricted to a subset of our frame.

PROPOSTION 2.9. *Let $\{\varphi_m\}_{m=1}^M$ be a frame for $\mathbb{H}_N$ and $I \subset \{1, 2, \ldots, M\}$. Then*

$$FP(\{\phi_m\}_{m\notin I}) = FP(\{\varphi_m\}_{m=1}^M) + FP(\{\phi_m\}_{m\in I}) - 2\sum_{n\in I}\sum_{m=1}^M |\langle \phi_m, \phi_m\rangle|^2.$$

PROOF. We compute:

$$FP(\{\varphi_m\}_{m\in I^c}) = \sum_{n,m\in I^c} |\langle \varphi_n, \varphi_m\rangle|^2$$

$$= \sum_{n,m=1}^M |\langle \varphi_n, \varphi_m\rangle|^2 - \sum_{n,m\in I} |\langle \varphi_n, \varphi_m\rangle|^2 - 2\sum_{n\in I}\sum_{m\in I^c} |\langle \varphi_n, \varphi_m\rangle|^2$$

$$= FP(\{\varphi_m\}_{m=1}^M) - FP(\{\varphi_n\}_{n\in I}) - 2\sum_{n\in I}\sum_{m\in I^c} |\langle \varphi_n, \varphi_m\rangle|^2 =$$

$$FP(\{\varphi_m\}_{m=1}^M) + FP(\{\varphi_n\}_{n\in I}) - 2\sum_{n\in I}\sum_{m\in I} |\langle \varphi_n, \varphi_m\rangle|^2 - 2\sum_{n\in I}\sum_{m\in I^c} |\langle \varphi_n, \varphi_m\rangle|^2 =$$

$$FP(\{\varphi_m\}_{m=1}^M) + FP(\{\varphi_m\}_{m\in I}) - 2\sum_{n\in I}\sum_{m=1}^M |\langle \varphi_n, \varphi_m\rangle|^2.$$

□

COROLLARY 2.10. *Let $\{\varphi_m\}_{m=1}^M$ be a tight frame for $\mathbb{H}_N$ with $\|\varphi_m\| = 1$, for all $1 \le m \le M$. If $I \subset \{1, 2, \ldots, M\}$ with $|I| = k$, then*

$$FP(\{\varphi_m\}_{m\in I^c}) = \frac{M^2}{N} - 2k\frac{M}{N} + FP(\{\varphi_m\}_{m\in I}).$$

Hence,

(1) *If $|I| > |I^c|$ then $FP(\{\varphi_m\}_{m\in I^c}) < FP(\{\varphi_m\}_{m\in I})$.*
(2) *If $|I| < |I^c|$ then $FP(\{\varphi_m\}_{m\in I^c}) > FP(\{\varphi_m\}_{m\in I})$.*
(3) *If M is even and $|I| = |I^c|$ then $FP(\{\varphi_m\}_{m\in I^c}) = FP(\{\phi_m\}_{m\in I})$.*

PROOF. Since $\|\phi_m\| = 1$, $FP(\{\varphi_m\}_{m=1}^M) = \frac{M^2}{N}$ and since the frame is a tight frame,

$$-2\sum_{n\in I}\sum_{m=1}^{M} |\langle\varphi_n, \varphi_m\rangle|^2 = -2\sum_{n\in I}\frac{M}{N} - 2k\frac{M}{N}.$$

□

3. A physical interpretation for finite tight frames

A problem which arose in communication theory is that of classifying those sequences $a_1 \geq a_2 \geq \cdots \geq a_m > 0$ so that there exists a tight frame $\{\phi_m\}_{m=1}^M$ for H_N for which $\|\phi_m\| = a_m$, for all $1 \leq m \leq N$. More importantly, if no tight frames exist for a given sequence $\{a_m\}_{m=1}^M$, can we identify those frames which are the closest to being tight and satisfy $\|\phi_m\| = a_m$? Casazza, Fickus, Kovačević, Leon and Tremain [**CF**] gave a complete solution to this question by building on earlier work of Benedetto and Fickus involving "frame-equivalent" notions to force and potential. In the process, the authors of [**CF**] arrived at the *Fundamental Inequality* for tight frames which governs the distribution of power among the frame vectors. All the results of this section come from [**CF**].

3.1. Coulomb's law and the frame force. The authors of [**CF**] investigated the common notions of what it means to equally distribute a collection of electrons upon a conductive spherical shell. Physically, in the absence of external forces the charged particles will repel each other according to the inverse-square Coulomb force law

$$CF(\phi_m, \phi_n) = \frac{\phi_m - \phi_n}{\|\phi_m - \phi_n\|^3}.$$

Intuitively, the corresponding optimal arrangements are those which minimize the internal pressure of the points upon each other. Specifically, given M distinct electrons located at points $\{\phi_m\}_{m=1}^M$, one seeks to minimize the total corresponding potential energy of the system,

$$CP(\{\phi_m\}_{m=1}^M) = \sum_{m=1}^{M}\sum_{n\neq m}\frac{1}{\|\phi_m - \phi_n\|}.$$

And though only a global minimizer corresponds to true equidistribution in this context, local minimizers are also of physical interest, in that they correspond to collections of points in equilibrium.

It is known that local minimizers of potential energy need not be global minimizers. For example, a dodecahedron inscribed in a sphere is a local minimizer of potential energy while there are different configurations with smaller total potential energy.

In [**BF**], Benedetto and Fickus defined a central force between real unit norm vectors ϕ_m and ϕ_n (called the *frame force*) by

$$FF(\phi_m, \phi_n) = 2\langle\phi_m, \phi_n\rangle(\phi_m - \phi_n).$$

The important point here is that the frame force for orthogonal vectors is 0. Also, vectors having an acute angle between them are repelling while vectors having an obtuse angle are attracting. So "charged particles" under the frame force are trying to reach equilibrium by becoming as orthogonal as possible. There are some other unnatural aspects to the frame force. For one, it needs a universal reference point—a fixed origin. Also, the force field generated by a point is not conservative. That is, the work required to get from one point to another depends upon the particular path taken. However (and this is all we need for our applications) the frame force is conservative when the points are constricted to lie on a sphere.

3.2. The weighted frame force. Since we need our vectors to lie on possibly different spheres, we need a corresponding *weighted frame force (potential)*. This makes the physical systems much more difficult to understand intuitively. That is, visualizing the movements of M charged particles restricted to M concentric spheres can challenge the imagination. However, this can be greatly simplified by "projecting" the dynamics down onto the unit sphere.

Consider two points, each of whose movement is restricted to a sphere of a given, yet arbitrary radius. That is, given $a_m, a_n > 0$, consider ϕ_m with $\|\phi_m\| = a_m$ and ϕ_n with $\|\phi_n\| = a_n$. Note that

$$\begin{aligned}\|\phi_m - \phi_n\|^2 &= \|\phi_m\|^2 - 2\langle\phi_m, \phi_n\rangle + \|\phi_n\|^2,\\ &= a_m^2 - 2\langle\phi_m, \phi_n\rangle + a_n^2,\end{aligned}$$

and so we may rewrite the frame force between these points as

$$\begin{aligned}FF(\phi_m, \phi_n) &= 2\langle\phi_m, \phi_n\rangle(\phi_m - \phi_n),\\ &= (a_m^2 + a_n^2 - \|\phi_m - \phi_n\|^2)(\phi_m - \phi_n).\end{aligned}$$

As in the Coulomb case, the pairwise potential between these points may be found by integrating the "magnitude" of this central force,

$$p(x) = -\int (a_m^2 + a_n^2 - x^2)x\,dx = \frac{1}{4}x^2[x^2 - 2(a_m^2 + a_n^2)],$$

and evaluating at $x = \|\phi_m - \phi_n\|$,

$$P(\phi_m, \phi_n) = p(\|\phi_m - \phi_n\|) = \langle\phi_m, \phi_n\rangle^2 - \frac{1}{4}(a_m^2 + a_n^2)^2.$$

The total potential contained within the physical system is the sum of all pairwise potentials,

$$TP(\{\phi_m\}_{m=1}^M) = \sum_{m,n} |\langle\phi_m, \phi_n\rangle|^2 - \frac{1}{4}\sum_{m,n}(a_m^2 + a_n^2)^2$$

However, we may disregard the additive constant, as it has no physical significance. This then leads to the *frame potential* definition of Benedetto and Fickus [Definition 2.1 above].

REMARK 3.1. Given two sequences $\{\phi_m\}_{m=1}^M$ and $\{\psi_m\}_{m=1}^M$ in $\mathbb{R}^N$ with $\|\phi_m\| = a_m = \|\psi_m\|$ for all m, $FP(\{\psi_m\}_{m=1}^M) - FP(\{\phi_m\}_{m=1}^M)$ is the work required to transform $\{\phi_m\}_{m=1}^M$ into $\{\psi_m\}_{m=1}^M$ while remaining on the spheres of radii $\{a_m\}_{m=1}^M$.

Though the frame potential is the potential energy contained within a system of points of equal weight on spheres of varying radii, we form an equivalent situation by considering points of varying weight all lying on a common sphere.

DEFINITION 3.2. Let S denote the unit sphere in $\mathbb{R}^N$, and $\psi_m \in \mathbb{R}^N$, $\|\psi_m\| = 1$ for all $1 \leq m \leq M$. We associate a weight w_m to each ψ_m.

- The *weighted frame force* between ψ_m and ψ_n is

$$WFF : [0,\infty) \times [0,\infty) \times S \times S \to \mathbb{R}^N,$$
$$WFF(w_m, w_n, \psi_m, \psi_n) = 2w_m w_n \langle \psi_m, \psi_n \rangle (\psi_m - \psi_n).$$

- The *weighted frame potential* of $\{\psi_m\}_{m=1}^M$ is

$$WFP(\{w_m\}_{m=1}^M, \{\psi_m\}_{m=1}^M) = \sum_{m=1}^M \sum_{n=1}^M w_m w_n |\langle \psi_m, \psi_n \rangle|^2.$$

For $\phi_m = a_m \psi_m$, we write,

$$WFP(\{a_m^2\}_{m=1}^M, \{\psi_m\}_{m=1}^M) = FP(\{\phi_m\}_{m=1}^M)$$

Thus, we shall no longer view the frame potential as the total potential energy of a system of points of mass 1 restricted to spheres of radius a_m. Rather, it shall be perceived more naturally as the energy of a system of points of masses a_m^2 restricted to a single sphere of radius 1.

In our situation, the points experiencing the frame force are constrained to move upon spheres. Therefore, only the components of the frame force acting upon a point which are tangent to the surface of the sphere at that point make a contribution to the frame potential. Thus, given $\phi_m, \phi_n \in \mathbb{R}^N$ with $\|\phi_m\| = a_m$, we wish to explicitly find the component of $FF(\phi_m, \phi_n)$ which lies tangent to the sphere of radius a_m at ϕ_m. Of course, to find the tangential component, we need only subtract the normal component from the whole. And since the surface in question is a sphere, the normal component of the force is simply the projection of the force onto the line passing through ϕ_m, i.e.,

$$\frac{\langle FF(\phi_m, \phi_n), \phi_m \rangle}{\langle \phi_m, \phi_m \rangle} \phi_m.$$

We therefore simplify

$$\begin{aligned} & FF(\phi_m, \phi_n) - \frac{\langle FF(\phi_m, \phi_n), \phi_m \rangle}{\langle \phi_m, \phi_m \rangle} \phi_m \\ &= \langle \phi_m, \phi_n \rangle (\phi_m - \phi_n) - \langle \phi_m, \phi_n \rangle \frac{\langle \phi_m, \phi_m \rangle - \langle \phi_m, \phi_n \rangle}{\langle \phi_m, \phi_m \rangle} \phi_m \\ &= \langle \phi_m, \phi_n \rangle \left(\frac{\langle \phi_m, \phi_n \rangle}{\langle \phi_m, \phi_m \rangle} \phi_m - \phi_n \right). \end{aligned}$$

This leads us to:

PROPOSTION 3.3. *Given a sequence* $\{\phi_m\}_{m=1}^M \subseteq \mathbb{R}^N$ *with* $\|\phi_m\| = a_m$, *the following are equivalent*

(1) $\{\phi_m\}_{m=1}^M$ *is a tight frame for* $\mathbb{R}^N$,
(2) $\sum_{m=1}^M FF(\phi, \phi_m) = 0$ *for all* $\phi \in \mathbb{R}^N$,
(3) $\sum_{m=1}^M FF(w, a_m^2, \phi, \phi_m) = 0$ *for all* $\phi \in S^{(N-1)}$.

3.3. The fundamental inequality and the minimizers of the weighted frame potential. To answer the question stated at the beginning of this section, we need an observation about decreasing sequences of positive numbers. Basically, this result compares the terms of the sequence to the "dimensional average" of the following terms.

PROPOSTION 3.4. *Given any sequence* $\{a_m\}_{m=1}^M \subset \mathbb{R}$ *with* $a_1 \geq \cdots \geq a_M \geq 0$, *and any* $N \leq M$, *there is a unique index* N_0 *with* $1 \leq N_0 \leq N$, *such that the inequality*

$$a_n > \frac{\sum_{m=n+1}^M a_m}{N-n} \tag{3.1}$$

holds for $1 \leq n < N_0$, *while the opposite inequality*

$$a_n \leq \frac{\sum_{m=n+1}^M a_m}{N-n} \tag{3.2}$$

holds for $N_0 \leq n < N$.

PROOF. Consider the set of indices n such that (3.2) holds, namely $\mathcal{I} = \{n : (N-n)a_n \leq \sum_{m=n+1}^M a_m\}$. Clearly $\mathcal{I} \neq \emptyset$, since $N \in \mathcal{I}$. Also, if $n \in \mathcal{I}$, then $n+1 \in \mathcal{I}$ since

$$\begin{aligned} [N-(n+1)]a_{n+1} &= -a_{n+1} + (N-n)a_{n+1}, \\ &\leq -a_{n+1} + (N-n)a_n, \\ &\leq -a_{n+1} + \sum_{m=n+1}^M a_m, \\ &= \sum_{m=n+2}^M a_m. \end{aligned}$$

The requisite properties upon N_o are then satisfied by letting N_0 be the minimum index in $\mathcal{I}$. □

The general answer to our original problem is contained in the following result from [**CF**]. We will denote by $S(a_1, \ldots, a_m)$ the product of spheres in $\mathbb{R}^N$ of radii $\{a_1, \ldots, a_m\}$.

THEOREM 3.5. *Given a sequence* $a_1 \geq a_2 \geq \cdots \geq a_M > 0$ *and any* $N \leq M$, *let* N_0 *be the smallest index* n *for which*

$$a_n^2 \leq \frac{\sum_{m=n+1}^M a_m^2}{N-n}$$

holds. Then any local minimizer of the frame potential

$$FP : S(a_1, \ldots, a_m) \to \mathbb{R}$$

is of the form

$$\{\phi_m\}_{m=1}^M = \{\phi_m\}_{m=1}^{N_0-1} \cup \{\phi_m\}_{m=N_0}^M$$

where $\{\phi_m\}_{m=1}^{N_0-1}$ *is an orthogonal set for whose orthogonal complement* $\{\phi_m\}_{m=N_0}^M$ *forms a tight frame.*

The proof of this theorem is a serious piece of work relying heavily on a deep intuitive understanding of the frame potential, and so we refer the reader to [**CF**] for the details. An important consequence of Theorem 3.5 is that we can only get tight frames for H_N when $N_0 = 1$. This case can be summarized as:

COROLLARY 3.6. *Fix* $N \leq M$ *and* $a_1 \geq a_2 \geq \cdots \geq a_M > 0$. *The following are equivalent:*

(1) *There is a tight frame* $\{\phi_m\}_{m=1}^M$ *for* H_N *satisfying* $\|\phi_m\| = a_m$, *for* $1 \le m \le M$.

(2) *For all* $1 \le n < N$ *we have*

$$a_n^2 \le \frac{\sum_{m=n+1}^M a_m^2}{N-n}.$$

(3) *We have:*

$$\sum_{m=1}^M a_m^2 \ge N a_1^2.$$

(4) *If*

$$\lambda = \sqrt{\frac{N}{\sum_{m=1}^M a_m^2}},$$

then $\lambda a_m \le 1$ *for all* $1 \le m \le M$.

PROOF. The equivalence of (1) and (3) is Theorem 3.5. The equivalence of (2) and (4) is immediate.

(2)⇒(3): Let $n = 1$ in (1) and get

$$\sum_{m=2}^M a_m^2 \ge (N-1)a_1^2.$$

Hence,

$$\sum_{m=1}^M a_m^2 \ge N a_1^2.$$

(3)⇒(2): We assume that

$$\sum_{m=1}^M a_m^2 \ge N a_1^2.$$

Since $a_1 \ge a_m$, for all $1 \le m \le M$ we have

$$\begin{aligned}\sum_{m=N+1}^M a_m^2 &\ge N a_1^2 - \sum_{m=1}^N a_m^2 \\ &= (N-n)a_1^2 - \sum_{m=n+1}^N a_m^2 + \sum_{m=1}^d (a_1^2 - a_m^2) \\ &\ge (N-n)a_n^2 - \sum_{m=n+1}^N a_m^2.\end{aligned}$$

Hence,

$$\sum_{m=n+1}^M a_m^2 \ge (N-n)a_n^2,$$

and thus

$$a_n^2 \le \frac{\sum_{m=n+1}^M a_m^2}{N-n}.$$

□

We refer to (3) in the corollary above as the *Fundamental Inequality* for tight frames. Note that what we have shown in Corollary 3.7 is that properties (2)-(4) are equivalent for any sequence of numbers.

As we discussed earlier, local minimizers of the Coulomb potential are not necessarily global minimizers. However, for the frame potential, we do have that local minimizers are global minimizers. Hence, the characterization of Theorem 3.5 applies to all minimizers and yields the following result from [**CF**].

COROLLARY 3.7. *Given a sequence* $\{a_m\}_{m=1}^M \subset \mathbb{R}$ *with* $a_1 \geq \cdots \geq a_M > 0$, *and any* $N \leq M$, *let* N_0 *denote the smallest index* n *such that*

$$a_n^2 \leq \frac{\sum_{m=n+1}^M a_n^2}{N-n}$$

holds. Then, for the frame potential $FP : S(a_1, \ldots, a_M) \to \mathbb{R}$,

(1) *The minimal value is* $\sum_{m=1}^{N_0-1} a_m^4 + \frac{(\sum_{m=N_0}^M a_m^2)^2}{N-N_0+1}$.
(2) *Any local minimizer is a global minimizer.*
(3) *The minimizers are precisely those sequences where* $\{\phi_m\}_{m=1}^{N_0-1}$ *is an orthogonal set for whose orthogonal complement* $\{\phi_m\}_{m=N_0}^M$ *forms a tight frame.*

PROOF. To begin, we compute the frame potential for any sequence of the form $\{\phi_m\}_{m=1}^{N_0-1} \cup \{\phi_m\}_{m=N_0}^M$, where $\{\phi_m\}_{m=1}^{N_0-1}$ is an orthogonal sequence for whose orthogonal complement the sequence $\{\phi_m\}_{m=N_0}^M$ is a tight frame. Thus, for $m = 1, \ldots, N_0 - 1$, $\langle \phi_m, \phi_n \rangle = 0$ for $n \neq m$, and so

$$\begin{aligned} FP(\{\phi_m\}_{m=1}^M) &= \sum_{m=1}^M \sum_{n=1}^M |\langle \phi_m, \phi_n \rangle|^2, \\ &= \sum_{m=1}^{N_0-1} |\langle \phi_m, \phi_m \rangle|^2 + \sum_{m=N_0}^M \sum_{n=N}^M |\langle \phi_m, \phi_n \rangle|^2, \\ &= \sum_{m=1}^{N_0-1} a_m^4 + FP(\{\phi_m\}_{m=N_0}^M). \end{aligned}$$

Now the fact that $\{\phi_m\}_{m=N_0}^M$ is a tight frame for the $(N - N_0 + 1)$-dimensional orthogonal complement implies that

$$FP(\{\phi_m\}_{m=1}^M) = \sum_{m=1}^{N_0-1} a_m^4 + \frac{(\sum_{m=N_0}^M a_m^2)^2}{N - N_0 + 1}. \tag{3.3}$$

In light of Theorem 3.5 we observe that any local minimizer of the frame potential is necessarily of this form, and consequently attains the potential value given in (3.3).

Next, we note that $S(a_1, \ldots, a_M)$ is compact, being a Cartesian product of spheres. The continuity of the frame potential upon this compact domain then guarantees that a global minimizer exists. As any global minimizer is also a local minimizer, the value given in (3.3) is indeed the global minimum, yielding the first claim.

Since all local minimizers attain the same value, the second claim follows.

For the third claim, note that any sequence of this form has already been shown to attain the minimal value, while Theorem 3.5 yields the converse implication. $\square$

4. Algorithmic constructions of frames

In this section we will look at algorithms for constructing frames with prescribed properties. This is critically important for applications where we need systematic methods for constructions which can be carried out, for e.g., in MatLab.

One serious inhibitor for algorithmic constructions of tight frames is that we cannot get tight frames at each stage of the construction. That is, if $\{\phi_m\}_{m=1}^M$ is a tight frame then $\{\phi_m\}_{m=1}^{M-K+1}$ cannot also be a tight frame for $1 \le K < N$. That is, assume for all $\phi \in H_N$ we have

$$\lambda\|\phi\|^2 = \sum_{m=1}^{M} |\langle \phi, \phi_m\rangle|^2.$$

Now, if $\{\phi_m\}_{m=1}^{M-K+1}$ is a δ tight frame,

$$\delta\|\phi\|^2 = \sum_{m=1}^{M-K+1} |\langle \phi, \phi_m\rangle|^2,$$

and so

$$(\lambda-\delta)\|\phi\|^2 = \sum_{m=M-K+2}^{M} |\langle \phi, \phi_m\rangle|^2.$$

But $\{\phi_m\}_{m=M-K+2}^M$ does not span H_N so this is impossible.

4.1. Nearly equal norm and nearly tight frames. In this subsection we will give algorithms for turning frames which are nearly tight (respectively, nearly equal norm) into tight frames (respectively, equal norm frames). We wish to thank Beata Randrianantoanina for sharing with us her notes on algorithmic constructions of frames. Propositions 4.1 and 4.3 of this section are an analgam of Randrianantoanina's notes and the author's.

Let $\{\phi_m\}_{m=1}^M$ be a frame with frame bounds A, B and satisfying $c-\epsilon \le \|\phi_m\| \le c+\epsilon$, for all $1 \le m \le M$. After a change of scale we may assume $1-\epsilon \le \|\phi_m\| \le 1+\epsilon$ and after a permutation we may as well assume that $1+\epsilon \ge \|\phi_1\| \ge \|\phi_2\| \ge \cdots \ge \|\phi_M\| \ge 1-\epsilon$. We call such a frame a ϵ *nearly unit norm frame*. We can turn $\{\phi_m\}_{m=1}^M$ into an equal norm frame by letting

$$\psi_m = x\frac{\phi_m}{\|\phi_m\|}, \quad \text{for } 0 < x.$$

Then

$$\max_{1\le m\le M} |x - \|\phi_m\||$$

and

$$\sum_{m=1}^{M} |\|\psi_m\| - \|\phi_m\|| = \sum_{m=1}^{M} |x - \|\phi_m\||$$

are minimized when

$$x = \frac{\|\phi_1\|^2 + \|\phi_M\|^2}{2}.$$

Also,

$$E = \sum_{m=1}^{M} |x - \|\phi_m\||^2$$

is minimized when

$$x = \frac{\sum_{m=1}^{M} \|\phi_m\|}{M}.$$

We need to check how close our new frame is to the old one.

$$\begin{aligned}\sum_{m=1}^{M} \|\psi_m - \phi_m\|^2 &= \sum_{m=1}^{M} \left\| x\frac{\phi_m}{\|\phi_m\|} - \phi_m \right\|^2 \\ &= \sum_{m=1}^{M} |x - \|\phi_m\||^2 = E.\end{aligned}$$

In this case we say $\{\psi_m\}_{m=1}^M$ is *E-close* to $\{\phi_m\}_{m=1}^M$. Also, for all $\phi \in H_N$,

$$\begin{aligned}\left(\frac{1-\epsilon}{1+\epsilon}\right)^2 A\|\phi\|^2 &\leq \left(\frac{1-\epsilon}{1+\epsilon}\right)^2 \sum_{m=1}^{M} |\langle \phi, \phi_m\rangle|^2 \\ &\leq \sum_{m=1}^{M} \frac{x^2}{\|\phi_m\|^2} |\langle \phi, \phi_m\rangle|^2 \\ &= \sum_{m=1}^{M} |\langle \phi, \psi_m\rangle|^2 \\ &\leq \left(\frac{1+\epsilon}{1-\epsilon}\right)^2 B\|\phi\|^2.\end{aligned}$$

We formulate this as:

PROPOSTION 4.1. *If $\{\phi_m\}_{m=1}^M$ is an ϵ nearly unit norm frame with frame bounds A, B, then $\{\phi_m\}_{m=1}^M$ is E-close to a unit norm frame with frame bounds $\left(\frac{1-\epsilon}{1+\epsilon}\right)^2 A$, and $\left(\frac{1+\epsilon}{1-\epsilon}\right)^2 B$.*

Now we look at nearly tight frames. After a change of scale, we may restrict our attention to nearly Parseval frames.

We say that a frame $\{\phi_m\}_{m=1}^M$ with frame bounds A, B is ϵ *nearly Parseval* if for all $\phi \in H_N$,

$$(1-\epsilon)\|\phi\|^2 \leq \sum_{m=1}^{M} |\langle \phi, \phi_m\rangle|^2 \leq (1+\epsilon)\|\phi\|^2.$$

Let S be the frame operator for a frame $\{\phi_m\}_{m=1}^M$ with eigenvalues $\lambda_1 \geq \lambda_2 \geq \cdots \geq \lambda_N > 0$ and a corresponding orthonormal set of eigenvectors $\{e_n\}_{n=1}^N$. The following result can be found in [**CF1**].

PROPOSTION 4.2. *For any $1 \leq n \neq k \leq N$,*

$$\sum_{m=1}^{M} \langle \phi_m, e_n\rangle \overline{\langle \phi_m, e_k\rangle} = 0.$$

That is, the column vectors of the matrix $(\langle \phi_m, e_n\rangle)_{m=1,n=1}^{M,N}$ are orthogonal.

PROOF. We compute:

$$\begin{aligned}\langle Se_n, e_k\rangle &= \langle \lambda_n e_n, e_k\rangle = \lambda_n \langle e_n, e_k\rangle = \lambda_n \delta_{nk} \\ &= \left\langle \sum_{m=1}^{M} \langle e_n, \phi_m\rangle \phi_m, e_k \right\rangle \\ &= \sum_{m=1}^{M} \overline{\langle \phi_m, e_n\rangle} \langle \phi_m, e_k\rangle\end{aligned}$$

□

Now we want to check that an ϵ nearly Parseval frame is close to a Parseval frame. Given $\{\phi_m\}_{m=1}^M$ ϵ nearly Parseval for H_N with frame bounds A, B, let $\lambda_1, \geq \lambda_2 \geq \cdots \geq \lambda_N > 0$ be the eigenvalues for the frame operator S with orthonormal eigenvectors $\{e_n\}_{n=1}^M$. It is known (see [**B**] and [**J**]) that the "closest" tight frame to $\{\phi_m\}$ is $\{S^{-\frac{1}{2}}\phi_m\}$. We have that $1+\epsilon \geq \lambda_1 \geq \lambda_2 \geq \cdots \geq \lambda_N \geq 1-\epsilon$. We construct a new frame $\{\psi_m\}_{m=1}^M$ by

$$S^{-\frac{1}{2}}\phi_m = \psi_m = \sum_{n=1}^{N} \frac{1}{\lambda_n} \langle \phi_m, e_n\rangle e_n.$$

Then

$$\begin{aligned}\sum_{m=1}^{M} \|\psi_m - \phi_m\|^2 &= \sum_{m=1}^{M}\sum_{n=1}^{N} |\langle \phi_m, e_n\rangle|^2 \left|\frac{1}{\lambda_n} - 1\right|^2 \\ &= \sum_{n=1}^{N} \left|\frac{1}{\lambda_n} - 1\right|^2 \sum_{m=1}^{M} |\langle \phi_m, e_n\rangle|^2 \\ &\leq \sum_{n=1}^{N} \left|\frac{1}{\lambda_n} - 1\right| (1+\epsilon) \\ &\leq N\frac{\epsilon}{1-\epsilon}(1+\epsilon).\end{aligned}$$

Also, for every $\phi \in H_N$,

$$\begin{aligned}\sum_{m=1}^{M} |\langle \phi, \psi_m\rangle|^2 &= \sum_{m=1}^{M} \left|\sum_{n=1}^{N} \frac{1}{\lambda_n} \langle \phi, e_n\rangle \overline{\langle \phi_m, e_n\rangle}\right|^2 \\ &= \sum_{m=1}^{M}\sum_{n=1}^{N}\sum_{k=1}^{M} \frac{1}{\lambda_n}\frac{1}{\lambda_k} \langle \phi, e_n\rangle \overline{\langle \phi, e_k\rangle}\ \overline{\langle \phi_m, e_n\rangle} \langle \phi_m, e_k\rangle \\ &= \sum_{n=1}^{N}\sum_{k=1}^{N} \frac{1}{\lambda_n}\frac{1}{\lambda_k} \langle \phi, e_n\rangle \overline{\langle \phi, e_k\rangle} \sum_{m=1}^{M} \overline{\langle \phi_m, e_n\rangle} \langle \phi_m, e_k\rangle \\ &= \sum_{n=1}^{N} \frac{1}{\lambda_n} |\langle \phi, e_n\rangle|^2 \sum_{m=1}^{M} |\langle \phi_m, e_n\rangle|^2 \\ &= \sum_{n=1}^{N} \frac{1}{\lambda_n} |\langle \phi, e_n\rangle|^2 \lambda_n = \|\phi\|^2.\end{aligned}$$

In the next to last equality above we used Proposition 4.2. We summarize this as:

PROPOSTION 4.3. *If* $\{\phi_m\}_{m=1}^M$ *is* ϵ *nearly Parseval, then* $\{\phi_m\}_{m=1}^M$ *is* $N\epsilon\frac{1+\epsilon}{1-\epsilon}$ *close to a Parseval frame.*

So, we have seen that nearly unit norm frames are close to unit norm frames and nearly Parseval frames are close to Parseval frames. However, each of these algorithms destroys the other property. Vern Paulsen has raised the question of whether there is a quantitative construction giving both of these properties at once.

PROBLEM 4.4 (Paulsen). If a frame is ϵ nearly equal norm and δ nearly Parseval, is the frame $f(\epsilon, \delta)$ close to a equal norm Parseval frame? The real point here is to identity the closeness function $f(\epsilon, \delta)$.

Related to this is the old and important question of identifying all (equal norm) Parseval frames.

PROBLEM 4.5. Is there an algorithm for constructing all (equal norm) Parseval frames?

There is a simple algorithm for turning a Parseval frame into an equal norm Parseval frame due to Holmes and Paulsen [**HP**]. Unfortunately, at this time we have no way to access the closeness of the new frame to the old under this construction.

PROPOSTION 4.6. *There is an algorithm for turning any frame into an equal norm frame without changing the frame operator.*

PROOF. Let $\{\phi_m\}_{m=1}^M$ be a frame for H_N with frame operator S and analysis operator F. Then

$$\sum_{m=1}^{M} \|\phi_m\|^2 = TrS,$$

so let $\lambda = \frac{TrS}{M}$. If $\|\phi_m\|^2 = \lambda$ for all m, we are done. Otherwise, there exists $1 \le i \ne j \le M$ with $\|\phi_i\|^2 > \lambda > \|\phi_j\|^2$. For any θ, replace the vectors ϕ_i, ϕ_j by the vectors $\psi_i = (\cos\theta)\phi_i - (\sin\theta)\phi_j$, $\psi_j = (\sin\theta)\phi_i + (\cos\theta)\phi_j$ and let $\psi_k = \phi_k$ for all other k. Then the analysis operator for $\{\psi_m\}_{m=1}^M$ is $F_1 = UF$ for a unitary operator U on ℓ_2^M given by the Givens rotation. Hence, $F_1^* F_1 = F^*U^*UF = F^*F = S$. So our frame operator is unchanged for any value of θ. By choosing θ appropriately, we can obtain $\|\phi_i\|^2 = \lambda$. Repeating this process at most $N-1$ times yields an equal norm frame with the same frame operator as $\{\phi_m\}_{m=1}^M$. Note that this gives an alternative proof of Corollary 4.11. □

D. Hadwin (see [**HP**]) has pointed out that a simple compactness argument shows that a solution to Problem 4.4 exists. Unfortunately, this argument does not give any quantitative estimates on the parameters involved. We want to show that for every $\epsilon > 0$, there exists a $\delta > 0$ such that if a frame has frame bounds $1-\delta$, $1+\delta$ and the lengths of the frame vectors differ by at most δ, then there is an equal norm tight frame whose vectors are at distance at most $\epsilon > 0$ from the original vectors [**HP**]. So, we assume not. Then there is an $\epsilon > 0$ so that for every $\delta = \frac{1}{n}$, there is a frame $\{\phi_m^n\}_{m=1}^M$ with frame bounds $1-\frac{1}{n}$, $1+\frac{1}{n}$ and whose vectors have lengths differing by at most $\frac{1}{n}$, but this frame is greater than $\epsilon > 0$ from any equal norm tight frame. By compactness (and switching to a subsequence) we may assume that

$$\lim_{n\to\infty} \phi_m^n = \phi_m, \quad \text{for all } 1 \le m \le M.$$

But now, $\{\phi_m\}_{m=1}^M$ is an equal norm tight frame, which is a contradiction.

4.2. Algorithmic constructions for tight frames. A problem which arose in Communication Theory is to identify those sequences of numbers $a_1 \geq a_2 \geq \cdots \geq a_M > 0$ so that there is a tight frame $\{\phi_m\}_{m=1}^M$ for H_N with $\|\phi_m\| = a_m$, for all $1 \leq m \leq M$. As we saw earlier, this problem was resolved in [**CF**]. But this proof is not implementable in practice. Casazza and Leon [**CL**] gave an algorithmic construction for this class of frames as well as a MatLab program [**CL1**] for carrying this out.

THEOREM 4.7. *For any $M \geq N$ and any sequence of numbers $1 \geq a_1 \geq a_2 \geq \cdots \geq a_M > 0$ satisfying $\sum_{m=1}^M a_m^2 = N$, there is a Parseval frame $\{\phi_m\}_{m=1}^M$ for H_N satisfying $\|\phi_m\| = a_m$, for all $1 \leq m \leq M$.*

PROOF. We sketch the algorithm here.

A *Givens rotation*

$$\theta(t,j,k) \text{ for } 1 \leq j < k \leq M$$

is

$$\theta(t,j,k) = \begin{pmatrix} I_{j-1,j-1} & 0 & 0 & 0 & 0 \\ 0 & \cos(t) & 0 & \sin(t) & 0 \\ 0 & 0 & I_{M-j-k-2,M-j-k-2} & 0 & 0 \\ 0 & -\sin(t) & 0 & \cos(t) & 0 \\ 0 & 0 & 0 & 0 & I_{k-1,k-1} \end{pmatrix}$$

It is clear that

$$\theta(t,j,k)^{-1} = \theta(-t,j,k).$$

To construct a tight frame $\{\phi_m\}_{m=1}^M$ with M elements in H_N with $\|\phi_m\| = a_m$, for all $1 \leq m \leq M$, we need [**CK, CF1**] to construct an $M \times M$ unitary matrix $(m_{ij})_{i,j=1}^M$ so that

$$\sum_{j=1}^N |m_{ij}|^2 = a_i^2, \quad \text{for all } 1 \leq i \leq M.$$

To do this, we will start with any block unitary matrix of the form

$$O = \begin{pmatrix} R_N & O \\ O & R_{M-N,M-N} \end{pmatrix}$$

where R_K denotes a $K \times K$ unitary matrix. Then by applying Givens notations, we will turn it into a solution to our problem.

For step I, let

$$t_1 = \arg(a_1 + i\sqrt{1-a_1^2})$$

and let

$$O^1 = \theta(t_1, 1, M)O$$

We will end up with a new unitary matrix O^1 with the values of the squared norms of the partial rows as described by the diagram:

$$\begin{pmatrix} a_1^2 & | & 1-a_1^2 \\ 1 & | & 0 \\ 1 & | & 0 \\ \vdots & | & \vdots \\ 1 & | & 0 \\ --- & | & --- \\ 0 & | & 1 \\ \vdots & | & \vdots \\ 0 & | & 1 \\ 1-a_1^2 & | & a_1^2 \end{pmatrix}$$

Now one of the following applies:

(CASE I) $1 \leq a_1^2 + a_2^2$, or

(CASE II) $1 > a_1^2 + a_2^2$

In CASE I, take t_2 a solution of

$$\sin(t_2)^2 = \frac{1-a_2^2}{a_1^2} \text{ and } O^2 = \theta(t_2, 2, M)O^1.$$

Then O^2 is also a unitary matrix with the values of the squared norms of the partial rows as described by the diagram:

$$\begin{pmatrix} a_1^2 & | & 1-a_1^2 \\ a_2^2 & | & 1-a_2^2 \\ 1 & | & 0 \\ \vdots & | & \vdots \\ 1 & | & 0 \\ ----- & | & ----- \\ 0 & | & 1 \\ \vdots & | & \vdots \\ 0 & | & 1 \\ 2-a_1^2-a_2^2 & | & a_1^2+a_2^2-1 \end{pmatrix}$$

In CASE II, take t_2 a solution of

$$\sin(t_2)^2 = \frac{a_2^2}{1-a_1^2} \text{ and let } O^2 = \theta(t_2, M-1, M)O^1.$$

Then O^2 is a unitary matrix with the values of the squared norms of the partial rows as described by the diagram:

$$\begin{pmatrix} a_1^2 & | & 1-a_1^2 \\ 1 & | & 0 \\ 1 & | & 0 \\ \cdots & | & \cdots \\ 1 & | & 0 \\ ------ & | & ------ \\ 0 & | & 1 \\ \cdots & | & \cdots \\ 0 & | & 1 \\ a_2^2 & | & 1-a_2^2 \\ 1-a_1^2-a_2^2 & | & a_1^2+a_2^2 \end{pmatrix}$$

Now it remains to prove that CASE I eventually takes place whenever $k < N-1$ and when $k = N-1$ only CASE II can apply. We refer the reader to [**CL**] for the details of this argument. □

The algorithm above actually proves more than we stated. A sequence $\{a_1 \geq a_2 \geq \cdots \geq a_M > 0\}$ is called *admissible* for N if $a_m \leq 1$, for $1 \leq m \leq M$ and $\sum_{m=1}^{M} a_m^2 = N$. Also, let $O(K)$ denote the family of $K \times K$ unitary matrices. We now have:

THEOREM 4.8. *Fix $N < M$ and $\{a_m\}_{m=1}^{M}$ an admissible sequence. Then*

(1) *There is a tight frame $\{\phi_m\}_{m=1}^{M}$ for H_N with $\|\phi_m\| = a_m$, for all $1 \leq m \leq M$.*
(2) *Moreover, there is an embedding from $O(N) \times O(M-N)$ into the space of all solutions for $\{a_j\}_{j=1}^{M}$. The embedding will take the pair of matrices $R_N \in O(N)$ and $R_{M-N} \in O(M-M)$ into*

$$\Psi \begin{pmatrix} R_N & 0 \\ 0 & R_{M-N,M-N} \end{pmatrix}$$

where the matrix Ψ is unitary and is determined by the admissible *sequence $\{a_j\}_{j=1}^{M}$.*

It must be pointed out that the embedding of $O(N) \times O(M-M)$ into $O(M)$ does not fill out the space of solutions. Using alternative algorithms it is possible to construct solutions which do no belong to the image of the embedding. That is, it can not be factorized as

$$\psi O = \psi \begin{pmatrix} R_N & 0 \\ 0 & R_{M-N} \end{pmatrix}$$

So we still have the problem:

PROBLEM 4.9. Find an algorithm which constructs all tight frames of the form $\|\phi_m\| = a_m$, $1 \leq m \leq M$, for H_N.

4.3. Frames with a given frame operator. Let S be a positive self-adjoint invertible operator on H_N. We want to find necessary and sufficient conditions on $a_1 \geq a_2 \geq \cdots \geq a_M > 0$ so that there is a frame $\{\phi_m\}_{m=1}^{M}$ for H_N with frame operator S and satisfying $\|\phi_m\| = a_m$, for all $1 \leq m \leq M$. Again, what we need is an algorithm for constructing such frames. Such an algorithm was given by Casazza and Leon [**CL**] along with a MatLab program for implementing it [**CL2**].

THEOREM 4.10. *Let S be a positive self-adjoint operator on a N-dimensional Hilbert space H_N. Let $\lambda_1 \geq \lambda_2 \geq \cdots \geq \lambda_N$ be the eigenvalues of S. Fix $M \geq N$ and real numbers $a_1 \geq a_2 \geq \cdots a_M > 0$. The following are equivalent:*

(1) *There is a frame $\{\varphi_m\}_{m=1}^M$ for H_N with frame operator S and $\|\varphi_m\| = a_m$, for all $m = 1, 2, \ldots, M$.*
(2) *For every $1 \leq k \leq N$,*

$$\sum_{m=1}^{k} a_i^2 \leq \sum_{i=1}^{k} \lambda_i, \quad \text{and} \quad \sum_{m=1}^{M} a_m^2 = \sum_{i=1}^{N} \lambda_i.$$

The construction of these frames is a variation of the algorithm given in Section 4.2 [**CL**].

In Section 4.4 we will see a class of equal norm Parseval frames called the *harmonic frames*. As a consequence of Theorem 4.10 above, we conclude that there are equal norm frames for any frame operator.

COROLLARY 4.11. *If S is a positive self-adjoint invertible operator on H_N, then for every $M \geq N$ there is an equal norm frame $\{\varphi_m\}_{m=1}^M$ for H_N whose frame operator equals S.*

PROOF. Let $\lambda_1 \geq \lambda_2 \geq \cdots \geq \lambda_N$ be the eigenvalues of S and let $\lambda = \sum_{i=1}^N \lambda_i$. Fix $M \geq N$. If $a_m = \sqrt{\lambda/M}$ for every m then $\sum_{m=1}^M a_m^2 = \lambda$, and for all $1 \leq k \leq N$,

$$\sum_{m=1}^{k} a_n^2 = \frac{k}{M}\lambda = \frac{k}{M}\sum_{i=1}^{N} \lambda_i = k\frac{N}{M}\frac{1}{N}\sum_{i=1}^{N} \lambda_i \leq k\frac{N}{M}\frac{1}{k}\sum_{i=1}^{k} \lambda_i \leq \sum_{i=1}^{k} \lambda_i.$$

In the next to last inequality above we have used the fact that deleting some of the smallest numbers from a set of numbers will increase the average of the numbers. Hence, by Theorem 4.10 there is a frame $\{\varphi_m\}_{m=1}^M$ with $\|\varphi_m\| = \sqrt{\lambda/M}$ for all $n = 1, 2, \ldots, M$ having frame operator S. □

However, as in Section 4.2, this algorithm does not construct all frames (or even all equal norm frames) having frame operator S. It would be invaluable to have such a construction.

PROBLEM 4.12. For a given positive, self-adjoint invertible operator S on H_N, and $a_1 \geq a_2 \geq \cdots \geq a_m > 0$ which satisfies Theorem 4.10, find an algorithm for constructing all frames $\{\phi_m\}_{m=1}^M$ for H_N with frame operator S and satisfying $\|\phi_m\| = a_m$, for all $1 \leq m \leq M$.

4.4. Grassmannian frames. An area which seriously needs algorithmic constructions is that of Grassmannian frames introduced by Heath and Strohmer [**HS**].

DEFINITION 4.13. For a given unit norm frame $\{\phi_m\}_{m=1}^M$ in H_N we define the *maximal frame correlation* $\mathcal{M}(\{\phi_m\}_{m=1}^M)$ by

$$\mathcal{M}(\{\phi_m\}_{m=1}^M) = \max_{1 \leq m \neq k \leq M} \{|\langle \phi_m, \phi_k \rangle|\}.$$

The assumption that our frames are unit norm is for convenience above since we could normalize the frame elements in the inner product above instead.

DEFINITION 4.14. A family $\{\phi_m\}_{m=1}^M$ in H_N is called a *Grassmannian frame* if it is a solution to

$$\min\{\mathcal{M}(\{\phi_m\}_{m=1}^M)\}$$

where the min is taken over all unit norm frames $\{\phi_m\}_{m=1}^M$ in H_N.

So a Grassmannian frame minimizes the maximum correlation between frame elements. Grassmannian frames are the best frames for many applications. Unfortunately, very few are known. Even the very useful *harmonic frames* are rarely Grassmannian. The elements of a harmonic tight frame $\{\phi_m\}_{m=1}^M$ for $\mathbb{C}^N$ are given by (see [**CK**]):

$$\phi_m = \frac{1}{\sqrt{N}}(w_1^m, w_2^m, \ldots, w_N^m)_{k=1}^N$$

where the w_k are M-th roots of unity. It is known that harmonic tight frames are always Grassmannian frames when $M = N$ or $N+1$. H. König [**K**] has given examples of harmonic tight frames in his work with spherical designs.

For p a prime number, set $k = p^\ell + 1$ for $\ell \in \mathbb{N}$ and let $K = n^2 - n + 1$. Then there exist integers $0 \leq d_1 < \cdots < d_k < M$ such that all numbers $1, 2, \ldots, M-1$ occur as residues $\mod M$ of the $n(n-1)$ differences $d_i - d_j$, $i \neq j$. For $m = 1, 2, \ldots, M$ we define

$$\phi_m = \left\{ \frac{1}{\sqrt{n}} e^{2\pi i m d_j / N} \right\}_{j=1}^k .$$

Then $\{\phi_m\}_{m=1}^M$ is a harmonic tight Grassmannian frame with $\mathcal{M}(\{\phi_m\}_{m=1}^M) = \frac{\sqrt{n-1}}{n}$.

DEFINITION 4.15. A unit norm frame $\{\phi_m\}_{m=1}^M$ for H_N is *equiangular* if there is a $c > 0$ so that

$$|\langle \phi_n, \phi_m \rangle| = c \text{ for all } n \neq m.$$

There are bounds for the maximal frame correlation function and the maximal number of elements in a Grassmannian frame [**HS**].

THEOREM 4.16. *Let $\{\phi_m\}_{m=1}^M$ be a unit norm frame for H_N. then*

$$\mathcal{M}(\{\phi_m\}_{m=1}^M) \geq \sqrt{\frac{M-N}{N(M-1)}}, \tag{4.1}$$

with equality if and only if $\{\phi_m\}_{m=1}^M$ is an equiangular tight frame.

Furthermore,

(1) *If $H_N = \mathbb{R}^N$, to have equality in (4.1) it is necessary that $M \leq \frac{N(N+1)}{2}$.*
(2) *If $H_N = \mathbb{C}^N$, to have equality in (4.1) it is necessary that $M \leq N^2$.*

We call unit norm fames which produce equality in (4.1) *optimal Grassmannian frames.*

There are few known examples of optimal Grassmannian frames. But they arise in a variety of areas such as Grassmannian spaces, spherical codes, graph theory and the study of equiangular line sets [**HS**]. It would be a major advance for all these subjects and more if we could find an algorithmic method for constructing Grassmannian frames. At this time we have no general methods for producing Grassmannian frames—with or without algorithms. Moreover, frames giving the optimal upper bound number of elements are known (or more accurately, are approximately given by computer programs) for $N \leq 15$.

The bound given in Theorem 4.16 can be generalized to non-unit norm frames. This was observed by Waldron [**W**] in relation to Welsh bounds in signal processing.

THEOREM 4.17. *If* $\{\phi_m\}_{m=1}^M$ *is a frame for* H_N *then*

$$\max_{1\le m\ne k\le M} |\langle \phi_m, \phi_k\rangle|^2 \ge \frac{\left(\sum_{m=1}^M \|\phi_m\|^2\right)^2 N - \sum_{m=1}^M \|\phi_m\|^4}{M(M-1)},$$

with equality if and only if $\{\phi_m\}_{m=1}^M$ *is an equiangular tight frame.*

5. Concluding remarks

Some of the results in this paper have been extended in [**CF2**] to infinite frames. First for infinite frames on finite dimensional Hilbert spaces.

THEOREM 5.1. *Given* $a_1 \ge a_2 \ge \cdots \ge 0$ *and a Hilbert space* $\mathbb{H}_N$*, the following are equivalent:*

(1) *There is a normalized tight frame* $\{\phi_m\}_{m=1}^\infty$ *for* $\mathbb{H}_N$ *with* $\|\varphi_m\| = a_m$*, for all* $m = 1, 2, 3, \ldots$.
(2) *We have that* $\sum_{m=1}^\infty a_m^2 < \infty$ *and*

$$\sum_{m=1}^\infty a_m^2 \ge N a_1^2.$$

Moreover, if (2) fails and $1 \le N_0 < N$ *is the largest number for which (2) fails, then the frame* $\{\psi_m\}_{m=1}^\infty$ *which is closest to being tight (in the sense of minimizing potential energy) with* $\|\varphi_m\| = a_m$ *is:*

$$\{c_m a_m e_m\}_{m=1}^{N_0} \cup \{\varphi_m\}_{m=N_0+1}^\infty$$

where $\{e_i\}_{i=1}^N$ *is an orthonormal basis for* $\mathbb{H}_N$*,* $|c_i| = 1$*, for all* $1 \le i \le N_0$*, and* $\{\varphi_m\}_{m=N_0+1}^\infty$ *is a tight frame for span* $\{e_i\}_{i=N_0+1}^N$.

Also in [**CF2**], we have the corresponding result for infinite dimensional Hilbert spaces.

THEOREM 5.2. *Let* $\mathbb{H}$ *be an infinite dimensional Hilbert space and* $a_m > 0$ *for all* $m = 1, 2, \ldots$. *The following are equivalent:*

(1) *There is a frame* $\{\psi_m\}_{m=1}^\infty$ *for* $\mathbb{H}$ *with* $\|\psi_m\| = a_m$*, for all* $m = 1, 2, \ldots$.
(2) *There is a* λ*-tight frame* $\{\varphi_m\}_{m=1}^\infty$ *for* H *with* $\|\varphi_m\| = a_m$*, for all* $m = 1, 2, \ldots$ *and* $\lambda = \sup_m a_m^2$.
(3) *The sequence* $\{a_m\}_{m=1}^\infty$ *is bounded and* $\sum_{m=1}^\infty a_m^2 = \infty$.

Moreover, λ *is the smallest real number satisfying (2). That is if* $\{\psi_m\}_{m=1}^\infty$ *is any* λ_1*-tight frame for* $\mathbb{H}$ *with* $\|\psi_m\| = a_m$*, for all* $m = 1, 2, \ldots$*, then* $\lambda_1 \ge \lambda$.

Finally, let us point out that the proofs in sections 2 and 3 can be done in much more generality showing that we really can *custom build* tight frames for most applications.

DEFINITION 5.3. If $\{\varphi_m\}_{m=1}^M$ is a frame for $\mathbb{H}_N$ with frame operator S and each φ_m is an eigenvector for S, we call $\{\varphi_m\}_{m=1}^M$ an *FF-critical sequence.*

DEFINITION 5.4. Let $\mathcal{F}$ be a family of frames and $\{\varphi_m\}_{m=1}^M \in \mathcal{F}$ have frame operator S.

(1) The family $\mathcal{F}$ *locally allows small perturbations* of $\{\varphi_m\}_{m=1}^M$ if for every $1 \le k \le M$ there is an $\epsilon > 0$ so that whenever $\psi_m = \varphi_m$ for all $m \ne k$, $\|\psi_k\| = \|\varphi_k\|$ and $\|\psi_k - \varphi_k\| < \epsilon$, then $\{\psi_m\}_{m=1}^M \in \mathcal{F}$.

(2) If $\{\varphi_m\}_{m=1}^M \in \mathcal{F}$ is an FF-critical sequence, we say the family $\mathcal{F}$ *allows eigenspace perturbations* of $\{\varphi_m\}_{m=1}^M$ if given any set $I \subset \{1, 2, \dots, M\}$ for which $\{\varphi_m\}_{m\in I}$ are in one eigenspace of S, there exists an $\epsilon > 0$ satisfying: Whenever $0 < \epsilon_m < \epsilon$, $\psi_m = \varphi_m$ for all $m \notin I$ and for $m \in I$, $\|\psi_m\| = \|\varphi_m\|$ and $\|\psi_m - \varphi_m\| < \epsilon_m$, then $\{\psi_m\}_{m=1}^M \in \mathcal{F}$.

Now, the proof of Proposition 2.8 yields:

THEOREM 5.5. *Let $\mathcal{F}$ be a family of frames for $\mathbb{H}_N$. Let $\{\varphi_m\}_{m=1}^M \in \mathcal{F}$ have frame operator S. If $\{\varphi_m\}_{m=1}^M$ is a local minimizer of the frame potential on $\mathcal{F}$ and $\mathcal{F}$ allows small perturbations of $\{\varphi_m\}_{m=1}^M$, then $\{\varphi_m\}_{m=1}^M$ is an FF-critical sequence.*

In general, if a family $\mathcal{F}$ allows small perturbations of its elements, it still may not contain a tight frame. For example, if $\mathcal{F}$ is the family of two element frames for $\mathbb{R}^2$, $\{\varphi_m\}_{m=1}^2$ with $\|\varphi_1\| = 1$ and $\|\varphi_2\| = 2$, then by Theorem 3.6 this family cannot contain any tight frames.

Now, the proof of Theorem 3.6 yields:

THEOREM 5.6. *Let $\mathcal{F}$ be a family of M-element frames for $\mathbb{H}_N$. Let $\{\varphi_m\}_{m=1}^M \in \mathcal{F}$ be an FF-critical sequence which is a minimizer of the frame potential on $\mathcal{F}$ and assume that $\mathcal{F}$ allows eigenspace perturbations of $\{\varphi_m\}_{m=1}^M$. Without loss of generality, assume $\|\varphi_1\| \ge \|\varphi_2\| \ge \cdots \ge \|\varphi_M\|$. Then there exists a natural number $0 \le N_0 < N$ so that $\{\varphi_m\}_{m=1}^M$ has the form:*

$$\{\varphi_m\}_{m=1}^M = \{c_m \|\varphi_m\| e_m\}_{m=1}^{N_0} \cup \{\varphi_m\}_{m=N_0+1}^M\},$$

where $\{e_i\}_{i=1}^N$ is an orthonormal basis for $\mathbb{H}_N$, $|c_m| = 1$ for $1 \le m \le N_0$ and $\{\varphi_m\}_{m=N_0+1}^M$ is a λ-tight frame for span $\{e_i\}_{i=N_0+1}^N$.

Finally, if $\{\|\varphi_m\|\}_{m=1}^M$ satisfies the fundamental inequality then $N_0 = 0$ and so $\{\varphi_m\}_{m=1}^M$ is a tight frame.

References

[B] R. Balan, *Equivalence relations and distances between Hilbert frames*, Proc. Amer. Math. Soc. **127** (1999), no. 8, 2353–2366.

[BF] J.J. Benedeto and M. Fickus, *Finite normalized tight frames*, Adv. Comput. Math. **18** (2003), no. 2-4, 357–385.

[C] P.G. Casazza, *The art of frame theory*, Taiwanese J. Math. **4** (2000), no. 2, 129–202.

[C1] P.G. Casazza, *Modern tools for Weyl-Heisenberg (Gabor) frame theory*, Adv. in Imag. and Electron. Physics, **115** (2000), 1–127.

[CF] P.G. Casazza, M. Fickus, J. Kovačević, M. Leon and J. Tremain, *A physical interpretation for tight frames*, preprint.

[CF1] P.G. Casazza, M. Fickus, J. Kovačević, M. Leon and J. Tremain, *Representations of frames*, preprint.

[CF2] P.G. Casazza, M. Fickus, J. Kovačević, M. Leon, and J. Tremain, *Constructing infinite tight frames*, preprint.

[CK] P.G. Casazza and J. Kovačević, *Equal norm tight frames with erasures*, Adv. Comput. Math. **18** (2003), no. 2-4, 387–430.

[CL] P.G. Casazza and M. Leon, *Existence and construction of finite tight frames*, preprint.

[CL1] P.G. Casazza and M. Leon, *Frames with a given frame operator*, preprint.

[CL2] P.G. Casazza and M. Leon, *Frameware*, 2002.

[Ch] O. Christensen, *An Introduction to Frames and Riesz Bases*, Birkhäuser, Boston, 2002.

[DGS] I. Daubechies, A. Grossman, and Y. Meyer, *Painless nonorthogonal expansions*, J. Math. Phys. **27** (November 1986), no. 5, 1271–1283.

[DS] R.J. Duffin and A.C. Schaeffer, *A class of nonharmonic Fourier series*, Trans. Amer. Math. Soc. **72** (1952), 341–366.

[GKK] V.K. Goyal, J. Kovačević and J.A. Kelner, *Quantized frame expansions with erasures*, Appl. Comput. Harmon. Anal. **10** (2001), no. 3, 203–233.

[G] K. Gröchenig, *Foundations of Time-Frequency Analysis*, Birkhäuser, Boston, 2000.

[HH] B. Hassibi, B. Hochwald, A. Shokrollahi, and W. Sweldens, *Representation theory for high-rate multiple-antenna code design*, IEEE Trans. Inform. Theory **47** (2001), no. 6, 2335–2367.

[HS] R. Heath and T. Strohmer, *Grassmannian frames with applications to coding and communication*, Appl. Comput. Harmon. Anal. **14** (2003), no. 3, 257–275.

[HM] B.M. Hochwarld, T.L. Marzetta, T.J. Richardson, W. Sweldens, and R. Urbanke, *Systematic design of unitary space-time constellations*, IEEE Transactions on Information Theory **46** (2000), no. 6, 1962–1973.

[HP] R.B. Homes and V.I. Paulsen, *Optimal frames for erasures*, preprint.

[J] A.J.E.M. Janssen, *Zak transforms with few zeroes and the tie*, Advances in Gabor Analysis, H.G. Feichtinger and T. Strohmer, eds., pages 31–70, Birkhäuser, Boston, 2003.

[K] H. König, *Cubature formulas on spheres*, Advances in Multivariate Approximation (Witten-Bommerholz, 1998), W. Haußmann, K. Jetter, and M. Reimer, eds., Math. Res., vol. 107, pages 201–211, Wiley-VCH, Berlin, 1999.

[W] S. Waldron, *Welch bound equality sequences are tight frames*, preprint.

[Y] R. Young, *An Introduction to Nonharmonic Fourier Series*, revised first edition, Academic Press, New York, 2001.

Department of Mathematics, University of Missouri, Coulombia, Missouri 65211
E-mail address: pete@math.missouri.edu

Contemporary Mathematics
Volume **345**, 2004

Frames of subspaces

Peter G. Casazza and Gitta Kutyniok

ABSTRACT. One approach to ease the construction of frames is to first construct local components and then build a global frame from these. In this paper we will show that the study of the relation between a frame and its local components leads to the definition of a frame of subspaces. We introduce this new notion and prove that it provides us with the link we need. It will also turn out that frames of subspaces behave as a generalization of frames. In particular, we can define an analysis, a synthesis and a frame operator for a frame of subspaces, which even yield a reconstruction formula. Also concepts such as completeness, minimality, and exactness are introduced and investigated. We further study several constructions of frames of subspaces, and also of frames and Riesz frames using the theory of frames of subspaces. An important special case are harmonic frames of subspaces which generalize harmonic frames. We show that wavelet subspaces coming from multiresolution analysis belong to this class.

1. Introduction

During the last 20 years the theory of frames has been growing rapidly, since several new applications have been developed. For example, besides traditional applications as signal processing, image processing, data compression, and sampling theory, frames are now used to mitigate the effect of losses in packet-based communication systems and hence to improve the robustness of data transmission [**7, 15**], and to design high-rate constellations with full diversity in multiple-antenna code design [**17**].

To handle these emerging applications of frames new methods have to be developed. One starting point is to first build frames "locally" and then piece them together to obtain frames for the whole space. One advantage of this idea is that it would facilitate the construction of frames for special applications, since we can first construct frames or choose already known frames for smaller spaces. And in a second step one would construct a frame for the whole space from them. Therefore

2000 *Mathematics Subject Classification.* Primary 42C15; Secondary 46C99.
Key words and phrases. Abstract frame theory, frame, harmonic frame, Hilbert space, resolution of the identity, Riesz basis, Riesz frame.
The first author was supported by NSF DMS 0102686.
The second author was supported by Forschungspreis 2003 der Universität Paderborn.

it is necessary to derive conditions for these components, so that there exists a construction, which yields a frame for the whole space with special properties. Various approaches to piecing together familes of vectors to get a frame for the whole space have been done over the years going back to Duffin and Schaeffer's original work [**12**]. One approach used in the wavelet as well as in the Gabor case [**10, 1**] is to start with non-frame sequences and piece them together to build frames for the whole space. Another is to build frames locally and piece them together orthogonally to get frames. We refer to Heil and Walnut [**18**] for an excellent introduction to these methods and Gabor frames in general. Recently, another approach was introduced by Fornasier [**13, 14**]. Fornasier uses subspaces which are quasi-orthogonal to construct local frames and piece them together to get global frames.

In this paper we will formulate a general method for piecing together local frames to get global frames. The importance of this approach is that it is both necessary and sufficient for the the construction of global frames from local frames. Some of these results are generalizations of Fornasier's work [**13, 14**] although they were done before his papers became available to us.

Another motivation comes from the theory of C*-algebras. Just recently Casazza, Christensen, Lindner, and Vershynin [**5**] proved that the so-called "Feichtinger conjecture" is equivalent to the weak Bourgain-Tzafriri conjecture. The Feichtinger conjecture states that each bounded frame is a finite union of Riesz basic sequences. Then, Casazza and Vershynin [**8**] showed that the Kadison-Singer problem is equivalent to the strong Bourgain-Tzafriri conjecture and that these two problems have a positive solution if and only if both the Feichtinger conjecture and the F_ϵ-conjecture have positive solutions. The F_ϵ-conjecture states: For every $\epsilon > 0$, every unit norm Riesz basis is a finite union of $(1+\epsilon)$-unconditional basic sequences. A unit norm sequence $\{f_i\}_{i\in I}$ is a $(1+\epsilon)$-basic sequence if for every sequence of scalars $\{a_i\}_{i\in I}$ we have

$$(1-\epsilon)\sum_{i\in I}|a_i|^2 \le \left\|\sum_{i\in I} a_i f_i\right\|^2 \le (1+\epsilon)\sum_{i\in I}|a_i|^2.$$

To attack these problems it is important to know into which components we can divide a frame. As we will see in this paper, the necessary divisions will form a frame of subspaces for the space. At this time, it is not even known how to divide a frame into two infinite frame sequences.

In this paper we want to answer the following two questions, which relate to the two different motivations:

- Let $\{W_i\}_{i\in I}$ be a collection of closed subspaces in a Hilbert space $\mathcal{H}$ in which we want to decompose our function, where each subspace W_i is equipped with a weight v_i, which indicate its importance. When can we find frames for W_i for each $i \in I$ so that the collection of all of them is a frame with special properties for the whole space $\mathcal{H}$?
- Let $\{f_i\}_{i\in I}$ be a frame for a Hilbert space $\mathcal{H}$, and let $I = \bigcup_{j\in\mathbb{Z}} I_j$ be a partition of I so that $\{f_i\}_{i\in I_j}$ is a frame sequence for each $j \in \mathbb{Z}$. Which relations exist between the closed linear spans of $\{f_i\}_{i\in I_j}$, $j \in \mathbb{Z}$?

We start our consideration by giving a brief review of the definitions and basic properties of frames and bases and stating some notation in Section 2.

In Section 3 it will turn out that both questions above lead to the definition of a frame of subspaces. In the first subsection we will state the definition of a frame of subspaces for a given family of closed subspaces $\{W_i\}_{i\in I}$ in a Hilbert space and a family of weights $\{v_i\}_{i\in I}$. Then it is shown that this definition leads to some answers to the above questions (see Theorem 3.2), since it shows that frames of subspaces behave as a link between local components of a frame and the global structure. This will also enlighten the advantage of our approach, since now we can choose the frames for the single subspaces W_i arbitrarily and always get a frame for the whole Hilbert space by just collecting them together. Thus it differs from previous approaches and is a generalization of the approach of Fornasier [**13, 14**]. It will turn out that frames of subspaces behave as a generalization of frames. We first give a definition of completeness of a family of subspaces and show that the relation between this property and the notion of a frame of subspaces is similar to the relation between the definition of completeness of a sequence and a frame. Further in Subsection 3.2 we introduce an analysis and a synthesis operator, a frame operator, and a dual frame of subspaces for a given frame of subspaces and prove that they behave in an analogous way as the corresponding objects in abstract frame theory. We even obtain a reconstruction formula using these ingredients (Proposition 3.16). The next subsection deals with Parseval frames of subspaces, which share several properties with Parseval frames. Finally in Subsection 3.4 we show that using the theory of frames of subspaces we can construct several useful resolutions of the identity.

Section 4 deals with Riesz decompositions, which are a generalization of the notion of Riesz bases to our general setting. We further define minimality for a family of subspaces and show that it behaves as expected. Also exactness is defined in a canonical way. However, it will turn out that this property is much weaker than exactness of a frame (compare Theorem 4.6).

Some constructions are given in Section 5. Here we first state some results which help constructing frames of subspaces. An extended example concerning the situation of Gabor frames is added. In Subsection 5.2 we then show how to construct frames and Riesz frames using a frame of subspaces.

Finally, Section 6 deals with harmonic frames of subspaces. These are a generalization of harmonic frames, which distinguish themselves by having an easy construction formula. In both the finite and the infinite dimensional cases we give the definition of a harmonic frame of subspaces, state some results, and give examples, e.g., subspaces coming from Gabor systems and subspaces coming from multiresolution analysis, for their occurance.

2. Review of frames and some notation

First we will briefly recall the definitions and basic properties of frames and bases. For more information we refer to the survey articles by Casazza [**3, 4**], the books by Christensen [**9**], Gröchenig [**16**], and Young [**22**] and the research-tutorial by Heil and Walnut [**18**].

Let $\mathcal{H}$ be a separable Hilbert space and let I be an indexing set. A family $\{f_i\}_{i\in I}$ is a *frame* for $\mathcal{H}$, if there exist $0 < A \leq B < \infty$ such that for all $h \in \mathcal{H}$,

$$A\,\|h\|^2 \leq \sum_{i\in I} |\langle h, f_i\rangle|^2 \leq B\,\|h\|^2 . \tag{2.1}$$

The constants A and B are called a *lower* and *upper frame bound* for the frame. Those sequences which satisfy only the upper inequality in (2.1) are called *Bessel sequences.* A frame is *tight,* if $A = B$. If $A = B = 1$, it is called a *Parseval frame.* We call a frame $\{f_i\}_{i\in I}$ *uniform* (or *equal norm*), if we have $\|f_i\| = \|f_j\|$ for all $i, j \in I$. A frame is *exact,* if it ceases to be a frame whenever any single element is deleted from the sequence $\{f_i\}_{i\in I}$. We say that two frames $\{f_i\}_{i\in I}$, $\{g_i\}_{i\in I}$ for $\mathcal{H}$ are *equivalent,* if there exists an invertible operator $U : \mathcal{H} \to \mathcal{H}$ satisfying $Uf_i = g_i$ for all $i \in I$. If U is a unitary operator, $\{f_i\}_{i\in I}$ and $\{g_i\}_{i\in I}$ are called *unitarily equivalent.* The *synthesis operator* $T_f : l^2(I) \to \mathcal{H}$ of a frame $f = \{f_i\}_{i\in I}$ is defined by $T_f(c) = \sum_{i\in I} c_i f_i$ for each sequence of scalars $c = \{c_i\}_{i\in I} \in l^2(I)$. The adjoint operator $T_f^* : \mathcal{H} \to l^2(I)$, the so-called *analysis operator* of $f = \{f_i\}_{i\in I}$, is given by $T_f^*(g) = \{\langle f_i, g\rangle\}_{i\in I}$. Then the *frame operator* $S_f(h) = T_f T_f^*(h) = \sum_{i\in I} \langle h, f_i\rangle f_i$ associated with $\{f_i\}_{i\in I}$ is a bounded, invertible, and positive operator mapping $\mathcal{H}$ onto itself. This provides the reconstruction formula

$$h = S_f^{-1} S_f(h) = \sum_{i\in I} \langle h, f_i\rangle \tilde{f}_i = \sum_{i\in I} \langle h, \tilde{f}_i\rangle f_i,$$

where $\tilde{f}_i = S_f^{-1} f_i$. The family $\{\tilde{f}_i\}_{i\in I}$ is also a frame for $\mathcal{H}$, called the *canonical dual frame* of $\{f_i\}_{i\in I}$. A sequence is called a *frame sequence,* if it is a frame only for its closed linear span. Moreover, we say that a frame $\{f_i\}_{i\in I}$ is a *Riesz frame,* if every subfamily of the sequence $\{f_i\}_{i\in I}$ is a frame sequence with uniform frame bounds A and B.

As important example of frames are the so-called *harmonic frames,* which are uniform Parseval frames of the form $\{U^i\varphi\}_{i\in I}$, where U is a unitary operator on $\mathcal{H}$ and $I = \{0, \ldots, N-1\}$, $N \in \mathbb{N}$ or $I = \mathbb{Z}$. Concerning a classification of harmonic frames we refer to the paper by Casazza and Kovaĉević [**7**].

Riesz bases are special cases of frames, and can be characterized as those frames which are biorthogonal to their dual frames. An equivalent definition is the following. A family $\{f_i\}_{i\in I}$ is a Riesz basis for $\mathcal{H}$, if there exist $0 < A \le B < \infty$ such that for all sequences of scalars $c = \{c_i\}_{i\in I}$,

$$A \left\|c\right\|_2 \le \left\| \sum_{i\in I} c_i f_i \right\| \le B \left\|c\right\|_2.$$

We define the *Riesz basis constants* for $\{f_i\}_{i\in I}$ to be the largest number A and the smallest number B such that this inequality holds for all sequences of scalars c. If $\{f_i\}_{i\in I}$ is a Riesz basis only for its closed linear span, we call it a *Riesz basic sequence.*

An arbitrary sequence $\{f_i\}_{i\in I}$ in $\mathcal{H}$ is *minimal,* if $f_i \notin \overline{\text{span}}_{j\in I, j\neq i}\{f_j\}$ for all $i \in I$, or equivalently if there exists a sequence $\{\tilde{f}_i\}_{i\in I}$, which is biorthogonal to $\{f_i\}_{i\in I}$. It is *complete,* if the span of $\{f_i\}_{i\in I}$ is dense in $\mathcal{H}$.

We conclude this section by giving some notation and remarks. Throughout this paper $\mathcal{H}$ shall always denote an arbitrary separable Hilbert space. Furthermore all subspaces are assumed to be closed although this is not stated explicitely. Moreover, for the remainder a sequence $\{v_i\}_{i\in I}$ always denotes a family of weights, i.e., $v_i > 0$ for all $i \in I$.

In addition we use the following notation. Dependent on the context I denotes an indexing set or the identity operator. If W is a subspace of a Hilbert space $\mathcal{H}$, we let π_W denote the orthogonal projection of $\mathcal{H}$ onto W. If $\{e_i\}_{i\in I}$ is an orthonormal basis for $\mathcal{H}$ and $J \subset I$, π_J is the orthogonal projection of $\mathcal{H}$ onto $\overline{\text{span}}_{i\in J}\{e_i\}$.

3. Frames of subspaces

3.1. Definition and basic properties. We start with the definition of a frame of subspaces. It will turn out that frames of subspaces share many of the properties of frames, and thus can be viewed as a generalization of frames.

DEFINITION 3.1. Let I be some index set, and let $\{v_i\}_{i\in I}$ be a family of weights, i.e., $v_i > 0$ for all $i \in I$. A family of closed subspaces $\{W_i\}_{i\in I}$ of a Hilbert space $\mathcal{H}$ is a *frame of subspaces with respect to* $\{v_i\}_{i\in I}$ *for* $\mathcal{H}$, if there exist constants $0 < C \le D < \infty$ such that

$$C\|f\|^2 \le \sum_{i\in I} v_i^2 \|\pi_{W_i}(f)\|^2 \le D\|f\|^2 \quad \text{for all } f \in \mathcal{H}. \tag{3.1}$$

We call C and D the *frame bounds* for the frame of subspaces. The family $\{W_i\}_{i\in I}$ is called a *C-tight frame of subspaces with respect to* $\{v_i\}_{i\in I}$, if in (3.1) the constants C and D can be chosen so that $C = D$, a *Parseval frame of subspaces with respect to* $\{v_i\}_{i\in I}$ provided that $C = D = 1$ and an *orthonormal basis of subspaces* if $\mathcal{H} = \bigoplus_{i\in I} W_i$. Moreover, we call a frame of subspaces with respect to $\{v_i\}_{i\in I}$ *v-uniform*, if $v := v_i = v_j$ for all $i, j \in I$. If we only have the upper bound, we call $\{W_i\}_{i\in I}$ a *Bessel sequence of subspaces with respect to* $\{v_i\}_{i\in I}$ with *Bessel bound* D.

Condition (3.1) states the necessary (and also sufficient) interaction between the subspaces so that taking frames from them and putting them together yields a frame for the whole space.

The importance of this definition is that it is both necessary and sufficient for us to be able to string together frames for each of the subspaces W_i (with uniformly bounded frame constants) to get a frame for $\mathcal{H}$. This is contained in the next theorem. The implication (3) $\Rightarrow$ (1) of the following result is [**7**, Proposition 4.5]. Fornasier [**13, 14**] obtains a similar result for quasi-orthogonal decompositions.

THEOREM 3.2. *For each $i \in I$ let $v_i > 0$ and let $\{f_{ij}\}_{j\in J_i}$ be a frame sequence in $\mathcal{H}$ with frame bounds A_i and B_i. Define $W_i = \overline{\text{span}}_{j\in J_i}\{f_{ij}\}$ for all $i \in I$ and choose an orthonormal basis $\{e_{ij}\}_{j\in J_i}$ for each subspace W_i. Suppose that $0 < A = \inf_{i\in I} A_i \le B = \sup_{i\in I} B_i < \infty$. The following conditions are equivalent.*

(1) *$\{v_i f_{ij}\}_{i\in I, j\in J_i}$ is a frame for $\mathcal{H}$.*
(2) *$\{v_i e_{ij}\}_{i\in I, j\in J_i}$ is a frame for $\mathcal{H}$.*
(3) *$\{W_i\}_{i\in I}$ is a frame of subspaces with respect to $\{v_i\}_{i\in I}$ for $\mathcal{H}$.*

PROOF. Since for each $i \in I$, $\{f_{ij}\}_{j\in J_i}$ is a frame for W_i with frame bounds A_i and B_i, we obtain

$$A\sum_{i\in I} v_i^2 \|\pi_{W_i}(f)\|^2 \le \sum_{i\in I} A_i v_i^2 \|\pi_{W_i}(f)\|^2 \le \sum_{i\in I}\sum_{j\in J_i} |\langle \pi_{W_i}(f), v_i f_{ij}\rangle|^2$$
$$\le \sum_{i\in I} B_i v_i^2 \|\pi_{W_i}(f)\|^2 \le B\sum_{i\in I} v_i^2 \|\pi_{W_i}(f)\|^2.$$

Now we observe that

$$\sum_{i\in I}\sum_{j\in J_i} |\langle \pi_{W_i}(f), v_i f_{ij}\rangle|^2 = \sum_{i\in I}\sum_{j\in J_i} |\langle f, v_i f_{ij}\rangle|^2.$$

This shows that provided $\{v_i f_{ij}\}_{i\in I, j\in J_i}$ is a frame for $\mathcal{H}$ with frame bounds C and D, the sets $\{W_i\}_{i\in I}$ form a frame of subspaces with respect to $\{v_i\}_{i\in I}$ for $\mathcal{H}$ with frame bounds $\frac{C}{B}$ and $\frac{D}{A}$. Moreover, if $\{W_i\}_{i\in I}$ is a frame of subspaces with respect to $\{v_i\}_{i\in I}$ for $\mathcal{H}$ with frame bounds C and D, the calculation above implies that $\{v_i f_{ij}\}_{i\in I, j\in J_i}$ is a frame for $\mathcal{H}$ with frame bounds AC and BD. Thus (1) $\Leftrightarrow$ (3).

To prove the equivalence of (2) and (3), note that we can now actually calculate the orthogonal projections in the following way

$$v_i^2 \|\pi_{W_i}(f)\|^2 = v_i^2 \left\| \sum_{j\in J_i} \langle f, e_{ij}\rangle e_{ij} \right\|^2 = \sum_{j\in J_i} |\langle f, v_i e_{ij}\rangle|^2 .$$

From this the claim follows immediately. □

The definition of completeness of a sequence gives rise to a definition of completeness for a sequence of subspaces.

DEFINITION 3.3. A family of subspaces $\{W_i\}_{i\in I}$ of $\mathcal{H}$ is called *complete*, if

$$\overline{\text{span}}_{i\in I}\{W_i\} = \mathcal{H}.$$

The next lemma possesses a well-known analog in the frame situation.

LEMMA 3.4. *Let $\{W_i\}_{i\in I}$ be a family of subspaces in $\mathcal{H}$, and let $\{v_i\}_{i\in I}$ be a family of weights. If $\{W_i\}_{i\in I}$ is a frame of subspaces with respect to $\{v_i\}_{i\in I}$ for $\mathcal{H}$, then it is complete.*

PROOF. Assume that $\{W_i\}_{i\in I}$ is not complete. Then there exists some $f \in \mathcal{H}$, $f \neq 0$ with $f \perp \overline{\text{span}}_{i\in I}\{W_i\}$. It follows that $\sum_{i\in I} v_i^2 \|\pi_{W_i}(f)\|^2 = 0$, hence $\{W_i\}_{i\in I}$ is not a frame of subspaces. □

To check completeness of a frame of subspaces, we derive the following useful characterization.

LEMMA 3.5. *Let $\{W_i\}_{i\in I}$ be a family of subspaces in $\mathcal{H}$, and for each $i \in I$ let $\{e_{ij}\}_{j\in J_i}$ be an orthonormal basis for W_i. Then the following conditions are equivalent.*

(1) *$\{W_i\}_{i\in I}$ is complete.*
(2) *$\{e_{ij}\}_{i\in I, j\in J_i}$ is complete.*

PROOF. The equivalence of (1) and (2) follows immediately from the definitions.

□

If we remove an element from a frame, we obtain either another frame or an incomplete set [**9**, Theorem 5.4.7]. A similar result holds in our situation.

PROPOSITION 3.6. *The removal of a subspace from a frame of subspaces with respect to some family of weights leaves either a frame of subspaces with respect to the same family of weights or an incomplete family of subspaces.*

PROOF. Let $\{W_i\}_{i\in I}$ be a frame of subspaces with respect to $\{v_i\}_{i\in I}$ for $\mathcal{H}$, and for each $i \in I$ let $\{e_{ij}\}_{j\in J_i}$ be an orthonormal basis for W_i. By Theorem 3.2, $\{v_i e_{ij}\}_{i\in I, j\in J_i}$ is a frame for $\mathcal{H}$. Let $i_0 \in I$. By [**9**, Theorem 5.4.7], $\{v_i e_{ij}\}_{i\in I\setminus\{i_0\}, j\in J_i}$ is either a frame or an incomplete set. If it is a frame, again by Theorem 3.2, also $\{W_i\}_{i\in I\setminus\{i_0\}}$ is a frame of subspaces with respect to $\{v_i\}_{i\in I}$ for $\mathcal{H}$.

Now suppose that $\{v_i e_{ij}\}_{i\in I\setminus\{i_0\}, j\in J_i}$ and hence $\{e_{ij}\}_{i\in I\setminus\{i_0\}, j\in J_i}$ is an incomplete set. By Lemma 3.5, also $\{W_i\}_{i\in I\setminus\{i_0\}}$ is incomplete. □

We further observe that the intersection of the elements of a frame of subspaces with a subspace still leaves a frame of subspaces for a smaller space.

LEMMA 3.7. *Let V be a subspace of $\mathcal{H}$ and let $\{W_i\}_{i\in I}$ be a frame of subspaces with respect to $\{v_i\}_{i\in I}$ for $\mathcal{H}$ with frame bounds C and D. Then $\{W_i \cap V\}_{i\in I}$ is a frame of subspaces with respect to $\{v_i\}_{i\in I}$ for V with frame bounds C and D.*

PROOF. For all $f \in V$ we have

$$\sum_{i\in I} v_i^2 \|\pi_{W_i}(f)\|^2 = \sum_{i\in I} v_i^2 \|\pi_{W_i\cap V}(f)\|^2.$$

From this the result follows at once. □

3.2. Frame properties. In this subsection we will show that a frame of subspaces behaves as a generalization of a frame, thus providing an associated analysis and synthesis operator, a frame operator and a dual object.

For the definition of an analysis and a synthesis operator for a frame of subspaces, we will need the following notation.

NOTATION 3.8. For each family of subspaces $\{W_i\}_{i\in I}$ of $\mathcal{H}$, we define the space $\left(\sum_{i\in I} \oplus W_i\right)_{\ell_2}$ by

$$\left(\sum_{i\in I} \oplus W_i\right)_{\ell_2} = \{\{f_i\}_{i\in I} | f_i \in W_i \text{ and } \sum_{i\in I} \|f_i\|^2 < \infty\}$$

with inner product given by

$$\langle \{f_i\}_{i\in I}, \{g_i\}_{i\in I}\rangle = \sum_{i\in I} \langle f_i, g_i\rangle.$$

We start with the definition of a synthesis operator for a frame of subspaces. To show that the series appearing in this formula converges unconditionally, we need the next lemma.

LEMMA 3.9. *Let $\{W_i\}_{i\in I}$ be a Bessel sequence of subspaces with respect to $\{v_i\}_{i\in I}$ for $\mathcal{H}$. Then, for each sequence $\{f_i\}_{i\in I}$ with $f_i \in W_i$ for each $i \in I$, the series $\sum_{i\in I} v_i f_i$ converges unconditionally.*

PROOF. Let $f = \{f_i\}_{i\in I} \in \left(\sum_{i\in I} \oplus W_i\right)_{\ell_2}$. Fix $J \subset I$ with $|J| < \infty$ and let $g = \sum_{i\in J} v_i f_i$. Then we compute

$$\|\sum_{i\in J} v_i f_i\|^4 = \left(\langle g, \sum_{i\in J} v_i f_i\rangle\right)^2 = \left(\sum_{i\in J} v_i \langle \pi_{W_i}(g), f_i\rangle\right)^2 \le \left(\sum_{i\in J} v_i \|\pi_{W_i}(g)\| \|f_i\|\right)^2$$

$$\le \sum_{i\in J} v_i^2 \|\pi_{W_i}(g)\|^2 \sum_{i\in J} \|f_i\|^2 \le D\|g\|^2 \sum_{i\in J} \|f_i\|^2 \le D\|\sum_{i\in J} v_i f_i\|^2 \|f\|^2.$$

Hence,

$$\|\sum_{i\in J} v_i f_i\|^2 \le D\|f\|^2.$$

It follows that $\sum_{i\in I} v_i f_i$ is weakly unconditionally Cauchy and hence unconditionally convergent in $\mathcal{H}$ (see [**11**], page 44, Theorems 6 and 8). □

DEFINITION 3.10. Let $\{W_i\}_{i\in I}$ be a frame of subspaces with respect to $\{v_i\}_{i\in I}$ for $\mathcal{H}$. Then the *synthesis operator* for $\{W_i\}_{i\in I}$ and $\{v_i\}_{i\in I}$ is the operator

$$T_{W,v} : \left(\sum_{i\in I} \oplus W_i\right)_{\ell_2} \longrightarrow \mathcal{H}$$

defined by

$$T_{W,v}(f) = \sum_{i\in I} v_i f_i \quad \text{for all } f = \{f_i\}_{i\in I} \in (\sum_{i\in I} \oplus W_i)_{\ell_2}.$$

We call the adjoint $T^*_{W,v}$ of the synthesis operator the *analysis operator.*

The following proposition will provide us with a concrete formula for the analysis operator.

PROPOSITION 3.11. *Let $\{W_i\}_{i\in I}$ be a frame of subspaces with respect to $\{v_i\}_{i\in I}$ for $\mathcal{H}$. Then the analysis operator $T^*_{W,v} : \mathcal{H} \to (\sum_{i\in I} \oplus W_i)_{\ell_2}$ is given by*

$$T^*_{W,v}(f) = \{v_i \pi_{W_i}(f)\}_{i\in I}.$$

PROOF. Let $f \in \mathcal{H}$ and $g = \{g_i\}_{i\in I} \in (\sum_{i\in I} \oplus W_i)_{\ell_2}$. Using the definition of $T_{W,v}$ we compute that

$$\langle T^*_{W,v}(f), g\rangle = \langle f, T_{W,v}(g)\rangle = \langle f, \sum_{i\in I} v_i g_i\rangle = \sum_{i\in I} v_i \langle f, g_i\rangle.$$

Since $g_i \in W_i$ for each $i \in I$, we can continue in the following way:

$$\sum_{i\in I} v_i \langle f, g_i\rangle = \sum_{i\in I} v_i \langle \pi_{W_i}(f), g_i\rangle = \langle \{v_i \pi_{W_i}(f)\}_{i\in I}, \{g_i\}_{i\in I}\rangle.$$

□

The well-known relations between a frame and the associated analysis and synthesis operator also holds in our more general situation.

THEOREM 3.12. *Let $\{W_i\}_{i\in I}$ be a family of subspaces in $\mathcal{H}$, and let $\{v_i\}_{i\in I}$ be a family of weights. Then the following conditions are equivalent.*

(1) *$\{W_i\}_{i\in I}$ is a frame of subspaces with respect to $\{v_i\}_{i\in I}$ for $\mathcal{H}$.*
(2) *The synthesis operator $T_{W,v}$ is bounded, linear and onto.*
(3) *The analysis operator $T^*_{W,v}$ is a (possibly into) isomorphism.*

PROOF. First we prove (1) $\Leftrightarrow$ (3). This claim follows immediately from the fact that for each $f \in \mathcal{H}$ we have

$$\|T^*_{W,v}(f)\|^2 = \|\{v_i \pi_{W_i}(f)\}_{i\in I}\|^2 = \sum_{i\in I} v_i^2 \|\pi_{W_i}(f)\|^2.$$

Further recall that (2) $\Leftrightarrow$ (3) holds in general for each operator on a Hilbert space.

□

In an analogous way as in frame theory we can define equivalence classes of frames of subspaces. Using the synthesis operator we can also characterize exactly the elements belonging to the same equivalence class.

DEFINITION 3.13. Let $\{W_i\}_{i\in I}$ and $\{\widetilde{W}_i\}_{i\in I}$ be frames of subspaces with respect to the same family of weights. We say that they are *(unitarily) equivalent*, if there exists an (unitary) invertible operator U on $\mathcal{H}$ such that $W_i = U(\widetilde{W}_i)$ for all $i \in I$.

LEMMA 3.14. *Let $\{W_i\}_{i\in I}$ and $\{\widetilde{W}_i\}_{i\in I}$ be frames of subspaces with respect to the same family of weights $\{v_i\}_{i\in I}$. The following conditions are equivalent.*

(1) *$\{W_i\}_{i\in I}$ and $\{\widetilde{W}_i\}_{i\in I}$ are (unitarily) equivalent.*
(2) *There exists an (unitary) invertible operator U on $\mathcal{H}$ such that $T_{W,v} = U^{-1}T_{\widetilde{W},v}U$, where U is applied to each component.*

PROOF. This follows immediately from the definition of the synthesis operator. □

As in the well-known frame situation, there also exists an associated frame operator for each frame of subspaces which satisfies similar properties as we will see in the next proposition. For instance we even obtain a reconstruction formula.

DEFINITION 3.15. Let $\{W_i\}_{i\in I}$ be a frame of subspaces with respect to $\{v_i\}_{i\in I}$ for $\mathcal{H}$. Then the frame operator $S_{W,v}$ for $\{W_i\}_{i\in I}$ and $\{v_i\}_{i\in I}$ is defined by

$$S_{W,v}(f) = T_{W,v}T^*_{W,v}(f) = T_{W,v}(\{v_i\pi_{W_i}(f)\}_{i\in I}) = \sum_{i\in I} v_i^2\pi_{W_i}(f).$$

The next proposition generalizes a result of Fornasier [**13, 14**].

PROPOSITION 3.16. *Let $\{W_i\}_{i\in I}$ be a frame of subspaces with respect to $\{v_i\}_{i\in I}$ with frame bounds C and D. Then the frame operator $S_{W,v}$ for $\{W_i\}_{i\in I}$ and $\{v_i\}_{i\in I}$ is a positive, self-adjoint, invertible operator on $\mathcal{H}$ with $CI \le S_{W,v} \le DI$. Further, we have the reconstruction formula*

$$f = T_{S^{-1}_{W,v}W,v}T^*_{W,v}(f) = \sum_{i\in I} v_i^2 S^{-1}_{W,v}\pi_{W_i}(f) \quad \text{for all } f \in \mathcal{H}.$$

PROOF. For any $f \in \mathcal{H}$, we have

$$\langle S_{W,v}(f), f\rangle = \langle \sum_{i\in I} v_i^2\pi_{W_i}(f), f\rangle = \sum_{i\in I} v_i^2\langle \pi_{W_i}(f), f\rangle = \sum_{i\in I} v_i^2\|\pi_{W_i}(f)\|^2,$$

which implies that $S_{W,v}$ is a positive operator. We further compute

$$\langle Cf, f\rangle = C\|f\|^2 \le \sum_{i\in I} v_i^2\|\pi_{W_i}(f)\|^2 = \langle S_{W,v}(f), f\rangle \le \langle Df, f\rangle.$$

This shows that $CI \le S_{W,v} \le DI$ and hence $S_{W,v}$ is an invertible operator on $\mathcal{H}$. Furthermore, for any $f, g \in \mathcal{H}$ we have

$$\langle S_{W,v}(f), g\rangle = \sum_{i\in I} v_i^2\langle \pi_{W_i}(f), g\rangle = \sum_{i\in I} v_i^2\langle f, \pi_{W_i}(g)\rangle.$$

Thus $S_{W,v}$ is self-adjoint. At last the reconstruction formula follows immediately from

$$f = S^{-1}_{W,v}S_{W,v}(f) = \sum_{i\in I} v_i^2 S^{-1}_{W,v}\pi_{W_i}(f).$$

□

The following result will show the connection between the frame operator for a frame of subspaces and the frame operator for the frame generated by orthonormal bases of the subspaces. Also the connection between the reconstruction formulas is exposed.

PROPOSITION 3.17. *Let $\{W_i\}_{i\in I}$ be a frame of subspaces with respect to $\{v_i\}_{i\in I}$ for $\mathcal{H}$ and $\{v_i f_{ij}\}_{j\in J_i}$ be a Parseval frame for W_i for each $i \in I$. Then the frame operator $S_{W,v}$ equals the frame operator S_{vf} for the frame $\{v_i f_{ij}\}_{i\in I, j\in J_i}$, and for all $g \in \mathcal{H}$ we have*

$$\sum_{i\in I} v_i^2 S_{W,v}^{-1} \pi_{W_i}(g) = \sum_{i\in I, j\in J_i} \langle g, v_i f_{ij}\rangle\, S_{vf}^{-1} v_i f_{ij}.$$

PROOF. Since $\{f_{ij}\}_{j\in J_i}$ is a Parseval frame for W_i for all $i \in I$, if $g \in \mathcal{H}$ then

$$\pi_{W_i}(g) = \sum_{j\in J_i} \langle \pi_{W_i}(g), f_{ij}\rangle f_{ij} = \sum_{j\in J_i} \langle g, f_{ij}\rangle f_{ij}.$$

Thus

$$S_{W,v}(g) = \sum_{i\in I} v_i^2 \pi_{W_i}(g) = \sum_{i\in I, j\in J_i} \langle g, v_i f_{ij}\rangle v_i f_{ij} = S_{vf}(g).$$

Moreover, we obtain

$$\sum_{i\in I} v_i^2 S_{W,v}^{-1} \pi_{W_i}(g) = \sum_{i\in I} S_{vf}^{-1} \sum_{j\in J_i} \langle g, v_i f_{ij}\rangle\, v_i f_{ij} = \sum_{i\in I, j\in J_i} \langle g, v_i f_{ij}\rangle\, S_{vf}^{-1} v_i f_{ij}.$$

□

Using the frame operator for a frame of subspaces for a special subspace yields an easy way to compute the orthogonal projection onto this subspace.

PROPOSITION 3.18. *Let $\{W_i\}_{i\in I}$ be a frame of subspaces with respect to $\{v_i\}_{i\in I}$ for a subspace V of $\mathcal{H}$. Then, the orthogonal projection π_V onto V is given by*

$$\pi_V(f) = \sum_{i\in I} v_i^2 S_{W,v}^{-1} \pi_{W_i}(f) \quad \textit{for all } f \in \mathcal{H}.$$

PROOF. The fact that $S_{W,v} : V \to V$ implies that $\pi_V(f) = 0$ for all $f \in V^\perp$. By Proposition 3.16, we have

$$f = \sum_{i\in I} v_i^2 S_{W,v}^{-1} \pi_{W_i}(f) \quad \text{for all } f \in V.$$

Thus $\pi_V^2 = \pi_V$, which finishes the proof. □

In the same manner as in frame theory we define a dual frame of subspaces associated with a frame of subspaces.

DEFINITION 3.19. Let $\{W_i\}_{i\in I}$ be a frame of subspaces with respect to $\{v_i\}_{i\in I}$ and with frame operator $S_{W,v}$. Then $\{S_{W,v}^{-1} W_i\}_{i\in I}$ is called the *dual frame of subspaces with respect to* $\{v_i\}_{i\in I}$.

The dual frame of subspaces is a frame of subspaces with the same weights. In fact, more is true.

PROPOSITION 3.20. *Let $\{W_i\}_{i\in I}$ be a frame of subspaces with respect to $\{v_i\}_{i\in I}$, and let $T : \mathcal{H} \to \mathcal{H}$ be an invertible operator on $\mathcal{H}$. Then $\{TW_i\}_{i\in I}$ is a frame of subspaces with respect to $\{v_i\}_{i\in I}$.*

PROOF. Since T is an invertible operator on $\mathcal{H}$, we have that $\pi_{TW_i} = T\pi_{W_i}T^{-1}$. Let $C, D > 0$ be the frame bounds for the frame of subspaces $\{W_i\}_{i\in I}$. Then for all $f \in \mathcal{H}$ we have

$$\sum_{i\in I} v_i^2 \|\pi_{TW_i}(f)\|^2 = \sum_{i\in I} v_i^2 \|T\pi_{W_i}T^{-1}(f)\|^2 \leq \|T\|^2 \sum_{i\in I} v_i^2 \|\pi_{W_i}T^{-1}(f)\|^2$$
$$\leq \|T\|^2 D \|T^{-1}(f)\|^2 \leq \|T\|^2 \|T^{-1}\|^2 D \|f\|^2.$$

Similarly, we obtain a lower frame of subspaces bound for $\{TW_i\}_{i\in I}$. □

3.3. Parseval frames of subspaces. Parseval frames play an important role in abstract frame theory, since they are extremely useful for applications. Therefore in this subsection we study characterizations of Parseval frames of subspaces and special cases of them.

The first result extends [**7**, Corollary 4.1].

COROLLARY 3.21. *For each $i \in I$ let $v_i > 0$ and let $\{f_{ij}\}_{j\in J_i}$ be a Parseval frame sequence in $\mathcal{H}$. Define $W_i = \overline{\text{span}}_{j\in J_i}\{f_{ij}\}$ for all $i \in I$, and choose for each subspace W_i an orthonormal basis $\{e_{ij}\}_{j\in J_i}$. Then the following conditions are equivalent.*

(1) *$\{v_i f_{ij}\}_{i\in I, j\in J_i}$ is a Parseval frame for $\mathcal{H}$.*
(2) *$\{v_i e_{ij}\}_{i\in I, j\in J_i}$ is a Parseval frame for $\mathcal{H}$.*
(3) *$\{W_i\}_{i\in I}$ is a Parseval frame of subspaces with respect to $\{v_i\}_{i\in I}$ for $\mathcal{H}$.*

PROOF. This follows immediately from Theorem 3.2. □

We can also characterize Parseval frames of subspaces in terms of their frame operators in a similar manner as in frame theory.

PROPOSITION 3.22. *Let $\{W_i\}_{i\in I}$ be a family of subspaces in $\mathcal{H}$, and let $\{v_i\}_{i\in I}$ be a family of weights. Then the following conditions are equivalent.*

(1) *$\{W_i\}_{i\in I}$ is a Parseval frame of subspaces with respect to $\{v_i\}_{i\in I}$ for $\mathcal{H}$.*
(2) *$S_{W,v} = I$.*

PROOF. For each $i \in I$, let $\{e_{ij}\}_{j\in J_i}$ be an orthonormal basis for W_i.

By Proposition 3.16, (1) implies (2). To prove the converse implication suppose that $S_{W,v} = I$. Then for all $f \in \mathcal{H}$ we have

$$f = S_{W,v}(f) = \sum_{i\in I} v_i^2 \pi_{W_i}(f) = \sum_{i\in I} v_i^2 \sum_{j\in J_i} \langle f, e_{ij}\rangle e_{ij}.$$

This yields

$$\|f\|^2 = \left\langle \sum_{i\in I} v_i^2 \sum_{j\in J_i} \langle f, e_{ij}\rangle e_{ij}, f \right\rangle = \sum_{i\in I} v_i^2 \|\pi_{W_i}(f)\|^2.$$

□

We also have the following characterization of orthonormal bases of subspaces, which reflects exactly the situation in frame theory.

PROPOSITION 3.23. *Let $\{W_i\}_{i\in I}$ be a family of subspaces in $\mathcal{H}$, and let $\{v_i\}_{i\in I}$ be a family of weights. Then the following conditions are equivalent.*

(1) *$\{W_i\}_{i\in I}$ is an orthonormal basis of subspaces for $\mathcal{H}$.*
(2) *$\{W_i\}_{i\in I}$ is a 1-uniform Parseval frame of subspaces for $\mathcal{H}$.*

PROOF. For each $i \in I$, let $\{e_{ij}\}_{j\in J_i}$ be an orthonormal basis for W_i.

If (1) is satisfied, then $\{e_{ij}\}_{i\in I, j\in J_i}$ is an orthonormal basis for $\mathcal{H}$. This implies

$$\|f\|^2 = \sum_{i\in I}\sum_{j\in J_i} |\langle e_{ij}, f\rangle|^2 = \sum_{i\in I} \|\pi_{W_i}(f)\|^2$$

for all $f \in \mathcal{H}$. Thus also (2) holds.

On the other hand suppose that (2) holds. Then for all $f \in \mathcal{H}$ we have

$$\|f\|^2 = \sum_{i\in I} \|\pi_{W_i}(f)\|^2 = \sum_{i\in I}\sum_{j\in J_i} |\langle e_{ij}, f\rangle|^2$$

and $\|e_{ij}\| = 1$ for all $i \in I, j \in J_i$, which shows that $\{e_{ij}\}_{i\in I, j\in J_i}$ is an orthonormal basis for $\mathcal{H}$. This immediately implies $\mathcal{H} = \bigoplus_{i\in I} W_i$, hence (1) follows. □

3.4. Resolution of the identity. Let $\{W_i\}_{i\in I}$ be a frame of subspaces with respect to $\{v_i\}_{i\in I}$ for $\mathcal{H}$ and let its frame operator be denoted by $S_{W,v}$. By Proposition 3.16, we have

$$f = \sum_{i\in I} v_i^2 S_{W,v}^{-1} \pi_{W_i}(f) \quad \text{for all } f \in \mathcal{H}.$$

This shows that the family of operators $\{v_i^2 S_{W,v}^{-1} \pi_{W_i}\}_{i\in I}$ is a resolution of the identity. But a frame of subspaces for $\mathcal{H}$ provides us with many more resolutions of the identity than only this one.

We start our consideration with the general definition of a resolution of the identity.

DEFINITION 3.24. Let I be an indexing set. A family of bounded operators $\{T_i\}_{i\in I}$ on $\mathcal{H}$ is called a *(unconditional) resolution of the identity* on $\mathcal{H}$, if for all $f \in \mathcal{H}$ we have

$$f = \sum_{i\in I} T_i(f)$$

(and the series converges unconditionally for all $f \in \mathcal{H}$).

Note that it follows from the definition and the uniform boundedness principle that $\sup_{i\in I} \|T_i\| < \infty$.

The following result shows another way to obtain a resolution of the identity from a frame of subspaces, which even satisfies an analog of (3.1).

PROPOSITION 3.25. *Let $\{v_i\}_{i\in I}$ be a family of weights, and for each $i \in I$ let $\{v_i f_{ij}\}_{j\in J_i}$ be a frame sequence in $\mathcal{H}$ with frame bounds A_i and B_i. Suppose that $\{W_i\}_{i\in I}$ is a frame of subspaces with respect to $\{v_i\}_{i\in I}$ for $\mathcal{H}$ with frame bounds C and D, where $W_i = \overline{\text{span}}_{j\in J_i}\{f_{ij}\}$ for all $i \in I$. Then $\{v_i f_{ij}\}_{i\in I, j\in J_i}$ is a frame for $\mathcal{H}$ with frame operator denoted by S_{vf}. Further, for each $i \in I$, let $T_i : \mathcal{H} \to W_i$ be given by*

$$T_i(f) = \sum_{j\in J_i} \langle f, S_{vf}^{-1} v_i f_{ij}\rangle v_i f_{ij}.$$

If $0 < A = \inf_{i\in I} A_i \le B = \sup_{i\in I} B_i < \infty$, then $\{T_i\}_{i\in I}$ is an unconditional resolution of the identity on $\mathcal{H}$ satisfying

$$\frac{AC}{B^2D^2}\|f\|^2 \le \sum_{i\in I} v_i^2 \|T_i(f)\|^2 \le \frac{B^2D^3}{A^2C^2}\|f\|^2 \quad \textit{for all } f \in \mathcal{H}.$$

PROOF. Recall that $\{v_i f_{ij}\}_{i\in I, j\in J_i}$ is a frame for $\mathcal{H}$ by Theorem 3.2 with frame bounds AC and BD. For any $f \in \mathcal{H}$ we have

$$f = \sum_{i\in I}\sum_{j\in J_i}\langle f, S_{vf}^{-1} v_i f_{ij}\rangle v_i f_{ij} = \sum_{i\in I} T_i(f).$$

Since this is convergence relative to a frame, the convergence is unconditional.

For each $i \in I$, let $S_{vf,i}$ be the frame operator for $\{v_i f_{ij}\}_{j\in J_i}$. Let $i \in I$ be fixed. Then we obtain

$$\begin{aligned}\|T_i(f)\|^2 &= \|\sum_{j\in J_i}\langle S_{vf}^{-1} f, v_i f_{ij}\rangle v_i f_{ij}\|^2\\ &= \|S_{vf,i}\pi_{W_i} S_{vf}^{-1}(f)\|^2\\ &\le \|S_{vf,i}\|^2\|\pi_{W_i} S_{vf}^{-1}(f)\|^2.\end{aligned}$$

To prove the upper bound, we compute

$$\sum_{i\in I} v_i^2\|T_i(f)\|^2 \le B^2D^2\sum_{i\in I} v_i^2\|\pi_{W_i} S_{vf}^{-1}(f)\|^2 \le B^2D^3\|S_{vf}^{-1}(f)\|^2 \le \frac{B^2D^3}{A^2C^2}\|f\|^2.$$

The lower bound follows from

$$\begin{aligned}\sum_{i\in I} v_i^2\|T_i(f)\|^2 &= \sum_{i\in I} v_i^2\|S_{vf,i}\pi_{W_i} S^{-1}(f)\|^2 \ge \sum_{i\in I} A_i v_i^2\|\pi_{W_i} S_{vf}^{-1}(f)\|^2\\ &\ge AC\|S_{vf}^{-1}(f)\|^2 \ge \frac{AC}{B^2D^2}\|f\|^2.\end{aligned}$$

□

We now give another method for obtaining an unconditional resolution of the identity from a frame of subspaces. A special case of this can be found in Fornasier [**13, 14**].

PROPOSITION 3.26. *Let $\{W_i\}_{i\in I}$ be a frame of subspaces with respect to $\{v_i\}_{i\in I}$ for $\mathcal{H}$ with frame bounds C and D, and let $S_{W,v}$ denote its frame operator. Then $\{T_i\}_{i\in I}$ defined by $T_i = \pi_{W_i} S_{W,v}^{-1}$, $i \in I$ satisfies that $\{v_i^2 T_i\}_{i\in I}$ is an unconditional resolution of the identity, and for all $f \in \mathcal{H}$ we have*

$$\frac{C}{D^2}\|f\|^2 \le \sum_{i\in I} v_i^2\|T_i(f)\|^2 \le \frac{D}{C^2}\|f\|^2.$$

PROOF. First, for any $f \in \mathcal{H}$ we have

$$\sum_{i\in I} v_i^2 \pi_{W_i} S_{W,v}^{-1}(f) = S_{W,v} S_{W,v}^{-1}(f) = f.$$

To prove the second claim we compute

$$\frac{C}{D^2}\|f\|^2 \le C\|S_{W,v}^{-1}(f)\|^2 \le \sum_{i\in I} v_i^2\|\pi_{W_i} S_{W,v}^{-1}(f)\|^2 \le D\|S_{W,v}^{-1}(f)\|^2 \le \frac{D}{C^2}\|f\|^2.$$

□

The next result will turn out to be useful for proving a lower bound for $\sum_{i\in I} v_i^2\|T_i(f)\|^2$ if $\{v_i^2 T_i\}_{i\in I}$ is a resolution of the identity.

LEMMA 3.27. *Let $\{W_i\}_{i\in I}$ be a frame of subspaces with respect to $\{v_i\}_{i\in I}$ with frame bounds C and D for $\mathcal{H}$, and let $T_i : \mathcal{H} \to W_i$ be such that $\{v_i^2 T_i\}_{i\in I}$ is a resolution of the identity on $\mathcal{H}$ (Note that a resolution of the identity need not be unconditional so the index set must have an ordering on it. In our case, the result will hold for any ordering so we do not specify the ordering here). For any $J \subset I$ we have*

$$\frac{1}{D}\|\sum_{j\in J} v_i^2 T_j(f)\|^2 \leq \sum_{j\in J} v_j^2 \|T_j(f)\|^2 \quad \text{for all } f \in \mathcal{H}.$$

PROOF. We may assume that $|J| < \infty$, since if our inequality holds for all finite subsets then it holds for all subsets. Let $f \in \mathcal{H}$ and set $g = \sum_{j\in J} v_j^2 T_j(f)$. Then, using the fact that $\{W_i\}_{i\in I}$ is a frame of subspaces with respect to $\{v_i\}_{i\in I}$ for $\mathcal{H}$, we compute

$$\begin{aligned}
\|g\|^4 &= \left(\langle g, \sum_{j\in J} v_j^2 T_j(f)\rangle\right)^2 \\
&= \left(\sum_{j\in J} v_j \langle g, v_j T_j(f)\rangle\right)^2 \\
&= \left(\sum_{j\in J} v_j \langle \pi_{W_j}(g), v_j T_j(f)\rangle\right)^2 \\
&\leq \left(\sum_{j\in J} v_j \|\pi_{W_j}(g)\| \|v_j T_j(f)\|\right)^2 \\
&\leq \sum_{j\in J} v_j^2 \|\pi_{W_j}(g)\|^2 \sum_{j\in J} \|v_j T_j(f)\|^2 \\
&\leq D\|g\|^2 \sum_{j\in J} \|v_j T_j(f)\|^2.
\end{aligned}$$

Dividing both sides of this inequality by $D\|g\|^2$ completes the proof.

□

Using this lemma, we obtain bounds for $\sum_{i\in I} v_i^2 \|T_i(f)\|^2$ for many resolutions of the identity $\{v_i^2 T_i\}_{i\in I}$.

PROPOSITION 3.28. *Let $\{W_i\}_{i\in I}$ be a frame of subspaces with respect to $\{v_i\}_{i\in I}$ for $\mathcal{H}$ with frame bounds C and D, let $T_i : \mathcal{H} \to W_i$ be such that $\{v_i^2 T_i\}_{i\in I}$ is a resolution of the identity on $\mathcal{H}$, and assume that $T_i \pi_{W_i} = T_i$. Then for all $f \in \mathcal{H}$*

$$\frac{1}{D}\|f\|^2 \leq \sum_{i\in I} v_i^2 \|T_i(f)\|^2 \leq DE\|f\|^2,$$

where $E = sup_i \|T_i\| < \infty$.

PROOF. By Lemma 3.27, for all $f \in \mathcal{H}$, we have

$$\frac{1}{D}\|f\|^2 = \frac{1}{D}\|\sum_{i\in I} v_i^2 T_i(f)\|^2 \leq \sum_{i\in I} v_i^2 \|T_i(f)\|^2 = \sum_{i\in I} v_i^2 \|T_i \pi_{W_i}(f)\|^2$$

$$\leq \sum_{i\in I} v_i^2 \|T_i\|^2 \|\pi_{W_i}(f)\|^2 \leq E \sum_{i\in I} v_i^2 \|\pi_{W_i}(f)\|^2 \leq DE\|f\|^2.$$

□

Obviously the condition $T_i \pi_{W_i} = T_i$ for all $i \in I$ is satisfied by the example $\{v_i^2 S_{W,v}^{-1} \pi_{W_i}\}_{i\in I}$ from the beginning of this subsection. This shows that this family of operators is not only a resolution of the identity but even satisfies an analog of (3.1).

The following definition provides us with a condition, which implies that a resolution of the identity $T_i : \mathcal{H} \to W_i$, where $\{W_i\}_{i\in I}$ is a frame of subspaces with respect to $\{v_i\}_{i\in I}$ for $\mathcal{H}$, automatically satisfies an analog of (3.1).

DEFINITION 3.29. A family of bounded operators $\{T_i\}_{i\in I}$ on $\mathcal{H}$ is called an *ℓ^2-resolution of the identity with respect to a family of weights* $\{v_i\}_{i\in I}$ on $\mathcal{H}$, if it is a resolution of the identity on $\mathcal{H}$ and there exists a constant $B > 0$ so that for all $f \in \mathcal{H}$ we have

$$\sum_{i\in I} v_i^{-2} \|T_i(f)\|^2 \leq B\|f\|^2.$$

THEOREM 3.30. *Let $\{W_i\}_{i\in I}$ be a frame of subspaces with respect to $\{v_i\}_{i\in I}$ for $\mathcal{H}$, and let $T_i : \mathcal{H} \to W_i$ be such that $\{v_i^2 T_i\}_{i\in I}$ is an ℓ^2-resolution of the identity with respect to $\{v_i\}_{i\in I}$ on $\mathcal{H}$. Then there exist constants $A, B > 0$ so that for all $f \in \mathcal{H}$ we have*

$$A\|f\|^2 \leq \sum_{i\in I} v_i^2 \|T_i(f)\|^2 \leq B\|f\|^2.$$

PROOF. This follows immediately from the definition of an ℓ^2-resolution of the identity on $\mathcal{H}$ and Lemma 3.27. □

4. Riesz decompositions

In this section we first study minimal frames of subspaces, which share similar properties with minimal frames.

DEFINITION 4.1. A family of subspaces $\{W_i\}_{i\in I}$ of $\mathcal{H}$ is called *minimal*, if for each $i \in I$

$$W_i \cap \overline{\operatorname{span}}_{j\in I, j\neq i}\{W_j\} = \{0\}.$$

Using orthonormal bases for the subspaces we obtain a useful characterization of minimal families of subspaces.

LEMMA 4.2. *Let $\{W_i\}_{i\in I}$ be a family of subspaces in $\mathcal{H}$, and for each $i \in I$ let $\{e_{ij}\}_{j\in J_i}$ be an orthonormal basis for W_i. Then the following conditions are equivalent.*

(1) *$\{W_i\}_{i\in I}$ is minimal.*
(2) *$\{e_{ij}\}_{i\in I, j\in J_i}$ is minimal.*

PROOF. The implication (2) ⇒ (1) is obvious. To prove (1) ⇒ (2) suppose that $\{c_{ij}\}_{j\in J_i} \in \ell_2(J_i)$ for all $i \in I$ and we have $f_i = \sum_{j\in J_i} c_{ij} e_{ij}$ and $f = \{f_i\}_{i\in I} \in \left(\sum_{i\in I} \oplus W_i\right)_{\ell_2}$. If $\sum_{i\in I} f_i = 0$, then by minimality of $\{W_i\}_{i\in I}$ we have that $f_i = 0$ for all $i \in I$ and so $c_{ij} = 0$ for all $i \in I, j \in J_i$. It follows that $\{e_{ij}\}_{i\in I, j\in J_i}$ is a minimal frame for $\mathcal{H}$. □

The following two propositions show that for families of subspaces we can also give a definition of biorthogonal families of subspaces, which possess similar properties compared to minimal frames of subspaces as in the situation of minimal frames (compare [**9**, Lemma 3.3.1]).

PROPOSITION 4.3. *Let $\{W_i\}_{i\in I}$ be a family of subspaces in $\mathcal{H}$. Then the following conditions are equivalent.*

(1) *$\{W_i\}_{i\in I}$ is minimal.*
(2) *There exists a unique maximal (up to containment) biorthogonal family of subspaces for $\{W_i\}_{i\in I}$, i.e., there exists a family of subspaces $\{V_i\}_{i\in I}$ with $W_i \perp V_j$ for all $i,j \in I$, $j \neq i$ and $f \not\perp V_i$ for all $f \in W_i$, $i \in I$.*

Moreover, if $\{W_i\}_{i\in I}$ is a minimal frame of subspaces with respect to $\{v_i\}_{i\in I}$ for $\mathcal{H}$, then $\{S_{W,v}^{-1/2}W_i\}_{i\in I}$ is an orthogonal family of subspaces in $\mathcal{H}$.

PROOF. Suppose that (2) holds and towards a contradiction assume that there exists $i \in I$ and $0 \neq f = \sum_{j\in I, j\neq i} g_j \in W_i$ with $g_j \in W_j$. By (2), we have $g_j \perp V_i$ for all $j \neq i$, hence $f \perp V_i$, but this is a contradiction. Thus (1) follows.

To prove the opposite direction suppose that $\{W_i\}_{i\in I}$ is minimal. For each $i \in I$, let P_i denote the orthogonal projection onto $\overline{\text{span}}_{j\in I, j\neq i}\{W_j\}$. Let $i \in I$ and let V_i be defined by $V_i = (I - P_i)\mathcal{H}$ for all $i \in I$. By the definition of V_i, we have $W_j \perp V_i$ for all $j \neq i$. Towards a contradiction assume that there exists $f \in W_i$ with $\langle f, g\rangle = 0$ for all $g \in V_i$. Then $f \in P_i\mathcal{H}$ and so $W_i \cap P_i\mathcal{H} \neq \{0\}$, which is a contradiction.

For the moreover part, let $\{v_i e_{ij}\}_{j\in J_i}$ be an orthonormal basis for W_i for each $i \in I$. By Proposition 3.17, $S_{W,v}$ is the frame operator for $\{v_i e_{ij}\}_{i\in I, j\in J_i}$. Since $\{W_i\}_{i\in I}$ is minimal, Lemma 4.2 implies that $\{v_i e_{ij}\}_{i\in I, j\in J_i}$ is a minimal frame for $\mathcal{H}$ and hence is a Riesz basis for $\mathcal{H}$. Thus $\{S_{W,v}^{-1/2} v_i e_{ij}\}_{i\in I, j\in J_i}$ is an orthonormal sequence in $\mathcal{H}$. Since we have

$$S_{W,v}^{-1/2}W_i = \overline{\text{span}}_{j\in J_i}\{S_{W,v}^{-1/2} v_i e_{ij}\},$$

it follows that $\{S_{W,v}^{-1/2}W_i\}_{i\in I, j\in J_i}$ is an orthogonal sequence in $\mathcal{H}$. □

The next definition transfers the definition of Riesz bases and exact sequences to families of subspaces in a canonical way. The so-called Riesz decomposition will share most properties with Riesz bases. However it will turn out that being an exact frame of subspaces is strictly weaker than being a Riesz decomposition, a fact which differs from the situation in abstract frame theory.

DEFINITION 4.4. We call a frame of subspaces $\{W_i\}_{i\in I}$ with respect to some family of weights for $\mathcal{H}$ a *Riesz decomposition* of $\mathcal{H}$, if for every $f \in \mathcal{H}$ there is a unique choice of $f_i \in W_i$ so that $f = \sum_{i\in I} f_i$. A frame of subspaces with respect to some family of weights is *exact*, if it ceases to be a frame of subspaces once one element is deleted.

LEMMA 4.5. *Let $\{W_i\}_{i\in I}$ be a family of subspaces in $\mathcal{H}$, and for each $i \in I$ let $\{e_{ij}\}_{j\in J_i}$ be an orthonormal basis for W_i.*

(1) *The following conditions are equivalent.*
   (a) *$\{W_i\}_{i\in I}$ is a Riesz decomposition of $\mathcal{H}$.*
   (b) *$\{e_{ij}\}_{i\in I, j\in J_i}$ is a Riesz basis for $\mathcal{H}$.*
   (c) *$\{e_{ij}\}_{i\in I, j\in J_i}$ is an unconditional basis for $\mathcal{H}$.*

(2) *Let* $\{W_i\}_{i\in I}$ *be a frame of subspaces with respect to* $\{v_i\}_{i\in I}$ *for* $\mathcal{H}$. *If* $\{v_ie_{ij}\}_{i\in I,j\in J_i}$ *is an exact frame for* $\mathcal{H}$, *then also* $\{W_i\}_{i\in I}$ *is an exact frame of subspaces with respect to* $\{v_i\}_{i\in I}$ *for* $\mathcal{H}$. *The opposite implication is not valid.*

PROOF. First we prove (1). The equivalence (b) $\Leftrightarrow$ (c) follows immediately from the fact that $\{v_ie_{ij}\}_{i\in I,j\in J_i}$ is bounded and that a Schauder basis is a Riesz basis if and only if it is a bounded unconditional basis. The implication (b) $\Rightarrow$ (a) is obvious. To prove (a) $\Rightarrow$ (c) assume that $\{e_{ij}\}_{i\in I,j\in J_i}$ is not an unconditional basis. Hence there exist $f \in \mathcal{H}$ and sequences $\{c_{ij}\}_{i\in I,j\in J_i}$ and $\{d_{ij}\}_{i\in I,j\in J_i}$ with $f = \sum_{i\in I,j\in J_i} c_{ij}e_{ij} = \sum_{i\in I,j\in J_i} d_{ij}e_{ij}$ such that $c_{i_0j_0} \neq d_{i_0j_0}$ for some $i_0 \in I$, $j_0 \in J_{i_0}$. By construction $\{e_{i_0j}\}_{j\in J_{i_0}}$ is an orthonormal basis for W_{i_0}, hence $\sum_{j\in J_{i_0}} c_{i_0j}e_{i_0j} \neq \sum_{j\in J_{i_0}} d_{i_0j}e_{i_0j}$, which implies that $\{W_i\}_{i\in I}$ is not a Riesz decomposition.

To prove (2) suppose that $\{W_i\}_{i\in I}$ is a frame of subspaces with respect to $\{v_i\}_{i\in I}$ for $\mathcal{H}$. By Theorem 3.2, $\{v_ie_{ij}\}_{i\in I,j\in J_i}$ is a frame for $\mathcal{H}$. If this is exact, then, by definition, deleting one element $v_{i_0}e_{i_0j_0}$ does not leave a frame. Thus also $\{v_ie_{ij}\}_{i\in I\setminus\{i_0\},j\in J_i}$ does not form a frame. Applying Theorem 3.2 once more yields the first claim. The fact that the opposite implication is not valid is demonstrated by Example 4.7. □

The next theorem is the analog of a well-known result in abstract frame theory (see [**9**, Theorem 6.1.1]), only the role of exact frames of subspaces differs from the frame situation.

THEOREM 4.6. *Let* $\{W_i\}_{i\in I}$ *be a frame of subspaces with respect to* $\{v_i\}_{i\in I}$ *for* $\mathcal{H}$. *Then the following conditions are equivalent.*

(1) $\{W_i\}_{i\in I}$ *is a Riesz decomposition of* $\mathcal{H}$.
(2) $\{W_i\}_{i\in I}$ *is minimal.*
(3) *The synthesis operator* $T_{W,v}$ *is one-to-one.*
(4) *The analysis operator* $T^*_{W,v}$ *is onto.*

Moreover, if $\{W_i\}_{i\in I}$ *is a Riesz decomposition of* $\mathcal{H}$, *then it is also an exact frame of subspaces for* $\mathcal{H}$. *The opposite implication is not valid.*

PROOF. First note that (3) $\Leftrightarrow$ (4) is always true for operators on a Hilbert space. Moreover, it is obvious that (1) implies (3).

Next we prove (4) $\Rightarrow$ (1). By Theorem 3.12, $T^*_{W,v}$ is an isomorphism. Therefore if it is onto, then it is invertible. Hence, $T_{W,v}$ is invertible. This implies that for every $f \in \mathcal{H}$ there exists a $\{f_i\}_{i\in I} \in \left(\sum_{i\in I}\oplus W_i\right)_{l^2}$ so that

$$f = T_{W,v}(\{f_i\}_{i\in I}) = \sum_{i\in I} v_if_i.$$

If we have $f = \sum_{i\in I} f_i = \sum_{i\in I} g_i$ with $\{f_i\},\{g_i\} \in \left(\sum_{i\in I}\oplus W_i\right)_{l^2}$, then it follows that $T_{W,v}(\{v_i^{-1}f_i\}_{i\in I}) = T_{W,v}(\{v_i^{-1}g_i\}_{i\in I})$. Since $T_{W,v}$ is one-to-one, $\{v_i^{-1}f_i\}_{i\in I} = \{v_i^{-1}g_i\}_{i\in I}$ and so $f_i = g_i$ for all $i \in I$. This shows the equivalence of (1), (3), and (4).

It remains to prove that (1) is equivalent to (2). If $\{W_i\}_{i\in I}$ is not a Riesz decomposition of $\mathcal{H}$, there exists an element $f \in \mathcal{H}$ and $f_i, g_i \in W_i$, $i \in I$ with $f_{i_0} \neq g_{i_0}$ for some $i_0 \in I$ and $f = \sum_{i\in I} f_i = \sum_{i\in I} g_i$. It follows $0 \neq g_{i_0} - f_{i_0} = \sum_{i\in I,i\neq i_0}(f_i - g_i)$ and $g_{i_0} - f_{i_0} \in W_{i_0}$. This proves

$$g_{i_0} - f_{i_0} \in W_{i_0} \cap \overline{\operatorname{span}}_{i\in I,i\neq i_0}\{W_i\},$$

which implies that $\{W_i\}_{i\in I}$ is not minimal.

To prove the converse implication assume that $\{W_i\}_{i\in I}$ is not minimal. Then, for some $i_0 \in I$, there exists $0 \neq f = \sum_{i\in I, i\neq i_0} f_i \in W_{i_0}$ with $f_i \in W_i$. Hence

$$0 = \sum_{i\in I, i\neq i_0} (f_i - f) = \sum_{i\in I} 0.$$

Thus $\{W_i\}_{i\in I}$ is not a Riesz decomposition of $\mathcal{H}$.

To prove the moreover part, suppose $\{W_i\}_{i\in I}$ is a frame of subspaces with respect to $\{v_i\}_{i\in I}$ for $\mathcal{H}$ and a Riesz decomposition of $\mathcal{H}$. We will prove that this implies that $\{W_i\}_{i\in I}$ is exact. For this, fix some $i_0 \in I$. Since $T_{W,v}^*$ is onto, there exists an element $f \in \mathcal{H}$ such that $\pi_{W_{i_0}}(f) \neq 0$, but $\pi_{W_i}(f) = 0$ for all $i \neq i_0$. Thus

$$\sum_{i\in I} v_i^2 \|\pi_{W_i}(f)\|^2 = v_{i_0}^2 \|\pi_{W_{i_0}}(f)\|^2.$$

Therefore it is not possible to delete W_{i_0} from the frame of subspaces yet leave a frame of subspaces. Since i_0 was chosen arbitrarily, the claim follows.

The fact that the opposite implication is not valid is demonstrated by Example 4.7. □

Note that we could have also proven the equivalence of (1) and (2) and the moreover part of the previous result by using Lemma 4.2, Lemma 4.5, and [**9**, Theorem 6.1.1]. We have chosen to add the extended proof here in order to enlighten the use of frames of subspaces.

Next we give an example for the different role exactness plays in the situation of families of subspaces.

EXAMPLE 4.7. Let $\{e_i\}_{i\in\mathbb{Z}}$ be an orthonormal basis for some Hilbert space $\mathcal{H}$ and define the subspaces W_1, W_2 by

$$W_1 = \overline{\operatorname{span}}_{i\geq 0}\{e_i\} \quad \text{and} \quad W_2 = \overline{\operatorname{span}}_{i\leq 0}\{e_i\}.$$

Then $\{W_1, W_2\}$ is a frame of subspaces with respect to weights $\{v_1, v_2\}$ with $v_1 = v_2 =: v > 0$, since

$$v_1\|\pi_{W_1}(f)\|^2 + v_2\|\pi_{W_2}(f)\|^2 = \sum_{i\in\mathbb{Z}} v\,|\langle f, e_i\rangle|^2 + v\,|\langle f, e_0\rangle|^2 = v\|f\|^2 + v\,|\langle f, e_0\rangle|^2$$

and

$$v\|f\|^2 \leq v\|f\|^2 + v\,|\langle f, e_0\rangle|^2 \leq 2v\|f\|^2.$$

It is also exact, since when we delete one subspace the remaining one does not span the space. But it is not a Riesz decomposition, because we can write the element e_0 as $e_0 = e_0 + 0$ and $e_0 = 0 + e_0$. Thus the decomposition is not unique. Also observe that the sequence $\{ve_i\}_{i\geq 0} \cup \{ve_i\}_{i\leq 0}$ is a frame, but is not exact.

We conclude this subsection by mentioning that orthonormal bases of subspaces are special cases of Riesz decompositions.

COROLLARY 4.8. *If $\{W_i\}_{i\in I}$ is an orthonormal basis of subspaces for $\mathcal{H}$, then it is also a Riesz decomposition of $\mathcal{H}$.*

PROOF. This follows immediately from the definition of a Riesz decomposition and Proposition 3.23. □

5. Several constructions

In this section we will discuss several constructions concerning frames of subspaces, frames, and Riesz frames. Recall that in addition to what follows we are already equipped with some constructions by Theorem 3.2, Corollary 3.21, and Lemma 4.5.

5.1. Constructions of frames of subspaces.

Dealing with Bessel families of subspaces is important, since there are easy ways to turn such a family into a frame of subspaces. One way is to just add the subspace $W_0 = \mathcal{H}$ to the family. Another more careful method is the following one: Take any orthonormal basis for $\mathcal{H}$ and partition its elements into the subspaces W_i, $i \in I$. Then add the subspaces spanned by the remaining elements to the Bessel family. This yields a frame of subspaces.

Using the synthesis operator $T_{W,v}$, we obtain a characterization of Bessel sequences of subspaces.

PROPOSITION 5.1. *Let $\{W_i\}_{i\in I}$ be a family of subspaces of $\mathcal{H}$, and let $\{v_i\}_{i\in I}$ be a family of weights. Then the following conditions are equivalent.*

(1) *$\{W_i\}_{i\in I}$ is a Bessel sequence of subspaces with respect to $\{v_i\}_{i\in I}$ for $\mathcal{H}$.*
(2) *The synthesis operator $T_{W,v}$ is bounded and linear.*

PROOF. First suppose that (1) holds. Then Lemma 3.9 shows that the series in the definition of the synthesis operator $T_{W,v}$ converges unconditionally. Moreover, we have

$$\sum_{i\in I} v_i^2 \|\pi_{W_i}(f)\|^2 \le B\|f\|^2.$$

By definition of the analysis operator $T^*_{W,v}$,

$$\|T^*_{W,v}(f)\|^2 = \sum_{i\in I} v_i^2 \|\pi_{W_i}(f)\|^2. \tag{5.1}$$

Since $\{W_i\}_{i\in I}$ is a Bessel sequence of subspaces with respect to $\{v_i\}_{i\in I}$, this implies that $T^*_{W,v}$ is bounded. Hence also $T_{W,v}$ is bounded, which shows (2).

If (2) holds, then also $T^*_{W,v}$ is a bounded operator. This fact together with (5.1) yields (1). □

One possible application for frames of subspaces is to the problem of classifying those $g \in L^2(\mathbb{R})$ and $0 < a, b \le 1$ so that (g, a, b) yields a Gabor frame (see Example 5.4 below). This is an exceptionally deep problem even in the case of characteristic functions [**6, 20**]. But we have simple classifications of when $\{e^{2\pi imbt}g(t)\}_{m\in\mathbb{Z}}$ is a frame sequence in $L^2(\mathbb{R})$ and when $\{e^{2\pi imbt}g(t-na)\}_{m,n\in\mathbb{Z}}$ has dense span in $L^2(\mathbb{R})$. By our results, this family will be a Gabor frame for $L^2(\mathbb{R})$ if and only if $\{W_n\}_{n\in\mathbb{Z}}$ is a frame of subspaces where W_n is the closed linear span of $\{e^{2\pi imbt}g(t-na)\}_{m\in\mathbb{Z}}$.

For some applications, we would like to take a frame for $\mathcal{H}$ and divide it into subsets so that the closed linear span of these subsets is a frame of subspaces for $\mathcal{H}$. This is not always possible. But the next proposition shows that one of the needed inequalities will always hold.

PROPOSITION 5.2. *Let $\{f_j\}_{j\in J}$ be a frame for $\mathcal{H}$ with frame bounds A and B. Let $\{J_i\}_{i\in I}$ be a partition of the indexing set J, and for all $i \in I$ let W_i denote the* *closed linear span of* $\{f_j\}_{j\in J_i}$. *Then for all* $f \in \mathcal{H}$ *we have*

$$\frac{A}{B}\|f\|^2 \le \sum_{i\in I}\|\pi_{W_i}(f)\|^2.$$

Hence, if $|I| < \infty$, *then* $\{W_i\}_{i\in I}$ *is a* 1*-uniform frame of subspaces for* $\mathcal{H}$.

PROOF. We compute

$$A\|f\|^2 \le \sum_{j\in J}|\langle f, f_j\rangle|^2 = \sum_{i\in I}\sum_{j\in J_i}|\langle f, f_j\rangle|^2 = \sum_{i\in I}\sum_{j\in J_i}|\langle \pi_{W_i}(f), f_j\rangle|^2.$$

Recall that if a family of vectors is a B-Bessel family then every subfamily is also B-Bessel. Thus

$$\sum_{i\in I}\sum_{j\in J_i}|\langle \pi_{W_i}(f), f_j\rangle|^2 \le \sum_{i\in I} B\|\pi_{W_i}(f)\|^2.$$

This proves the first claim.

If $|I| < \infty$, then we have

$$\sum_{i\in I}\|\pi_{W_i}(f)\|^2 \le |I|\cdot\|f\|^2.$$

Hence in this case $\{W_i\}_{i\in I}$ is always a frame of subspaces for $\mathcal{H}$, in particular a 1-uniform frame of subspaces. □

An easy way to obtain a frame of subspaces is provided by the next result.

PROPOSITION 5.3. *Let* $\{f_j\}_{j\in J}$ *be a frame for* $\mathcal{H}$, *let* $J = J_1 \cup \ldots \cup J_n$ *be a finite partition of* J, *and let* $\{v_i\}_{i=1}^n$ *be a family of weights. Then* $\{W_i\}_{i=1}^n$ *is a frame of subspaces with respect to* $\{v_i\}_{i=1}^n$ *for* $\mathcal{H}$, *where* $W_i = \overline{\text{span}}_{j\in J_i}\{f_j\}$.

PROOF. Let $f \in \mathcal{H}$. Obviously, $\|\pi_{W_i}(f)\|^2 \le \|f\|^2$ for all $1 \le i \le n$, which implies that

$$\sum_{i=1}^n v_i^2\|\pi_{W_i}(f)\|^2 \le \max_{i=1,\ldots,n}\{v_i^2\}\cdot\|f\|^2.$$

Thus $\{W_i\}_{i=1}^n$ is a Bessel sequence of subspaces with respect to $\{v_i\}_{i=1}^n$. The lower bound follows from an application of Proposition 5.2. That is, for any $f \in \mathcal{H}$ we have

$$\frac{A}{B}\|f\|^2 \le \sum_{i\in I}\|\pi_{W_i}(f)\|^2 \le \frac{1}{\max_{i=1,\ldots,n}\{v_i^2\}}\sum_{i\in I} v_i^2\|\pi_{W_i}(f)\|^2.$$

□

This partition of the frame elements is not always a partition into frame sequences. Let us consider the case of Gabor systems. In the following example we will show that a large class of Gabor systems can be written as a frame of subspaces. Moreover, we can characterize those Gabor atoms, for which this partition is a partition into frame sequences.

EXAMPLE 5.4. For each $a \in \mathbb{R}$, let the unitary operators E_a, T_a on $L^2(\mathbb{R})$ be defined by

$$E_a f(x) = e^{2\pi i a x} f(x) \quad \text{and} \quad T_a f(x) = f(x-a).$$

Given a function $g \in L^2(\mathbb{R})$ and $a, b > 0$, the Gabor system determined by g and a, b is defined by

$$\mathcal{G}(g, a, b) = \{E_{ma}T_{nb}g : m, n \in \mathbb{Z}\}.$$

Let

$$Z : L^2(\mathbb{R}) \to L^2([0,1)^2), \quad Zf(x,y) = \sum_{k\in\mathbb{Z}} f(x+k)e^{2\pi iky}.$$

denote the Zak transform on $L^2(\mathbb{R})$ (compare [**19**]).

Let $h \in L^2(\mathbb{R})$ and $q \in \mathbb{N}$. In the following we will consider some Gabor system $\mathcal{G}(h,a,b)$ with $a, b > 0$, $ab = \frac{1}{q}$. Using a metaplectic transform it can be shown that this system is unitarily equivalent to $\mathcal{G}(g, \frac{1}{q}, 1)$ for some $g \in L^2(\mathbb{R})$ [**16**, Proposition 9.4.4], hence it suffices to consider this system. Now the Gabor system $\mathcal{G}(g, \frac{1}{q}, 1)$ in turn can be decomposed using the partition

$$\mathcal{G}(g, \tfrac{1}{q}, 1) = \bigcup_{j=0}^{q-1} \{E_{\frac{1}{q}(mq+j)} T_n g\}_{m,n\in\mathbb{Z}}. \tag{5.2}$$

By Proposition 5.3, the set of the subspaces $W_j := \overline{\text{span}}_{m,n\in\mathbb{Z}}\{E_{\frac{1}{q}(mq+j)} T_n g\}$, $j = 0, \ldots, q-1$ is indeed a frame of subspaces.

In a second step we will investigate, whether the sequences $\{E_{\frac{1}{q}(mq+j)} T_n g\}_{m,n\in\mathbb{Z}}$ are frame sequences. It will turn out that this will not happen unless the Zak transform is discontinuous.

We first observe that the following conditions are equivalent.

(1) The sequence $\{E_{\frac{1}{q}(mq+j)} T_n g\}_{m,n\in\mathbb{Z}}$ is a frame sequence for each $0 \le j \le q-1$.

(2) There exist $0 < A \le B < \infty$ such that

$$A \le |Zg(x,y)|^2 \le B \quad \text{for almost all } (x,y) \in [0,1)^2 \backslash V,$$

where $V = \{(x,y) \in [0,1)^2 : Zg(x,y) = 0\}$.

This can be proven as follows. First notice that since

$$E_{\frac{1}{q}}(\{E_{\frac{1}{q}(mq+j)} T_n g\}_{m,n\in\mathbb{Z}}) = \{E_{\frac{1}{q}(mq+j+1)} T_n g\}_{m,n\in\mathbb{Z}} \quad \text{for all } 0 \le j < q-1$$

and

$$E_{\frac{1}{q}}(\{E_{\frac{1}{q}(mq+q-1)} T_n g\}_{m,n\in\mathbb{Z}}) = \{E_{\frac{1}{q}(mq)} T_n g\}_{m,n\in\mathbb{Z}} = \{E_m T_n g\}_{m,n\in\mathbb{Z}},$$

the fact that $E_{\frac{1}{q}}$ is a unitary operator implies that condition (1) holds if and only if $\{E_m T_n g\}_{m,n\in\mathbb{Z}}$ is a frame sequence. Since $Z : L^2(\mathbb{R}) \to L^2([0,1)^2)$ is an isomorphism [**19**] and it is an easy calculation to show that $Z(E_m T_n g)(x,y) = E_m(x)E_n(y)Zg(x,y)$, condition (1) holds if and only if $\{E_m(x)E_n(y)Zg(x,y)\}_{m,n\in\mathbb{Z}}$ is a frame sequence. Since $\{E_m E_n\}_{m,n\in\mathbb{Z}}$ is an orthonormal basis for $L^2([0,1)^2)$, for each $f \in L^2([0,1)^2)$ we obtain

$$\sum_{m,n\in\mathbb{Z}} |\langle f, E_m E_n Zg\rangle|^2 = \left\|f \cdot \overline{Zg}\right\|^2.$$

This implies that (1) is equivalent to

$$A\|f\|^2 \le \left\|f \cdot \overline{Zg}\right\|^2 \le B\|f\|^2 \quad \text{for all } f \in \overline{\text{span}}_{m,n\in\mathbb{Z}}\{E_m(x)E_n(y)Zg(x,y)\},$$

which holds if and only if

$$A\|f \cdot Zg\|^2 \le \left\|f \cdot |Zg|^2\right\|^2 \le B\|f \cdot Zg\|^2 \quad \text{for all } f \in L^2([0,1)^2).$$

It is easy to check that this is equivalent to (2), which proves the claim.

Finally, we consider $g \in L^2(\mathbb{R})$ with Zg being continuous. By [**19**], this implies that Zg has a zero. Hence condition (2) can never be fulfilled. This shows that the sequences $\{E_{\frac{1}{q}(mq+j)}T_n g\}_{m,n\in\mathbb{Z}}$ can only be frame sequences, if the Zak transform Zg is discontinuous.

5.2. Constructions of frames and Riesz frames. If we have a frame for $\mathcal{H}$, using a frame of subspaces for $\mathcal{H}$ we can construct new frames for $\mathcal{H}$ from these components.

PROPOSITION 5.5. *Let $\{W_i\}_{i\in I}$ be a frame of subspaces with respect to $\{v_i\}_{i\in I}$ for $\mathcal{H}$ and let $\{f_j\}_{j\in J}$ be a frame for $\mathcal{H}$. Then there exist $A, B > 0$ so that $\{\pi_{W_i} S_{W,v}^{-1}(f_j)\}_{j\in J}$ is a frame for W_i with frame bounds A and B for each $i \in I$. Hence $\{\pi_{W_i} S_{W,v}^{-1}(f_j)\}_{i\in I, j\in J}$ is also a frame for $\mathcal{H}$.*

PROOF. Since $S_{W,v}^{-1}$ is an invertible operator on $\mathcal{H}$ and $\{f_j\}_{j\in J}$ is a frame for $\mathcal{H}$, we have that $\{S_{W,v}^{-1} f_j\}_{j\in J}$ is a frame for $\mathcal{H}$ with frame bounds A and B. Therefore $\{\pi_{W_i} S_{W,v}^{-1}(f_j)\}_{j\in J}$ is a frame for W_j with frame bounds A and B for every $i \in I$. Since $\{W_i\}_{i\in I}$ is a frame of subspaces for $\mathcal{H}$, we have that $\{\pi_{W_i} S_{W,v}^{-1}(f_j)\}_{i\in I, j\in J}$ is a frame for $\mathcal{H}$ by Theorem 3.2. □

To construct Riesz frames for $\mathcal{H}$ we first need to give an analog definition for families of subspaces.

DEFINITION 5.6. We call a frame of subspaces $\{W_i\}_{i\in I}$ a *Riesz frame of subspaces with respect to* $\{v_i\}_{i\in I}$, if there exist constants $C, D > 0$ so that every subfamily $\{W_i\}_{i\in J}$ with $J \subset I$ is a frame of subspaces with respect to $\{v_i\}_{i\in J}$ for its closed linear span with frame bounds C and D.

First we may ask whether subfamilies of a frame of subspaces are automatically frames of subspaces for their closed linear spans. The following example shows that this is not always the case.

EXAMPLE 5.7. In general, if $\{W_i\}_{i\in I}$ is a 1-uniform frame of subspaces and $J \subset I$, then $\{W_i\}_{i\in J}$ need not be a frame of subspaces for its closed linear span. For example, let $\{e_i\}_{i=1}^{\infty}$ be an orthonormal basis for $\mathcal{H}$ and for each $i \in I$ define the subspaces W_i^1, W_i^2, and W_i^3 by

$$W_i^1 = \text{span}\{e_{2i} + \tfrac{1}{i} e_{2i+1}\}, \quad W_i^2 = \text{span}\{e_{2i}\}, \quad \text{and} \quad W_i^3 = \text{span}\{e_{2i+1}\}.$$

Then it is easily checked that $\{W_i^1, W_i^2, W_i^3\}_{i=1}^{\infty}$ is a frame of subspaces for $\mathcal{H}$. Also observe that $\overline{\text{span}}_{i=1,\ldots,\infty}\{W_i^1, W_i^2\} = \mathcal{H}$. Since for all positive integers i we have

$$\pi_{W_i^1}(e_{2i+1}) = \frac{1}{i\sqrt{1+\frac{1}{i^2}}}(e_{2i} + \tfrac{1}{i} e_{2i+1}) \quad \text{and} \quad \pi_{W_i^2}(e_{2i+1}) = 0,$$

it follows that $\{W_i^1, W_i^2\}_{i=1}^{\infty}$ is not a frame of subspaces for its closed linear span.

Using a Riesz frame of subspaces and Riesz frames for the single subspaces, we can construct a Riesz frame for $\mathcal{H}$ by just taking all elements of the Riesz frames.

PROPOSITION 5.8. *Let $\{W_i\}_{i\in I}$ be a Riesz frame of subspaces with respect to $\{v_i\}_{i\in I}$ for $\mathcal{H}$, and let $\{f_{ij}\}_{j\in I_i}$ be a Riesz frame for W_i with Riesz frame bounds A and B for all $i \in I$. Then $\{v_i f_{ij}\}_{i\in I, j\in I_i}$ is a Riesz frame for $\mathcal{H}$. Also, for any $J \subset I$, $\{W_j\}_{j\in J}$ is a Riesz frame of subspaces with respect to $\{v_i\}_{i\in I}$ for its closed linear span.*

PROOF. Let C and D be the Riesz frame of subspaces bounds for $\{W_i\}_{i\in I}$. For every $i \in I$ choose $J_i \subset I_i$ and define $\widetilde{W}_i$ by

$$\widetilde{W}_i = \overline{\text{span}}_{j\in J_i}\{f_{ij}\} \subset W_i.$$

Let $f \in \overline{\text{span}}_{i\in I, j\in J_i}\{f_{ij}\}$. Then we have

$$\sum_{i\in I, j\in J_i} |\langle f, v_i f_{ij}\rangle|^2 = \sum_{i\in I} v_i^2 \sum_{j\in J_i} |\langle \pi_{\widetilde{W}_i}(f), f_{ij}\rangle|^2 \leq \sum_{i\in I} Bv_i^2 \|\pi_{\widetilde{W}_i}(f)\|^2 \leq BD\|f\|^2.$$

Concerning the lower bounds, we compute

$$\sum_{i\in I, j\in J_i} |\langle f, v_i f_{ij}\rangle|^2 = \sum_{i\in I} v_i^2 \sum_{j\in J_i} |\langle \pi_{\widetilde{W}_i}(f), f_{ij}\rangle|^2 \geq \sum_{i\in I} Av_i^2 \|\pi_{\widetilde{W}_i}(f)\|^2 \geq CA\|f\|^2.$$

□

6. Harmonic frames of subspaces

Harmonic frames of subspaces are a special case of frames of subspaces, which are equipped with a natural structure and which occur in several situations, e.g., in Gabor analysis or in wavelet analysis in the form of multiresolution analysis.

6.1. The finite case.

We start by giving the definition of a harmonic frame of subspaces for a finite family of subspaces.

DEFINITION 6.1. A frame of subspaces $\{W_i\}_{i\in I}$ with respect to $\{v_i\}_{i\in I}$ for $\mathcal{H}$ is a *finite harmonic frame of subspaces with respect to* $\{v_i\}_{i\in I}$, if $I = \{0, \ldots, N-1\}$, $N \in \mathbb{N}$ and there exists a unitary operator U on $\mathcal{H}$ so that

$$UW_{N-1} = W_0 \quad \text{and} \quad UW_i = W_{i+1} \quad \text{for all } 0 \leq i \leq N-2.$$

If $\{W_i\}_{i\in I}$ is a uniform Parseval frame of subspaces, $UW_{N-1} = W_0$ follows automatically as the following proposition shows. This result equals the corresponding result in the frame situation [**7**, Theorem 4.1].

THEOREM 6.2. *Let $\{W_i\}_{i\in I}$, $|I| = N$, be a uniform Parseval frame of subspaces such that there exists a unitary operator U on $\mathcal{H}$ so that*

$$UW_i = W_{i+1} \quad \textit{for all } 0 \leq i \leq N-2.$$

Then

$$UW_{N-1} = W_0.$$

PROOF. Without loss of generality we can assume that $v_i = 1$ for all $i \in I$. Let $\{g_j\}_{j\in J}$ be an orthonormal basis for W_0. By hypothesis, $\{U^i g_j\}_{j\in J}$ is an orthonormal basis for W_i for all $0 \leq i \leq N-1$. Now let $f \in \mathcal{H}$. By Proposition 3.23 and since $\{W_i\}_{i\in I}$ was assumed to be uniform, we have

$$f = \sum_{i=0}^{N-1} \sum_{j\in J} \langle f, U^i g_j\rangle U^i g_j. \tag{6.1}$$

Applying U, this leads to

$$\begin{aligned} Uf &= \sum_{i=0}^{N-1}\sum_{j\in J}\langle Uf, U^i g_j\rangle U^i g_j \\ &= \sum_{i=0}^{N-1}\sum_{j\in J}\langle f, U^{i-1} g_j\rangle U^i g_j \\ &= U\left[\sum_{i=0}^{N-1}\sum_{j\in J}\langle f, U^{i-1} g_j\rangle U^{i-1} g_j\right] \\ &= U\left[\sum_{i=-1}^{N-2}\sum_{j\in J}\langle f, U^{i} g_j\rangle U^{i} g_j\right]. \end{aligned}$$

Since U is unitary and by (6.1), we obtain

$$f = \sum_{i=-1}^{N-2}\sum_{j\in J}\langle f, U^i g_j\rangle U^i g_j = \sum_{i=0}^{N-1}\sum_{j\in J}\langle f, U^i g_j\rangle U^i g_j.$$

This implies

$$\sum_{j\in J}\langle f, U^{-1} g_j\rangle U^{-1} g_j = \sum_{j\in J}\langle f, U^{N-1} g_j\rangle U^{N-1} g_j.$$

Now we apply U, which yields

$$\sum_{j\in J}\langle f, U^{-1} g_j\rangle g_j = \sum_{j\in J}\langle f, U^{N-1} g_j\rangle U^{N} g_j.$$

Using $U^{-1}f$ instead of f gives

$$\sum_{j\in J}\langle f, g_j\rangle g_j = \sum_{j\in J}\langle f, U^{N} g_j\rangle U^{N} g_j,$$

which shows that $\pi_{W_0} = \pi_{\overline{\text{span}}_{j\in J}\{U^N g_j\}} = \pi_{UW_{N-1}}$. This completes the proof. □

The following result is [**7**, Theorem 4.2 and Theorem 4.3], which we add for completeness in a reformulated version and with a proof using our results.

PROPOSITION 6.3. *Let $\varphi \in \mathcal{H}$ and let V be a unitary operator on $\mathcal{H}$ such that $\{V^j\varphi\}_{j=0}^{M-1}$ is a uniform Parseval frame sequence. Define the subspace W_0 by $W_0 = \text{span}_{j=0,\dots,M-1}\{V^j\varphi\}$. Further let U be a unitary operator on $\mathcal{H}$. Then the following conditions are equivalent.*

(1) *$\{U^iV^j\varphi\}_{i=0,\ j=0}^{L-1,\ M-1}$ is a uniform Parseval frame for $\mathcal{H}$.*
(2) *$\{U^iW_0\}_{i=0}^{L-1}$ is a uniform Parseval frame of subspaces for $\mathcal{H}$.*

PROOF. This follows immediately from Theorem 3.2 with setting $W_i = U^iW_0$ for all $i = 1, \dots, L-1$. □

We conclude this subsection by giving an example for this type of frames of subspaces.

EXAMPLE 6.4. Let $a \in \mathbb{R}$ and $g \in L^2(\mathbb{R})$ be such that $\{E_{am}T_n g\}_{m,n\in\mathbb{Z}}$ is a Parseval Gabor frame. Fix some $N \in \mathbb{N}$ and define W_i, $0 \le i \le N-1$, by

$$W_i = \overline{\text{span}}_{m,n\in\mathbb{Z}}\{E_{a(mN+i)}T_n g\}.$$

Then $\{W_i\}_{i=0}^{N-1}$ is a finite harmonic frame of subspaces. To see this let $U := E_a$. Then

$$UW_{N-1} = W_0 \quad \text{and} \quad UW_i = W_{i+1} \quad \text{for all } 0 \le i \le N-2.$$

Also,

$$U^i(E_{amN}T_n g) = E_{a(mN+i)}T_n g.$$

Hence the sequences $\{E_{a(mN+i)}T_n g\}_{m,n\in\mathbb{Z}}$ are even unitarily equivalent to each other.

Notice that this construction generalizes the one in Example 5.4.

6.2. The infinite case. We can also define a harmonic frame of subspaces for an infinite family of subspaces.

DEFINITION 6.5. A frame of subspaces $\{W_i\}_{i\in I}$ with respect to $\{v_i\}_{i\in I}$ for $\mathcal{H}$ is an *infinite harmonic frame of subspaces with respect to* $\{v_i\}_{i\in I}$, if $I = \mathbb{Z}$ and there exists a unitary operator U on $\mathcal{H}$ so that

$$UW_i = W_{i+1} \quad \text{for all } i \in I.$$

An interesting example for this type of frame of subspaces is the generalized frame multiresolution analysis in the sense of Papadakis [**21**], which generalizes the classical multiresolution analysis.

EXAMPLE 6.6. Here we consider the generalized frame multiresolution analysis in the sense of Papadakis [**21**], whose approach includes all classical multiresolution analysis (MRAs) in one and higher dimensions as well as the FMRAs of Benedetto and Li [**2**].

Let $\mathcal{H}$ be a Hilbert space, $U : \mathcal{H} \to \mathcal{H}$ be a unitary operator, and let G be a unitary abelian group acting on $\mathcal{H}$. A sequence $\{V_i\}_{i\in\mathbb{Z}}$ of closed subspaces of $\mathcal{H}$ is a *generalized frame multiresolution analysis* of $\mathcal{H}$, if it satisfies the following properties.

(1) $V_i \subseteq V_{i+1}$ for all $i \in \mathbb{Z}$.
(2) $V_i = U^i(V_0)$ for all $i \in \mathbb{Z}$.
(3) $\bigcap_i V_i = \{0\}$ and $\overline{\bigcup_i V_i} = \mathcal{H}$.
(4) There exists a countable subset B of V_0 such that the set $G(B) = \{g\phi : g \in G, \phi \in B\}$ is a frame for V_0.

Now $\{W_i\}_{i\in\mathbb{Z}}$ is defined by $W_0 := V_1 \cap V_0^\perp$ and $W_i := U^i(W_0)$ for every $i \in \mathbb{Z}$. Then we have

$$W_i = U^i(V_1 \cap V_0^\perp) = U^i(V_1) \cap U^i(V_0)^\perp = V_{i+1} \cap V_i^\perp.$$

In the case $\mathcal{H} = L^2(\mathbb{R})$, $G = \{T_k : k \in \mathbb{Z}\}$, B containing only one element, and $U = D_2$ being the dilation operator $D_2 f(t) = \sqrt{2}f(2t)$ this definition reduces to the well-known MRA.

In the general situation we have the following relations:

(i) If $\{V_i\}_{i\in\mathbb{Z}}$ is a GFMRA of some Hilbert space $\mathcal{H}$, then $\{W_i\}_{i\in\mathbb{Z}}$ is an infinite 1-uniform harmonic Parseval frame of subspaces, since $\mathcal{H} = \bigoplus_{i\in\mathbb{Z}} W_i$, with unitary operator U.

(ii) Let $\{W_i\}_{i\in\mathbb{Z}}$ be an infinite 1-uniform harmonic Parseval frame of subspaces with unitary operator denoted by U. By Proposition 3.23, we have $\mathcal{H} = \bigoplus_{i\in\mathbb{Z}} W_i$. Then we can define $\{V_i\}_{i\in\mathbb{Z}}$ by

$$V_i = \bigoplus_{m\leq i-1} W_m.$$

Then (1)-(3) are obviously satisfied. Therefore $\{W_i\}_{i\in\mathbb{Z}}$ is a GFMRA if and only if (4) is satisfied.

Acknowledgments

The majority of the research for this paper was performed while the second author was visiting the Departments of Mathematics at the University of Missouri and Washington University in St. Louis. This author thanks these departments for their hospitality and support during these visits.

The second author is indebted to Damir Bakić, Ilya Krishtal, Demetrio Labate, Guido Weiss, and Ed Wilson for helpful discussions.

References

[1] A. Aldroubi, C. Cabrelli, and U.M. Molter, *Wavelets on irregular grids and arbitrary dilation matrices, and frame atoms for* $L^2(\mathbb{R}^d)$, preprint, 2003.

[2] J. Benedetto and S. Li, *The theory of multiresolution analysis frames and applications to filter banks*, Appl. Comput. Harmon. Anal. **5** (1998), 389–427.

[3] P.G. Casazza, *Modern tools for Weyl-Heisenberg (Gabor) Frame Theory*, Adv. in Imag. and Electron. Physics **115** (2001), 1–127.

[4] P.G. Casazza, *The art of frame theory*, Taiwanese J. Math. **4** (2000), 129–201.

[5] P.G. Casazza, O. Christensen, A.M. Lindner, and R. Vershynin, *Frames and the Feichtinger conjecture*, preprint, 2003.

[6] P.G. Casazza and N.J. Kalton, *Roots of complex polynomials and Weyl-Heisenberg frame sets*, Proc. Amer. Math. Soc. **130** (2002), 2313–2318.

[7] P.G. Casazza and J. Kovačević, *Equal-norm tight frames with erasures*, Adv. Comput. Math. **18** (2003), 387–430.

[8] P.G. Casazza and R. Vershynin, *Kadison-Singer meets Bourgain-Tzafriri*, preprint, 2003.

[9] O. Christensen, *An Introduction to Frames and Riesz Bases*, Birkhäuser, Boston, 2003.

[10] I. Daubechies, A. Grossman and Y. Meyer, *Painless nonorthogonal expansions*, J. Math. Phys. **27** (1986), 1271–1283.

[11] J. Diestel, *Sequences and Series in Banach Spaces*, Springer-Verlag, New York, 1984.

[12] R.J. Duffin and A.C. Schaeffer, *A class of nonharmonic Fourier series*, Trans. Amer. Math. Soc. **72** (1952), 341–366.

[13] M. Fornasier, *Decompositions of Hilbert spaces: Local construction of global frames*, Proc. Int. Conf. on Constructive function theory, Varna (2002) B. Bojanov Ed., DARBA, Sofia (2003), 275–281.

[14] M. Fornasier, *Quasi-orthogonal decompositions of structured frames*, preprint, 2003.

[15] V.K. Goyal, J. Kovačević, and J.A. Kelner, *Quantized frame expansions with erasures*, Appl. Comput. Harmon. Anal. **10** (2001), 203–233.

[16] K. Gröchenig, *Foundations of Time-Frequency Analysis*, Birkhäuser, Boston, 2001.

[17] B. Hassibi, B. Hochwald, A. Shokrollahi, and W. Sweldens, *Representation theory for high-rate multiple-antenna code design*, IEEE Trans. Inform. Theory **47** (2001), 2335–2367.

[18] C.E. Heil and D.F. Walnut, *Continuous and discrete Wavelet transforms*, SIAM Rev. **31** (1989), 628–666.

[19] A.J.E.M. Janssen, *The Zak transform: A signal transform for sampled time-continuous signals*, Philips J. Res. **43** (1988), 23–69.

[20] A.J.E.M. Janssen, *Zak transforms with few zeroes and the tie*, in Advances in Gabor Analysis, H.G. Feichtinger and T. Strohmer, eds., Birkhäuser, Boston, 2003, pp. 31–70.

[21] M. Papadakis, *Generalized frame multiresolution analysis of abstract Hilbert spaces*, in Sampling, Wavelets, and Tomography, Birkhäuser, 2003.
[22] R. Young, *An Introduction to Nonharmonic Fourier Series*, Academic Press, New York, 1980.

Department of Mathematics, University of Missouri, Columbia, Missouri 65211 USA
E-mail address: pete@math.missouri.edu

Institute of Mathematics, University of Paderborn, 33095 Paderborn, Germany
E-mail address: gittak@uni-paderborn.de

Contemporary Mathematics
Volume **345**, 2004

The local trace function for super-wavelets

Dorin Ervin Dutkay

ABSTRACT. We define an affine structure on $L^2(\mathbb{R})\oplus\cdots\oplus L^2(\mathbb{R})$ and, following some ideas developed in [**Dut1**], we construct a local trace function for this situation. This trace function is a complete invariant for a shift invariant subspace and it has a variety of properties which make it easily computable. The local trace is then used to give a characterization of super-wavelets and to analyze their multiplicity function, dimension function and spectral function. The "$n\times$" oversampling result of Chui and Shi [**CS**] is refined to produce super-wavelets.

CONTENTS

1. Introduction

The wavelet theory involves the study of an affine structure existent on a Hilbert space H, given by two unitary operators, T the translation, and U the dilation that satisfy the commutation relation:

$$UTU^{-1} = T^N,$$

where $N \geq 2$ is an integer called the scale. For the classical wavelet theory, the Hilbert space is $L^2(\mathbb{R})$, the translation is

$$T_0 f(x) = f(x-1), \quad (x \in \mathbb{R},\ f \in L^2(\mathbb{R})),$$

and the dilation operator is

$$U_0 f(x) = \frac{1}{\sqrt{N}} f\left(\frac{x}{N}\right), \quad (x \in \mathbb{R},\ f \in L^2(\mathbb{R})),$$

where $N \geq 2$ is an integer.

Consider an affine structure U, T on a Hilbert space H. A wavelet is a set $\Psi = \{\psi_1, \ldots, \psi_L\}$ such that the affine system

$$\{ U^m T^k \psi \,|\, m, k \in \mathbb{Z},\ \psi \in \Psi \}$$

is an orthonormal basis for H. Often the requirement is weaker, and one looks for the affine system to be only a frame. We recall that a family $(e_i)_{i\in I}$ of vectors in a Hilbert space H is called a frame with constants $A > 0$ and $B > 0$ if

$$A\|f\|^2 \leq \sum_{i\in I} |\langle f \,|\, e_i \rangle|^2 \leq B\|f\|^2, \quad (f \in H).$$

When only the second inequality is satisfied, the family is called Bessel. When $A = B = 1$, the frame is called a normalized tight frame.

Each orthonormal wavelet has a generalized multiresolution attached to it.

A generalized multiresolution analysis (GMRA) is a collection $(V_j)_{j\in\mathbb{Z}}$ of subspaces of H that satisfy:

(i) $V_j \subset V_{j+1}$ for all j.
(ii) $UV_j = V_{j-1}$ for all j.
(iii) $\bigcup V_j$ is dense in H and $\bigcap V_j = \{0\}$.
(iv) V_0 is invariant under T_k for all $k \in \mathbb{Z}$.

If, in addition, there is a vector $\varphi \in V_0$ such that $\{ T^k\varphi \,|\, k \in \mathbb{Z} \}$ is an orthonormal basis for V_0, then $(V_j)_j$ is called a multiresolution analysis (MRA) and φ is called a scaling vector for this MRA. For more information on GMRA's we refer to [**BMM**].

The GMRA associated to a wavelet Ψ is defined by

$$V_j := \{ U^m T^k \psi \,|\, m > -j,\ k \in \mathbb{Z} \}, \quad (j \in \mathbb{Z}).$$

If the GMRA associated to Ψ is actually a MRA, then Ψ is called a MRA wavelet.

As it is shown in [**HL**] and [**Dut2**], it is possible to construct an affine structure and wavelets on the larger Hilbert space $L^2(\mathbb{R}) \oplus \cdots \oplus L^2(\mathbb{R})$, i.e., "super-wavelets". One way to do this is by considering the operators:

$$T := T_0 \oplus \cdots \oplus T_0, \quad U := U_0 \oplus \cdots \oplus U_0,$$

and as shown in [**HL**], wavelets can be constructed for this affine structure, but none of them are MRA wavelets.

The affine structure described in [**Dut2**] takes into consideration the cycles of the map $z \mapsto z^N$. A set $\{z_1, \ldots, z_p\}$ of points on the unit circle $\mathbb{T} := \{ z \in \mathbb{C} \,|\, |z| = 1 \}$ is called a cycle if the points are distinct and $z_1^N = z_2, \ldots, z_{p-1}^N = z_p, z_p^N = z_1$. Then the translation is defined by

$$T(f_1, \ldots, f_p) = (z_1 T_0 f_1, \ldots, z_p T_0 f_p), \quad (f_1, \ldots, f_p \in L^2(\mathbb{R})),$$

and the dilation is defined by

$$U(f_1, \ldots, f_p) = (U_0 f_2, U_0 f_3, \ldots, U_0 f_p, U_0 f_1), \quad (f_1, \ldots, f_p \in L^2(\mathbb{R})).$$

Note that the dilation operator permutes cyclically the components $f_1, \ldots, f_p$. Then one can do finite direct sums of these representations for different cycles to obtain new affine structures.

The advantage of the second type of affine structure is that it possesses MRA-wavelets (see [**Dut2**]).

The objective of this paper is to analyze these affine structures (see section 2 for the definition) and their "super-wavelets", their characterization, their dimension function, multiplicity function and spectral function, the relation between super-wavelets and super-scaling functions. We follow the ideas introduced in [**Dut1**], and the main tool will be the local trace function.

The subspaces of $L^2(\mathbb{R}) \oplus \cdots \oplus L^2(\mathbb{R})$ which are invariant under all integer translations T^k are called shift invariant. The local trace function is a complex valued map defined on $\mathbb{R}$ which is associated to a shift invariant space and an operator on $l^2(\mathbb{Z}) \oplus \cdots \oplus l^2(\mathbb{Z})$ (definition 4.1). Varying the operator, we obtain a lot of information about the shift invariant subspace. Taking the operator to be the identity yields the dimension function (definition 6.7), the spectral function is obtained by taking the operator to be a certain canonical one-dimensional projection (definition 6.3).

The central results are theorem 4.4 and 4.5. Their power is twofold: they give a formula for computing the local trace in terms of normalized tight frame generators and also they show that this formula is independent of the choice of the normalized tight frame generator. Particular instances of these theorems give fundamental results about wavelets and shift invariant subspaces:

(i) Take the shift invariant subspace to be $L^2(\mathbb{R}) \oplus \cdots \oplus L^2(\mathbb{R})$ and the consequence is the theorem that characterizes normalized tight frames generated by translations: theorem 4.9. This in turn gives a characterization result for super-wavelets, theorem 5.4.

(ii) Take the shift invariant subspace to be V_0, the core space of the GMRA of a wavelet and take the operator to be I, then the result is the equality between the dimension function and the multiplicity function (proposition 6.10).

(iii) For V_0, take the operator to be a canonical one-dimensional projection and the result is the Gripenberg-Weiss formula of corollary 6.6.

Also, using the spectral function we are able to give a lower bound estimate on the dimension function (corollary 6.13) which shows not only that some of our affine structures do not have MRA wavelets, but also we can give a lower bound on the number of scaling functions any such wavelet needs.

Another interesting application, which follows from the characterization theorem 5.4, is a refinement of an oversampling result of Chui and Shi [**CS**], which states that when ψ is a NTF wavelet for $N = 2$, then $\eta := (1/p)\psi(\cdot/p)$ is also a NTF wavelet, p being an odd number. We prove in theorem 5.8 that much more is true: η is part of a super-wavelet $\vec{\eta} := (\eta, \ldots, \eta)$ in $L^2(\mathbb{R})^p$ and there is also a converse: if $\vec{\eta}$ is a NTF super-wavelet then ψ is a wavelet. Moreover, going from the wavelet ψ to the super-wavelet $\vec{\eta}$ preserves the orthogonality.

Some notations:

$\mathbb{T}$ is the unit circle in $\mathbb{C}$. We will often identify it with the interval $[-\pi, \pi)$ and the functions on $\mathbb{T}$ with functions on $[-\pi, \pi)$ and with 2π-periodic functions on $\mathbb{R}$. The identification is done via $z = e^{-i\theta}$.

The Fourier transform is given by

$$\widehat{f}(\xi) = \frac{1}{\sqrt{2\pi}} \int_{\mathbb{R}} f(x) e^{-i\xi x}\, dx, \quad (\xi \in \mathbb{R}).$$

If T in an operator on $L^2(\mathbb{R})$ we denote by $\widehat{T}$ its conjugate by the Fourier transform $\widehat{T}\widehat{f} = \widehat{Tf}$, for $f \in L^2(\mathbb{R})$. On $L^2(\mathbb{R}) \oplus \cdots \oplus L^2(\mathbb{R})$ the Fourier transform is done componentwise.

For $f \in L^{\infty}(\mathbb{T})$, denote by $\pi_0(f)$ the operator on $L^2(\mathbb{R})$ defined by

$$\widehat{\pi_0(f)\varphi} = f\varphi, \quad (f \in L^{\infty}(\mathbb{T}));$$

note that $T_0 = \pi_0(z)$ (here z indicates the identity function on $\mathbb{T}$).

2. Permutative wavelet representations

We define here the affine structure on $L^2(\mathbb{R}) \oplus \cdots \oplus L^2(\mathbb{R})$ that we will work with. This structure contains the ones used in [**HL**] and [**Dut2**] which we mentioned in the introduction.

Let σ be a permutation of the set $\{1, \ldots, n\}$ and $z_k = e^{-i\theta_k} \in \mathbb{T}$, $\theta_k \in [-\pi, \pi)$, with the property that $z_k^N = z_{\sigma(k)}$ for all $k \in \{1, \ldots, n\}$. We denote by $Z := (z_1, \ldots, z_n)$.

We define a wavelet representation on the Hilbert space

$$L^2(\mathbb{R})^n = \underbrace{L^2(\mathbb{R}) \oplus \cdots \oplus L^2(\mathbb{R})}_{n \text{ times}}.$$

For $f \in L^{\infty}(\mathbb{T})$ define the operator $\pi_{\sigma,Z}(f)$ on $L^2(\mathbb{R})^n$ by:

$$\pi_{\sigma,Z}(f)(\varphi_1, \ldots, \varphi_n) = (\pi_0(f(zz_1))\varphi_1, \ldots, \pi_0(f(zz_n))\varphi_n),$$

for $\varphi_i \in L^2(\mathbb{R})$, $(i \in \{1, \ldots, n\})$, which means that the Fourier transform of this operator is

$$\widehat{\pi}_{\sigma,Z}(f)(\varphi_1, \ldots, \varphi_n) = (f(\cdot + \theta_1)\varphi_1, \ldots, f(\cdot + \theta_n)\varphi_n).$$

Define the dilation $U_{\sigma,Z}$ on $L^2(\mathbb{R})^n$ by

$$U_{\sigma,Z}(\varphi_1, \ldots, \varphi_n) = (U_0\varphi_{\sigma(1)}, \ldots, U_0\varphi_{\sigma(n)}).$$

Then $(L^2(\mathbb{R})^n, \pi_{\sigma,Z}, U_{\sigma,Z})$ is a normal wavelet representation, i.e., it satisfies the following conditions:

(i) $U_{\sigma,Z}$ is unitary;
(ii) $\pi_{\sigma,Z}$ is a unital representation of the C^*-algebra $L^{\infty}(\mathbb{T})$;
(iii) $U_{\sigma,Z}\pi_{\sigma,Z}(f)U_{\sigma,Z}^{-1} = \pi_{\sigma,Z}(f(z^N))$ for all $f \in L^{\infty}(\mathbb{T})$;
(iv) For every uniformly bounded sequence $(f_n)_{n\in\mathbb{N}}$ in $L^{\infty}(\mathbb{T})$ which converges a.e to a function $f \in L^{\infty}(\mathbb{T})$, one has that $\pi_{\sigma,Z}(f_n)$ converges to $\pi_{\sigma,Z}(f)$ in the strong operator topology.

Checking these properties requires just some simple computations. The translation (or shift) of this wavelet representation is

$$T_{\sigma,Z} := \pi_{\sigma,Z}(z),$$

where z indicates the identity function on $\mathbb{T}$, $z \mapsto z$. Also note that $T_{\sigma,Z}^n = \pi_{\sigma,Z}(z^n)$.

REMARK 2.1. Each permutation σ can be decomposed into a finite product of disjoint cycles. Since $z_i^N = z_{\sigma(i)}$ for all i, this implies that all z_i have a finite orbit under the map $z \mapsto z^N$, i.e., they are cycles, as defined in the introduction.

It may be that the length of the cycle given by z_i is shorter then the length of the corresponding cycle of σ (for example, when all $z_i = 1$ and σ is a product of disjoint transpositions, then the length of the cycle of z_i is 1, but the cycles of σ have length 2). However when these lengths coincide we obtain the representations defined in [**Dut2**].

When $z_i = 1$ for all i and σ is the identity we have the amplification of the standard representation on $L^2(\mathbb{R})$, amplification which was studied in [**HL**].

For the rest of the paper we will consider a fixed permutation σ and the points $z_1, z_2, \ldots, z_n$, and we will omit the subscripts.

3. Shift invariant subspaces

In this section we present some structural theorem for subspaces of $L^2(\mathbb{R})^n$ which are invariant under the integer translations T^k.

DEFINITION 3.1. A subspace V of $L^2(\mathbb{R})^n$ is called shift invariant (SI) if $T^k V \subset V$ for all $k \in \mathbb{Z}$.

If $\mathcal{A}$ is a subset of $L^2(\mathbb{R})^n$, we denote by $S(\mathcal{A})$ the shift invariant subspace generated by $\mathcal{A}$.

DEFINITION 3.2. Let V be a shift invariant subspace of $L^2(\mathbb{R})^n$. A subset Φ of V is called a normalized tight frame generator (or NTF generator) for V if

$$\{ T_k \varphi \mid k \in \mathbb{Z},\ \varphi \in \Phi \}$$

is a NTF for V.

We use also the notation $S(\varphi) := S(\{\varphi\})$. φ is called a quasi-orthogonal generator for $S(\varphi)$ if

$$\{ T_k \varphi \mid k \in \mathbb{Z} \}$$

is a NTF for $S(\varphi)$ and for all $\xi \in \mathbb{R}$,

$$\operatorname{Per} |\widehat{\varphi}|^2(\xi) := \sum_{k \in \mathbb{Z}} |\widehat{\varphi}|^2(\xi + 2k\pi) \in \{0, 1\}.$$

(actually, the second condition is a consequence of the first but we include it anyway).

We will need the "fiberization" techniques based on the range function of Helson [**H**], which were used in the classical case in the work of de Boor, DeVore and Ron [**BDR**] and further developed by Ron and Shen [**RS1**]-[**RS4**], Bownik [**Bo1**] and others.

For $\varphi = (\varphi_1, \ldots, \varphi_n) \in L^2(\mathbb{R})^n$, the map $\mathcal{T}_{per}\varphi$ assigns to each point $\xi \in \mathbb{R}$ a vector in $l^2(\mathbb{Z})^n$, called the fiber of φ at ξ:

$$\begin{aligned}\mathcal{T}_{per}&(\varphi_1, \ldots, \varphi_n)(\xi) \\ &= \left((\widehat{\varphi}_1(\xi - \theta_1 + 2k\pi))_{k \in \mathbb{Z}}, \ldots, (\widehat{\varphi}_n(\xi - \theta_n + 2k\pi))_{k \in \mathbb{Z}} \right), \quad (\xi \in \mathbb{R}),\end{aligned}$$

for all $(\varphi_1, \ldots, \varphi_n) \in L^2(\mathbb{R})^n$.

This map transforms the translations into multiplications by scalars in each fiber:

PROPOSITION 3.3. *For all $f \in L^\infty(\mathbb{T})$ and $(\varphi_1, \ldots, \varphi_n) \in \mathbb{R}$,*

$$\mathcal{T}_{per}(\pi(f)(\varphi_1, \ldots, \varphi_n))(\xi) = f(\xi)\mathcal{T}_{per}(\varphi_1, \ldots, \varphi_n)(\xi), \quad (\xi \in \mathbb{R}).$$

In particular, for all $k \in \mathbb{Z}$,

$$\mathcal{T}_{per}(T^k(\varphi_1, \ldots, \varphi_n)(\xi) = e^{-ik\xi}\mathcal{T}_{per}(\varphi_1, \ldots, \varphi_n)(\xi), \quad (\xi \in \mathbb{R}).$$

Note also the periodicity property of $\mathcal{T}_{per}$ (which justifies the subscript "per"):

$$\mathcal{T}_{per}\varphi(\xi + 2s\pi) = \lambda(s)^*(\mathcal{T}_{per}f(\xi)), \quad (\xi \in \mathbb{R},\ s \in \mathbb{Z}),$$

where, for $s \in \mathbb{Z}$, $\lambda(s)$ is the shift on $l^2(\mathbb{Z})^n$:

$$(\lambda(s)\alpha)(k,i) = \alpha(k-s,i), \quad (k \in \mathbb{Z},\ i \in \{1,\dots,N\}).$$

For a shift invariant subspace the fibers $\mathcal{T}_{per}\varphi(\xi)$ at a fixed point ξ form a subspace of $l^2(\mathbb{Z})^n$, which will be denoted by $J_{per}(\xi)$. Thus J_{per} will be a bundle map which assigns to each point $\xi \in \mathbb{R}$ a subspace of $l^2(\mathbb{Z})^n$. This is the idea behind the range function.

DEFINITION 3.4. A range function is a measurable mapping

$$J : [-\pi, \pi) \to \{\text{closed subspaces of } l^2(\mathbb{Z})^n\}.$$

Measurable means weakly operator measurable, i.e., $\xi \mapsto \langle P_{J(\xi)}a \,|\, b \rangle$ is measurable for any choice of vectors $a, b \in l^2(\mathbb{Z})^n$.

A periodic range function is a measurable function

$$J_{per} : \mathbb{R} \to \{\text{closed subspaces of } l^2(\mathbb{Z})^n\},$$

with the periodicity property:

$$J_{per}(\xi + 2k\pi) = \lambda(k)^*(J_{per}(\xi)), \quad (k \in \mathbb{Z},\ \xi \in \mathbb{R}).$$

Sometimes we will use the same letter to denote the subspace $J_{per}(\xi)$ and the projection onto $J_{per}(\xi)$. In terms of projections, the periodicity can be written as:

$$J_{per}(\xi + 2k\pi) = \lambda(k)^* J_{per}(\xi)\lambda(k), \quad (k \in \mathbb{Z},\ \xi \in \mathbb{R}).$$

There is a one-to-one correspondence between shift invariant subspaces and range functions. This correspondence is made precise in the next theorem due to Helson [**H**]. The proof given in [**Bo1**], proposition 1.5, carries over here, without any significant modification.

THEOREM 3.5. *A closed subspace V of $L^2(\mathbb{R})^n$ is shift invariant if and only if*

$$V = \{ f \in L^2(\mathbb{R})^n \,|\, \mathcal{T}_{per}f(\xi) \in J_{per}(\xi) \text{ for a.e. } \xi \in \mathbb{R}\},$$

for some measurable periodic range function J_{per}. The correspondence between V and J_{per} is bijective under the convention that range functions are identified if they are equal a.e. Furthermore, if $V = S(\mathcal{A})$ for some countable $\mathcal{A} \subset L^2(\mathbb{R})^n$, then

$$J_{per}(\xi) = \overline{\text{span}}\{\mathcal{T}_{per}\varphi(\xi) \,|\, \varphi \in \mathcal{A}\}, \quad \text{for a.e. } \xi \in \mathbb{R}.$$

Another fundamental fact about the range function is that it transforms the frame property into a local property: more precisely, a family of vectors generate a normalized tight frame by translations if and only if the fibers form normalized tight frames at each point in $\mathbb{R}$. This is given in the next theorem, for a proof look in [**Bo1**], theorem 2.3.

THEOREM 3.6. *Let V be a SI subspace of $L^2(\mathbb{R})^n$, J_{per} its periodic range function and Φ a countable subset of V. $\{T^k\varphi \,|\, k \in \mathbb{Z},\ \varphi \in \Phi\}$ is a frame with constants A and B for V (Bessel family with constant B) if and only if $\{\mathcal{T}_{per}\varphi(\xi) \,|\, \varphi \in \Phi\}$ is a frame with constants A and B for $J_{per}(\xi)$ (Bessel sequence with constant B) for almost every $\xi \in \mathbb{R}$.*

Another approach, with a more representation-theoretic flavor, is proposed in [**BMM**], [**BM**], [**B**]. The group $\mathbb{Z}$ has a unitary representation on a shift invariant space and the abstract harmonic analysis arguments given in the papers by Baggett et al. work here with minor changes and yield the following results.

THEOREM 3.7. *If V_0 is a shift invariant subspace of $L^2(\mathbb{R})^n$ then there exists a multiplicity function $m : \mathbb{T} \to \{0, 1, \ldots, \infty\}$ such that:*

(i) *If*

$$S_j := \{\, \xi \in \mathbb{T} \,|\, m(\xi) \geq j \,\},$$

then there is a unitary

$$\mathcal{J} : V_0 \to \bigoplus_{j=1}^{\infty} L^2(S_j)$$

which intertwines the representations

$$\mathcal{J}T^k = \rho_k \mathcal{J}, \quad (k \in \mathbb{Z}), \tag{3.1}$$

where ρ_k is multiplication by the function $\xi \mapsto e^{-ik\xi}$ on each component:

$$\rho_k(f_1, f_2, \ldots) = (e^{-ik\cdot} f_1, e^{-ik\cdot} f_2, \ldots), \quad (k \in \mathbb{Z}).$$

(ii) *If*

$$\phi_i := \mathcal{J}^{-1}(\chi_{S_i}), \quad (i \in \{1, 2, \ldots\}),$$

(where χ_{S_i} stands for the element of $\oplus L^2(S_j)$ whose only nonzero component is χ_{S_i} in $L^2(S_i)$), then

$$\{\, \phi_i \,|\, i \in \{1, 2, \ldots\} \,\} \quad \text{is a NTF generator for } V_0, \tag{3.2}$$

$$T^k \phi_i \perp T_l \phi_j \quad (i \neq j, k, l \in \mathbb{Z}), \tag{3.3}$$

and

$$\langle\, \mathcal{T}_{per}\phi_i(\xi) \,|\, \mathcal{T}_{per}\phi_j(\xi) \,\rangle = \begin{cases} \chi_{S_i} & \text{if } i = j, \\ 0 & \text{if } i \neq j \end{cases} \quad (\xi \in [-\pi, \pi)). \tag{3.4}$$

(iii) *V_0 can be decomposed as the orthogonal sum of the shift invariant subspaces generated by the quasi-orthogonal generators ϕ_i:*

$$V_0 = \bigoplus_{i=1}^{\infty} S(\phi_i). \tag{3.5}$$

PROOF. We follow the arguments given in [**BMM**] which use the spectral theory of Stone and Mackey for the commutative group $\mathbb{Z}$ that has as dual $\widehat{\mathbb{Z}} = \mathbb{T}$.

For the representation of $\mathbb{Z}$ by translations T^k on V_0, there is a unique projection-valued measure p on $\mathbb{T}$ for which

$$T^k|_{V_0} = \int_{\mathbb{T}} e^{-ik\xi}\, dp(\xi), \quad (k \in \mathbb{Z}).$$

The projection-valued measure p is completely determined by a measure class $[\mu]$ on $\mathbb{T}$ and a measurable multiplicity function $m : \mathbb{T} \to \{0, 1, \ldots, \infty\}$.

As in [**BMM**] proposition 1.2, we will show that the measure class $[\mu]$ is absolutely continuous with respect to the Haar measure on $\mathbb{T}$.

Indeed, the Fourier transform establishes an equivalence between the translation T on $L^2(\mathbb{R})^n$ and its Fourier version

$$\widehat{T}^k(f_1,\dots,f_n) = (e^{-ik(\cdot\,+\theta_i)}f_i)_{i\in\{1,\dots,n\}}, \quad (k\in\mathbb{Z}).$$

Define $W : L^2(\mathbb{R})^n \to L^2(\mathbb{R})^n$ by

$$W(f_1,\dots,f_n) = (f_i(\cdot - \theta_i))_{i\in\{1,\dots,n\}}.$$

This intertwines the given representation with the one defined by

$$\tilde{T}^k(f_1,\dots,f_n) = (e^{-ik\cdot}f_i)_{i\in\{1,\dots,n\}}.$$

But $\mathbb{R}$ can be decomposed as

$$\mathbb{R} = \bigcup_{k\in\mathbb{Z}}[(2k-1)\pi,(2k+1)\pi),$$

thus, the representation of $\mathbb{Z}$ on the entire space $L^2(\mathbb{R})^n$ is equivalent to n times an infinite multiple of the regular representation of $\mathbb{Z}$ (which is multiplication by $e^{-ik\xi}$ on $L^2(\mathbb{T})$.

Therefore, the projection-valued measure associated to the representation of $\mathbb{Z}$ on the entire space $L^2(\mathbb{R})^n$ is equivalent to the Haar measure on $\mathbb{T}$, so the projection-valued measure associated to any subrepresentation of this representation must have a measure which is absolutely continuous w.r.t. the Haar measure.

For (3.1), (3.2) and (3.3) the argument is the one used on [**BM**] lemma 1.1.

(3.4) can be proved as in lemma 1.2 from [**BM**]. It follows from the fact that $\mathcal{J}$ is unitary and intertwining so, for $k\in\mathbb{Z}$, $i,j\in\{1,2,\dots\}$,

$$\langle T^k\phi_i \,|\, \phi_j\rangle = \int_{\mathbb{T}} e^{-ik\xi}\chi_{S_i}\delta_{ij}\,d\xi,$$

which implies

$$\int_{\mathbb{T}} e^{-ik\xi}\,\langle \mathcal{T}_{per}\phi_i(\xi) \,|\, \mathcal{T}_{per}\phi_j(\xi)\rangle\,d\xi = \int_{\mathbb{T}} e^{-ik\xi}\chi_{S_i}\delta_{ij}\,d\xi.$$

(3.5) is just a rephrasing of the previous statements. □

4. The local trace function

We are able now to define the local trace function. The proofs from [**Dut1**] work here in most of the instances and, when a result is presented without a proof, the reader can look in [**Dut1**].

Each shift invariant subspace V, can be seen as a bundle of vector subspaces of $l^2(\mathbb{Z})^n$ by means of a range function J_{per}. The local trace function associated to V and an operator T on $l^2(\mathbb{Z})^n$ is the function which associates to each point ξ the value of the canonical trace of the restriction of T to the subspace $J_{per}(\xi)$. We recall that the trace of a positive or trace-class operator T on some Hilbert space H, can be computed by

$$\mathrm{Trace}(T) = \sum_{i\in I}\langle Tf_i \,|\, f_i\rangle,$$

where $(f_i)_{i\in I}$ is any normalized tight frame of the Hilbert space H (see [**Dut1**]).

DEFINITION 4.1. Let V be a SI subspace of $L^2(\mathbb{R})^n$, T a positive (or trace-class) operator on $l^2(\mathbb{Z})^n$ and let J_{per} be the range function associated to V. We define the local trace function associated to V and T as the map from $\mathbb{R}$ to $[0,\infty]$ (or $\mathbb{C}$) given by the formula

$$\tau_{V,T}(\xi) = \text{Trace}\,(TJ_{per}(\xi)), \quad (\xi \in \mathbb{R}).$$

We define the restricted local trace function associated to V and a vector f in $l^2(\mathbb{Z})^n$ by

$$\tau_{V,f}(\xi) = \text{Trace}\,(P_f J_{per}(\xi))\,(= \tau_{V,P_f}(\xi)), \quad (\xi \in \mathbb{R}),$$

where P_f is the operator on $l^2(\mathbb{Z})^n$ defined by $P_f(v) = \langle\, v \mid f \,\rangle\, f$.

REMARK 4.2. Even though, $TJ_{per}(\xi)$ is not necessarily positive when T is, when computing the trace, we can consider instead of $TJ_{per}(\xi)$ the operator $J_{per}(\xi)TJ_{per}(\xi)$, or we can use normalized tight frames just for the subspace $J_{per}(\xi)$ (not for the entire $l^2(\mathbb{Z})^n$), in any case the expression Trace$(TJ_{per}(\xi))$ makes sense and is a positive number.

When T is trace-class, since the trace class operators form an ideal, $TJ_{per}(\xi)$ is also trace class so Trace$(TJ_{per}(\xi))$

Here are some ways to compute the local trace function.

PROPOSITION 4.3. *For all* $f \in l^2(\mathbb{Z})^n$,

$$\tau_{V,f}(\xi) = \|J_{per}(\xi)(f)\|^2, \quad (\text{ for a.e. } \xi \in \mathbb{R}).$$

The next result is central for the paper. As we have mentioned in the introduction, this theorem can be used for two purposes: it gives a method to compute the local trace function and secondly, and maybe even more important, it shows that the formula on the right does not depend on the choice of the normalized tight frame. So one can look at the local trace function as an invariant for the SI subspace, and the theorem is an index theorem.

THEOREM 4.4. *Let V be a SI subspace of $L^2(\mathbb{R})^n$ and $\Phi \subset V$ a NTF generator for V. Then for every positive (or trace-class) operator T on $l^2(\mathbb{Z})^n$ and any $f \in l^2(\mathbb{Z})^n$,*

$$\tau_{V,T}(\xi) = \sum_{\varphi\in\Phi} \langle\, T\mathcal{T}_{per}\varphi(\xi) \mid \mathcal{T}_{per}\varphi(\xi) \,\rangle, \quad (\text{ for a.e. } \xi \in \mathbb{R}); \tag{4.1}$$

$$\tau_{V,f}(\xi) = \sum_{\varphi\in\Phi} |\langle\, f \mid \mathcal{T}_{per}\varphi(\xi) \,\rangle|^2, \quad \text{for a.e. } \xi \in \mathbb{R}). \tag{4.2}$$

The converse is also true: if a family of vectors can be used to compute the local trace then they form a NTF generator for the SI space. Moreover it is sufficient that this is satisfied only for some very particular vectors (see (iii) in the next result):

THEOREM 4.5. *Let V be a SI subspace of $L^2(\mathbb{R})^n$, J_{per} its periodic range function and Φ a countable subset of $L^2(\mathbb{R})^n$. Then following affirmations are equivalent:*

(i) *$\phi \subset V$ and ϕ is a NTF generator for V;*
(ii) *For every $f \in l^2(\mathbb{Z})^n$*

$$\sum_{\varphi\in\Phi} |\langle\, f \mid \mathcal{T}_{per}\varphi(\xi) \,\rangle|^2 = \|J_{per}(\xi)(f)\|^2, \quad \text{for a.e. } \xi \in \mathbb{R} \tag{4.3}$$

(iii) *For every* $0 \neq k \in \mathbb{Z}$, $l, j \in \{1, \dots, n\}$ *and* $\alpha \in \{0, 1, i = \sqrt{-1}\}$,

$$\sum_{\varphi \in \Phi} |\widehat{\varphi}_l(\xi) + \overline{\alpha}\widehat{\varphi}_j(\xi + 2k\pi)|^2 = \|J_{per}(\xi)(\delta_{0l} + \alpha\delta_{kj})\|^2, \quad \text{for a.e. } \xi \in \mathbb{R}. \tag{4.4}$$

The local trace function is related in a very simple way to another important invariant, the dual Gramian. This was introduced by A. Ron and Z. Shen and successfully used for the analysis of the structure of SI spaces and of frames generated by translations in [**RS1**]-[**RS4**].

DEFINITION 4.6. Let V be a shift invariant subspace of $L^2(\mathbb{R})^n$ and Φ, a NTF generator for V. The dual Gramian of Φ is the function $\tilde{G}^{\Phi}$ which assigns to each point $\xi \in \mathbb{R}$ the matrix

$$\tilde{G}^{\Phi}_{ki,lj}(\xi) = \sum_{\varphi \in \Phi} \widehat{\varphi}_i(\xi - \theta_i + 2k\pi)\overline{\widehat{\varphi}}_j(\xi - \theta_j + 2l\pi), \quad (k, l \in \mathbb{Z},\ i, j \in \{1, \dots, n\}).$$

We can recuperate the Gramian from the local trace function just by computing it at some rank-one operators.

PROPOSITION 4.7. *If V is a SI space, J_{per} its range function, and Φ a NTF generator for it then, for all $k, l \in \mathbb{Z}$, $i, j \in \{1, \dots, n\}$,*

$$\tilde{G}^{\Phi}_{ki,lj}(\xi) = \tau_{V, P_{ki,lj}}(\xi) = \langle J_{per}(\xi)\delta_{lj} \,|\, J_{per}(\xi)\delta_{ki} \rangle, \quad (\xi \in \mathbb{R}),$$

where $P_{ki,lj}$ is the rank-one operator defined by

$$P_{ki,lj}v = \langle v \,|\, \delta_{ki} \rangle \, \delta_{lj}, \quad (v \in l^2(\mathbb{Z})^n).$$

PROOF. The first part is a consequence of theorem 4.4. For the last equality we use the following argument: if $(f_q)_{q \in Q}$ is an orthonormal basis for $J_{per}(\xi)$ then

$$\begin{aligned}
\mathrm{Trace}(P_{ki,lj}J_{per}(\xi)) &= \sum_{q \in Q} \langle P_{ki,lj} f_q \,|\, f_q \rangle \\
&= \sum_{q \in Q} \langle f_q \,|\, \delta_{ki} \rangle \langle \delta_{lj} \,|\, f_q \rangle \\
&= \sum_{q \in Q} \langle f_q \,|\, J_{per}(\xi)\delta_{ki} \rangle \langle J_{per}(\xi)\delta_{lj} \,|\, f_q \rangle \\
&= \langle J_{per}(\xi)\delta_{lj} \,|\, J_{per}(\xi)\delta_{ki} \rangle .
\end{aligned}$$

□

Having these it is clear that the dual Gramian is an invariant and in fact it is a complete one (see [**Dut1**], theorem 4.6 for the proof):

THEOREM 4.8. *Let V be a SI subspace of $L^2(\mathbb{R})^n$, Φ_1 a NTF generator for V and Φ_2 a countable family of vectors from $L^2(\mathbb{R})^n$. The following affirmations are equivalent:*

(i) *$\Phi_2 \subset V$ and Φ_2 is a NTF generator for V;*
(ii) *The dual Gramians $\tilde{G}^{\Phi_1}$ and $\tilde{G}^{\Phi_2}$ are equal almost everywhere.*

If we particularize this result to the case when the shift invariant subspace is $L^2(\mathbb{R})^n$, then since the range function for it is constant $l^2(\mathbb{Z})^n$ at each point, we obtain the following characterization of families that generate normalized tight frames by translations:

THEOREM 4.9. *Let Φ be a countable subset of $L^2(\mathbb{R})^n$. The following affirmations are equivalent:*

(i) *Φ is a NTF generator for $L^2(\mathbb{R})^n$;*

(ii)

$$\sum_{\varphi\in\Phi}\widehat{\varphi}_i(\xi-\theta_i)\overline{\widehat{\varphi}}_j(\xi-\theta_j+2k\pi)=\delta_{ij}\delta_k,\quad (\xi\in\mathbb{R},\ i\in\{1,\dots,n\},\ k\in\mathbb{Z}).$$

Next, we present some elementary properties of the local trace function. The proofs of these results are simple and can be found in [**Dut1**] for the classical case.

PROPOSITION 4.10. [**Periodicity**] *Let V be a SI subspace, T a positive or trace-class operator on $l^2(\mathbb{Z})^n$, $f\in l^2(\mathbb{Z})^n$. Then, for $k\in\mathbb{Z}$*

$$\tau_{V,T}(\xi+2k\pi)=\tau_{V,\lambda(k)T\lambda(k)^*}(\xi),\quad (\text{ for a.e. } \xi\in\mathbb{R}); \tag{4.5}$$

$$\tau_{V,f}(\xi+2k\pi)=\tau_{V,\lambda(k)f}(\xi),\quad (\text{ for a.e. } \in\mathbb{R}), \tag{4.6}$$

where $\lambda(k)$ is the shift on $l^2(\mathbb{Z})^n$, i.e., the unitary operator on $l^2(\mathbb{Z})^n$ defined by $(\lambda(k)\alpha)(l,i)=\alpha(l-k,i)$ for all $l\in\mathbb{Z}$, $i\in\{1,\dots,n\}$, $\alpha\in l^2(\mathbb{Z})^n$.

PROPOSITION 4.11. [**Additivity**] *Suppose $(V_i)_{i\in I}$ are mutually orthogonal SI subspaces (I countable) and let $V=\oplus_{i\in I}V_i$. Then, for every positive or trace-class operator T on $l^2(\mathbb{Z})^n$ and every $f\in l^2(\mathbb{Z})^n$:*

$$\tau_{V,T}=\sum_{i\in I}\tau_{V_i,T},\quad a.e.\ on\ \mathbb{R}; \tag{4.7}$$

$$\tau_{V,f}=\sum_{i\in I}\tau_{V_i,f},\quad a.e.\ on\ \mathbb{R}. \tag{4.8}$$

PROPOSITION 4.12. [**Monotony and injectivity**] *Let V,W be SI subspaces.*

(i) *$V\subset W$ iff $\tau_{V,T}\le\tau_{W,T}$ a.e. for all positive operators T iff $\tau_{V,f}\le\tau_{W,f}$ a.e. for all $f\in l^2(\mathbb{Z})^n$.*

(ii) *$V=W$ iff $\tau_{V,f}=\tau_{W,f}$ a.e for all $f\in l^2(\mathbb{Z})^n$.*

THEOREM 4.13. *Let $(V_j)_{j\in\mathbb{N}}$ be an increasing sequence of SI subspaces of $L^2(\mathbb{R})^n$,*

$$V:=\overline{\bigcup_{j\in\mathbb{N}}V_j},$$

T a positive (trace-class) operator on $l^2(\mathbb{Z})^n$ and $f\in l^2(\mathbb{Z})^n$. Then, for a.e. $\xi\in\mathbb{R}$, $\tau_{V_j,T}(\xi)$ increases (converges) to $\tau_{V,T}(\xi)$ and $\tau_{V_j,f}(\xi)$ increases to $\tau_{V,f}(\xi)$.

We will use the local trace function for the analysis of wavelets. For this reason we have to see how it behaves under dilation.

THEOREM 4.14. *Let V be a shift invariant subspace of $L^2(\mathbb{R})^n$ and T a positive or trace-class operator on $l^2(\mathbb{Z})^n$. For $l\in\{0,\dots,N-1\}$ define the operators D_l on $l^2(\mathbb{Z})^n$ by*

$$D_lv(k,i)=\begin{cases}v(s,i) & \text{if}\quad 2k\pi=2l\pi+\theta_i-N\theta_{\sigma^{-1}(i)}+2Ns\pi,\ \text{with } s\in\mathbb{Z},\\ 0 & \text{otherwise,}\end{cases}$$

and the unitary operator $\mathcal{S}$ on $l^2(\mathbb{Z})^n$, by

$$\mathcal{S}(v_1,\dots,v_n)=(v_{\sigma^{-1}(1)},\dots,v_{\sigma^{-1}(n)}),\quad ((v_1,\dots,v_n)\in l^2(\mathbb{Z})^n).$$

Then $U^{-1}V$ is shift invariant and its local trace function is given by

$$\tau_{U^{-1}V,T}(\xi) = \sum_{l=0}^{N-1} \tau_{V,\mathcal{S}^* D_l^* T D_l \mathcal{S}}\left(\frac{\xi + 2l\pi}{N}\right), \quad (\xi \in \mathbb{R}).$$

PROOF. By theorem 3.7, V can be decomposed into an orthogonal sum of singly-generated SI spaces. Also the local trace is additive and U is unitary so we may assume that $V = S(\varphi)$ with φ a quasi-orthogonal generator.

Due to the commutation relation $UTU^{-1} = T^N$, the vectors

$$\{\, \varphi_{-1,r} := U^{-1}T^r\varphi \,|\, r \in \{0, \dots, N-1\}\,\}$$

form a NTF generator for $U^{-1}V$. Their Fourier transforms are

$$\widehat{\varphi}_{-1,r}(\xi) = \left(\frac{1}{\sqrt{N}} e^{-ir(\xi/N + \theta_{\sigma^{-1}(i)})} \widehat{\varphi}_{\sigma^{-1}(i)}(\xi/N)\right)_{i \in \{1,\dots,n\}}, \quad (\xi \in \mathbb{R}).$$

Then a short computation shows that, if we consider the vectors $v_l(\xi)$, $l \in \{0, \dots, N-1\}$ in $l^2(\mathbb{Z})^n$,

$$v_l(\xi)(k,i) = \begin{cases} \widehat{\varphi}_{\sigma^{-1}(i)}\left(\frac{\xi - \theta_i + 2k\pi}{N}\right) & \text{if } \frac{-\theta_i + 2k\pi}{N} \in \frac{2l\pi}{N} - \theta_{\sigma^{-1}(i)} + 2\pi\mathbb{Z}, \\ 0 & \text{otherwise,} \end{cases}$$

then

$$\mathcal{T}_{per}\varphi_{-1,r}(\xi) = \frac{1}{\sqrt{N}} e^{-ir\frac{\xi}{N}} \sum_{l=0}^{N-1} e^{-ir\frac{2\pi l}{N}} v_l(\xi), \quad (r \in \{0, \dots, N-1\}).$$

The matrix

$$\left(\frac{1}{\sqrt{N}} e^{-ir\frac{2\pi l}{N}}\right)_{l,r \in \{0,\dots,N-1\}}$$

is unitary, so the vectors $\{\, v_l(\xi) \,|\, l \in \{0, \dots, N-1\}\,\}$ span the same subspace of $l^2(\mathbb{Z})^n$ as the vectors $\{\, \mathcal{T}_{per}\varphi_{-1,r}(\xi) \,|\, r \in \{0, \dots, N-1\}\,\}$, namely $J_{U^{-1}V}(\xi)$.

Note that

$$v_l(\xi) = D_l \mathcal{S} \mathcal{T}_{per}\varphi\left(\frac{\xi + 2l\pi}{N}\right), \quad (l \in \{0, \dots, N-1\}),$$

D_l are isometries and $\mathcal{S}$ is unitary. In addition φ is quasi-orthogonal so $\|\mathcal{T}_{per}\varphi(\xi)\| \in \{0, 1\}$ for almost every $\xi \in \mathbb{R}$, hence $\|v_l(\xi)\| \in \{0, 1\}$ for all l. It is also clear that the vectors $v_l(\xi)$ are mutually orthogonal for a fixed ξ. Having all these, it follows that $\{v_l(\xi) \,|\, l \in \{0, \dots, N-1\}\}$ is a NTF for $J_{U^{-1}V}(\xi)$, therefore we can compute the trace with them.

$$\begin{aligned} \tau_{U^{-1}V,T}(\xi) &= \sum_{l=0}^{N-1} \langle\, T v_l(\xi) \,|\, v_l(\xi)\,\rangle \\ &= \sum_{l=0}^{N-1} \left\langle T D_l \mathcal{S} \mathcal{T}_{per}\varphi\left(\frac{\xi + 2l\pi}{N}\right) \,\middle|\, D_l \mathcal{S} \mathcal{T}_{per}\varphi\left(\frac{\xi + 2l\pi}{N}\right)\right\rangle \\ &= \sum_{l=0}^{N-1} \left\langle \mathcal{S}^* D_l^* T D_l \mathcal{S} \mathcal{T}_{per}\varphi\left(\frac{\xi + 2l\pi}{N}\right) \,\middle|\, \mathcal{T}_{per}\varphi\left(\frac{\xi + 2l\pi}{N}\right)\right\rangle, \end{aligned}$$

which proves the formula. □

5. A characterization of super-wavelets

We apply now the local trace function to super-wavelets. Our first goal is to obtain a characterization of NTF super-wavelets. This is done in theorem 5.4.

DEFINITION 5.1. A finite subset $\Psi = \{\psi^1, \dots, \psi^L\}$ of $L^2(\mathbb{R})^n$ is a NTF wavelet if the affine system

$$X(\Psi) := \{\, \psi_{j,k} := U^{-j}T^k\psi \mid j \in \mathbb{Z},\ k \in \mathbb{Z},\ \psi \in \Psi \,\}$$

is a NTF for $L^2(\mathbb{R})^n$.

The quasi-affine system $X^q(\Psi)$ is

$$X^q(\Psi) := \{\, \tilde{\psi}_{j,k} \mid j \in \mathbb{Z},\ k \in \mathbb{Z},\ \psi \in \Psi \,\},$$

with the convention

$$\tilde{\psi}_{j,k} := \begin{cases} U^{-j}T^k\psi & \text{if } j \geq 0,\ k \in \mathbb{Z}, \\ N^{j/2}T^kU^{-j}\psi & \text{if } j < 0,\ k \in \mathbb{Z}. \end{cases}$$

There is an equivalence between affine and quasi-affine frames. This was proved in full generality for the $L^2(\mathbb{R})$-case in [**CSS**]. Again the result will work here with some slight modifications in the proof.

THEOREM 5.2. *Let $\Psi = \{\psi^1, \dots, \psi^L\}$ be a finite subset of $L^2(\mathbb{R})^n$.*

(i) *$X(\Psi)$ is a Bessel family if and only if $X^q(\Psi)$ is a Bessel family. Furthermore, their exact frame bounds are equal.*

(ii) *$X(\Psi)$ is an affine frame with constants A and B if and only if $X^q(\Psi)$ is a quasi-affine frame with constants A and B.*

PROOF. The proof of theorem 2 in [**CSS**] works here word for word once we have the following lemma, which is the analogue of lemma 4 in [**CSS**]:

LEMMA 5.3. *Let $\Psi = \{\psi_1, \dots, \psi_L\} \subset L^2(\mathbb{R})^n$ and let $f \in L^2(\mathbb{R})^n$ be a vector such that $f_1, \dots, f_n$ have all compact supports. Then*

$$\lim_{M \to \infty} \sum_{j<0} \sum_{\psi \in \Psi} \sum_{k \in \mathbb{Z}} \left| \left\langle U^{-M} f \mid \tilde{\psi}_{j,k} \right\rangle \right|^2 = 0,$$

$$\lim_{M \to \infty} N^{-M} \sum_{j \leq -M} \sum_{\nu=0}^{N^M-1} \sum_{\psi \in \Psi} \sum_{k \in \mathbb{Z}} |\langle T^\nu f \mid \psi_{j,k} \rangle|^2 = 0.$$

Lemma 4 from [**CSS**] shows that this is true for the classical wavelet representation on $L^2(\mathbb{R})$ (i.e., $n = 1$, $\sigma =$ identity, $\theta_1 = 0$). We will show that the general case follows from this particular one.

For $M \geq 0$, $j < 0$, $\psi \in \Psi$, $k \in \mathbb{Z}$ we evaluate

$$\begin{aligned} \left| \left\langle U^{-M} f \mid \tilde{\psi}_{j,k} \right\rangle \right|^2 &= \left| \sum_{i=1}^n \int_{\mathbb{R}} U_0^{-M} f_{\sigma^{-M}(i)}(x) N^{j/2}\, \overline{z_i^k T_0^k U_0^{-j} \psi_{\sigma^{-j}(i)}(x)}\, dx \right|^2 \\ &\leq \left| \sum_{i=1}^n \int_{\mathbb{R}} \left| U_0^{-M} f_{\sigma^{-M}(i)}(x) \right| N^{j/2} \left| T_0^k U_0^{-j} \psi_{\sigma^{-j}(i)}(x) \right| dx \right|^2 \\ &\leq \left| \sum_{i=1}^n \int_{\mathbb{R}} U_0^{-M} F_f(x) N^{j/2} T_0^k U_0^{-j} P_\psi(x)\, dx \right|^2, \end{aligned}$$

where

$$F_f := \max\{\, |f_i| \mid i \in \{1, \dots, n\} \,\},\ P_\psi := \max\{\, |\psi_i| \mid i \in \{1, \dots, n\} \,\},$$

and the last term in the previous inequality is actually

$$= n^2 |\left\langle U_0^{-M} F_f \mid (P_\psi)\tilde{}_{j,k} \right\rangle|^2.$$

Apply now lemma 4 of [**CSS**] for $f \leftrightarrow F_f$ and $\Psi \leftrightarrow \{\, P_\psi \mid \psi \in \Psi \,\}$ and with the previous inequality we obtain the first limit.

The second one can be obtain by a similar argument. □

Theorem 5.2 is one of the crucial steps in obtaining a characterization for wavelets. The only thing that remains to do is to use theorem 4.9 for the quasi-affine system and carefully compute the Gramians to obtain the next result which gives the characterizing equation for NTF wavelets in $L^2(\mathbb{R})^n$.

THEOREM 5.4. *Let $\Psi = \{\psi_1, \dots, \psi_p\}$ be a finite subset of $L^2(\mathbb{R})^n$. The following affirmations are equivalent:*

(i) *The affine system $X(\Psi)$ is a NTF for $L^2(\mathbb{R})^n$;*

(ii) *The following equation hold for almost every $\xi \in \mathbb{R}$, $i, j \in \{1, \dots, n\}$: For all indices i, j with $z_i = z_j$:*

$$\sum_{\psi\in\Psi} \sum_{m=-\infty}^{\infty} \widehat{\psi}_{\sigma^m(i)}(N^m\xi)\overline{\widehat{\psi}}_{\sigma^m(j)}(N^m\xi) = \delta_{ij}. \tag{5.1}$$

For all i, j and for all $s \in \mathbb{Z} \setminus N\mathbb{Z}$:

$$\sum_{\psi\in\Psi} \sum_{m\geq 0} \widehat{\psi}_{\sigma^m(i)}(N^m\xi)\overline{\widehat{\psi}}_{\sigma^m(j)}(N^m(\xi + N(\theta_{\sigma^{-1}(i)} - \theta_{\sigma^{-1}(j)}) + 2s\pi)) = 0. \tag{5.2}$$

PROOF. According to theorem 5.2, the affine system $X(\Psi)$ is a normalized tight frame if and only if the quasi-affine $X^q(\Psi)$ is. This can be reformulated as

$$\{\, \tilde{\psi}_{m,0} \mid m \leq 0,\ \psi \in \Psi \,\} \cup \{\, \tilde{\psi}_{m,r} \mid m > 0,\ r \in \{0, \dots, N^m - 1\},\ \psi \in \Psi \,\}$$

is a NTF generator for $L^2(\mathbb{R})^n$. We will use theorem 4.9 for this family.

We have to compute the Fourier transforms:

$$\widehat{\tilde{\psi}}_{m,0}(\xi) = \left(\widehat{\psi}_{\sigma^{-m}(i)}(N^{-m}\xi)\right)_{i\in\{1,\dots,n\}}, \quad (m \leq 0),$$

and

$$\widehat{\tilde{\psi}}_{m,r}(\xi) = N^{-m/2}\left(e^{-i(N^{-m}\xi+\theta_{\sigma^{-m}(i)})r}\widehat{\psi}_{\sigma^{-m}(i)}(N^{-m}\xi)\right)_{i\in\{1,\dots,n\}},$$

when $m > 0, r \in \{0, \dots, N^m - 1\}$. Change m into $-m$ and the equations of theorem 4.9 are

$$\begin{aligned}
&\sum_{\psi\in\Psi} \sum_{m\geq 0} \widehat{\psi}_{\sigma^m(i)}(N^m(\xi - \theta_i))\overline{\widehat{\psi}}_{\sigma^m(i)}(N^m(\xi - \theta_j + 2k\pi)) \\
&+ \sum_{\psi\in\Psi} \sum_{m<0} N^m \left(\sum_{r=0}^{N^{-m}-1} e^{-ir(\theta_{\sigma^m(i)}+N^m(\xi-\theta_i)-(\theta_{\sigma^m(j)}+N^m(\xi-\theta_j+2k\pi)))} \right. \\
&\left. \cdot\, \widehat{\psi}_{\sigma^m(i)}(N^m(\xi - \theta_i))\overline{\widehat{\psi}}_{\sigma^m(j)}(N^m(\xi - \theta_j + 2k\pi)) \right) = \delta_{ij}\delta_l,
\end{aligned} \tag{5.3}$$

$i, j \in \{1, \dots, n\}$, $k \in \mathbb{Z}$ for almost all $\xi \in \mathbb{R}$.

The rest of the proof will be just the attempt to rewrite the equation (5.3) in the nicer form given in the statement of the theorem.

We evaluate the inner sum, for fixed ξ, i, j, k. Denote by

$$\alpha(m) := \theta_{\sigma^m(i)} - \theta_{\sigma^m(j)} - N^m(\theta_i - \theta_j + 2k\pi), \quad (m \leq 0).$$

$$S_m := N^m \sum_{r=0}^{N^{-m}-1} e^{-ir(\theta_{\sigma^m(i)} + N^m(\xi - \theta_i) - (\theta_{\sigma^m(j)} + N^m(\xi - \theta_j + 2k\pi)))}$$
$$= \begin{cases} N^m \frac{1-e^{-i\left(N^{-m}(\theta_{\sigma^m(i)} - \theta_{\sigma^m(j)}) - (\theta_i - \theta_j)\right)}}{1-\alpha(m)} & \text{if } e^{-i\alpha(m)} \neq 1, \\ 1 & \text{if } e^{-i\alpha(m)} = 1. \end{cases}$$

But $m \leq 0$ so $N^{-m}\theta_{\sigma^m(i)} \equiv \theta_i \mod 2\pi$ and similarly for j so that

$$S_m = \begin{cases} 0 & \text{if } \alpha(m) \not\equiv 0 \mod 2\pi, \\ 1 & \text{if } \alpha(m) \equiv 0 \mod 2\pi. \end{cases}$$

Also note that, if for some $m < 0$, $\alpha(m) \in 2\pi\mathbb{Z}$ then $\alpha(m+1) \in 2\pi\mathbb{Z}$.

Indeed,

$$2N\pi\mathbb{Z} \ni N\alpha(m) = N\theta_{\sigma^m(i)} - N\theta_{\sigma^m(j)} - N^{m+1}(\theta_i - \theta_j + 2k\pi).$$

But

$$N\theta_{\sigma^m(i)} \equiv \theta_{\sigma^{m+1}(i)} \mod 2\pi$$

and similarly for j therefore we see that $\alpha(m+1)$ is in $2\pi\mathbb{Z}$ also.

Thus, if for some $m_0 \leq 0$ we have $\alpha(m_0) \in 2\pi\mathbb{Z}$ then all $\alpha(m)$ are in $2\pi\mathbb{Z}$ for $0 \geq m \geq m_0$, and so $S_m = 1$ for $m \geq m_0$.

Define

$$m_0 := \min\{\, m \leq 0 \,|\, \alpha(m) \in 2\pi\mathbb{Z}\,\} \in \{-\infty, \dots, -1, 0\}.$$

Rewrite (5.3):

$$\text{(5.4)} \qquad \sum_{\psi \in \Psi} \sum_{m \geq m_0} \widehat{\psi}_{\sigma^m(i)}(N^m(\xi - \theta_i))\overline{\widehat{\psi}}_{\sigma^m(j)}(N^m(\xi - \theta_j + 2k\pi)) = \delta_{ij}\delta_k.$$

Of course, m_0 depends on i, j, k and we have to make this dependence more precise.

If $z_i = z_j$ and $k = 0$ then $\theta_{\sigma^m(i)} = \theta_{\sigma^m(j)}$ for all $m \in \mathbb{Z}$ so $\alpha(m) = 0$ for all $m \leq 0$ and therefore $m_0 = -\infty$. In this case the equation (5.4) is

$$\sum_{m=-\infty}^{\infty} \widehat{\psi}_{\sigma^m(i)}(N^m(\xi - \theta_i))\overline{\widehat{\psi}}_{\sigma^m(j)}(N^m(\xi - \theta_j)) = \delta_{ij},$$

which after the change of variable $\xi - \theta_i \mapsto \xi$ becomes (5.1).

If $z_i \neq z_j$ or $k \neq 0$ then m_0 is finite. When $z_i \neq z_j$ we have $\theta_{\sigma^m(i)} \neq \theta_{\sigma^m(j)}$ for all m, by cyclicity $\theta_{\sigma^m(i)} - \theta_{\sigma^m(j)}$ will assume only a finite number of nonzero values in $(-2\pi, 2\pi)$ as m varies. For m close to $-\infty$ the quantity $-N^m(\theta_i - \theta_j + 2k\pi)$ is too small to cover the distance from $\theta_{\sigma^m(i)} - \theta_{\sigma^m(j)}$ to any point in $2\pi\mathbb{Z}$ so that $\alpha(m) \notin 2\pi\mathbb{Z}$.

We change the variable in (5.4), $\xi' - \theta_{\sigma^{m_0}(i)} = N^{m_0}(\xi - \theta_i)$. Then

$$N^{m_0}(\xi - \theta_j + 2k\pi) = \xi' - \theta_{\sigma^{m_0}(j)} - \alpha(m_0),$$

so (5.4) is equivalent to

$$(5.5)\qquad \sum_{m\geq 0}\widehat{\psi}_{\sigma^{m+m_0}(i)}(N^m(\xi-\theta_{\sigma^{m_0}(i)}))\overline{\widehat{\psi}}(N^m(\xi-\theta_{\sigma^{m_0}(j)}-\alpha(m_0)))=0,$$

We prove that (5.5) is equivalent to:
For all i,j and $s\in\mathbb{Z}\setminus N\mathbb{Z}$:

$$(5.6)\qquad \sum_{\psi\in\Psi}\sum_{m\geq 0}\widehat{\psi}_{\sigma^m(i)}(N^m(\xi-\theta_i))\overline{\widehat{\psi}}_{\sigma^m(j)}(N^m(\xi-\theta_j+A(i,j)+2s\pi))=0,$$

where $A(i,j)=N(\theta_{\sigma^{-1}(i)}-\theta_{\sigma^{-1}(j)})-(\theta_i-\theta_j)$.

First, assume (5.6) holds. Fix $i,j\in\{1,\dots,n\}$, $k\in\mathbb{Z}$ with $z_i\neq z_j$ or $k\neq 0$. We prove that $-\alpha(m_0)$ is of the form $A(\sigma^{m_0}(i),\sigma^{m_0}(j))-2s\pi$ with $s\in\mathbb{Z}\setminus N\mathbb{Z}$.

We have

$$\alpha(m_0)=\theta_{\sigma^{m_0}(i)}-\theta_{\sigma^{m_0}(j)}-N^{m_0}(\theta_i-\theta_j+2k\pi).$$

By the definition of m_0, $\alpha(m_0-1)\notin 2\pi\mathbb{Z}$. But observe that

$$\begin{aligned}N\alpha(m_0-1)&=N(\theta_{\sigma^{m_0-1}(i)}-\theta_{\sigma^{m_0-1}(j)})-N^{m_0}(\theta_i-\theta_j+2k\pi)\\&=\alpha(m_0)+A(\sigma^{m_0}(i),\sigma^{m_0}(j)).\end{aligned}$$

Also $A(\sigma^{m_0}(i),\sigma^{m_0}(j))\in 2\pi\mathbb{Z}$ and $\alpha(m_0)\in 2\pi\mathbb{Z}$ so $N\alpha(m_0-1)\in 2\pi\mathbb{Z}$ which means that $\alpha(m_0-1)\in\frac{2\pi}{N}\mathbb{Z}$.

We also know that $\alpha(m_0-1)\notin 2\pi\mathbb{Z}$, hence $\alpha(m_0-1)=2\pi\frac{s}{N}$ with $s\in\mathbb{Z}\setminus N\mathbb{Z}$. Then

$$-\alpha(m_0)=-2s\pi+A(\sigma^{m_0}(i),\sigma^{m_0}(j)).$$

Now using (5.6) for $\sigma^{m_0}(i)$, $\sigma^{m_0}(j)$ and $-s$, we obtain (5.5).

For the converse, suppose (5.5) holds and fix i,j and $s\in\mathbb{Z}\setminus N\mathbb{Z}$. Let $k:=2s\pi+A(i,j)$ and we show that the m_0 associated to i,j and k is 0.

Indeed, we compute: $\alpha(-1)=-2\frac{s}{N}\pi\notin 2\pi\mathbb{Z}$. Since $\alpha(0)=-2s\pi-A(i,j)$, using (5.5), the relation (5.6) is obtained.

Equation (5.6) is clearly equivalent to (5.2) after a change of variable. □

REMARK 5.5. Note that there is some redundancy in the equations of theorem 5.4. Indeed, the relations (5.1) and (5.2) for $i>j$ follow from the corresponding relations for $i\leq j$ after a conjugation and a change of variable.

Also, if $i_1,\dots,i_l$ form a cycle for σ then, once we have equation (5.1) for $i=j=i_1$, the equation (5.1) will be also true for $i_2,\dots,i_l$, because one needs only to change the variable $\xi\leftrightarrow N^p\xi$ for some appropriate p.

We can apply theorem 5.4 to the particular case of the amplification of the classic representation on $L^2(\mathbb{R})$. Our result complements the equations obtained for orthogonal super-wavelets in [**HL**], theorem 5.13.

COROLLARY 5.6. *Let σ be the identity permutation and $z_1=\dots=z_n=1$. Then Ψ is a NTF wavelet if and only if the following equations are satisfied for almost all $\xi\in\mathbb{R}$:*

$$\sum_{\psi\in\Psi}\sum_{m=-\infty}^{\infty}\widehat{\psi}_i(N^m\xi)\overline{\widehat{\psi}}_j(N^m\xi)=\delta_{i,j},\quad (i,j\in\{1,\dots,n\});$$

$$\sum_{\psi\in\Psi}\sum_{m=0}^{\infty}\widehat{\psi}_i(N^m\xi)\overline{\widehat{\psi}}_j(N^m(\xi+2s\pi))=0,\quad (i,j\in\{1,\dots,n\},\ s\in\mathbb{Z}\setminus N\mathbb{Z}).$$

In [**HL**], two normalized tight frames $(e_i)_{i\in I}$ for H and $(f_i)_{i\in I}$ for K are called strongly disjoint if their direct sum $(e_i \oplus f_i)_{i\in I}$ is a normalized tight frame for $H\oplus K$. Two normalized tight frame wavelets $\Psi = \{\psi_1, \ldots, \psi_L\}$, $\Psi' = \{\psi'_1, \ldots, \psi'_L\}$ for $L^2(\mathbb{R})$ (note that the cardinality is the same) are strongly disjoint if their affine systems $X(\psi)$ and $X(\Psi')$ are strongly disjoint.

If we look at corollary 5.6, we see that the fact that each component $\{\psi_i \mid \psi \in \Psi\}$, $(i \in \{1, \ldots, n\})$ is a NTF wavelet for $L^2(\mathbb{R})$ is equivalent to the equations with $i = j$ (and we reobtain in fact the well known characterization of wavelets in $L^2(\mathbb{R})$, see [**Bo2**], [**Ca**] or [**HW**]). Therefore the disjointness part is covered by the equations with $i \neq j$, hence we can use the corollary 5.6 to produce a characterization of disjointness:

COROLLARY 5.7. *Two NTF wavelets* $\Psi := \{\psi^1, \ldots, \psi^L\}$, $\Psi' := \{\psi'^1, \ldots, \psi'^L\}$ *for* $L^2(\mathbb{R})$ *are strongly disjoint if and only if*

$$\sum_{l=1}^{L} \sum_{m=-\infty}^{\infty} \widehat{\psi^l}(N^m \xi)\overline{\widehat{\psi'^l}}(N^m \xi) = 0,$$

$$\sum_{l=1}^{L} \sum_{m=0}^{\infty} \widehat{\psi^l}(N^m \xi)\overline{\widehat{\psi'^l}}(N^m (\xi + 2s\pi)) = 0, \quad (s \in \mathbb{Z} \setminus N\mathbb{Z}).$$

In the remainder of this section we present another application of the characterization theorem. The following oversampling result is proved in [**CS**] for the scaling factor $N = 2$: if one has a NTF wavelet ψ (for $L^2(\mathbb{R})$), then also $\eta := \frac{1}{p}\psi\left(\frac{x}{p}\right)$ is a NTF wavelet, where p is any odd number.

We will refine this result here: not only that η is a NTF wavelet, but it is part of a super-wavelet $(\eta, \ldots, \eta)$ for $L^2(\mathbb{R})^p$, and the starting wavelet is an orthogonal one if and only if the super-wavelet is also orthogonal. Here is the precise formulation of this result:

Let p be a positive integer which is prime with N. For $\psi \in L^2(\mathbb{R})$, define

$$\eta(\psi)(x) = \frac{1}{p}\psi\left(\frac{x}{p}\right), \quad (x \in \mathbb{R}),$$

$$\vec{\eta}(\psi) = (\eta(\psi), \ldots, \eta(\psi)) \in L^2(\mathbb{R})^p, \quad (\psi \in \Psi).$$

Let $\rho := e^{-2\pi i/p}$, $z_k = \rho^k$, $k \in \{1, \ldots, p\}$ and σ the permutation of $\{1, \ldots, p\}$ defined by $(\rho^i)^N = \rho^{\sigma(i)}$ for all $i \in \{1, \ldots, p\}$, and let $\mathfrak{R}_{\sigma,Z}$ be the corresponding affine structure on $L^2(\mathbb{R})^p$.

THEOREM 5.8. *Let* $\Psi := \{\psi_1, \ldots, \psi_p\}$ *in* $L^2(\mathbb{R})$. Ψ *is a NTF wavelet for* $L^2(\mathbb{R})$ *if and only if*

$$\vec{\eta}(\Psi) := \{\vec{\eta}(\psi) \mid \psi \in \Psi\}$$

is a NTF wavelet on $L^2(\mathbb{R})^p$ *for the affine structure* $\mathfrak{R}_{\sigma,Z}$.

Moreover, Ψ *is an orthogonal wavelet for* $L^2(\mathbb{R})$ *if and only if* $\vec{\eta}(\Psi)$ *is an orthogonal wavelet for* $L^2(\mathbb{R})^p$.

PROOF. We have to check that the equations of theorem 5.4 are satisfied.

The Fourier transform of $\eta(\psi)$ is

$$\widehat{\eta(\psi)}(\xi) = \widehat{\psi}(px), \quad (\xi \in \mathbb{R}).$$

Then, we can rewrite the equations (5.1) and (5.2) for $\vec{\eta}(\psi)$:

$$\sum_{\psi\in\Psi}\sum_{m=-\infty}^{\infty}|\widehat{\psi}|^2(N^m p\xi)=1,$$

$$\sum_{\psi\in\Psi}\sum_{m\geq 0}\widehat{\psi}(N^m p\xi)\overline{\widehat{\psi}}(N^m(p\xi+a))=0,$$

for all $a\in\{\,p(N\theta_{\sigma^{-1}(i)}-N\theta_{\sigma^{-1}(j)}+2s\pi)\,|\,i,j\in\{1,\dots,p\},\ s\in\mathbb{Z}\setminus N\mathbb{Z}\,\}=:A$. We want to see what this set is and we will show that it is equal to $\{\,2q\pi\,|\,q\in\mathbb{Z}\setminus N\mathbb{Z}\,\}$.

Note that θ_i takes all the values of the form $2k\pi/p$ inside $[-\pi,\pi)$ with k integer. Then $p\theta_i$ takes all values of the form $2k\pi$ inside $[-p\pi,p\pi)$ and therefore $p\theta_{\sigma^{-1}(i)}-p\theta_{\sigma^{-1}(j)}$ will cover all values of the form $2k\pi$ within $(-2p\pi,2p\pi)$. Therefore

$$A=\{\,2\pi(Nl+ps)\,|\,l\in\{-p+1-p+2,\dots,-1,0,1,\dots,p-1\},\ s\in\mathbb{Z}\setminus N\mathbb{Z}\,\}.$$

Since p is prime with N it is clear that all numbers in A are of the form $2\pi q$ with $q\in\mathbb{Z}\setminus N\mathbb{Z}$.

For the converse, take q not divisible by N. Since N and p are prime we can write $q=Nl_1+ps_1$ with l_1, s_1 integers. Clearly s_1 cannot be divisible by N. Also we can write $l_1=pr_1+l$ with $r_1\in\mathbb{Z}$ and $l\in\{0,\dots,p-1\}$. So

$$q=Nl+p(Nr_1+s_1)$$

with $l\in\{0,\dots,p-1\}$, Nr_1+s_1 not divisible by N and therefore $2\pi q$ is in A.

Finally, if we change the variable $p\xi\leftrightarrow\xi$, we see that the equations are equivalent to the fact that Ψ is a NTF wavelet for $L^2(\mathbb{R})$.

The orthogonal case follows by an inspection of the norms: a normalized tight frame is an orthonormal basis if and only if all the norms of the vectors are 1. But observe that

$$\|\vec{\eta}(\psi)\|^2=p\|\eta(\psi)\|^2=\|\psi\|^2,$$

and therefore the last equivalence is clear. □

6. The spectral function and the dimension function

DEFINITION 6.1. A finite subset $\Psi=\{\psi^1,\dots,\psi^L\}$ of $L^2(\mathbb{R})^n$ is called a semi-orthogonal wavelet if the affine system

$$\{\,U^jT^k\psi\,|\,j\in\mathbb{Z},\ k\in\mathbb{Z},\ \psi\in\Psi\,\}$$

is a NTF for $L^2(\mathbb{R})^n$ and $W_i\perp W_j$ for $i\neq j$, where

$$W_j=V_{j+1}\ominus V_j=\overline{\operatorname{span}}\{\,U^{-j}T^k\psi\,|\,k\in\mathbb{Z},\ \psi\in\Psi\,\}=U^{-j}(S(\Psi)),\quad (j\in\mathbb{Z}).$$

and $(V_j)_j$ is the GMRA associated to Ψ.

If Φ is a NTF generator for V_0 then Φ is called a set of scaling functions for Ψ.

For a semi-orthogonal wavelet, theorem 5.2 and the orthogonality conditions given in the definition imply the following:

PROPOSITION 6.2. *Let Ψ be a semi-orthogonal wavelet and $(V_j)_j$ its GMRA. Then the vectors*

$$\{\,\tilde{\psi}_{m,0}=N^{-m/2}U^m\psi\,|\,m>0,\ \psi\in\Psi\,\}$$

form a NTF generator for V_0.

This proposition will be one of the essential ingredients for our results. In V_0 we have two NTF generators: one given by the wavelet Ψ as in proposition 6.2 and another given by a set of scaling functions for V_0 which always exists according to theorem 3.7. Computing the local trace for various operators in two ways will give rise to some fundamental equalities about wavelets.

The spectral function was introduced for the classical case on $L^2(\mathbb{R})$ by M. Bownik and Z. Rzeszotnik in [**BoRz**]. It contains a lot of information about the shift invariant subspace and it has several nice features. In some sense, the spectral function corresponds to the diagonal entries of the dual Gramian.

DEFINITION 6.3. Let V be a shift invariant subspace of $L^2(\mathbb{R})^n$. For $i \in \{1, \dots, n\}$, the i-th spectral function of V is

$$\sigma_{V,i}(\xi) := \tau_{V,P_{0,i}}(\xi), \quad (\xi \in \mathbb{R}),$$

where $P_{0,i}$ is the projection onto the 0-th component of the i-th vector

$$(P_{0,i}v)(k, j) = v_{0,i}\delta_{0k}\delta_{ij}, \qquad \text{for all } j \in \{1, \dots, n\},\ k \in \mathbb{Z}.$$

The next proposition is just a consequence of theorem 4.4

PROPOSITION 6.4. *Let V be a shift invariant subspace and Φ be a NTF generator for V. Then*

$$\sigma_{V,i}(\xi) = \sum_{\varphi \in \Phi} |\widehat{\varphi}_i|^2(\xi - \theta_i), \quad (\xi \in \mathbb{R},\ i \in \{1, \dots, n\}).$$

Using the NTF generator given in proposition 6.2, we obtain a formula for the spectral functions associated to wavelets:

PROPOSITION 6.5. *Let Ψ be a semi-orthogonal wavelet and $(V_j)_j$ its GMRA. Then the spectral functions of the core space V_0 can be computed by:*

$$\sigma_{V_0,i}(\xi) = \sum_{\psi \in \Psi} \sum_{m \geq 1} |\widehat{\psi}_{\sigma^m(i)}|^2(N^m(\xi - \theta_i)), \quad (\xi \in \mathbb{R},\ i \in \{1, \dots, n\}).$$

Combining proposition 6.4 and 6.5 we obtain a Gripenberg-Weiss-type formula:

COROLLARY 6.6. *If Ψ is a semi-orthogonal wavelet and Φ is a set of scaling function for it, then*

$$\sum_{\varphi \in \Phi} |\widehat{\varphi}_i|^2(\xi - \theta_i) = \sum_{\psi \in \Psi} \sum_{m \geq 1} |\widehat{\psi}_{\sigma^m(i)}|^2(N^m(\xi - \theta_i)), \quad (\xi \in \mathbb{R},\ i \in \{1, \dots, n\}).$$

The dimension function was extensively used on $L^2(\mathbb{R})$ to decide whether a wavelet is a MRA wavelet or not, see [**HW**]. It also gives information about the multiplicity of the wavelets ([**B**], [**BM**], [**BMM**], [**Web**]).

DEFINITION 6.7. Let V be a shift invariant subspace of $L^2(\mathbb{R})^n$ and J_{per} its range function. The dimension function of V is

$$\dim_V(\xi) = \dim(J_{per}(\xi)), \quad (\xi \in \mathbb{R}).$$

Since the trace of a projection is the dimension of its range, using the definition of the local trace we have the following equality:

PROPOSITION 6.8. *Let V be a shift invariant subspace of $L^2(\mathbb{R})^n$. Then*

$$\dim_V(\xi) = \tau_{V,I}(\xi),$$

where I is the identity operator on $l^2(\mathbb{Z})^n$.

Then, with theorem 4.4, and proposition 6.4 we have

PROPOSITION 6.9. *If Φ is a NTF generator for a shift invariant subspace V then*

$$\dim_V(\xi) = \sum_{\varphi\in\Phi}\sum_{i=1}^{n}\sum_{k\in\mathbb{Z}}|\widehat{\varphi}_i|^2(\xi-\theta_i+2k\pi) = \sum_{i=1}^{n}\operatorname{Per}(\sigma_{V,i})(\xi), \quad (\xi\in\mathbb{R}).$$

where

$$\operatorname{Per}(f)(\xi) = \sum_{k\in\mathbb{Z}} f(\xi+2k\pi), \quad (\xi\in\mathbb{R},\ f\in L^1(\mathbb{R})).$$

We can choose the scaling functions to be as in theorem 3.7, and the consequence is that the dimension function and the multiplicity function are the same (a generalization of the result of E. Weber [**Web**]).

THEOREM 6.10. *If Ψ is a semi-orthogonal wavelet and V_0 the core space of the associated GMRA. Then the multiplicity function and the dimension function of V_0 are equal and they are given by*

$$m_{V_0}(\xi) = \dim_{V_0}(\xi) = \sum_{\psi\in\Psi}\sum_{m\geq 1}\sum_{i=1}^{n}\sum_{k\in\mathbb{Z}}|\widehat{\psi}_{\sigma^m(i)}|^2(N^m(\xi-\theta_i+2k\pi) =: D_\Psi(\xi),$$

for almost all $\xi\in[-\pi,\pi)$.

Next we want to establish some dilation formulas for the spectral function and for the dimension function. We use theorem 4.14.

PROPOSITION 6.11. *Let V be a SI subspace of $L^2(\mathbb{R})^n$. Then for all $j\in\{1,\dots,n\}$,*

$$\sigma_{U^{-1}V,j}(\xi) = \sigma_{V,\sigma^{-1}(j)}\left(\frac{\xi-\theta_j+N\theta_{\sigma^{-1}(j)}}{N}\right), \quad (\xi\in\mathbb{R}),$$

$$\dim_{U^{-1}V}(\xi) = \sum_{l=0}^{N-1}\dim_V\left(\frac{\xi+2l\pi}{N}\right), \quad (\xi\in\mathbb{R}).$$

PROOF. With theorem 4.14, we have to compute for $l\in\{0,\dots,N-1\}$, $\mathcal{S}^*D_l^*P_{0,j}D_l\mathcal{S}$ is the rank-one operator given by the vector $\mathcal{S}^*D_l^*e_{0,j}$. We have for $k\in\mathbb{Z}$, $i\in\{0,\dots,N-1\}$:

$$\mathcal{S}^*D_l^*e_{0,j}(k,i) = (D_l^*e_{0,j})(k,\sigma(i)) = e_{0,j}(2l\pi+\theta_{\sigma(i)}-N\theta_i+2kN\pi),$$

and this shows that

$$\mathcal{S}^*D_l^*e_{0,j} = \begin{cases} e_{k_j,\sigma^{-1}(j)} & \text{if } l=l_j,\\ 0 & \text{otherwise,}\end{cases}$$

where $l_j\in\{0,\dots,N-1\}$ and $k_j\in\mathbb{Z}$ are the only such numbers that verify the equation $2l_j\pi+\theta_j-N\theta_{\sigma^{-1}(j)}+2k_jN\pi=0$. Therefore

$$\mathcal{S}^*D_l^*P_{0,j}D_l\mathcal{S} = \begin{cases} P_{k_j,\sigma^{-1}(j)} = \lambda(k_j)P_{0,\sigma^{-1}(j)}\lambda(k_j)^* & \text{if } l=l_j,\\ 0 & \text{otherwise.}\end{cases}$$

Using theorem 4.14 and then the periodicity property given in proposition 4.10, we obtain:

$$\begin{aligned}\sigma_{U^{-1}V,j}(\xi) &= \tau_{V,P_{k_j,\sigma^{-1}(j)}}\left(\frac{\xi + 2l_j\pi}{N}\right)\\ &= \sigma_{V_0,\sigma^{-1}(j)}\left(\frac{\xi + 2l_j\pi}{N} + 2k_j\pi\right),\end{aligned}$$

and, with the definition of k_j and l_j the equation follows.

The equation for the dimension functions follows from the fact that $\mathcal{S}^* D_l^* I D_l \mathcal{S} = I$ for all $l \in \{0, \dots, N-1\}$. □

PROPOSITION 6.12. *Let $(V_j)_j$ be a GMRA and Φ a NTF generator for V_0. Then, for almost every $\xi \in \mathbb{R}$ and all $i \in \{1, \dots, n\}$,*

$$\lim_{m\to\infty} \sum_{\varphi\in\Phi} |\widehat{\varphi}_i|^2 \left(\frac{\xi}{N^m}\right) = 1.$$

PROOF. With the monotone convergence theorem 4.13, we have that $\sigma_{U^{-m}V,i}(\xi)$ converges pointwise a.e. to $\sigma_{L^2(\mathbb{R})^n,i}(\xi) = 1$. With proposition 6.11, by induction we get:

$$\begin{aligned}\sigma_{U^{-m}V,i}(\xi) &= \sigma_{V_0,\sigma^{-m}(i)}\left(\frac{\xi - \theta_i + N^m\theta_{\sigma^{-m}(i)}}{N^m}\right)\\ &= \sum_{\varphi\in\Phi} |\widehat{\varphi}_{\sigma^{-m}(i)}|^2 \left(\frac{\xi - \theta_i}{N^m}\right).\end{aligned}$$

Changing $\xi - \theta_i$ to ξ we obtain

$$\lim_{m\to\infty} \sum_{\varphi\in\Phi} |\widehat{\varphi}_{\sigma^{-m}(i)}|^2 \left(\frac{\xi}{N^m}\right) = 1.$$

But σ is a finite permutation so $\sigma^p = \mathrm{id}$ for some p. Thus, for the subsequence mp, $\sigma^{-mp}(i) = i$ for all m, hence

$$\lim_{m\to\infty} \sum_{\varphi\in\Phi} |\widehat{\varphi}_i|^2 \left(\frac{\xi}{N^{mp}}\right) = 1,$$

and if we apply this to ξ, ξ/N, ..., ξ/N^{p-1}, we obtain the desired limit. □

The limit given in proposition 6.12 enables us to give a lower bound for the dimension function of a wavelet. We can conclude that, if there are two distinct indices i, j such that $z_i = z_j$ then the dimension function of any wavelet is strictly bigger then 1 at some points so there are no MRA wavelets, and for any wavelet one needs at least 2 scaling functions. If there are three indices for which the points z_i are the same, then any wavelet needs at least three scaling functions, and so on. This result generalizes and refines proposition 5.16 in [**HL**].

COROLLARY 6.13. *If Ψ is a semi-orthogonal wavelet then, for all $\alpha_0 \in [-\pi, \pi)$,*

$$\limsup_{\xi\to\alpha_0} D_\Psi(\xi) \geq \operatorname{card}\{\, i \in \{1, \dots, n\} \mid \theta_i = \alpha_0 \,\}.$$

In particular, if there are two distinct indices $i \neq j$ with $\theta_i = \theta_j$ then there are no MRA wavelets.

PROOF. Let Φ be a set of scaling functions associated to Ψ and compute D_Ψ at $\xi/N^m + \alpha_0$ using proposition 6.9. Then take the limsup and use proposition 6.12.

For a MRA wavelet D_Ψ is constant 1, and the second statement follows by contradiction. □

References

[B] L. Baggett, *An abstract interpretation of the wavelet dimension function using group representations*, J. Funct. Anal. **173** (2000), 1–20.

[BDR] C. de Boor, R.A. Devore, A. Ron, *The structure of finitely generated shift-invariant subspaces in $L_2(\mathbb{R}^d)$*, J. Funct. Anal. **119** (1994), 37–78.

[BM] L. Baggett, K. Merrill, *Abstract harmonic analysis and wavelets in $\mathbb{R}^n$*, The Functional and Harmonic Analysis of Wavelets and Frames (San Antonio, TX, 1999) (L.W. Baggett and D.R. Larson, eds.), Contemporary Mathematics, vol. 247, American Mathematical Society, Providence, RI, 1999, pp. 17–27.

[BMM] L. Baggett, H. Medina, K. Merrill, *Generalized multi-resolution analyses and a construction procedure for all wavelet sets in $\mathbb{R}^n$*, J. Fourier Anal. Appl. **5** (1999), 563–573.

[Bo1] M. Bownik, *The structure of shift-invariant subspaces of $L^2(\mathbb{R}^n)$*, J. Funct. Anal. **177** (2000), 282–309.

[Bo2] M. Bownik, *On characterizations of multiwavelets in $L^2(\mathbb{R}^n)$*, Proc. Amer. Math. Soc. **129** (2001), 3265–3274.

[BoRz] M. Bownik, Z. Rzeszotnik, *The spectral function of shift-invariant spaces*, Michigan Math. J. **51** (2003), 387–414.

[Ca] A. Calogero, *A characterization of wavelets on general lattices*, J. Geom. Anal. **10** (2000), no. 4, 597–622.

[CS] C.K. Chui, X. Shi, *$n\times$ oversampling preserves any tight affine frame for odd n*, Proc. Amer. Math. Soc. **121** (1994), no. 2, 511–517.

[CSS] C.K. Chui, X. Shi, J. Stoeckler, *Affine frames, quasi-affine frames, and their duals*, Adv. Comput. Math. **8** (1998), 1–17.

[Dut1] D. Dutkay, *The local trace function of shift invariant subspaces*, to appear in Journal of Operator Theory.

[Dut2] D. Dutkay, *MRA super-wavelets*, preprint.

[HL] D. Han, D. Larson, *Frames, bases and group representations*, Mem. Amer. Math. Soc. **147** (2000), no. 697.

[H] H. Helson, *Lectures on Invariant Subspaces*, Academic Press, New York, 1964.

[HW] E. Hernandez, G. Weiss, *A First Course on Wavelets*, CRC Press, Boca Raton, FL, 1996.

[RS1] A. Ron, Z. Shen, *Frames and stable bases for shift-invariant subspaces of $L_2(\mathbb{R}^d)$*, Canad. J. Math. **47** (1995), 1051–1094.

[RS2] A. Ron, Z. Shen, *Affine systems in $L_2(\mathbb{R}^d)$: the analysis of the analysis operator*, J. Funct. Anal. **148** (1997), 408–447.

[RS3] A. Ron, Z. Shen, *Affine systems in $L_2(\mathbb{R}^d)$, II: Dual systems*, J. Fourier Anal. Appl. **3** (1997), 617–637.

[RS4] A. Ron, Z. Shen, *The wavelet dimension function is the trace function of a shift-invariant system*, Proc. Amer. Math. Soc. **131** (2003), no. 5, 1385–1398.

[Web] E. Weber, *Applications of the wavelet multiplicity function*, The Functional and Harmonic Analysis of Wavelets and Frames (San Antonio, TX, 1999) (L.W. Baggett and D.R. Larson, eds.), Contemporary Mathematics, vol. 247, American Mathematical Society, Providence, RI, 1999, pp. 297–306.

DEPARTMENT OF MATHEMATICS, THE UNIVERSITY OF IOWA, 14 MACLEAN HALL, IOWA CITY, IA 52242-1419, U.S.A.

E-mail address: ddutkay@math.uiowa.edu

Contemporary Mathematics
Volume **345**, 2004

Recovery of band-limited functions on manifolds by an iterative algorithm

Hans Feichtinger and Isaac Pesenson

Abstract. The main goal of the paper is to extend some results of traditional Sampling Theory in which one considers signals that propagate in Euclidean space to more general spaces.

It is our strong believe that there exist many real life problems in Signal Analysis and Information Theory which would require non-Euclidean models.

It is known for a long time that band-limited functions on R^d can be reconstructed from their irregular samples, if only the density of sampling points is high enough. For some time there exist iterative methods to do this constructively. In this paper we present a family of iterative algorithms which allow to perform a similar task for band-limited functions over a Riemannian manifold M. We make use of a suitable concept of band-limitedness which is based on the existence of the Laplace- Beltrami operator on M. It is shown that for sufficiently dense metric lattices a geometric rate of convergence can be guaranteed as long as the sampling density is high enough compared to the "band-width" of the sampled function. This claim applies to the L^2-norm, but also the sup-norm or some Sobolev space norms.

1. Introduction and Main Results

The main goal of the paper is to extend some results of traditional Sampling Theory in which one considers signals that propagate in Euclidean space to more general spaces. Although the vast majority of problems in Signal Analysis usually take place in Euclidean space, there are situations which are genuinely non-Euclidean. The first example of such kind could be a Signal Analysis on 2-dimensional sphere.

It is our strong believe that there exist many real life problems in Signal Analysis and Information Theory which would require non-Euclidean models.

A Theory which will unify the ideas of the Classical Sampling Theorem, one of the most beautiful and applied results of the Euclidean Fourier Analysis, with ideas of Differential Geometry and non-Euclidean Harmonic Analysis would be of great interest and importance.

2000 *Mathematics Subject Classification.* 42C05; Secondary 41A17, 41A65, 43A85, 46C99.

Key words and phrases. Riemannian manifold, Laplace-Beltrami operator, band-limited functions, iteration algorithm.

We consider the present paper as a foundation for future papers in which we are planning to investigate in details more specific examples such as:

1) Spheres, projective spaces and general compact manifolds.
2) Hyperboloids and general non-compact symmetric spaces.
3) Different Lie groups.

The classical sampling theorem says, that if $f \in L^2(R)$ is from the Paley-Wiener space PW_ω, i.e. if its Fourier transform $\hat{f}$ has support supp$(\hat{f})$ in $[-\omega, \omega]$, then f is completely determined by its values at points $n\Omega$, where $\Omega = \pi/\omega$ and in L^2-sense

$$f(t) = \sum f(n\Omega)\frac{\sin(\pi(t - n\Omega))}{\pi(t - n\Omega)}.$$

The Paley-Wiener theorem states that $f \in PW_\omega$ if and only if f is an entire function of exponential type ω.

Although the group structure of the "sampling lattice" makes the reconstruction very easy it is nevertheless true that band-limited functions are also uniquely determined and can be recovered in a stable way from their values on specific irregular sets of points $X = \{x_n\}$. An early result in this direction is [**14**], due to Paley and Wiener, where they observe that it is enough to assume that functions $\exp ix_n t, n \in Z$ form a Riesz basis for $L^2([-\pi, \pi])$. More recently frame theory was giving the perspective that one indeed only has to verify certain (frame) estimates in order to grant stable reconstruction. The irregular sampling problem has inspired many mathematicians and culminated in the work of A.Beurling, P.Malliavin [**2**], [**3**], and H.Landau [**12**].

In the early ninety's H. Feichtinger and K. Gröchenig have introduced iterative reconstruction methods (cf. the series of papers [**7**]-[**10**]) which allow to recover band-limited functions (from a variety of function spaces) from their irregular samples. Their approach was very closely related to their work on coorbit spaces (see [**6**]), which in fact was partially inspired by the work of R.Coifman and R.Rochberg [**4**], who prove irregular sampling theorems for spaces of analytic functions. See also the paper by K. Gröchenig [**10**] for applications to the short time Fourier transforms and wavelet transforms. On the other hand version of an irregular sampling theory in which band-limited functions are reconstructed as limits of variational splines was developed by I. Pesenson in the case of $L_2(R^d)$ [**20**], in the case of stratified Lie groups [**18**], [**19**], in the case of manifolds [**17**], [**21**], and in the case of a general Hilbert space [**22**]. Note that the notion of band-limited functions in Hilbert spaces was developed by I. Pesenson in [**11**], [**15**] and [**16**] in connection with approximation theory and Besov spaces on manifolds.

The goal of the present article is to show that it is possible to adapt the iterative reconstruction schemes to the context of Banach spaces of band-limited functions over manifolds and thus to merge the approaches taken by the two authors of this paper in their earlier papers.

The following is a brief description of main results. Although our results hold true for a general Riemannian manifold of bounded geometry, we will assume for the sake of simplicity that M be a C^∞-homogeneous manifold i. e. the group of isometries is transitive on M. The Laplace-Beltrami operator Δ of the corresponding Riemannian metric $dist(x, y), x, y, \in M$ and the operator $D = \Delta^{1/2}$ are self adjoint positive definite operators in the corresponding Hilbert space $L^2(M)$, which is constructed with the help of the invariant measure on M.

To introduce our notion of band-limited functions we will apply the spectral theorem for a self-adjoint positive definite operator D in the space $L_2(M)$.

According to the spectral theory [**13**] for any self-adjoint positive definite operator S in a Hilbert space H there exist a direct integral of Hilbert spaces $X = \int X(\lambda)\, dm(\lambda)$ and a unitary operator F from H onto X, which transforms domain of $S^\mu, \mu \geq 0$, onto $X_\mu = \{x \in X | \lambda^\mu x \in X\}$ with norm

$$\|x(\lambda)\|_{X_\mu} = \left(\int_0^\infty \lambda^{2\mu} \|x(\lambda)\|^2_{X(\lambda)}\, dm(\lambda) \right)^{1/2}$$

besides $F(S^\mu f) = \lambda^\mu (Ff)$, if f belongs to the domain of S^μ. As known, X is the set of all m-measurable functions $\lambda \to x(\lambda) \in X(\lambda)$, for which the norm

$$\|x\|_X = \left(\int_0^\infty \|x(\lambda)\|^2_{X(\lambda)}\, dm(\lambda) \right)^{1/2}$$

is finite.

We will say that a vector f from H is ω-band-limited if its "Fourier transform" Ff has support in $[0, \omega]$. The $E_\omega(S)$ will denote the set of all ω-band-limited vectors.

In our case we apply the above construction to the positive square root $S = D$ from the Laplace-Beltrami operator Δ in the space $L_2(M)$. In this situation the corresponding spaces of ω-band-limited functions are denoted as $E_\omega(M)$.

The next theorem is taken from [**21**] and can be considered as an abstract version of the Paley-Wiener theorem.

THEOREM 1.1. *The following conditions are equivalent:*

1) $f \in E_\omega(M)$

2) *for all $s > 0$ the following Bernstein inequality holds true*

$$\|D^s f\| \leq \omega^s \|f\|, f \in L_2(M), \tag{1.1}$$

3) *for every $g \in L_2(M)$ the scalar-valued function $\langle e^{itD} f, g\rangle$ that is bounded on the real line has an extension to the complex plane as an entire function of exponential type ω.*

4) *the vector-valued function $t \mapsto e^{itD} f$ that is bounded on the real line has an extension to the complex plane as an entire function of the exponential type ω.*

We will prove this theorem in section 3. Throughout the paper norm $\|f\|$ means $L_2(M)$-norm.

Here are some examples of band-limited functions on manifolds.

1. If $M = R^d$ and $D = \Delta^{1/2}$, where Δ is the usual Laplace operator then we can show that $E_\omega(R^d)$ is just the set of all band-limited functions in $L_2(R^d)$ whose Fourier transform has support in the ball of radius ω.

2. If M is a compact manifold then $\bigcup_{\omega > 0} E_\omega(M) = E(M)$ is the linear span of all eigen-functions of $D = \Delta^{1/2}$, where Δ is the Laplace-Beltrami operator of the invariant metric of M. If M is equivariantly imbedded in an R^n, then $E(M)$ can be identified with restrictions of polynomials in R^n onto M, [**11**], [**16**].

3. In the case of non-compact symmetric space M and corresponding Laplace-Beltrami operator the space $E(M)$ consists of all $L^2(M)$ functions whose Fourier-Helgason transform has compact support [**11**], [**16**].

A discrete set $X = (x_i)$ will be called a *metric lattice* if there exist a strictly positive number $\rho > 0$ and a natural number N such that

a) the balls $B(x_i, \rho/4)$ are disjoint, i.e. are a *pavement* over M;
b) the balls $B(x_i, \rho/2)$ form a cover of M;
c) the balls $B(x_i, \rho)$, form a cover of M of *uniform height* $\leq N$.

We will call, more specifically such a family X a (ρ, N)-lattice and use the symbol $\Xi(\rho, N)$ for the collection of all such lattices. In short, instead of saying that X satisfies (a)–(c) above we will write $X \in \Xi(\rho, N)$.

The later means that every ball has non-empty intersections with no more than $N - 1$ other balls from this family.

LEMMA 1.2. *For every manifold M as described above there exists an integer $N_M = N(M) > 0$ and $\rho_M = \rho(M) > 0$ such that for any $\rho < \rho(M)$ there exist metric lattices in $X \in \Xi(\rho, N_M)$.*

In the rest of section 2 we discuss the analysis on manifolds. We need a notion of an elliptic operator and corresponding regularity theory in the Sobolev scale of spaces $H^k(M), k \in \mathbb{Z}$.

For $X \in \Xi(\rho, N_M)$ one can construct corresponding partitions of unity $\theta = \{\theta_i\}$ (we will call it *associated* to the metric lattice X) with the properties

(1) $\mathrm{supp}(\theta_i) \subseteq B(x_i, \rho/2)$,
(2) for each multi-index α there exists $C_\theta(\alpha)$ such that

$$\sup_i \sup_x |\partial^{|\alpha|}\theta_i| \leq C_\theta(\alpha).$$

In analogy to established terminology for the group case we call such a partition a *bounded uniform partition of unity* (for short a BUPU), or more precisely a (ρ, N)-BUPU, if the specific set X is not relevant. For any continuous function f we define the (Voronoi-type) function $V_X f$ by:

$$V_X(f) = \sum_i f(x_i)\, \theta_i, \quad \theta_j \in C_0^\infty(B(x_j, \rho/2)). \tag{1.2}$$

For the case that θ_j is just the indicator function of those points which are closer to x_j than to any other point the resulting function is constant on Voronoi domains and is simply the nearest neighborhood interpolator.

It will be shown that for any lattice $X \in \Xi(\rho, N_M)$ the function $V_X(f)$ belongs to $L_2(M)$ as long as f belongs to a $H^k(M)$, for some $k > d/2$, i.e. that V_X is a linear operator from $H^k(M)$ into $L^2(M)$.

The (approximation) operator A will then be defined as

$$A_X = P_\omega \circ V_X, \qquad \text{resp.} \quad A_X f = P_\omega V_X(f). \tag{1.3}$$

where P_ω is the orthonormal projector from $L_2(M)$ onto $E_\omega(M)$.

Note that the notation A_X is not perfect in the sense that operator A depends not just on the lattice X but also on the corresponding partition of unity. We hope the notation A_X will not cause any confusion.

The main result of the paper is the following theorem.

THEOREM 1.3. *For a given $k > d/2$ there exist constants $C(M, k) > 0$ and $\rho(M, k) > 0$ such that for any lattice $X \in \Xi(\rho, N_M), \rho < \rho(M, k)$, and any $\omega > 0$*

$$\|f - A_X f\| \leq C(M, k)\rho(1 + \omega^2)^{k/2}\, \|f\|. \quad \text{for all } f \in E_\omega(M). \tag{1.4}$$

Hence, if $k > d/2$ and $\omega > 0$ are given we can choose

$$\rho < \left(C(M,k)(1+\omega^2)^{k/2}\right)^{-1},$$

which implies that

$$\|f - A_X f\| \;\leq\; \gamma\;\|f\| \text{ with } \gamma = C(M,k)^{-1}\rho(1+\omega^2)^{-k/2} < 1.$$

As a consequence we obtain that $f \in E_\omega(M)$ can be recovered from $A_X f$ by the following iterative procedure. Starting from $f_0 = A_X f$ and defining inductively

$$f_{n+1} = f_n + A_X(f - f_n) \tag{1.5}$$

one has

$$\lim_{n\to\infty} f_n = f \tag{1.6}$$

with the error estimate

$$\|f - f_n\| \leq \gamma^{n+1}\|f\|. \tag{1.7}$$

We will also show that convergence of iterations takes place not only in the $L_2(M)$ norm but also in Sobolev and uniform norms on the manifold.

2. Analysis on homogeneous manifolds

Let M be a C^∞-homogeneous Riemannian manifold with $\dim(M) = d$. The symbol $B(x,\rho)$ will denote a ball whose center is $x \in M$ and radius is $\rho > 0$. The measure of this ball $B(x,\rho)$ is independent of x and will be denoted by $v(\rho)$.

Denote by $T_x(M)$ the tangent space of M at a point $x \in M$ and let $exp_x : T_x(M) \to M$ be the exponential geodesic map i. e. $exp_x(u) = \gamma(1), u \in T_x(M)$ where $\gamma(t)$ is the geodesic starting at x with the initial vector $u : \gamma(0) = x, \frac{d\gamma(0)}{dt} = u$. Writing $\mathrm{inj}(M)$ for the injectivity radius of M we recall that the exponential map is a diffeomorphism from a ball of radius $\rho < inj(M)$ in the tangent space $T_x(M)$ onto the ball $B(x,\rho)$.

Fix a point $x \in M$ and consider ball $B(x,\rho/4), \rho \leq inj(M)$. Using isometries we can construct a family of disjoint balls $B(x_i,\rho/4)$ such that there is no ball $B(x,\rho/4), x \in M$ which has non-empty intersections with balls from our family. The same property implies that the family $B(x_i,\rho/2)$ is a cover for M. Now, if $x \in B(x_i,\rho)$ then $B(x_i,\rho/4) \subset B(x,2\rho)$. Since any two balls from the family $B(x_i,\rho/4)$ are disjoint, it gives the estimate for the index of multiplicity N of the cover $B(x_i,\rho)$ i.e. $N \leq v(2\rho)/v(\rho/4)$ and using the doubling property we obtain the estimate

$$N \leq \frac{v(2\rho)}{v(\rho/4)} \leq N(M).$$

So, we have proved the Lemma 1.2 stated in the introduction. Note that the argument here is a more concrete version of a general principle described in [**5**].

We fix an arbitrary (λ, N_M)-lattice $Y = \{y_\nu\}$ for some $\lambda < \rho_M$ and for the corresponding cover $\{B(y_\nu,\lambda)\}$ we choose a partition of unity satisfying:

a) $\operatorname{supp}\varphi_\nu \subset B(y_\nu,\lambda)$, for all ν,

b) $|\varphi_\nu^{(\alpha)}(x)| \leq C(\alpha)$, where $C(\alpha)$ is independent of ν for every multi-index α in the coordinate system defined by exp.

By means of such partition of unity is possible to construct the Sobolev spaces $H^k(M)$ of any order, i.e. for $k \in \mathbb{Z}_+$. On $C_0^\infty(M)$ the following norm is considered

$$\|g\|_k = \left(\sum_{\nu=1}^{\infty} \sum_{1 \le |\alpha| \le k} \|\partial^{|\alpha|} \varphi_\nu g\|^2_{L_2(B(y_\nu, \lambda))} \right)^{1/2}, \tag{2.1}$$

where we have partial derivatives in canonical coordinates on the right hand side. Then the Sobolev space $H^k(M)$ is introduced as the completion of $C_0^\infty(M)$ with respect to this norm (2.1). The spaces $H^k(M), k \in \mathbb{Z}$ have the same properties as the standard Sobolev spaces on R^d.

The following facts about the Laplace-Beltrami operator Δ can be found in [**25**]. The closure from $C_0^\infty(M)$ of Δ in $L^2(M)$ is self-adjoint and positive definite operator and domain of $D^k, k \in \mathbb{Z}, D = \Delta^{1/2}$ is exactly the Sobolev space $H^k(M)$.

The operator Δ has the form $\sum_{|\alpha| \le 2} a_\alpha(x)\partial^\alpha$ in any geodesic coordinate system in the neighborhood $B(x_i, \rho)$ and since it is an invariant operator the estimate $|\partial^\gamma a_\alpha| \le C_\gamma$ takes place uniformly with respect to i. The invariance of this operator also implies its uniform ellipticity.

For such uniformly elliptic differential operator on a homogeneous manifold M the following regularity properties are simple consequences of the corresponding results in R^d and can be proved using partition of unity from the definition of the Sobolev norm (2.1) (see [**23**], [**25**]).

LEMMA 2.1. *The following holds true.*
a) *If $k > 0$ is an integer then on $H^k(M)$ norms*

$$\|(I + \Delta)^{k/2} f\|, \quad or \ \|f\| + \|\Delta^{k/2} f\| \ or \ \|f\|_{H^k(M)}$$

which is defined by (2.1) *are equivalent.*

b) *The map $\Delta : H^k(M) \to H^{k-2}(M), k \in \mathbb{Z}$ is continuous.*

c) *For any $k, m \in \mathbb{Z}$ there exists $c > 0$ such that*

$$\|g\|_k \le c \ (\|\Delta g\|_{k-2} + \|g\|_m), \ g \in C_0^\infty(M).$$

d) *If $g \in \bigcup_{l \in \mathbb{Z}} H^l(M)$ and $\Delta g \in H^{k-2}(M), k \in Z$, then $g \in H^k(M)$.*

3. Band limited vectors in a Hilbert space

In this section we discuss basic properties of band-limited vectors on manifolds. Some of them we formulate in terms of an abstract self-adjoint operator S in a Hilbert space H. Corresponding notations were introduced in the Introduction.

The next theorem is hence evident.

THEOREM 3.1. a) *The set $\bigcup_{\omega > 0} E_\omega(S)$ is dense in H;*
b) *the $E_\omega(S)$ is a linear closed subspace in H.*

We will also need the following lemma.

LEMMA 3.2. *Let S be a generator of an one parameter group of operators e^{tS} in a Hilbert space H and $\|e^{tS}f\| = \|f\|$. If for some $f \in H$ there exists an $\omega > 0$ such that the quantity*

$$\sup_{k\in N} \|S^k f\|\omega^{-k} = R(f)$$

is finite, then $R(f) \leq \|f\|$.

PROOF. By assumption $\|S^r f\| \leq R(f)\omega^r, r \in N$. Now for any complex number z we have

$$\|e^{zS}g\| = \left\|\sum_0^\infty (z^r S^r g)/r!\right\| \leq R(f)\sum_0^\infty |z|^r\omega^r/r! = R(f)e^{|z|\omega}.$$

It implies that for any functional $h \in H^*$ the scalar function $(e^{zS}f, h)$ is an entire function of exponential type ω which is bounded on the real axis R^1 by the constant $\|h\|\,\|f\|$. An application of the Bernstein inequality gives

$$\|(e^{tS}S^k f, h)\|_{C(R^1)} = \left\|\left(\frac{d}{dt}\right)^k (e^{tS}f, h)\right\|_{C(R^1)} \leq \omega^k\|h\|\,\|f\|.$$

The last one gives for $t = 0$

$$|(S^k f, h)| \leq \omega^k\|h\|\,\|f\|.$$

Choosing h such that $\|h\| = 1$ and $(S^k f, h) = \|S^k f\|$ we obtain the inequality $\|S^k f\| \leq \omega^k\|f\|, k \in N$, which gives

$$R(f) = \sup_{k\in N}(\omega^{-k}\|S^k f\|) \leq \|f\|.$$

Lemma is proved. □

The next theorem is an abstract version of the Paley-Wiener theorem.

THEOREM 3.3. *The following conditions are equivalent:*
1) $f \in E_\omega(S)$
2) *for all $s > 0$ the following Bernstein inequality holds true*

$$\|S^s f\| \leq \omega^s\|f\| \tag{3.1}$$

3) *for every $g \in H$ the scalar-valued function $\langle e^{itS}f, g\rangle$ that is bounded on the real line has an extension to the complex plane as an entire function of the exponential type ω.*
4) *the vector-valued function $e^{itS}f$ that is bounded on the real line has an extension to the complex plane as an entire function of the exponential type ω.*

PROOF. Let f belongs to the space $E_\omega(S)$ and $Ff = x \in X$. Then

$$\left(\int_0^\infty \lambda^{2\mu}\|x(\lambda)\|^2_{X(\lambda)}\,dm(\lambda)\right)^{1/2} = \left(\int_0^\omega \lambda^{2\mu}\|x(\lambda)\|^2_{X(\lambda)}\,dm(\lambda)\right)^{1/2} \leq \omega^\mu\|x\|_X,$$

which gives Bernstein inequality for f.

Conversely, if f satisfies Bernstein inequality then $x = Ff$ satisfies $\|x\|_{X_\mu} \leq \omega^\mu\|x\|_X$. Suppose that there exists a set $\sigma \subset [0, \infty] \setminus [0, \omega]$ whose m-measure is not zero and $x|_\sigma \neq 0$. We can assume that $\sigma \subset [\omega+\epsilon, \infty)$ for some $\epsilon > 0$. Then for any real μ we have

$$\int_\sigma \|x(\lambda)\|^2_{X(\lambda)}\, dm(\lambda) \leq \int_{\omega+\epsilon}^{\infty} \lambda^{-2\mu}\|\lambda^\mu x(\lambda)\|^2_{X(\lambda)}\, d\mu \leq \|x\|^2_X\, (\omega/\omega+\epsilon)^{2\mu}$$

which shows that either $x(\lambda)$ is zero on σ or σ has measure zero.

The implications 2) $\rightarrow$ 4) $\rightarrow$ 3) of the Theorem 3.3 are evident. So it is enough to show the implication 3) $\rightarrow$ 2) which is a consequence of the previous lemma. □

In the case when $S = D = \Delta^{1/2}$ and $H = L_2(M)$ we obtain all the above properties for band-limited functions in $L_2(M)$. In this situation the corresponding spaces of ω-band-limited functions are denoted as $E_\omega(M)$.

Note, that in the classical case of R^d the space $E_\omega(R^d)$ coincides with the space of all functions from $L_2(R^d)$ whose Fourier transform has support in the ball of radius ω. It is well known, that in this case there are following continuous embeddings

$$E_\omega(R^d) \subset L_p(R^d)$$

for all $2 \leq p \leq \infty$.

Such continuous embeddings hold also true in the case when manifold M is compact. But in general situation we have just the following result.

THEOREM 3.4. *The following continuous embeddings hold true*

$$E_\omega(M) \subset W_p^m(M), \tag{3.2}$$

where $W_p^m(M)$, *is the Sobolev space on* M, $p > 2$, $\frac{1}{p} = \frac{1}{2} - \frac{k-m}{d}$, $0 \leq m < k$, k *is any natural number and*

$$E_\omega(M) \subset C_b^k(M), \tag{3.3}$$

where $C_b^k(M)$ *is the space of* k*-differentiable bounded functions on* M *and* $k > d/2$.

PROOF. Since the norm of the Sobolev space $H^k(M)$ is equivalent to the norm

$$\|f\| + \|\Delta^{k/2} f\|$$

we have for any function f from $E_\omega(M)$, and for any natural k

$$\|f\|_{H^k(M)} \leq C_M \left(\|f\| + \|\Delta^{k/2} f\|\right) \leq C_M(1+\omega^k)\|f\|.$$

The last inequality along with Sobolev embeddings

$$H^k(M) \subset W_p^m(M), \quad p > 2,\ 0 \leq m < k,\ \frac{1}{p} = \frac{1}{2} - \frac{k-m}{d},$$

and

$$H^k(M) \subset C_b^k(M), \quad k > d/2,$$

imply our theorem. □

Let us also to note, that on the space $E_\omega(M)$ the $L_2(M)$-norm is equivalent to the norm of the space $H^k(M)$ for any k. Indeed, if $f \in E_\omega(M)$, then

$$\|f\| \leq \|f\| + \|\Delta^{k/2} f\| \leq (1+\omega^k)\|f\|.$$

4. Iterative reconstruction algorithm

In this section we describe our first iterative algorithms for reconstruction of band-limited functions on manifold. The following Lemma 4.1 presents a generic idea which is used for reconstruction of different classes of analytic functions. This Theorem basically says that if a bounded operator is close to identity operator then it can be inverted by a Neumann series.

The goal of section is to realize this idea in the case of band-limited functions on manifolds.

LEMMA 4.1. *Let A be a bounded operator in a Hilbert space H such that for some $\gamma < 1$ and for all $f \in H$*

$$\|f - Af\| \leq \gamma \|f\|. \tag{4.1}$$

Then A is invertible and f can be recovered from Af by the following iterative procedure. If $f_0 = Af$ and

$$f_{n+1} = f_n + A(f - f_n) \tag{4.2}$$

then

$$\lim_{n\to\infty} f_n = f \tag{4.3}$$

with the error estimate

$$\|f - f_n\| \leq \gamma^{n+1}\|f\|. \tag{4.4}$$

PROOF. Since the norm of the operator $I - A$ is less than 1, operator A is invertible and the inverse A^{-1} can be represented by a Neumann series

$$A^{-1} = \sum_{n=0} (I - A)^n.$$

Thus, every $f \in H$ is determined by Af because

$$f = A^{-1}Af = \sum_{n=0}^{\infty} (I - A)^k Af = \lim_{n\to\infty} \left(\sum_{n=0}^{n} (I - A)^n Af \right). \tag{4.5}$$

The sequence f_n of partial sums

$$f_n := \sum_{k=0}^{n} (I - A)^k Af$$

satisfies the stated recursion relation follows from

$$f_{n+1} = Af + (I - A)\sum_{k=0}^{n} (I - A)^k Af = Af + (I - A)f_n = f_n + A(f - f_n).$$

Moreover, the identity

$$\sum_{k=n+1}^{\infty} (I - A)^k = (I - A)^{n+1}A^{-1},$$

implies that we have for all $f \in E_\omega(M)$

$$\|f - f_n\| = \left\| \sum_{k=n+1}^{\infty} (I - A)^k Af \right\| = \left\|(I - A)^{n+1}A^{-1}Af\right\| \leq \gamma^{n+1}\|f\|.$$

□

We will apply this theorem in the following situation.

Consider a ρ-lattice $X = \{x_i\} \in \Xi(\rho, N_M)$, where $\rho < \lambda/2 < \rho_M/2$ and where λ is from the definition of the Sobolev norm (2.1). For the cover $\{B(x_i, \rho/2)\}$ construct a partition of unity θ_i with the properties:

a) $\operatorname{supp} \theta_i \subset B(x_i, \rho/2)$,
b) $\sup_x |\partial^{|\alpha|}\theta_i| \leq C_\theta(\alpha)$, where $C(\alpha)$ is independent of i.

Note that the assumptions that $\rho < \lambda/2$ and that the multiplicity of the cover $\{B(y_\nu, \lambda)\}$ is not greater than an N_M imply that each ball $B(x_i, \rho)$ has non-empty intersections with no more than N_M balls of the family $\{B(y_\nu, \lambda)\}$.

Given a function $f \in H^k(M), k > d/2$, we consider an operator

$$V_X(f) = \sum_i f(x_i)\theta_i, \quad \theta_j \in C_0^\infty(B(x_j, \rho/2)). \tag{4.6}$$

It will be shown that for any lattice $X(\rho, N_M), \rho < \lambda/2$, the function $V_X(f)$ belongs to $L_2(M)$ as long as f belongs to a $H^k(M)$, where $k > d/2$.

The operator A will be defined as

$$A_X f = P_\omega V_X(f), \tag{4.7}$$

where P_ω is the orthonormal projector from $L_2(M)$ onto $E_\omega(M)$.

Our next goal is to provide a uniform estimate for the norms of the operator $Id - A_X$ on the subspace $E_\omega(M)$ for all (ρ, N)-BUPUs, if only $\rho < \lambda/2$.

The following lemma provides a step in this direction.

LEMMA 4.2. *For any lattice $X(\rho, N_M), \rho < \lambda/2$, the map $V_X(f)$ is a continuous operator from every $H^k(M), k > d/2$, into $L_2(M)$ in other words, for any lattice $X(\rho, N_M), \rho < \lambda/2$, and a partition of unity $\theta = \{\theta_i\}$ with properties* a), b) *and any natural k there exists a constant $C = C(d, N_M, \theta, k)$ such that*

$$\left\| \sum_i f(x_i)\theta_i \right\|_{L_2(M)} \leq C \|f\|_{H^k(M)} \tag{4.8}$$

for all $f \in H^k(M)$.

PROOF. We have

$$\|V_X(f)\|_{L_2(M)} = \left\| \sum_i f(x_i)\theta_i \right\|_{L_2(M)} \leq \sum_i \|f(x_i)\theta_i\|_{L_2(M)} \leq C_1 \sum_i \sup_{x \in B(x_i, \rho/2)} |f(x_i)\theta(x)|,$$

for all $\rho < \rho_M, C_1 = C_1(d, \rho_M)$. By the Sobolev Embedding Theorem

$$\sup_{x \in B(x_j, \rho/2)} |\theta_j(x) f(x)| \leq C_2(d) \|\theta_j f\|_{H^k(B(x_j, \rho))}, \quad k > d/2.$$

Thus we obtain for some $C_3 = C_3(d, N_M, \rho_M)$:

$$\|V_X(f)\|_{L_2(M)} \leq C_3 \sum_i \|\theta_i f\|_{H^k(B(x_i, \rho))}.$$

Our goal is to estimate the last term through the Sobolev norm $H^k(M)$, $k > d/2$, which was introduced in (2.1). In order to do so we use the partition of unity $\{\varphi_\nu\}$, with $\text{supp}(\varphi_\nu) \subset B(x_\nu, \lambda)$, from definition (2.1). We obtain

$$\begin{aligned}\sum_i \|\theta_i f\|_{H^k(B(x_i,\rho)} &= \sum_i \sum_{|\alpha|\le k} \|\partial^{|\alpha|} \sum_\nu \varphi_\nu \theta_i f\|_{L_2(B(x_i,\rho))} \\ &\le C_4 \sum_\nu \sum_i \|\varphi_\nu f\|_{H^k(B(x_i,\rho))},\end{aligned}$$

where $C_4 = C_4(d, N_M, \rho_M, \theta, k)$.

For a fixed index ν let $n_{\nu,1}, \ldots, n_{\nu,K}$ be all the indexes for which balls $B(x_{n_{\nu,l}}, \rho)$, $l = 1, \ldots, K \le N_M$, have non-empty intersections with the ball $B(x_\nu, \lambda)$.

Since the balls $B(x_i, \rho)$ form a cover of M of multiplicity N_M we have

$$\sum_{l=n_{\nu,1}}^{n_{\nu,K}} \|\varphi_\nu f\|_{H^k(B(x_l,\rho))} \le N_M \|\varphi_\nu f\|_{H^k(B(x_\nu,\lambda))}$$

and finally for any $f \in H^k(M), k > d/2$, we have

$$\|V_X(f)\| \le \sum_i \|\theta_i f\|_{L_2(B(x_i,\rho))} \le C\|f\|_{H^k(M)},$$

where $C = C(d, N_M, \rho_M, \theta, k), \rho < \lambda/2$, that means that $V_X(f)$ belongs to $L_2(M)$. □

We will need the following lemma.

LEMMA 4.3. *For any $k > d/2$ there exist constants $C(M,k) > 0$, $\rho(M,k) > 0$ such that for any $\rho < \rho(M,k)$ and any $X \in \Xi(\rho, N_M)$*

$$\sum_i \|f - f(x_i)\|_{B(x_i,\rho/2)} \le C(M,k)\rho\|(I + \Delta)^{k/2} f\|, \quad k > d/2. \tag{4.9}$$

PROOF. For the given (ρ, N) lattice $X = \{x_i\}$ let $\{\varphi_\nu\}$ be an associated BUPU according to the above construction. For any $f \in C^\infty(M)$ one has

$$\begin{aligned} f(x) = f(x_i) &+ \sum_{1\le|\alpha|\le k-1} \frac{1}{\alpha!}\partial^{|\alpha|} f(x_i)(x - x_i)^\alpha \\ &+ \sum_{|\alpha|=k} \frac{1}{\alpha!} \int_0^\tau t^{k-1}\partial^{|\alpha|} f(x_i + t\vartheta)\vartheta^\alpha \, dt,\end{aligned} \tag{4.10}$$

for each $B(x_i, \rho)$ and every $x \in B(x_i, \rho/2)$, where $x = (x_1, \ldots, x_d)$, $x_i = (x_1^i, \ldots, x_d^i)$, $\alpha = (\alpha_1, \ldots, \alpha_d)$, $(x - x_i)^\alpha = (x_1 - x_1^i)^{\alpha_1} \cdots (x_d - x_d^i)^{\alpha_d}$, $\tau = \|x - x_i\|$, $\vartheta = (x - x_i)/\tau$.

From the last formula we obtain

$$\begin{aligned} &\|f(x) - f(x_i)\|_{L_2(B(x_i,\rho/2))} \\ &\le \sum_{1\le|\alpha|\le k-1} \left\|\partial^{|\alpha|} f(x_i)(x - x_i)^\alpha\right\|_{L_2(B(x_i,\rho/2))} \\ &\quad + \sum_{|\alpha|=k} \left\|\int_0^\tau t^{k-1}\partial^{|\alpha|} f(x_i + t\vartheta)\vartheta^\alpha \, dt\right\|_{L_2(B(x_i,\rho/2))}.\end{aligned} \tag{4.11}$$

We have for sufficiently small ρ

$$|f(x_i)| \leq C_{d,m} \sum_{|\mu| \leq m} \rho^{-d/2} \|\partial^{|\mu|} f\|_{L_2(B(x_i,\rho))}, \tag{4.12}$$

where $\mu = (\mu_1, \mu_2, \ldots, \mu_d), m > d/2$. Using this inequality we can estimate the first term on the right side of (4.11) as follows:

$$\sum_{1 \leq |\alpha| \leq k-1} \left(\int_{B(x_i,\rho/2)} |\partial^{|\alpha|} f(x_i)(x - x_i)^\alpha|^2 \, dx \right)^{1/2} \leq C_{d,k} \sum_{1 \leq |\alpha| \leq k-1} \rho^{|\alpha|} \left\| \partial^{|\alpha|} f \right\|_{L_2(B(x_i,\rho))}. \tag{4.13}$$

Next, to estimate the second term in (4.11) we use the Schwartz inequality and the assumption $k > d/2$:

$$\left| \int_0^\tau t^{k-1} \partial^{|\alpha|} f(x_i + t\vartheta)\vartheta^\alpha \, dt \right|^2 \leq \left(\int_0^\tau t^{k-d/2-1/2} \left| t^{d/2-1/2} \partial^{|\alpha|} f(x_i + t\vartheta) \right| dt \right)^2 \leq C_{d,k} \tau^{2k-d} \int_0^\tau t^{d-1} \left| \partial^{|\alpha|} f(x_i + t\vartheta) \right|^2 dt.$$

We integrate both sides of this inequality over the ball $B(x_i, \rho/2)$ using the spherical coordinate system (τ, ϑ).

$$\begin{aligned} &\int_0^{\rho/2} \tau^{d-1} \int_0^{2\pi} \left| \int_0^\tau t^{k-1} \partial^{|\alpha|} f(x_i + t\vartheta)\vartheta^\alpha \, dt \right|^2 d\vartheta \, d\tau \\ &\leq C_{d,k} \int_0^{\rho/2} t^{d-1} \left(\int_0^{2\pi} \int_0^{\rho/2} \tau^{2k-d} \left| \partial^{|\alpha|} f(x_i + t\vartheta) \right|^2 \tau^{d-1} \, d\tau \, d\vartheta \right) dt \\ &\leq C_{d,k} \rho^{2k} \left\| \partial^{|\alpha|} f \right\|^2_{L_2(B(x_i,\rho))}, \end{aligned} \tag{4.14}$$

where $\tau = \|x - x_i\| \leq \rho/2$, $|\alpha| = k$.

Substitution of (4.13) and (4.14) into (4.11) and summation over all i gives

$$\sum_i \|f(x) - f(x_i)\|_{L_2(B(x_i,\rho/2))} \leq C_{d,k} \sum_i \sum_{1 \leq |\alpha| \leq k} \rho^{|\alpha|} \left\| \partial^{|\alpha|} f \right\|_{L_2(B(x_i,\rho))}, \quad k > d/2. \tag{4.15}$$

Our goal is to obtain the Sobolev norm $H^k(M)$ on the right side of this inequality which was defined in (2.1) in terms of the balls $B(y_\nu, \lambda)$ and corresponding partition of unity (φ_ν). For this partition of unity we have

$$\begin{aligned} \sum_i \|f - f(x_i)\|_{L_2(B(x_i,\rho/2))} &\leq \sum_i \sum_\nu \|\varphi_\nu f - (\varphi_\nu f)(x_i)\|_{L_2(B(x_i,\rho/2))} \\ &\leq C_1 \sum_i \sum_\nu \sum_{1 \leq |\alpha| \leq k} \rho^{|\alpha|} \left\| \partial^{|\alpha|} (\varphi_\nu f) \right\|_{L_2(B(x_i,\rho/2))}, \end{aligned}$$

where $C_1 = C_1(N_M, d, k)$. Since the coverings $\{B(y_\nu, \lambda)\}$ and $\{B(x_i, \rho)\}$ have finite height N_M and $\rho < \lambda/2$, we obtain

$$\sum_i \sum_\nu \sum_{1\le|\alpha|\le k} \rho^{2|\alpha|} \left\| \partial^{|\alpha|}(\varphi_\nu f) \right\|^2_{L_2(B(x_i,\rho))}$$
$$\le C_2(N_M) \sum_\nu \sum_{1\le|\alpha|\le k} \rho^{2|\alpha|} \left\| \partial^{|\alpha|}(\varphi_\nu f) \right\|^2_{L_2(B(y_\nu,\lambda))}$$

and therefore it follows that for $k > d/2$ and some $C_3 = C_3(N_M, d, k)$

$$\sum_i \|f(x) - f(x_i)\|_{L_2(B(x_i,\rho/2))} \le C_3 \sum_i \sum_{j=1}^{k} \rho^j \|f\|_{H^j(M)}.$$

Using this and the regularity theorem for the Laplace-Beltrami in the form

$$\|f\|_{H^k(M)} \le C(M,k)\, \|(I+\Delta)^{k/2} f\|,$$

we derive, again for $k > d/2$ and some $C_4 = C_4(M,k)$

$$\sum_i \|f(x) - f(x_i)\|_{L_2(B(x_i,\rho/2))} \le C_4 \sum_{j=1}^{k} \rho^j \|(I+\Delta)^{j/2} f\|. \tag{4.16}$$

For the self-adjoint $I + \Delta$ for any $\rho > 0$, $0 \le j \le k$, we have the following interpolation inequality:

$$\rho^j \|(I+\Delta)^{j/2} f\| \le \rho^k \|(I+\Delta)^{k/2} f\| + c_k \|f\|. \tag{4.17}$$

Substituting this inequality into (4.16) and using the inequality

$$\|f\| \le C(M,k)\, \|(I+\Delta)^{k/2} f\|$$

we obtain inequality (4.9). □

Our next goal is to show that

THEOREM 4.4. *For a given $k > d/2$ there exist constants $C(M,k) > 0$ and $\rho(M,k)$ such that for any lattice $X(\rho, N_M)$, $\rho < \rho(M,k)$, and for any $f \in E_\omega(M)$*

$$\|f - A_X f\| \le C(M,k)\, \rho (1+\omega^2)^{k/2}\, \|f\|. \tag{4.18}$$

Consequently, for fixed $k > d/2$ and $\omega > 0$ it is true that for any ρ satisfying

$$\rho < \left(C(M,k)(1+\omega^2)^{k/2} \right)^{-1}$$

one has

$$\|f - A_X f\| \le \gamma \|f\| \quad \textit{where} \quad \gamma = C(M,k)\rho(1+\omega^2)^{k/2} < 1.$$

PROOF. For any $f \in E_\omega(M)$ we have

$$\|f - A_X f\| = \|P_\omega f - P_\omega V_X(f)\| \le \|f - V_X(f)\|$$
$$\le \sum_i \|f - V_X(f)\|_{L_2(B(x_i,\rho/2))}$$
$$\le \sum_i \left\| f - \sum_{j=j_1}^{j_{N_M}} \theta_j f(x_j) \right\|_{L_2(B(x_i,\rho/2))},$$

where $j_1, \ldots, j_{N_M}$ are the indices for which corresponding balls have non-empty intersections with the ball $B(x_i, \rho/2)$. Since for $x \in B(x_i, \rho/2)$ the following identity holds true

$$f(x) = \sum_{j=j_1}^{j_{N_M}} \theta_j(x) f(x)$$

and $0 \le \theta_j \le 1$ we obtain

$$\begin{aligned} \|f - A_X f\|^2 &\le \sum_i \sum_{j=j_1}^{j_{N_M}} \|f - f(x_j)\|^2_{L_2(B(x_i,\rho/2)} \\ &\le \sum_i \sum_{j=j_1}^{j_{N_M}} \|f - f(x_i)\|^2_{L_2(B(x_i,\rho/2))} \\ &\quad + \sum_i \sum_{j=j_1}^{j_{N_M}} \|f(x_i) - f(x_j)\|^2_{L_2(B(x_i,\rho/2))} . \end{aligned}$$

The first double sum here can be estimated by using our inequality (4.9). The second double sum can be estimated by using the Sobolev embedding theorem. We have

$$\|f - A_X f\| \le C(M,k)\rho \|(I + \Delta)^{k/2} f\|,$$

for any $k > d/2$.

Because for $f \in E_\omega(M)$ the Bernstein inequality

$$\|\Delta^k f\| \le \omega^{2k} \|f\|$$

holds true it gives the inequality

$$\|f - A_X f\| \le C(M,k)\rho(1 + \omega^2)^{k/2} \|f\|.$$

Since in the last inequality the constant $C(M,k)$ is independent on a lattice $X \in \Xi(\rho)$ it is obvious that for fixed ω and a $k > d/2$ one can choose

$$\rho < \left(C(M,k)(1 + \omega^2)^{k/2} \right)^{-1}$$

such that corresponding

$$\gamma = C(M,k)\rho(1 + \omega^2)^{k/2}$$

will be less than 1. Theorem is proved. □

Combining this result with Theorem 3.4 we obtain

THEOREM 4.5. *For a given $k > d/2$ and $\omega > 0$ choose a $\rho > 0$ such that*

$$\rho < (C(M,k))^{-1}(1 + \omega^2)^{-k/2},$$

where the constant $C(M,k)$ taken from Theorem 4.4. Then, for any lattice $X \in \Xi(\rho, N_M)$ and for any corresponding $A = A_{X,\theta}$ in the space $E_\omega(M)$ define inductively, starting from $f_0 = Af$:

$$f_{n+1} = f_n + A(f - f_n).$$

Then in the space $L_2(M)$ the following convergence holds true:

$$\lim_{n \to \infty} f_n = f \ in \ L_2(M),$$

with the error estimate

$$\|f - f_n\| \leq \gamma^{n+1}\|f\|,$$

where $\gamma = C(M,k)\rho(1+\omega^2)^{k/2} < 1$.

Convergence also holds in the Sobolev spaces

$$W_p^m(M),\ p > 2,\ \frac{1}{p} = \frac{1}{2} - \frac{k-m}{d},\ 0 \leq m < k,$$

where k *is any natural number, and in the uniform norms of the spaces* $C_b^m(M)$.

References

[1] J. Benedetto, *Irregular sampling and frames*, Wavelets: A Tutorial in Theory and Applications (C.K. Chui, ed.), Academic Press, Boston, 1992, pp. 445–507.

[2] A. Beurling, *Local harmonic analysis with some applications to differential operators*, Some Recent Advances in the Basic Sciences, vol. 1 (New York, 1962–1964) (A. Gelbart, ed.), Belfer Graduate School of Science, Yeshiva University, distributed by Academic Press, New York, 1966, pp. 109–125.

[3] A. Beurling and P. Malliavin, *On the closure of characters and the zeros of entire functions*, Acta Math. **118** (1967), 79–95.

[4] R. Coifman and R. Rochberg, *Representation theorems for holomorphic and harmonic functions in* L^p, Astérisque **77** (1980), 11–66.

[5] H.G. Feichtinger, *Banach spaces of distributions defined by decomposition methods, II*, Math. Nachr. **132** (1987), 207–237.

[6] H.G. Feichtinger and K. Gröchenig, *Banach spaces related to integrable group representations and their atomic decompositions, I*, J. Funct. Anal. **86** (1989), 307–340.

[7] H. Feichtinger and K. Gröchenig, *Iterative reconstruction of multivariate band-limited functions from irregular sampling values*, SIAM J. Math. Anal. **23** (1992), 244–261.

[8] H. Feichtinger and K. Gröchenig, *Irregular sampling theorems and series expansions of band-limited functions*, J. Math. Anal. Appl. **167** (1992), 530–556.

[9] H. Feichtinger and K. Gröchenig, *Theory and practice of irregular sampling*, Wavelets: Mathematics and Applications (J.J. Benedetto and M.W. Frazier, eds.), Stud. Adv. Math., CRC Press, Boca Raton, FL, 1994, pp. 305–363.

[10] K. Gröchenig, *Irregular sampling of wavelet and short time Fourier transforms*, Constr. Approx. **9** (1993), no. 2–3 (special issue on wavelet theory), 283–297.

[11] S. Krein and I. Pesenson, *Interpolation Spaces and Approximation on Lie Groups*, The Voronezh State University, Voronezh, 1990 (Russian).

[12] H. Landau, *Necessary density conditions for sampling and interpolation of certain entire functions*, Acta. Math. **117** (1967), 37–52.

[13] J.-L. Lions and E. Magenes, *Non-Homogeneous Boundary Value Problems and Applications*, Springer-Verlag, New York, Heidelberg, 1975.

[14] R.E.A.C. Paley and N. Wiener, *Fourier Transforms in the Complex Domain*, Amer. Math. Soc. Colloq. Publ., vol. 19, American Mathematical Society, Providence, RI, 1934.

[15] I. Pesenson, *The best approximation in a representation space of a Lie group*, Dokl. Acad. Nauk SSSR **302** (1988), no. 5, 1055–1059 (English translation in Soviet Math. Dokl. **38** (1989), no. 2, 384–388).

[16] I. Pesenson, *The Bernstein inequality in the space of representation of Lie group*, Dokl. Acad. Nauk SSSR **313** (1990), no. 4, 803–806 (English translation in Soviet Math. Dokl. **42** (1991), no. 1, 87–90).

[17] I. Pesenson, *Lagrangian Splines, Spectral Entire Functions and Shannon-Whittaker Theorem on Manifolds*, Temple University Research Report 95-87, 1995, pp. 1–28.

[18] I. Pesenson, *Reconstruction of Paley-Wiener functions on the Heisenberg group*, Voronezh Winter Mathematical Schools (P. Kuchment and V. Lin, eds.), Amer. Math. Soc. Transl. Ser. 2, vol. 184, American Mathematical Society, Providence, RI, 1998, pp. 207–216.

[19] I. Pesenson, *Sampling of Paley-Wiener functions on stratified groups*, J. Fourier Anal. Appl. **4** (1998), 271–281.

[20] I. Pesenson, *A reconstruction formula for band limited functions in* $L_2(R^d)$, Proc. Amer. Math. Soc. **127** (1999), no. 12, 3593–3600.

[21] I. Pesenson, *A sampling theorem on homogeneous manifolds*, Trans. Amer. Math. Soc. **352** (2000), no. 9, 4257–4269.
[22] I. Pesenson, *Sampling of band-limited vectors*, J. Fourier Anal. Appl. **7** (2001), no. 1, 93–100.
[23] J. Roe, *An index theorem on open manifolds, I, II*, J. Differential Geom. **27** (1988), no. 1, 87–113, 115–136.
[24] I. Schoenberg, *Cardinal Spline Interpolation*, CBMS Regional Conf. Ser. in Appl. Math., vol. 12, SIAM, Philadelphia, 1973.
[25] R. Strichartz, *Analysis of the Laplacian on the complete Riemannian manifold*, J. Funct. Anal. **52** (1983), 48–79.

Department of Mathematics, Vienna University, AUSTRIA
E-mail address: hans.feichtinger@univie.ac.at

Department of Mathematics, Temple University, Philadelphia, PA 19122
E-mail address: pesenson@@math.temple.edu

Contemporary Mathematics
Volume **345**, 2004

On a characterization of the local Hardy space by Gabor frames

John E. Gilbert and Joseph D. Lakey

ABSTRACT. It is shown that under a Lipschitz condition on the window g, a Gabor expansion $f = \sum_{mn} c_{mn} e^{2\pi i m\beta x} g(x - n\alpha)$ belongs to Goldberg's local Hardy space $h^1(\mathbb{R})$ if and only if its Gabor coefficients satisfy

$$\sum_n \| \sum_m c_{mn} e^{2\pi i m\beta x} \|_{H^1(\mathbb{T}_\beta)} < \infty$$

where $H^1(\mathbb{T}_\beta)$ denotes the periodic atomic Hardy space in the sense of Coifman and Weiss. This extends a Gabor characterization of L^p-spaces given by Gröchenig and Heil.

1. Gabor frames and the local Hardy space

The 'natural' spaces for Gabor theory are the modulation spaces ([**3**]), while Besov and Triebel-Lizorkin spaces are the 'natural' spaces for wavelet analysis, natural meaning that frame expansions converge unconditionally. These two scales of spaces are decidely different, however. All L^p-spaces and Hardy H^p-spaces belong to the latter scale, for instance, but only L^2 is a modulation space. In fact, Gabor expansions do not converge unconditionally in $L^p(\mathbb{R})$ if $p \neq 2$. Nonetheless, as Gröchenig and Heil have shown, Gabor frames can still provide meaningful discrete characterizations of membership in $L^p(\mathbb{R})$, $1 < p < \infty$ [**4**] (*cf.* also Grafakos and Lennard [**5**]). They also suggested that their characterization would extend to the case $p = 1$, provided one had the right Hardy space for that endpoint. The primary aim of this paper is to show that the *right* Hardy space is the *local Hardy space* $h^1(\mathbb{R})$ as defined in Goldberg's thesis [**6**], not the much more familiar Hardy H^1-space. Crucially, the standard real Hardy space $H^1(\mathbb{R})$ is not preserved under multiplication by the Gabor window as is the local Hardy space, at least when the window enjoys some regularity. Identifying where this regularity comes into play is all that is really needed to push the techniques of Gröchenig and Heil through. By contrast, Grafakos and Lennard [**5**] interpreted $L^1(\mathbb{R})$ as the endpoint space, taking convergence of Gabor expansions in the sense of Césaro averages, not convergence in norm.

2000 *Mathematics Subject Classification.* Primary 42A45, 42B30, 42C40.

All results extend routinely to $\mathbb{R}^n$ but for simplicity we restrict attention to one dimension and use the same notation as in [**4**]. Standard facts from time-frequency analysis will also be taken for granted [**3**]. Let $T_af(x) = f(x-a)$ and $E_bf(x) = e^{2\pi ibx}f(x)$ denote the respective unitary translation and modulation operators. Then the family $\mathcal{G}(g,\alpha,\beta)$ of functions

$$g_{mn}(x) = E_{m\beta}T_{n\alpha}g(x) = e^{2\pi im\beta x}g(x-n\alpha)$$

is said to form a *Gabor frame* for $L^2(\mathbb{R})$ provided there are constants $A, B > 0$ such that

$$A\|f\|_2^2 \le \sum_{m,n\in\mathbb{Z}} |\langle f, g_{mn}\rangle|^2 \le B\|f\|_2^2. \tag{1.1}$$

Because of the Balian-Low theorem we shall assume throughout that α, β are fixed with $\alpha\beta < 1$. Associated with $\mathcal{G}(g,\alpha,\beta)$ are a *coefficient mapping*

$$C_g : L^2(\mathbb{R}) \to \ell^2(\mathbb{Z}\times\mathbb{Z}), \quad f \mapsto \{\langle f, g_{mn}\rangle\}$$

and a *reconstruction operator*

$$R_g : \ell^2(\mathbb{Z}\times\mathbb{Z}) \to L^2(\mathbb{R}), \quad \{c_{mn}\} \mapsto \sum_{m,n} c_{mn}g_{mn}.$$

Then $\mathcal{G}(g,\alpha,\beta)$ is a frame precisely when the *frame operator* $S_g = R_gC_g$ is bounded and invertible. Setting $\gamma = S_g^{-1}(g)$, we obtain the dual frame $\mathcal{G}(\gamma,\alpha,\beta)$ having the property $I = R_\gamma C_g = R_gC_\gamma$. Each f in $L^2(\mathbb{R})$ can then be written as

$$f = \sum_{n\in\mathbb{Z}} f_n(x)g(x-n\alpha), \qquad f_n = \sum_{m\in\mathbb{Z}} \langle f, \gamma_{mn}\rangle e^{2\pi im\beta x}$$

with f_n a Fourier series having period $1/\beta$, *i.e*, a function in $L^1(\mathbb{T}_\beta)$.

When g, γ generate dual Gabor frames and satisfy a certain *Wiener criterion*, Heil and Gröchenig's characterization of $L^p(\mathbb{R})$, $1 < p < \infty$, in terms of Gabor frames now reduces to two basic facts:

$$\|f\|^p_{L^p(\mathbb{R})} \sim \sum_n \|f_n\|^p_{L^p}, \qquad \|f_n\|_{L^p(\mathbb{R})} \sim \|\sum_m \langle f, \gamma_{mn}\rangle e^{2\pi im\beta x}\|_{L^p(\mathbb{T}_\beta)}.$$

The main result of the present paper is the end-point case $p = 1$; it runs parallel to the range $1 < p < \infty$, replacing $L^p(\mathbb{R})$ with $h^1(\mathbb{R})$ and $L^p(\mathbb{T}_\beta)$ with the periodic Hardy space $H^1(\mathbb{T}_\beta)$ (precise definitions given in detail later).

Denote by S_1 the doubly-indexed sequence space of all Fourier coefficients of functions in $\ell^1(H^1(\mathbb{T}_\beta))$, *i.e.*, all sequences $\{c_{mn}\}$ such that $\sum_m c_{mn}e^{2\pi im\beta x}$ defines a function in $H^1(\mathbb{T}_\beta)$ for each n and

$$\|\{c_{mn}\}\|_{S_1} = \sum_n \Big\{ \|\sum_m c_{mn}e^{2\pi im\beta x}\|_{H^1(\mathbb{T}_\beta)} \Big\} < \infty. \tag{1.2}$$

To introduce the requisite smoothness on g and γ, let $W(\mathrm{Lip}_s, \ell^1)$, $0 < s \le 1$, be the amalgam space of those functions ϕ for which $\sum_{k\in\mathbb{Z}} \|\phi\|_{\mathrm{Lip}_s(k-1,k+1)}$ is finite, where $\|f\|_{\mathrm{Lip}_s(a,b)} = \sup_{x,y\in(a,b),\ x\ne y} \frac{|f(x)-f(y)|}{|x-y|^s}$. Here is the main result of this paper:

THEOREM 1.1. *Let g, γ be functions in $W(\mathrm{Lip}_s, \ell^1)(\mathbb{R})$ for some $s \in (0,1)$ such that $\mathcal{G}(g,\alpha,\beta)$ is a Gabor frame for $L^2(\mathbb{R})$, and let $\mathcal{G}(\gamma,\alpha,\beta)$ be the dual frame. Then*

$$\|f\|_{h^1} \sim \|\{\langle f, g_{mn}\rangle\}\|_{S_1} \sim \|\{\langle f, \gamma_{mn}\rangle\}\|_{S_1}$$

up to norm equivalence.

Once the Hardy spaces have been introduced in Section 2 along with some of their key properties, the proof of Theorem 1.1 is given in Section 3, while several ancillary results are then proved in Section 4. Section 5 concludes the paper with some remarks concerning a wavelet characterization of $H^1(\mathbb{T})$ that brings out the overlapping roles of Gabor functions and wavelets in characterizing the local Hardy space in terms of magnitudes of coefficients. A few simple observations concerning the role of these coefficient characterizations in providing criteria for boundedness of multipliers on $h^1(\mathbb{R})$ are made also, but no attempt is made here to extend Theorem 1.1 to $p < 1$.

2. Local and periodic Hardy spaces

Atomic decompositions provide the most convenient definitions of Hardy spaces for us. A *periodic* function a is said to be an $H^1(\mathbb{T})$-*atom* provided a is supported in a subinterval I of $\mathbb{T}$ such that (i) $|a(\theta)| \leq 1/|I|$ and (ii) $\int a(\theta)\, d\theta = 0$ ([**1**] p. 573).

DEFINITION 2.1. A 1-periodic function h belongs to the real Hardy space $H^1(\mathbb{T})$ if h can be written as $h = \sum_{j=0}^{\infty} \lambda_j a_j$ where $a_0 = 1$ and each a_j is an $H^1(\mathbb{T})$-atom for $j \geq 1$. The norm of h is defined by $\|h\|_{H^1(\mathbb{T})} = \inf \sum_{j=0}^{\infty} |\lambda_j|$, the infimum being taken over all possible such representations of h.

To define the local Hardy space on $\mathbb{R}$ a corresponding notion of atom is needed. We say that a function $b(x)$ on $\mathbb{R}$ is an $h^1(\mathbb{R})$-atom provided b is supported in an interval $I \subset \mathbb{R}$ such that (i) $|b(x)| \leq 1/|I|$ and (ii) if $|I| \leq 1$ then $\int b(x)\, dx = 0$. No moment condition is imposed on b if $|I| > 1$. To distinguish these two types of atoms one says that b is *type I* if $|I| \leq 1$ and is *type II* if $|I| > 1$.

DEFINITION 2.2. A function f on $\mathbb{R}$ belongs to the local Hardy space $h^1(\mathbb{R})$ if f can be written as $f = \sum_{j=1}^{\infty} \mu_j b_j$ with each b_j an $h^1(\mathbb{R})$-atom. The norm of f is $\|f\|_{h^1(\mathbb{R})} = \inf \sum_{j=1}^{\infty} |\mu_j|$, the infimum again being taken over all such representations of f.

Imposing the moment condition on intervals I with $|I| < 1$ is not essential in that one could define an equivalent local Hardy space in terms of α-type I and type II atoms by imposing vanishing moments when $|I| < \alpha$; this becomes important when α is the Gabor frame shift parameter. Both of $H^1(\mathbb{T})$ and $h^1(\mathbb{R})$ are Banach spaces. The fundamental property distinguishing $h^1(\mathbb{R})$ from $H^1(\mathbb{R})$ is that h^1 is invariant under multiplication by Hölder continuous functions *cf.* [**6**] or [**2**]:

LEMMA 2.3. *If* $\sigma(x) \in \mathrm{Lip}_s(\mathbb{R})$ *for some* $s \in (0, 1]$ *then* $f \mapsto M_\sigma(f)(x) = \sigma(x)f(x)$ *is bounded on* $h^1(\mathbb{R})$ *with norm depending linearly on* $\|\sigma\|_{\mathrm{Lip}_s}$.

The crucial point is that functions in $H^1(\mathbb{R})$ have integral zero – a property that will usually be destroyed after pointwise multiplication by a Lipschitz function.

3. Proof of the Main Theorem

Several preliminary facts are required for Theorem 1.1. In order to get to its proof quickly, we collect proofs of some of the preliminary facts in Section 4. The first says that a sufficiently regular cut-off defines a bounded mapping from $H^1(\mathbb{T})$ (thought of as periodic functions on $\mathbb{R}$) to $h^1(\mathbb{R})$.

PROPOSITION 3.1. *If a is an $H^1(\mathbb{T})$ atom, thought of as a periodic function on $\mathbb{R}$, and if η is a compactly supported Lip_1-function, then pointwise multiplication $a \mapsto M_\eta(a)(x) = \eta(x)a(x)$ is bounded on $h^1(\mathbb{R})$ and satisfies the inequality*

$$\|M_\eta(a)\|_{h^1} \leq C\|\eta\|_{\mathrm{Lip}_1}.$$

Here C depends on the length of the support of η.

COROLLARY 3.2. *$f \mapsto M_\eta(f)$ is continuous from $H^1(\mathbb{T})$ to $h^1(\mathbb{R})$.*

The corollary follows immediately from the atomic definition of $H^1(\mathbb{T})$. The corresponding localization for L^p is trivial and requires only essential boundedness and compact support of η, so Proposition 3.1 brings out a critical technicality not present in the L^p case.

Inequalities showing how Gabor expansions provide a relationship between $h^1(\mathbb{R})$ and the sequence space S_1 are also required.

PROPOSITION 3.3. *Let $\{p_n\}$ be a sequence in $H^1(\mathbb{T}_\beta)$ such that $\sum_n \|p_n\|_{H^1} < \infty$ and let $g \in W(\mathrm{Lip}_s, \ell^1)$ for some $s \in (0,1)$. Then*

$$\Big\|\sum_n p_n T_{\alpha n} g\Big\|_{h^1(\mathbb{R})} \leq C\|g\|_{W(\mathrm{Lip}_s,\ell^1)} \sum_n \|p_n\|_{H^1(\mathbb{T}_\beta)}.$$

The proof of Theorem 1.1 requires a dual version of this result. The duals of $H^1(\mathbb{T})$ and $h^1(\mathbb{R})$, denoted respectively by $\mathrm{BMO}(\mathbb{T})$ and $\mathrm{bmo}(\mathbb{R})$, are the spaces of functions of bounded mean oscillation on the circle and of bounded local mean oscillation on $\mathbb{R}$, see [**6**]. Recall that a function $b \in \mathrm{bmo}$ satisfies $\frac{1}{|I|}\int_I |b - b_I| \leq C$ when $|I| \leq 1$ and $\frac{1}{|I|}\int_I |b| \leq C$ when $|I| > 1$; the smallest such constant C is the bmo-norm of b. The dual pairings of the BMO spaces with the corresponding Hardy spaces are given by integrations over $\mathbb{T}$ and $\mathbb{R}$ respectively. The required result dual to Proposition 3.3 is:

PROPOSITION 3.4. *If $g \in W(\mathrm{Lip}_s, \ell^1)$ and $\{q_n\} \subset \mathrm{BMO}(\mathbb{T}_\beta)$ then*

$$\Big\|\sum_n q_n T_{\alpha n} g\Big\|_{\mathrm{bmo}} \leq C\|g\|_{W(\mathrm{Lip}_s,\ell^1)} \sup_n \|q_n\|_{\mathrm{BMO}}.$$

Armed with these preliminary facts we can now start the proof of Theorem 1.1.

Suppose that $\alpha > 0, \beta > 0$ are fixed as before, and that $\mathcal{G}(g,\alpha,\beta)$ generates a Gabor frame with dual γ such that both $g, \gamma \in W(\mathrm{Lip}_s, \ell^1)$ for some $r \in (0,1)$. As before, let

$$C_g f = \{\langle f, E_{\beta m} T_{\alpha n} g\rangle\} \quad \text{and} \quad R_g\{c\} = \sum_{m,n} c_{mn} e^{2\pi i m\beta x} g(x - n\alpha) \tag{3.1}$$

denote the Gabor coefficient mapping and Gabor reconstruction operator respectively. Recall that these operators satisfy $R_g C_\gamma = I = R_\gamma C_g$ when $\mathcal{G}(g,\alpha,\beta)$ and $\mathcal{G}(\gamma,\alpha,\beta)$ are dual Gabor frames. More precisely, the identities apply, a-priori, to h^1 atoms and extend by respective limiting or duality arguments to the spaces in question. The sequence $c = \{c_{mn}\}$ belongs to the space S_1 provided the sequence $\{p_n\}$ defined by $p_n(x) = \sum_m c_{mn} e^{2\pi i m\beta x}$ belongs to the space $\ell^1(\alpha\mathbb{Z}, H^1(\mathbb{T}_\beta))$ and the norm on S_1 is the sum of the H^1 norms of the p_n. Having recalled these conventions, Proposition 3.3 is then nothing other than the inequality,

$$\|R_g c\|_{h^1(\mathbb{R})} \leq C\|g\|_{W(\mathrm{Lip}_s,\ell^1)}\|c\|_{S_1}. \tag{3.2}$$

Similarly, Proposition 3.4 says that the reconstruction mapping is bounded from S_∞ to bmo. Here, S_∞ consists of those sequences d_{mn} such that $q_n = \sum_m d_{mn} e^{2\pi i m\beta x} \in \mathrm{BMO}(\mathbb{T}_\beta)$. The norm on the space is the $\ell^\infty(\mathrm{BMO})$ norm. The fundamental roles of duality along with Theorem 1.7 appear in the following:

LEMMA 3.5. *Let* $f \in h^1(\mathbb{R})$. *Then*

$$\|C_\gamma f\|_{S_1} \leq C\|\gamma\|_{W(\mathrm{Lip}_s, \ell^1)}\|f\|_{h^1}.$$

PROOF OF LEMMA 3.5. Since $C_\gamma R_g = I$ and $R_g = C_g^*$ it follows that, for any $f \in h^1$ and sequence d in S_∞,

$$\big|\langle C_\gamma f, C_g R_\gamma d\rangle\big| = \big|\langle R_g C_\gamma f, R_\gamma d\rangle\big| \leq \|f\|_{h^1}\|R_\gamma d\|_{\mathrm{bmo}} \leq C\|f\|_{h^1}\|d\|_{S_\infty}$$

by Proposition 3.4. Since S_∞ is the dual of S_1 the lemma now follows. □

Now we can complete the proof of Theorem 1.1. Combining (3.2) and Lemma 3.5 proves that $\|f\|_{h^1} \sim \|\{\langle f, g_{mn}\rangle\}\|_{S_1}$. The corresponding norm equivalence with g replaced by γ follows from the same argument. This proves the Main Theorem.

4. Proofs of preliminary facts

It remains to prove Proposition 3.1 and Propositions 3.3 and 3.4. It will be convenient to assume $\beta = 1$.

PROOF OF PROPOSITION 3.1. To prove Proposition 3.1, first note that if $a = a_0 = 1$ then $M_\eta(a_0) = \eta \in h^1(\mathbb{R})$ because $h^1(\mathbb{R})$ contains Lipschitz functions with compact support. To simplify the argument slightly, assume that supp $\eta \subset [-\gamma, 1+\gamma]$ for some $0 < \gamma < 1/2$. Suppose then that a is supported in some circular interval I and satisfies $|a| \leq 1/|I|$ and $\int_{\mathbb{T}} a = 0$. The only potential problem occurs when a has short support so, to fix ideas, assume that the periodic support of a lies in $[1-\delta, \delta]$ where $\gamma > \delta > 0$.

Under the *worst case* hypotheses that we have put on a, $M_\eta(a)$ will be localized in two disjoint regions $[-\delta, \delta] \cup [1-\delta, 1+\delta]$. Fix $b \in$ bmo such that $\|b\|_{\mathrm{bmo}} = 1$. The support hypotheses on η and moment condition on a allow one to write

$$\begin{aligned}
\langle M_\eta(a), b\rangle &= \int_{-\delta}^{\delta} \eta ab + \int_{1-\delta}^{1+\delta} \eta ab \\
&= \int_{-\delta}^{\delta} (\eta(\cdot) - \eta(0))ab + \eta(0)\int_{-\delta}^{\delta} a(b - b_\delta) + \dots \\
&= \mathrm{I} + \mathrm{II} + \dots
\end{aligned}$$

where "..." denotes a similar breakdown of the integral centered at $x = 1$ to which corresponding estimates can be applied. Here, b_δ is the average of b over $[-\delta, \delta]$. Then,

$$|\mathrm{I}| = \left|\int_{-\delta}^{\delta} (\eta(\cdot) - \eta(0))ab\right| \leq \delta\|\eta\|_{\mathrm{Lip}}\|a\|_\infty \int_{-\delta}^{\delta} |b| \leq \frac{1}{2}\|\eta\|_{\mathrm{Lip}} \int_{-1}^{1} |b| \leq \|\eta\|_{\mathrm{Lip}}.$$

Here we have used the size estimate on a, the Lipschitz condition on η, and the assumption $\|b\|_{\mathrm{bmo}} = 1$. On the other hand,

$$|\mathrm{II}| \leq \|\eta\|_\infty\|a\|_\infty \int_{-\delta}^{\delta} |b - b_\delta| \leq \|\eta\|_\infty \frac{1}{2\delta}\int_{-\delta}^{\delta} |b - b_\delta| \leq \|\eta\|_\infty.$$

Here we have used the size and support condition on a and, again, the definition of the norm on bmo. Similar estimates apply to the term "..." The Corollary then follows from writing down the atomic decomposition of $f = \sum_{j=0}^{\infty} \lambda_j a_j \in H^1(\mathbb{T})$ and applying the argument above to each atom individually to conclude that if $b \in$ bmo then $|\langle M_\eta(f), b\rangle| \le C\|\eta\|_{\mathrm{Lip}}\|b\|_{\mathrm{bmo}} \sum_{j=0}^{\infty} |\lambda_j|$. Taking infima in the appropriate order then yields Proposition 3.1 and Corollary 3.2. □

In what follows, let $\eta \ge 0$ be a Lipschitz function having compact support such that $\sum_{l\in\mathbb{Z}} \eta(x - \alpha l) \equiv 1$ for fixed $\alpha > 0$. Let $\tilde\eta$ be a smooth bump function as well, chosen such that $\eta\tilde\eta = \eta$.

Proof of Proposition 3.3. With $\eta, \tilde\eta$ as above, $T_{\alpha n}\eta(x) = T_{\alpha n}\eta(x) T_{\alpha n}\tilde\eta(x)$ and $\sum_n T_{\alpha n}\eta(x) = 1$, and so

$$\begin{aligned}\sum_n p_n(x) T_{\alpha n} g(x) &= \sum_n p_n(x) T_{\alpha n} g(x) \sum_l T_{\alpha(n+l)}\eta(x)\\ &= \sum_n p_n(x) \sum_l T_{\alpha(n+l)}\eta(x) T_{\alpha n} g(x) T_{\alpha(n+l)}\tilde\eta(x)\\ &= \sum_{n,l} p_n T_{\alpha(n+l)}\eta \Big[T_{\alpha n} g T_{\alpha(n+l)}\tilde\eta\Big].\end{aligned}$$

Consequently, by the triangle inequality, Lemma 2.3, and translation invariance of the Lip_s norm,

$$\begin{aligned}\|\sum_n p_n T_{\alpha n} g\|_{h^1(\mathbb{R})} &\le C\sum_l\sum_n \|T_{\alpha n} g T_{\alpha(n+l)}\tilde\eta\|_{\mathrm{Lip}_s(\mathbb{R})} \|p_n T_{\alpha(n+l)}\eta\|_{h^1(\mathbb{R})}\\ &= C\sum_l \|g T_{\alpha l}\tilde\eta\|_{\mathrm{Lip}_s(\mathbb{R})} \sum_n \|p_n T_{\alpha(n+l)}\eta\|_{h^1(\mathbb{R})}\\ &\le C'\sum_l \|g T_{\alpha l}\tilde\eta\|_{\mathrm{Lip}_s(\mathbb{R})} \sum_n \|p_n\|_{H^1(\mathbb{T})},\end{aligned}$$

using Corollary 3.2 in the final step. But

$$\|g T_{\alpha l}\tilde\eta\|_{\mathrm{Lip}_s} \le C\|\tilde\eta\|_{\mathrm{Lip}_1}\|g\|_{\mathrm{Lip}_s(I_{\alpha l})}$$

with $I_{\alpha l} = \mathrm{supp}\ \tilde\eta + \alpha l$, since $s < 1$. Hence, summing over l, we see that

$$\|\sum_n p_n T_{\alpha n} g\|_{h^1(\mathbb{R})} \le C''\|g\|_{W(\mathrm{Lip}_s,\ell^1)} \sum_n \|p_n\|_{H^1(\mathbb{T})}.$$

Note that C'' depends on α because it reflects how many shifts $T_{\alpha l}\tilde\eta$ have support intersecting $[-1, 1)$. This proves Proposition 3.3. □

Proof of Proposition 3.4. One must show that for any h^1 atom a,

$$\Big|\int a \sum_n q_n T_{\alpha n} g\Big| \le C\|g\|_{W(\mathrm{Lip}_s,\ell^1)} \sup_n \|q_n\|_{\mathrm{BMO}}.$$

The result then follows from the α-atomic decomposition of $h^1(R)$.

First suppose that a is an α-type I atom. In this case one may assume without loss of generality that supp $a \subset \{x : \eta(x) = 1\}$. In general, the sum of a fixed finite number of shifts of η will add to one on supp a. Therefore

$$\begin{aligned}|\int a \sum_n q_n T_{\alpha n} g| &= |\sum_n \int \eta a \; q_n \tilde{\eta} T_{\alpha n} g \;| \\ &\leq \sum_n \|\eta a\|_{h^1} \|q_n \tilde{\eta} T_{\alpha n} g\|_{\mathrm{bmo}} \\ &\leq C \sum_n \|\tilde{\eta} T_{\alpha n} g\|_{\mathrm{Lip}_s} \|q_n\|_{\mathrm{BMO}} \leq C \|g\|_{W(\mathrm{Lip}_s, \ell^1)} \sup_n \|q_n\|_{\mathrm{BMO}}\end{aligned}$$

where the penultimate inequality essentially follows from the same arguments as given in the previous proposition, using now the fact that Lip_s is a bmo module under pointwise multiplication (*cf.* [**6**]).

Next suppose that a is an α-type II atom. Then $a = \sum_{n: aT_{\alpha n}\eta \neq 0} aT_{\alpha n}\eta$ and since, essentially, $aT_{\alpha j}\eta$ can be regarded as a type II atom divided by the length of the atomic support I of a, one has $\|aT_{\alpha j}\eta\|_{h^1} \leq C/|I|$. Applying the same reasoning as above one again obtains the desired inequality. This proves the proposition and completes the proof of the ancillary results for Theorem 1.1. □

5. Concluding remarks

5.1. Gabor meets Littlewood-Paley and wavelets. As Gröchenig and Heil noted, $\| \sum_m c_m e^{2\pi i m x}\|_p$ can be characterized in terms of the Littlewood-Paley square function $S(\{c_m\})(x) = \left(\sum_{j=0}^{\infty} |\sum_{m \in \Lambda_j} c_m e^{2\pi i m x}|^2\right)^{1/2}$ which satisfies $\|S(f)\|_{L^p(\mathbb{T})} \sim \|f\|_{L^p(\mathbb{T})}$ when $1 < p < \infty$ and $f = \sum_m c_m e^{2\pi i m x}$. Here $\Lambda_j = \{m : |m| \leq 2^j\} \setminus \{m : |m| \leq 2^{j-1}\}$. To extend such a characterization to $H^1(\mathbb{T})$ one needs a slightly mollified square function. One of the early mathematical observations in wavelet theory was that wavelets could provide a corresponding Littlewood-Paley characterization of $L^p(\mathbb{R})$ in terms of a certain *wavelet square function*. Although several basic aspects of Triebel-Lizorkin spaces have been worked out in the periodic case (*e.g,* [**8**], [**9**]), the role of wavelet square functions in providing a discrete norm for $H^1(\mathbb{T})$ seems not to have been reported. We will describe such a characterization here without proofs, as the methods – essentially periodic versions of basic estimates of Frazier and Jawerth [**2**], (*cf.* [**7**]) – would lead us astray from the central focus of this work.

In the following remarks it will be convenient to suppose that ψ is a *bandlimited wavelet.* In particular, one assumes that $\psi_{jk}(x) = 2^{j/2}\psi(2^j x - k)$ forms an orthonormal basis and that $\hat{\psi}$ has compact support. Moreover one imposes that $\hat{\psi}$ is Lipschitz. This rules out the case of the Shannon wavelet $\psi(x) = (\sin 2\pi x - \sin \pi x)/(\pi x)$ that, in what follows, corresponds to the square function S above. It is a simple matter to check that if ψ comes from a multiresolution analysis for $L^2(\mathbb{R})$ with orthogonal scaling function ϕ then the periodizations $\pi(\phi)$ and $\pi(\psi_{jk})$, $j = 0, 1, \ldots$ and $k = 0, \ldots, 2^j - 1$ together form an orthonormal basis for $L^2(\mathbb{T})$. Here $\pi(f)(x) = \sum_{l \in \mathbb{Z}} f(x + l)$. One defines the *periodic Treibel-Lizorkin space* (*cf.* [**8**]) $F^1(\mathbb{T})$ to consist of those periodic functions h such that

$$\|h\|_{F^1(\mathbb{T})} = \|h * \pi(\phi)\|_{L^1(\mathbb{T})} + \left\| \left(\sum_{j=0}^{\infty} |h * \pi(\psi_j)(\cdot)|^2 \right)^{1/2} \right\|_{L^1(\mathbb{T})} < \infty.$$

The convolution of course is on $\mathbb{T}$. Ostensibly, the definition of $F^1(\mathbb{T})$ depends on ψ but it can be shown that different, but sufficiently regular, wavelets give rise to equivalent norms. One has the following Littlewood-Paley type of characterization of $H^1(\mathbb{T})$:

THEOREM 5.1. *A periodic function h belongs to $H^1(\mathbb{T})$ if and only if $h \in F^1(\mathbb{T})$. Moreover,* $\|h\|_{H^1} \sim \|h\|_{F^1}$.

Notice that $e^{2\pi imx} * \pi(\psi_j) = 2^{-j/2}e^{2\pi imx}\hat{\psi}(m/2^j)$. This suggests defining

$$S_\psi(\{c_m\}) = \Big|\sum_{m\in\mathbb{Z}} \hat{\phi}(m)c_m e^{2\pi im\beta x}\Big| + \Big(\sum_{j=0}^{\infty}\Big|\sum_{m\in\mathbb{Z}} \hat{\psi}_j(m)c_m e^{2\pi im\beta x}\Big|^2\Big)^{1/2}$$

as a square function substitute for $S(\{c_m\})$ above. Using an appropriate rescaling to take into account the parameter β, one then has the following characterization of S_1:

COROLLARY 5.2. If $\{c_{mn}\} \in S_1$ and ψ is as above then

$$\|\{c_{mn}\}\|_{S_1} \sim \sum_{n\in\mathbb{Z}} \|S_\psi(\{c_{mn}\})\|_{L^1(\mathbb{T}_\beta)}.$$

Consequently, one obtains via Theorem 1.1 a sort of wavelet-Gabor characterization of $h^1(\mathbb{R})$ in the spirit of Gröchenig and Heil's Gabor characterization of $L^p(\mathbb{R})$.

5.2. Multipliers. A Fourier multiplier for $H^1(\mathbb{T}_\beta)$ is a sequence $d = \{d_m\}$ such that the mapping

$$M_d : f = \sum_m c_m e^{2\pi im\beta x} \mapsto \sum_m c_m d_m e^{2\pi im\beta x}$$

is bounded on $H^1(\mathbb{T}_\beta)$. Let $d_n = \{d_{mn}\}_m$ define a sequence of $H^1(\mathbb{T}_\beta)$ multipliers whose operator norms $\|M_{d_n}\|_{H^1\to H^1}$ are uniformly bounded. It follows immediately from Theorem 1.1 that the induced mapping

$$M : \{c_{mn}\} \mapsto \{c_{mn}d_{mn}\}$$

is bounded on S_1 with $\|M\|_{S_1\to S_1} \leq \sup_n \|M_{d_n}\|_{H^1\to H^1}$ and so:

COROLLARY 5.3. Under the hypotheses of Theorem 1.1 and the assumption that $\sup_n \|M_{d_n}\|_{H^1\to H^1} < \infty$, the mapping

$$M_{\gamma,d} : f = \sum_{mn} \langle f, \gamma_{mn}\rangle g_{mn} \mapsto \sum_{mn} \langle f, \gamma_{mn}\rangle d_{mn} g_{mn}$$

is bounded on $h^1(\mathbb{R})$.

One can take advantage of Theorem 5.1 to obtain useful conditions for defining H^1-multipliers.

References

[1] R. Coifman and G. Weiss, *Extensions of Hardy spaces and their use in analysis*, Bull. Amer. Math. Soc. **83** (1977), 569–645.

[2] M. Frazier and B. Jawerth, *A discrete transform and decomposition of distribution spaces*, J. Funct. Anal. **93** (1990), 34–170.

[3] K. Gröchenig, *Foundations of Time-Frequency Analysis*, Birkhäuser, Boston, 2000.

[4] K. Gröchenig and C. Heil, *Gabor meets Littlewood-Paley: Gabor expansions in $L^p(\mathbb{R}^d)$*, Studia Math. **146** (2001), 15–33.

[5] L. Grafakos and C. Lennard, *Characterization of $L^p(\mathbb{R}^n)$ using the Gabor frame*, J. Fourier Anal. Appl. **7** (2001), 101–126.

[6] D. Goldberg, *A local version of real Hardy spaces*, Duke Math. J. **46** (1979), 27–42.
[7] E. Hernández and G. Weiss, *A First Course on Wavelets*, CRC Press, Boca Raton, 1996.
[8] H. Triebel, *Theory of Function Spaces*, Birkhäuser, Basel, 1983.
[9] H.-J. Schmeisser and H. Triebel, *Topics in Fourier Analysis and Function Spaces*, J. Wiley, New York, 1987.

DEPARTMENT OF MATHEMATICS, THE UNIVERSITY OF TEXAS AT AUSTIN, AUSTIN, TX 78712
E-mail address: gilbert@math.utexas.edu

DEPARTMENT OF MATHEMATICAL SCIENCES, NEW MEXICO STATE UNIVERSITY, LAS CRUCES, NM 88001
E-mail address: jlakey@nmsu.edu

Contemporary Mathematics
Volume **345**, 2004

Riesz bases, multiresolution analyses, and perturbation

Alfredo L. González and Richard A. Zalik

ABSTRACT. This article is partly a review of known results and partly a research paper. We study affine sequences with dilation factor 2, as well as sequences of translates, generated by a single function in $L^2(\mathbb{R})$.

In the first section we introduce our notation, some definitions, and some basic theorems. In Section 2 we discuss properties of frame and Riesz bases of translates and clarify the proofs of several theorems. Section 3 reviews results on perturbations of the Haar wavelet. In Section 4 we study the characterization of MRA orthonormal and Riesz wavelets. We also discuss the relationship between functions associated with an MRA, and functions generated by an MRA. Finally, Section 5 contains old and new results on the convolution of Riesz bases, scaling functions, and multiresolution analyses.

Many of the results discussed here have been obtained for the univariate case only. Their extension to a more general setting is a source of interesting open problems.

1. Introduction

In this section we will introduce our notation, some definitions, and some basic theorems.

In what follows $\mathbb{Z}$ will denote the integers, $\mathbb{Z}^+$ the nonnegative integers, and $\mathbb{R}$ the real numbers; t and x will always denote real variables. The Fourier transform of a function f will be denoted by $\widehat{f}$. If $f \in L(\mathbb{R})$,

$$\widehat{f}(x) := \int_{\mathbb{R}} e^{-2\pi txi} f(t)\, dt.$$

Other definitions of the Fourier transform are common in the literature. Since all these definitions are equivalent modulo a linear change of variable and, in some cases, multiplication by a suitable constant, we have modified, where needed, all the statements quoted herein, so as to conform to the definition given above.

Let $\mathbb{H}$ be a (separable) Hilbert space with inner product $\langle \cdot, \cdot \rangle$ and norm $\| \cdot \| := \langle \cdot, \cdot \rangle^{1/2}$, let $\Lambda \subset \mathbb{Z}$, and let $\mathbf{F} := \{ f_k \,;\, k \in \Lambda \} \subset \mathbb{H}$. $\mathbf{F}$ is called a *frame* if there are

2000 *Mathematics Subject Classification.* 42C40.

Key words and phrases. Bessel wavelets, frame wavelets, Riesz wavelets, biorthogonal wavelets, orthogonal wavelets, smooth compactly supported wavelets, convolution, multiresolution analysis.

constants $0 < A \le B$ such that for every $f \in \mathbb{H}$

$$A\|f\|^2 \le \sum_{k\in\Lambda} |\langle f, f_k\rangle|^2 \le B\|f\|^2. \tag{1.1}$$

The constants A and B are called (lower and upper) *bounds* of the frame. The operator $S : \mathbb{H} \to \mathbb{H}$ defined by

$$Sf := \sum_{k\in\Lambda} \langle f, f_k\rangle f_k$$

is called the *frame operator*. It is invertible and self–adjoint, and for every $f \in \mathbb{H}$

$$f = \sum_{k\in\Lambda} \langle f, S^{-1} f_k\rangle f_k = \sum_{k\in\Lambda} \langle f, f_k\rangle S^{-1} f_k.$$

If only the right-hand inequality in (1.1) is satisfied for all $f \in \mathbb{H}$, then $\mathbf{F}$ is called a *Bessel sequence* with bound B. If $\mathbf{a} := \{ a_k \ ; \ k \in \Lambda \}$, then $\mathbf{F}$ is a Bessel sequence with Bessel bound B if and only if the *analysis operator*

$$T : \mathbf{a} \longrightarrow \sum_{k\in\Lambda} a_k f_k$$

is a well-defined bounded operator from $\ell^2(\Lambda)$ into $\mathbb{H}$ and $||T|| \le \sqrt{B}$. T is also called the *pre–frame operator*

$\mathbf{F}$ is called a *Riesz basis* if its linear span is dense in $\mathbb{H}$ and there are constants A and B, $A > 0$, such that for every sequence $\mathbf{a} \in \ell^2(\Lambda)$

$$\sqrt{A}||\mathbf{a}|| \le ||T\mathbf{a}|| \le \sqrt{B}||\mathbf{a}||. \tag{1.2}$$

The constants A and B are called (lower and upper) *bounds* of the Riesz basis. Every orthonormal basis is a Riesz basis with bounds $A = B = 1$, every Riesz basis is a frame, and Riesz bounds and frame bounds coincide. A sequence is a Riesz basis if and only if it is a frame having the additional property that upon the removal of any element from the sequence, it ceases to be a frame.

$\mathbf{F}$ is called a *frame sequence* (resp. a *Riesz sequence*) if it is a frame (resp. a Riesz basis) in the closure of its linear span. Riesz sequences are also called *exact frame sequences*. Note that if $\mathbf{F}$ is an orthogonal sequence, then it is an orthogonal basis of the closure of its linear span.

Two sequences $\mathbf{F} := \{ f_k \ ; \ k \in \Lambda \}$ and $\mathbf{G} := \{ g_k \ ; \ k \in \Lambda \}$ in $\mathbb{H}$ are said to be *biorthogonal* if $\langle f_m, g_k\rangle = \delta_{m,k}$ for every m and k; we also say that $\mathbf{G}$ is biorthogonal to $\mathbf{F}$. If $\mathbf{F}$ is a Riesz basis then it has a unique biorthogonal sequence $\mathbf{G}$, and this sequence is a Riesz basis. Moreover, if A and B are Riesz bounds for $\mathbf{F}$, then $1/B$ and $1/A$ are Riesz bounds for $\mathbf{G}$. For a discussion of the theory of frames and Riesz bases see [**14**], and also [**7, 43**].

Here is a characterization of frame and Riesz sequences that will be useful in the subsequent discussion:

THEOREM 1.1.
(a) *A sequence* $\mathbf{F} \in \mathbb{H}$ *is a frame sequence if and only if the pre-frame operator* T *satisfies* (1.2) *for every* $\mathbf{a} \in (\ker T)^{\perp}$.
(b) *In particular,* $\mathbf{F}$ *is a Riesz sequence if and only if* T *is one–to–one.*

The proof of (a) follows from [**14**, Lemma 5.5.4]. Since T is one–to–one if and only if $\ker T = \{0\}$, (b) follows trivially from the definition of Riesz sequence.

In this paper the underlying Hilbert space will be $L^2(\mathbb{R})$ with the usual inner product and norm, and we will study affine sequences with dilation factor 2 generated by a single function, i.e., sequences of the form $\{\psi_{j,k}\ ;\ j,k \in \mathbb{Z}\}$, where $\psi \in L^2(\mathbb{R})$ and $\psi_{j,k}(t) := 2^{j/2}\psi(2^j t - k)$, as well as sequences of translates generated by a single function, i.e., sequences of the form $\{\varphi_k\ ;\ k \in \mathbb{Z}\}$, where $\varphi \in L^2(\mathbb{R})$ and $\varphi_k := \varphi_{0,k} = \varphi(t-k)$.

Although the rest of the theorems in this article are stated for functions of a real variable, we direct the reader to generalizations to the multivariate case, whenever we are aware of any.

A function $\psi \in L^2(\mathbb{R})$ will be called a frame wavelet, a Bessel wavelet, a Riesz wavelet, or an orthonormal wavelet, if the affine sequence $\{\psi_{j,k}\ ;\ j,k \in \mathbb{Z}\}$ it generates is respectively a frame, a Bessel sequence, a Riesz basis, or an orthonormal basis of $L^2(\mathbb{R})$. When we refer to the bounds of a wavelet ψ we mean the frame, Bessel, or Riesz bounds, as the case may be, of the affine sequence generated by ψ. A Riesz wavelet ψ is said to be biorthogonal if there is a function $\tau \in L^2(\mathbb{R})$ such that $\{\tau_{j,k}\ ;\ k \in \mathbb{Z}\}$ is biorthogonal to $\{\psi_{j,k}\ ;\ k \in \mathbb{Z}\}$. The discussion of the preceding paragraph implies that τ is a Riesz wavelet. It is also easy to see that τ must be unique. We call τ the (Riesz) wavelet biorthogonal to ψ.

A *multiresolution analysis* (MRA) is a sequence $\{V_j\ ;\ j \in \mathbb{Z}\}$ of closed linear subspaces of $L^2(\mathbb{R})$ such that:

(i)
$$V_j \subset V_{j+1} \text{ for every } j \in \mathbb{Z}.$$

(ii)
$$\text{For every } j \in \mathbb{Z},\ f(t) \in V_j \text{ if and only if } f(2t) \in V_{j+1}.$$

(iii)
$$\bigcup_{j\in\mathbb{Z}} V_j \text{ is dense in } L^2(\mathbb{R}).$$

(iv)
There is a function φ such that $\{\varphi_k\ ;\ k \in \mathbb{Z}\}$ is an orthonormal basis of V_0.

For any function $\varphi \in L^2(\mathbb{R})$, let
$$\Phi(x) := \sum_{k\in\mathbb{Z}} |\widehat{\varphi}(x+k)|^2.$$
If $\{\varphi_k\ ;\ k \in \mathbb{Z}\}$ is a Riesz basis of V_0, and $\sigma(x) := \widehat{\varphi}(x)[\Phi(x)]^{-1/2}$, it is well known that $\{\sigma_k\ ;\ k \in \mathbb{Z}\}$ is an orthonormal basis of V_0 (see, e.g., [**17**, p. 139], and also Theorem 2.2 below). Thus, condition (iv) can be replaced by the following:

(iv′)
There is a function φ such that $\{\varphi_k\ ;\ k \in \mathbb{Z}\}$ is a Riesz basis of V_0.

Traditionally, the definition of MRA includes an additional condition:

(v)
$$\bigcap_{j\in\mathbb{Z}} V_j = \{0\}.$$

It was later discovered that (iv′) (and therefore (iv)) implies (v) [**17**, p. 141, Proposition 5.3.1] (see also [**28**, p. 45, Theorem 1.6] for a proof that (i), (ii), and (iv) imply (v)).

It follows from the definition of MRA that there is a 1–periodic function $p \in L^2[0,1]$ such that

$$\widehat{\varphi}(2x) = p(x)\widehat{\varphi}(x) \qquad \text{a.e.} \tag{1.3}$$

The function φ is called a *scaling function* for the MRA, (1.3) is called a *two scale equation*, and p is called the *low pass filter* associated with φ. If $\{\,\varphi_k\ ;\ k \in \mathbb{Z}\,\}$ is an orthonormal basis of V_0, then φ will be called an *orthonormal scaling function* for the MRA. This definition is consistent with [**28**, p. 53]. For a thorough study of scaling functions see [**38**].

A function will be called a (orthonormal) scaling function, if it is a (orthonormal) scaling function for some MRA.

The following theorem shows that in certain cases, condition (iii) may be replaced by the assumption that the function φ does not vanish at the origin. It will be used in the proof of Theorem 5.8 below.

THEOREM 1.2. [**28**, p. 46, Theorem 1.7]. *Let $\{\,V_j\ ;\ j \in \mathbb{Z}\,\}$ be a sequence of closed linear subspaces of $L^2(\mathbb{R})$ satisfying* (i), (ii), *and* (iv)*; assume that the scaling function φ of condition* (iv) *is such that $|\widehat{\varphi}|$ is continuous at* 0*. Then* (iii) *is satisfied if and only if $\widehat{\varphi}(0) \neq 0$. If either is the case, $|\widehat{\varphi}(0)| = 1$.*

The basic characterization of orthonormal wavelets is due to Gripenberg [**24**] and Wang [**41**]. See the comments at the end of Chapter 7 of [**28**] for an account of the history of this result.

THEOREM 1.3. [**28**, p. 332]. *A function $\psi \in L^2(\mathbb{R})$ is an orthonormal wavelet if and only if the following three conditions are satisfied:*

(a) $\|\psi\| = 1$.

(b) $\displaystyle\sum_{j\in\mathbb{Z}} |\widehat{\psi}(2^j x)|^2 = 1 \qquad a.e.$

(c) $\displaystyle\sum_{j\in\mathbb{Z}^+} \widehat{\psi}(2^j x)\,\overline{\widehat{\psi}(2^j(x+m))} = 0, \qquad a.e., \qquad m \in 2\mathbb{Z}+1.$

Theorem 1.3 has been extended to several variables by M. Frazier, G. Garrigós, K. Wang, and G. Weiss [**19**] (see also [**18**]). This theorem allows us to determine whether a given function is an orthonormal wavelet, but it does not tell us how to actually construct a wavelet. This problem, in its full generality, is still open. A constructive method for MRA orthogonal wavelets will be discussed in Section 4.

2. Frames and Riesz bases of translates

In what follows we will use the following notation: $I := [0,1]$, and, for $\varphi \in L^2(\mathbb{R})$,

$$r(x) := [\Phi(x)]^{1/2} = \left[\sum_{k\in\mathbb{Z}} |\widehat{\varphi}(x+k)|^2\right]^{1/2}. \tag{2.1}$$

If $\Lambda \subset \mathbb{Z}$, define

$$H_\Lambda := \overline{\operatorname{span}}\,\{\, e^{2\pi kxi}\ ;\ k \in \Lambda \,\},$$

and let E_Λ be the closed subspace of H_Λ of all f such that $f(x)\Phi(-x) = 0$ a.e.

The following is part of the statement of [**10**, Theorem 2.1]. The proof is essentially the original one.

THEOREM 2.1. *Suppose* $\varphi \in L^2(\mathbb{R})$*, let* $\Lambda \subset \mathbb{Z}$*, and let* $\mathbf{S} := \{\, \varphi_k\ ;\ k \in \Lambda \,\}$*. Then:*
(a) $\mathbf{S}$ *is a frame sequence with frame bounds* A *and* B *if and only if for every* $f \in H_\Lambda \cap E_\Lambda^\perp$*,*

$$A\,\|f\|^2 \le \int_I |f(x)|^2 \Phi(-x)\, dx \le B\,\|f\|^2. \tag{2.2}$$

(b) *Furthermore, if the condition of* (a) *is satisfied, then* $\mathbf{S}$ *is a Riesz sequence with the same bounds if and only if* $E_\Lambda = \{0\}$*.*

PROOF. (a) Note that $\mathbf{a} := \{\, a_k\ ;\ k \in \Lambda \,\} \in \ell^2(\Lambda)$ if and only if

$$f(x) := \sum_{k \in \Lambda} a_k e^{2\pi kxi} \in H_\Lambda.$$

If T denotes the analysis operator, then, using Plancherel's identity we have:

$$\begin{aligned}
\|T\mathbf{a}\|^2 &= \Big\| \sum_{k\in\Lambda} a_k \varphi_k \Big\|^2 = \int_{\mathbb{R}} \Big| \sum_{k\in\Lambda} a_k e^{-2\pi kxi} \widehat{\varphi}(x) \Big|^2 \, dx \\
&= \int_{\mathbb{R}} \Big| \sum_{k\in\Lambda} a_k e^{2\pi kxi} \widehat{\varphi}(-x) \Big|^2 \, dx \\
&= \int_I \Big| \sum_{k\in\Lambda} a_k e^{2\pi kxi} \Big|^2 \sum_{r\in\mathbb{Z}} |\widehat{\varphi}(-x-r)|^2 \, dx \\
&= \int_I \Big| \sum_{k\in\Lambda} a_k e^{2\pi kxi} \Big|^2 \sum_{r\in\mathbb{Z}} |\widehat{\varphi}(-x+r)|^2 \, dx = \int_I |f(x)|^2 \Phi(-x)\, dx.
\end{aligned}$$

This implies that $\mathbf{a} \in \ker T$ if and only if $f(x)\Phi(-x) = 0$ a.e., and the assertion follows from Theorem 1.1 (a).
(b) Recall first, that for a Riesz sequence frame bounds and Riesz bounds coincide. Since T is one–to–one if and only if $\ker T = \{0\}$ (here "0" denotes a vector in $\ell^2(\Lambda)$), and this condition in turn is satisfied if and only if $E_\Lambda = \{0\}$ (here "0" denotes a function), the assertion follows from Theorem 1.1 (b). □

The following result was originally proved by Benedetto and Li with a mild restriction [**5, 7**]. The more general version that we state here, was proved by Di–Rong Chen [**11**], Casazza, Christensen and Kalton [**10**], and Kim, Kim and Li [**31**] (see also [**6**]). It can be derived by arguments similar to those that were used to prove Theorem 2.1. For a detailed proof see [**14**, Theorem 7.2.3].

THEOREM 2.2. *Let* $\varphi \in L^2(\mathbb{R})$ *and* $\mathbf{S} := \{\, \varphi_k\ ;\ k \in \mathbb{Z} \,\}$*. For any* $A, B > 0$ *the following characterizations hold:*

(a) **S** *is a Bessel sequence with bound* B *if and only if*

$$\Phi(x) \leq B \qquad \text{a.e. } x \in I.$$

(b) **S** *is an orthonormal sequence if and only if*

$$\Phi(x) = 1 \qquad \text{a.e. } x \in I.$$

(c) **S** *is a Riesz sequence with bounds* $A \leq B$ *if and only if*

$$A \leq \Phi(x) \leq B \qquad \text{a.e. } x \in I.$$

(d) **S** *is a frame sequence with bounds* $A \leq B$ *if and only if*

$$A \leq \Phi(x) \leq B \qquad \text{a.e. } x \in I \setminus N,$$

where $N := \{\, x \,;\, x \in I,\ \Phi(x) = 0 \,\}$.

The proof of the next theorem relies on the properties of the Hardy class H^2. Let $p > 0$. A function $f(z)$ is said to *belong to the class* H^p if it is analytic in the unit disc $|z| < 1$, and the set of integrals

$$\int_0^{2\pi} |f(re^{xi})|^p \, dx$$

is bounded for $0 < r < 1$.

The case $p = 2$ is particularly simple, since if

$$f(z) := \sum_{k \in \mathbb{Z}^+} a_k z^k = \sum_{k \in \mathbb{Z}^+} a_k r^k e^{kxi}, \tag{2.3}$$

Bessel's identity

$$\int_0^{2\pi} |f(re^{xi})|^2 \, dx = \frac{1}{2\pi} \sum_{k \in \mathbb{Z}^+} |a_k|^2 r^{2k}$$

shows that $f \in H^2$ if and only if $\{\, a_k \,;\, k \in \mathbb{Z}^+ \,\} \in \ell^2(\mathbb{Z}^+)$ (cf., e.g., [**46**]).

Here is an interesting result of Casazza, Christensen and Kalton (it appears as a remark following the proof of [**10**, Theorem 2.4]). The argument we will use can be found there. We have added some details to make the proof more transparent. Note that if $\varphi(t) = 0$ a.e., then trivially $\{\, \varphi_k \,;\, k \in \Lambda \,\}$ is a frame, but not a Riesz basis.

THEOREM 2.3. *Let* $\varphi \in L^2(\mathbb{R})$, *and assume that* $\varphi(t)$ *does not vanish a.e. Then* $\mathbf{S_0} := \{\, \varphi_k \,;\, k \in \mathbb{Z}^+ \,\}$ *is a frame sequence if and only if it is a Riesz sequence.*

PROOF. Assume $\mathbf{S_0}$ is a frame sequence with bounds $0 < A \leq B$. From Theorem 2.1 (a) we know that (2.2) is satisfied for every $g \in H_{\mathbb{Z}^+} \cap E_{\mathbb{Z}^+}^{\perp}$. If $g \in H_{\mathbb{Z}^+}$, then

$$g(x) = \sum_{k \in \mathbb{Z}^+} a_k e^{2\pi k x i}, \qquad \{\, a_k \,;\, k \in \mathbb{Z}^+ \,\} \in \ell^2(\mathbb{Z}^+).$$

Now, the function $f(z)$ defined by (2.3) is clearly in the Hardy class H^2. Thus, if in particular $g \neq 0$ (whence $f \neq 0$), we deduce from [**29**, p. 90] or [**46**, Vol 1., Chapter VII, Theorem 7.25] that $f(e^{xi}) \neq 0$ a.e. Since $g(x) = f(e^{2\pi xi})$ and $\Phi(-x)$ cannot vanish a.e. (for otherwise $\varphi(t)$ would vanish a.e.), we conclude that $E_{\mathbb{Z}^+} = \{0\}$, and the assertion follows from Theorem 2.1 (b). □

We also have:

THEOREM 2.4. *Let* $\varphi \in L^2(\mathbb{R})$ *and* $\mathbf{S_0} := \{ \varphi_k \,;\, k \in \mathbb{Z}^+ \}$. *If* $\mathbf{S_0}$ *is a frame sequence then* $\mathbf{S} := \{ \varphi_k \,;\, k \in \mathbb{Z} \}$ *is a frame sequence.*

PROOF. Assume first that f is a trigonometric polynomial in $H_{\mathbb{Z}^+} \cap E_{\mathbb{Z}^+}^{\perp}$. Then Theorem 2.2 implies that f satisfies (2.2). Suppose now that f is any trigonometric polynomial. Then for large enough n we have that $e^{2\pi n x i} f \in H_{\mathbb{Z}^+}$. Thus (2.2) holds for all trigonometric polynomials in $H_{\mathbb{Z}} \cap E_{\mathbb{Z}}^{\perp}$, and therefore for all functions in $H_{\mathbb{Z}} \cap E_{\mathbb{Z}}^{\perp}$. Applying Theorem 2.2 (a) we conclude that $\mathbf{S}$ is a frame sequence. □

The preceding proof has been culled from [**10**, Theorem 2.4].

3. Perturbations of the Haar wavelet

Let ψ be a frame wavelet in $L^2(\mathbb{R})$; for $j \in \mathbb{Z}$, let P_j denote the closure of the linear span of $\{ \psi_{j,k} \,;\, k \in \mathbb{Z} \}$, and let $V_j := \sum_{r<j} P_r$. Note that $\psi \in V_1$. We say that ψ is *associated* with an MRA, or that ψ is an MRA wavelet, if $M := \{ V_j \,;\, j \in \mathbb{Z} \}$ is a multiresolution analysis; we also say that ψ is associated with M. This definition is apparently due to Wang [**41**].

A feature of MRA orthonormal wavelets is the existence of the *fast wavelet transform algorithm*, which allows for the computation of the wavelet coefficients in $O(n)$ steps, where n is the sample size [**34**]. This compares favorably with the *fast Fourier transform algorithm*, which allows for the computation of the Fourier coefficients in $O(n \log n)$ steps [**9**]. If, in addition, the wavelet is of compact support and smooth, it has the additional advantages of having good localization in both time and frequency. Not everything is perfect, though, since most MRA orthonormal wavelets do not have a closed form representation and have to be obtained recursively using the cascade algorithm [**17, 40**]. Moreover, compactly supported MRA orthonormal wavelets with dilation factor 2 lack symmetry: Recall that the Haar function is defined by

$$\psi_H(t) := \begin{cases} 1 & \text{if } 0 \le t < 1/2, \\ -1 & \text{if } 1/2 \le t \le 1, \\ 0 & \text{otherwise.} \end{cases}$$

We have:

THEOREM 3.1. [**17**, Theorem 8.1.4] *Suppose that* φ *and* ψ, *the scaling function and orthonormal wavelet (with dilation factor* 2*) associated with a multiresolution analysis, are both real and compactly supported. If* ψ *has either a symmetry or an antisymmetry axis, then* ψ *is the Haar function.*

In view of this theorem, we know that we cannot have an orthonormal wavelet with dilation factor 2 that is symmetric, smooth, and of compact support. We cannot have all four properties, but we can sacrifice orthonormality in order to obtain symmetry: Given $\varepsilon > 0$ and an arbitrary positive integer m, we can perturb the Haar wavelet into a Riesz wavelet $\psi_H^{\{\varepsilon,m\}} \in C^m(-\infty, \infty)$ with support in $[-\varepsilon, 1+\varepsilon]$ (thus having good time and frequency localization), and known Riesz bounds. This perturbation is based on the observation that adding a Bessel sequence with a small bound to a Riesz basis transforms the original basis into another Riesz basis. Moreover, Riesz bounds for the new basis can be expressed in terms of the Riesz bounds of the original basis:

THEOREM 3.2. [**20**, Theorem 5]. *Let* $\{f_n\}$ *be a Riesz basis in a Hilbert space* $\mathbb{H}$ *with bounds* A *and* B. *Assume* $\{\, g_n \,;\, n \in \mathbb{Z}\} \subset \mathbb{H}$ *is such that* $\{\, f_n - g_n \,;\, n \in \mathbb{Z}\}$ *is a Bessel sequence with bound* $M < A$. *Then* $\{\, g_n \,;\, n \in \mathbb{Z}\}$ *is a Riesz basis with bounds* $[1-(M/A)^{1/2}]^2 A$ *and* $[1+(M/B)^{1/2}]^2 B$. *Conversely, if* $\{\, f_n \,;\, n \in \mathbb{Z}\}$ *and* $\{\, g_n \,;\, n \in \mathbb{Z}\}$ *are Riesz bases in* $\mathbb{H}$ *with bounds* A_1, B_1 *and* A_2, B_2 *respectively, and* U *is a linear homeomorphism such that* $U f_n = g_n, n \in \mathbb{Z}$, *then* $\{\, f_n - g_n \,;\, n \in \mathbb{Z}\}$ *is a Bessel sequence with bound* $M := \min\{B_1 \, \|I - U\|^2, B_2 \, \|I - U^{-1}\|^2\}$.

A similar theorem for frames was obtained by Christensen [**12, 13**]. A very nice discussion of frame and Riesz basis perturbations in Hilbert spaces can be found in Chapter 15 of [**14**].

We now describe how the perturbed Haar wavelets $\psi_H^{\{\varepsilon,m\}}$ are defined. The basic idea is to excise intervals around the discontinuities of the Haar function and replace them with functions based on B–splines of designated smoothness.

Let $N_m(t)$ denote the B–spline of order m ($m \geq 2$) [**15**, Chapter 4], $\chi_{[0,m-1]}(t)$ the characteristic function of $[0, m-1]$,

$$g_m(t) := \chi_{[0,m-1]}(t) \sum_{k=0}^{m-2} N_m(t-k),$$

$$g_{m,1}(t) := g_m(t-m+1), \quad h_m(t) := (1/2) \sum_{k=0}^{m-2} N_m(t-k),$$

and

$$q_m(t) := g_{m,1}(t) - h_m(t).$$

For $0 < \varepsilon < 1/2$, let

$$\alpha_{\varepsilon,m,1} := -\alpha_{\varepsilon,m,2} := -\alpha_{\varepsilon,m,3} := \alpha_{\varepsilon,m,4} := 2(m-1)/\varepsilon,$$
$$\beta_{\varepsilon,m,1} := 2(m-1), \qquad \beta_{\varepsilon,m,2} := 2(m-1)(1+\varepsilon)/\varepsilon,$$
$$\beta_{\varepsilon,m,3} = -\beta_{\varepsilon,m,4} = (m-1)/\varepsilon,$$
$$p_{\varepsilon,m}^{\{i\}}(t) := (-1)^{i-1} q_m(\alpha_{\varepsilon,m,i} t + \beta_{\varepsilon,m,i}), \qquad i = 1,2,3,4,$$
$$p_\varepsilon^{\{5\}}(t) := \chi_{[1/2,1/2+\varepsilon)}(t) - \chi_{[1/2-\varepsilon,1/2)}(t)$$

and

$$\delta^{\{\varepsilon,m\}}(t) := \sum_{i=1}^{4} p_{\varepsilon,m}^{\{i\}}(t) + p_\varepsilon^{\{5\}}(t)$$

Applying Theorem 3.2 the following can be obtained:

THEOREM 3.3. [**21**] *Let* $m \geq 2$ *be an integer, and let* $\varepsilon > 0$ *be such that*

$$B_\varepsilon := \{4[\frac{\sqrt{2}\,\varepsilon}{\sqrt{2}-1}\{(\varepsilon+1)(2m-2) + \frac{3}{\sqrt{2}}\}]^{\frac{1}{2}} + [\frac{\sqrt{2}\,\varepsilon}{\sqrt{2}-1}\{8\varepsilon + 4 + \frac{3}{\sqrt{2}}\}]^{\frac{1}{2}}\}^2 < 1.$$

Let $\psi_H^{\{\varepsilon,m\}} := \psi_H + \delta^{\{\varepsilon,m\}}$. *Then*

(a) $\psi_H^{\{\varepsilon,m\}} \in C^{m-2}(-\infty, \infty)$.

(b) $\operatorname{supp} \psi_H^{\{\varepsilon,m\}} \subseteq [-\varepsilon, 1+\varepsilon]$.

(c) $\psi_H^{\{\varepsilon,m\}}$ *is a Riesz wavelet with Riesz bounds* $(1-\sqrt{B_\varepsilon})^2$ *and* $(1+\sqrt{B_\varepsilon})^2$.

(d) *If* $0 < \delta < \infty$, *then*

$$||\psi_H^{\{\varepsilon,m\}} - \psi_H||_{L^\delta(\mathbb{R})} \leq (1+2^\delta)^{1/\delta}(2\varepsilon)^{1/\delta}.$$

The functions $\psi_H^{\{\varepsilon,m\}}$ are called *perturbed Haar wavelets*. Since they are not orthonormal wavelets, the fast wavelet transform cannot be used to compute the wavelet coefficients. However, in [**44**] an algorithm is given for the fast computation of the wavelet coefficients, provided that $\varepsilon = 2^{-k}, k \in \mathbb{Z}^+$. The underlying justification for the method is that, as explained above, the perturbations are crafted by excising intervals around the discontinuities of the Haar functions and replacing them with functions based on B–splines of designated smoothness. The resulting functions can be calculated in $O(n)$ steps by relying on multilevel relations among the B–splines and interval integrals on the sections of the Haar functions which have not been excised. Due to the lack of orthogonality, there is still no completely satisfactory solution to the problem of reconstructing the original function. Since Riesz bounds are given explicitly, the frame algorithm or other algorithms can be used to reconstruct a function from its wavelet coefficients ([**17, 22, 23**]). The approach used by Schuler for the sine–Gabor frame wavelet that he introduced in [**39**] is to simply treat the wavelet as if it were orthogonal. Since his frames are nearly tight, this seems to work fine. But of course, one has no control over the error.

The remaining question about the perturbed Riesz wavelets is whether they are associated with an MRA. This was the motivation for some of the work cited in the next section.

4. Characterizations of orthonormal and Riesz wavelets.

Recall that MRA wavelets were defined at the beginning of Section 3.

If we know that a function is an orthonormal wavelet, how can we tell whether it is associated with an MRA? This question is answered using the *dimension function* $D_\psi(x)$, defined by

$$D_\psi(x) := \sum_{j=1}^{\infty}\sum_{k\in\mathbb{Z}} |\widehat{\psi}(2^j(x+k))|^2.$$

Let $\mathbf{F}_\psi$ denote the closure in $\ell^2(\mathbb{Z})$ of the linear span of $\{\, \Psi_j(x) \,;\, j \in \mathbb{Z}\,\}$, where $\Psi_j(x) := \{\, \widehat{\psi}(2^j(x+k)) \,;\, k \in \mathbb{Z}\,\} \in \ell^2(\mathbb{Z})$.

THEOREM 4.1. [**28**, p. 363] *Suppose* ψ *is an orthonormal wavelet. Then the following assertions are equivalent:*

(a) ψ *is an MRA wavelet.*

(b) $D_\psi(x) > 0$ *a.e.*

(c) $D_\psi(x) = 1$ *a.e.*

(d) $\dim \mathbf{F}_\psi(x) = 1$ *a.e.*

Motivated by Theorem 4.1 and by the question of whether the perturbed Haar wavelets are associated with an MRA, the second author [**45**] began the study of conditions that determine whether a Riesz wavelet ψ is associated with an MRA.

We will say that ψ is *obtained* by a multiresolution analysis, if there is some multiresolution analysis $M = \{ U_j \, ; \, j \in \mathbb{Z} \}$ such that $\psi \in U_1$. In this case we also say that ψ is obtained by M. A set A is *essentially contained* in a set B, if $A \setminus B$ has Lebesgue measure 0, and that A is *essentially equal* to B, if both $A \setminus B$ and $B \setminus A$ have Lebesgue measure 0. The main result of [**45**] is the following:

THEOREM 4.2. *Let ψ be a Riesz wavelet. Then the following propositions are equivalent:*

(a) *ψ is obtained by an MRA and* $\text{supp}\{\widehat{\psi}(2x)\}$ *is essentially equal to* $\text{supp}\{\widehat{\psi}(x)\}$.

(b) *ψ is obtained by an MRA and* $\text{supp}\{\widehat{\psi}(2x)\}$ *is essentially contained in* $\text{supp}\{\widehat{\psi}(x)\}$.

(c) *There is a 2–periodic function g such that $g(x) \neq 0$ and*

$$\widehat{\psi}(2x) = g(x)\widehat{\psi}(x) \qquad a.e. \tag{4.1}$$

(d)

$$\sum_{k \in \mathbb{Z}} |\widehat{\psi}(x + 2k)|^2 > 0 \qquad a.e.,$$

and there is a 2–periodic function g such that (4.1) *is satisfied a.e.*

A multivariate version of Theorem 4.2 was obtained by Bownik [**8**].

We can now answer the question of whether the perturbed Haar wavelets $\psi_H^{\{\varepsilon,m\}}$ are obtained by an MRA. A nontrivial application of Theorem 4.2 yields:

THEOREM 4.3. [**45**, Theorem 4.3] *Let B_ε be defined as in Theorem* 3.3, *let $m \geq 2$ be an integer, and let $\varepsilon > 0$ be such that $B(\varepsilon, m) < 1$. Then $\psi_H^{\{\varepsilon,m\}}$ cannot be obtained by an MRA.*

The question now is what relationship there is between MRA wavelets and wavelets obtained by an MRA. The definitions trivially imply that if ψ is an MRA Riesz wavelet, then it is a Riesz wavelet obtained by the same MRA. The question then, is whether the converse is true. This has been studied by Kim et al. [**32**] and completely solved by Bownik [**8**], who studied the problem in the multidimensional setting and showed that a Riesz wavelet obtained by an MRA is an MRA Riesz wavelet. The univariate version of his result is the following:

THEOREM 4.4. [**8**, Corollary 3.8] *Suppose ψ is a Riesz wavelet obtained by an MRA. Then ψ is a biorthogonal wavelet associated with the same MRA.*

But since an MRA Riesz wavelet is obtained by the same MRA, it follows that every MRA Riesz wavelet is biorthogonal. However, this result is already known: it is a consequence of results of Kim et al. [**30**] in one dimension, and Larson et al. [**33**] in several dimensions.

In summation: a Riesz wavelet is associated with an MRA if and only if it is obtained by an MRA. Moreover, every MRA Riesz wavelet is biorthogonal.

Basically, Theorem 4.2 gives necessary and sufficient conditions for a Riesz wavelet to be an MRA Riesz wavelet if $\widehat{\psi}(x) \neq 0$ a. e. For arbitrary Riesz wavelets another condition was given in [**45**, Proposition 2.2]. Kim et al. [**32**, Proposition 1.1] noted that, since a Riesz wavelet is an MRA Riesz wavelet if and only if it is associated with an MRA, this result can be expressed as follows:

PROPOSITION 4.1. *A Riesz wavelet ψ is associated with an MRA if and only if there exist $p, q \in L^2(-1/2, 1/2)$ and 1–periodic, and $\varphi \in L^2(\mathbb{R})$ such that:*

(a) $\{ \varphi_k \, ; \, k \in \mathbb{Z} \}$ *is a Riesz sequence.*

(b) $\widehat{\varphi}(2x) = p(x)\widehat{\varphi}(x) \qquad a.e.$

(c) $\widehat{\psi}(2x) = q(x)\widehat{\varphi}(x) \qquad a.e.$

We say that a closed subspace S of $L^2(\mathbb{R})$ is *shift invariant*, if $\varphi \in S$ implies that $\varphi_k \in S$ for every $k \in \mathbb{Z}$. The theory of invariant subspaces is discussed in [**25**].

If $\{ V_j \, ; \, j \in \mathbb{Z} \}$ is an MRA, W_r will denote the orthogonal complement of V_r in V_{r+1}. Thus, $V_{r+1} = V_r \oplus W_r$.

Kim and his collaborators have also shown that Theorem 4.2 is a corollary of the following generalization of Theorem 4.1:

THEOREM 4.5. [**30**, Theorem 2.6] *Suppose ψ is a Riesz wavelet. For $k \in \mathbb{Z}$, let $W_k := \overline{\operatorname{span}}\{ \psi_{k,l} \, ; \, l \in \mathbb{Z} \}$. Then the following assertions are equivalent:*

(a) *ψ is an MRA Riesz wavelet.*

(b) $\sum_{k<0} W_k$ *is shift–invariant and* $\dim \mathbf{F}_\psi(x) = 1 \quad a.e.$

(c) $\sum_{k<0} W_k$ *is shift–invariant and* $\dim \mathbf{F}(x) \geq 1 \quad a.e.$

(d) $\sum_{k<0} W_k$ *is shift–invariant and* $D_\psi(x) > 0 \quad a.e.$

The definitions of Riesz wavelets *associated* with an MRA and of Riesz wavelets *obtained* by an MRA can be carried over *verbatim* to frame wavelets. However in [**36**], Paluszyński et al. have introduced a different definition of MRA frame wavelet involving so–called *pseudo–scaling functions*, whereas in [**37**] they introduced the concept of MSA tight frame wavelets, also involving pseudo–scaling functions. Since frame theory is beyond the scope of this article, suffice it to say that the exact relationship between these different types of wavelet frames is still unclear.

The next result, due to Auscher, shows that an orthonormal wavelet that is continuous and satisfies certain decay conditions, is associated with an MRA having a continuous scaling function that also satisfies some decay conditions:

THEOREM 4.6. *Let h be an orthonormal wavelet. Assume that $\widehat{h}$ is continuous on $\mathbb{R}$ and that there are constants $\alpha, C > 0$, such that*

$$|\widehat{h}(x)| \leq C|x|^{\alpha}(1 + |x|)^{-1/2-\alpha}. \tag{4.2}$$

Then h is associated with an MRA. Moreover, this MRA has a scaling function φ such that $\widehat{\varphi}$ is continuous everywhere on $\mathbb{R}$, and there is a constant $D > 0$ such that

$$|\widehat{\varphi}(x)| \leq D(1+|x|)^{-1/2-\alpha}. \tag{4.3}$$

A proof of this theorem is sketched in [**2**]. It is a particular case of [**3**, Theorem 1.3], which is a much more general multivariate result. Whether the results of [**3**] can be generalized to Riesz bases still seems to be an open question.

So far in this section, we have studied the following problem: If we know that a function ψ is a Riesz wavelet, how do we determine whether it is associated with an MRA? We will now focus on a slightly different issue: given an arbitrary function in $L^2(\mathbb{R})$, how do we determine whether it is an MRA Riesz wavelet? In other words, we need to determine whether the function satisfies two properties: that it is a Riesz wavelet *and* that it is associated with an MRA.

Gripenberg [**24**], Hernández, Wang and Weiss [**27**], and Wang [**41**] (see also [**42**]) have characterized the set of all orthonormal wavelets associated with a given MRA.

THEOREM 4.7. [**28**, p. 57]. *If φ is an orthonormal scaling function for an MRA $\{V_j\ ;\ j \in \mathbb{Z}\}$ and p is the associated low pass filter, then h is an orthonormal wavelet in W_0 if and only if there is a measurable unimodular and 1-periodic function $\nu(x)$, such that*

$$\widehat{h}(2x) = e^{2\pi x i}\,\nu(2x)\,\overline{p(x+1/2)}\,\widehat{\varphi}(x) \qquad a.e.$$

To determine whether a given function is an orthonormal wavelet associated with an MRA, we can use

THEOREM 4.8. *A function $\psi \in L^2(\mathbb{R})$ is an MRA orthonormal wavelet if and only if the following four conditions are satisfied:*

(a) $\|\psi\| = 1$.

(b) $\displaystyle\sum_{j\in\mathbb{Z}} |\widehat{\psi}(2^j x)|^2 = 1, \qquad a.e.$

(c) $\displaystyle\sum_{j\in\mathbb{Z}^+} \widehat{\psi}(2^j x)\,\overline{\widehat{\psi}(2^j(x+m))} = 0, \qquad a.e., \qquad m \in 2\mathbb{Z}+1.$

(d) $D_\psi(x) = 1 \qquad a.e.$

Here $D_\psi(x)$ is the dimension function defined at the beginning of this section.

Note that conditions (a), (b) and (c) are the same as in Theorem 1.3; they guarantee that ψ is an orthonormal wavelet. Adding condition (d) guarantees that ψ is associated with an MRA (cf. [**28**, p. 355, Theorem 3.2]).

Recent work by Hernández, Labate and Weiss [**26**] greatly extends Theorem 4.8, not only to a multivariate setting, but also to a more general class of functions that encompasses affine and Gabor wavelets as particular cases.

5. Perturbation in the frequency domain

Since the perturbed Haar wavelets are not associated with an MRA, the question remains as to whether it is possible to find an MRA Riesz wavelet having the same properties as the perturbed Haar wavelets. Further insight has been provided by Aimar et al., who have pointed out that Theorem 3.3 is a consequence of the following theorem.

THEOREM 5.1. [**1**] *Let m be a nonnegative integer and let φ be such that its mth derivative belongs to $L(\mathbb{R})$ and $\int_{\mathbb{R}} \varphi(t)\,dt = 1$. Let $\varepsilon > 0$ be such that*

$$M_\varepsilon := 2\sqrt{3}\,\varepsilon(1 + ||\varphi||_1)^2(4\sqrt{2}+1)(\sqrt{2}-1)^{-1}(4\varepsilon+1) < 1,$$

and let $\varphi_\varepsilon(t) := (1/\varepsilon)\varphi(t/\varepsilon)$. Then $\psi_H^\varepsilon := \psi_H * \varphi_\varepsilon$ belongs to $C^m(\mathbb{R})$, has support in $[-\varepsilon, 1+\varepsilon]$, and is a Riesz wavelet with bounds $(1-\sqrt{M_\varepsilon})^2$ and $(1+\sqrt{M_\varepsilon})^2$.

From Theorem 5.1 we see that adding a Bessel wavelet to the Haar wavelet ψ_H to obtain a Riesz wavelet is equivalent to multiplying $\widehat{\psi_H}$ by a suitable function μ. This suggests the idea of trying to multiply the Fourier transform of an orthonormal wavelet by a suitable function, so as to obtain a Riesz wavelet associated with a *different* MRA.

We begin with two auxiliary propositions. the first one is a direct consequence of Theorem 2.2:

PROPOSITION 5.1. [**14**, Proposition 7.3.6 (i)]. *Let $\varphi \in L^2(\mathbb{R})$, and assume that $\{\varphi_k\,;\,k \in \mathbb{Z}\}$ is a frame sequence but not a Riesz sequence (i.e., it is overcomplete). Then $\Phi(x)$ is discontinuous.*

The following assertion is proved by an argument similar to the one used in the proof of [**14**, Proposition 7.3.6 (ii)].

PROPOSITION 5.2. *Assume that $\varphi \in L^2(\mathbb{R})$, that $\widehat{\varphi}$ is continuous on $\mathbb{R}$, and that there are constants $D > 0$ and $\alpha > 0$ for which (4.3) is satisfied. If $\{\varphi_k\,;\,k \in \mathbb{Z}\}$ is a Riesz sequence, then $\Phi(x)$ is continuous in $\mathbb{R}$.*

PROOF. Assume that (4.3) is satisfied, and let $x \in I$. Then, for all $n \in \mathbb{Z}^+$,

$$\begin{aligned}\left|\Phi(x) - \sum_{|k|\le n} |\widehat{\varphi}(x+k)|^2\right| &= \sum_{|k|>n} |\widehat{\varphi}(x+k)|^2 \\ &\le D^2 \sum_{|k|>n} (1+|x+k|)^{-1-2\alpha} \\ &\le 2D^2 \sum_{k>n} k^{-1-2\alpha}.\end{aligned}$$

This implies that the series

$$\sum_{k\in\mathbb{Z}} |\widehat{\varphi}(x+k)|^2$$

is uniformly convergent. Thus $\Phi(x)$ is the uniform limit of continuous functions, and the assertion follows. □

Motivated by a paper of Aldroubi and Unser [**4**] we are now going to develop the idea of convolving Riesz bases, scaling functions, and multiresolution analyses. The first result involves convolving two suitably behaved functions, only one of which generates a Riesz basis of translates.

THEOREM 5.2. *Assume that* $\varphi \in L^2(\mathbb{R})$, *that* $\widehat{\varphi}$ *is continuous on* $\mathbb{R}$, *and that there are numbers* $D > 0$ *and* $\alpha > 0$ *for which* (4.3) *is satisfied. Let* $u \in L^2(\mathbb{R})$ *be such that* $\widehat{u}$ *is continuous and bounded on* $\mathbb{R}$, *and* $\widehat{u}(x) \neq 0$ *whenever* $\widehat{\varphi}(x) \neq 0$. *Assume, moreover, that* $\mathbf{S} := \{ \varphi_k \; ; \; k \in \mathbb{Z} \}$ *is a Riesz sequence. If* $\sigma := u * \varphi$, *then* $\mathbf{T} := \{ \sigma_k \; ; \; k \in \mathbb{Z} \}$ *is a Riesz sequence.*

Theorem 5.2 is a corollary of the following generalization of [**14**, Proposition 7.3.9].

THEOREM 5.3. *Assume that* $\varphi \in L^2(\mathbb{R})$, *that* $\widehat{\varphi}$ *is continuous on* $\mathbb{R}$, *and that there are numbers* $D > 0$ *and* $\alpha > 0$ *for which* (4.3) *is satisfied. Let* v *be a continuous and bounded function in* $\mathbb{R}$ *such that* $v(x) \neq 0$ *whenever* $\widehat{\varphi}(x) \neq 0$. *Assume, moreover, that* $\mathbf{S} := \{ \varphi_k \; ; \; k \in \mathbb{Z} \}$ *is a Riesz sequence. If* $\widehat{\sigma}(x) := v(x)\widehat{\varphi}(x)$, *then* $\mathbf{T} := \{ \sigma_k \; ; \; k \in \mathbb{Z} \}$ *is a Riesz sequence.*

PROOF. Let A and B be a lower and an upper Riesz bound for S respectively. From Proposition 5.2 we know that $\Phi(x)$ is continuous. Applying Theorem 2.2 we therefore conclude that

$$A \leq \Phi(x) \leq B \qquad \text{for every } x \in I. \tag{5.1}$$

Thus, for every $x \in I$

$$\begin{aligned} \Phi_1(x) := \sum_{k \in \mathbb{Z}} |\widehat{\sigma}(x+k)|^2 &= \sum_{k \in \mathbb{Z}} |v(x+k)\widehat{\varphi}(x+k)|^2 \\ &\leq ||v||_\infty^2 \Phi(x) \leq ||v||_\infty^2 B. \end{aligned} \tag{5.2}$$

To find a lower bound for $\Phi_1(x)$ we proceed as follows: For $m \in \mathbb{Z}$ and $\delta > 0$, let

$$U(m, \delta) := \{ x \; ; \; -1 < x < 2 \text{ and } |\widehat{\varphi}(x+m)|^2 > \delta \},$$

and let $\mathfrak{F}$ denote the family of all these sets $U(m,\delta)$. From (5.1) we deduce that for every $x \in I$ there are numbers $m \in \mathbb{Z}$ and $\delta > 0$ such that $x \in U(m,\delta)$. Thus $\mathfrak{F}$ is a an open covering of I, and Heine–Borel's lemma implies that there is a finite subcollection of $\mathfrak{F}$, say $\{ U(m_n, \delta_n) \; ; \; 1 \leq n \leq N \}$, that also covers $\mathfrak{F}$. Let F_n denote the closure of $U(m_n, \delta_n)$. Since $F_n \subset [-1, 2]$, we see that $\{ F_n \; ; \; 1 \leq n \leq N \}$ is a family of compact sets that covers I.

If $x \in F_n$, then $|\widehat{\varphi}(x+m_n)|^2 \geq \delta_n$. Since F_n is compact, the hypotheses imply there is a number $\varepsilon_n > 0$ such that $|v(x+m_n)|^2 \geq \varepsilon_n$. Clearly

$$\delta := \min\{ \delta_n \; ; \; 1 \leq n \leq N \} > 0 \qquad \text{and} \qquad \varepsilon := \min\{ \varepsilon_n \; ; \; 1 \leq n \leq N \} > 0.$$

Let $x \in I$. Then there is an integer n, $1 \leq n \leq N$, such that $x \in F_n$. Thus

$$\Phi_1(x) \geq |\widehat{\sigma}(x+m_n)|^2 = |v(x+m_n)\widehat{\varphi}(x+m_n)|^2 \geq \delta\varepsilon. \tag{5.3}$$

Combining (5.2), (5.3), and Theorem 2.2, the assertion follows. □

The convolution of scaling functions was studied as early as 1987 by Micchelli and Prautzsch [**35**]. The most general result is due to Wang:

THEOREM 5.4. [**41**, Theorem 2.47] *Let* φ_1 *and* φ_2 *be orthonormal scaling functions (not necessarily for the same MRA), with low pass filters* p_1 *and* p_2 *respectively. Let* $\varphi(x) := \varphi_1 * \varphi_2$, *and let* $r(x)$ *be defined by* (2.1). *Assume that*

$$\sum_{k \in \mathbb{Z}} |\widehat{\varphi_1}(x+k)\widehat{\varphi_2}(x+k)| > 0, \qquad a.e., \tag{5.4}$$

and let

$$\widehat{\sigma}(x) := \frac{\widehat{\varphi_1}(x)\widehat{\varphi_2}(x)}{r(x)}. \tag{5.5}$$

Then σ is an orthonormal scaling function with low pass filter

$$p(x) := p_1(x)p_2(x)\frac{r(x)}{r(2x)}. \tag{5.6}$$

Condition (5.4) may be difficult to verify directly, since it involves an infinite sum. Here is a theorem that may be easier to apply.

THEOREM 5.5. *Let φ_1 and φ_2 be orthonormal scaling functions with low pass filters p_1 and p_2 respectively. Assume, moreover that both $\widehat{\varphi_1}$ and $\widehat{\varphi_2}$ are continuous on $\mathbb{R}$, that $\widehat{\varphi_2}$ satisfies an inequality of the form (4.3) for some $D > 0$ and $\alpha > 0$, and that $\widehat{\varphi_1}(x) \neq 0$ whenever $\widehat{\varphi_2}(x) \neq 0$. Let σ and p be defined by (5.5) and (5.6) respectively. Then σ is an orthonormal scaling function with low pass filter p.*

PROOF. Proceeding as in the proof of Theorem 5.2 we see that $\Phi(x)$ satisfies an inequality of the form (5.3), i.e, there is a $C > 0$ such that $\Phi(x) > C$, and the assertion follows from Theorem 5.4. □

We now turn to the convolution of multiresolution analyses. We begin with a consequence of Theorem 5.4.

THEOREM 5.6. *Let h_1 and h_2 be MRA orthonormal wavelets with orthonormal scaling functions φ_1 and φ_2, and low pass filters p_1 and p_2 respectively. Let $\varphi(x) := \varphi_1 * \varphi_2$, assume that (5.4) is satisfied, let $r(x)$ be given by (2.1), and*

$$\mu(x) := e^{-\pi x i}\frac{r(x/2 + 1/2)}{r(x/2)r(x)}, \tag{5.7}$$

and let σ and p be defined by (5.5) and (5.6) respectively. If

$$\widehat{h}(x) := \widehat{h_1}(x)\widehat{h_2}(x)\mu(x), \tag{5.8}$$

then h is an MRA orthonormal wavelet with orthonormal scaling function σ, and low pass filter p.

PROOF. Applying Theorem 4.7 we see that there are unimodular 1–periodic functions $\nu_1(x)$ and $\nu_2(x)$ such that

$$\widehat{h_1}(2x) = e^{2\pi x i}\,\nu_1(2x)\,\overline{p_1(x + 1/2)}\,\widehat{\varphi_1}(x) \qquad \text{a.e.}$$

and

$$\widehat{h_2}(2x) = e^{2\pi x i}\,\nu_2(2x)\,\overline{p_2(x + 1/2)}\,\widehat{\varphi_2}(x) \qquad \text{a.e.}$$

Since $r(x)$ is 1–periodic, a straightforward computation yields

$$\widehat{h_1}(2x)\widehat{h_2}(2x) = e^{2\pi x i}\,\nu(2x)\,\overline{p(x + 1/2)}\,\widehat{\varphi}(x)\frac{e^{2\pi x i}r(2x)}{r(x + 1/2)} \qquad \text{a.e.,}$$

where $\nu(x) = \nu_1(x)\nu_2(x)$. Thus

$$\widehat{h}(2x) = e^{2\pi x i}\,\nu(2x)\,\overline{p(x + 1/2)}\,\widehat{\sigma}(x) \qquad \text{a.e..}$$

From Theorem 5.4 we know that σ is an orthonormal scaling function with low pass filter p, and the assertion follows by another application of Theorem 4.7. □

Since the hypotheses of Theorem 5.5 imply that (5.4) is satisfied, Theorem 5.6 yields

THEOREM 5.7. *Let h_1 and h_2 be MRA orthonormal wavelets with orthonormal scaling functions φ_1 and φ_2, and low pass filters p_1 and p_2 respectively. Assume, moreover that both $\widehat{\varphi_1}$ and $\widehat{\varphi_2}$ are continuous on $\mathbb{R}$, that $\widehat{\varphi_2}$ satisfies an inequality of the form (4.3) for some $D > 0$ and $\alpha > 0$, and that $\widehat{\varphi_1}(x) \neq 0$ whenever $\widehat{\varphi_2}(x) \neq 0$. Let σ, p and h be defined by (5.5), (5.6) and (5.8) respectively. Then h is an MRA orthonormal wavelet with orthonormal scaling function σ and low pass filter p.*

To determine whether h is an MRA orthonormal wavelet using Theorem 5.7, we need certain information on the scaling functions φ_1 and φ_2. The following theorem bypasses this requirement.

THEOREM 5.8. *Let h_1 and h_2 be orthonormal wavelets. Assume that $\widehat{h_1}$ and $\widehat{h_2}$ are continuous on $\mathbb{R}$, and that there are constants $\alpha_1, \alpha_2, C_1, C_2 > 0$ such that*

$$|\widehat{h_i}(x)| \leq C_i|x|^{\alpha_i}(1+|x|)^{-1/2-\alpha_i}, \qquad i = 1,2.$$

Assume, moreover that

$$\sum_{k\in\mathbb{Z}+} |\widehat{h_1}(2^j x)|^2 \neq 0 \quad \text{whenever} \quad \sum_{k\in\mathbb{Z}+} |\widehat{h_2}(2^j x)|^2 \neq 0.$$

Then h_1 and h_2 are MRA wavelets. If φ_1 and φ_2 are their scaling functions and h, σ and p are defined by (5.5), (5.6) and (5.8), respectively, then h is an MRA orthonormal wavelet with orthonormal scaling function σ, and low pass filter p.

PROOF. By Theorem 4.6 h_1 and h_2 are MRA wavelets, $\widehat{\varphi_1}$ and $\widehat{\varphi_2}$ have continuous Fourier transforms, and there are constants $D_1, D_2 > 0$, such that

$$|\widehat{\varphi}_i(x)| \leq D_i(1+|x|)^{-1/2-\alpha_i}, \qquad i = 1,2.$$

We want to establish a connection between the zeros of the scaling functions φ_i and the zeros of the wavelets h_i, so we can apply Theorem 5.5. We first need to establish a representation for the scaling functions in terms of the wavelets.

From([**28**, p. 61]) we know that

$$|\widehat{\varphi}_i(x)|^2 = \sum_{j\in\mathbb{Z}+} |\widehat{h}_i(2^j x)|^2 \qquad \text{a.e} \tag{5.9}$$

Define

$$u_i(x) := \left(\sum_{j\in\mathbb{Z}+} |\widehat{h}_i(2^j x)|^2\right)^{1/2}, \qquad i = 1,2.$$

From the hypotheses we know that

$$|\widehat{h}_i(2^j x)| \leq C_i|2^j x|^{\alpha_i}(1+|2^j x|)^{-1/2-\alpha_i}, \qquad i = 1,2.$$

Thus, if $|x| \geq \delta > 0$ we see that

$$|\widehat{h}_i(2^j x)| \leq C_i\delta^{-1/2}2^{-j/2}, \qquad i = 1,2,$$

and proceeding as in the proof of Proposition 5.2, we readily see that the functions $u_i(x)$ are continuous on $|x| \geq \delta$. Since δ is arbitrary, this implies that the functions $u_i(x)$ are continuous on $\mathbb{R} \setminus \{0\}$. Since the functions $\widehat{\varphi}_i$ are continuous on $\mathbb{R}$, (5.9) implies that

$$|\widehat{\varphi}_i(x)| = u_i(x), \qquad x \in \mathbb{R} \setminus \{0\}, \qquad i = 1,2.$$

Since Theorem 1.2 implies that

$$\widehat{\varphi}_i(0) \neq 0, \qquad i = 1, 2,$$

we conclude that $\widehat{\varphi_1}(x) \neq 0$ whenever $\widehat{\varphi_2}(x) \neq 0$, and the assertion follows from Theorem 5.7. □

Example. The following example illustrates the application of Theorem 5.5, Theorem 5.6, Theorem 5.7, and Theorem 5.8.

Let φ_1 denote the characteristic function of the interval $[0, 1)$. The following facts are well known [**17, 28**]: The Haar function ψ_H is an MRA orthonormal wavelet, and there is an MRA associated with ψ_H that has φ_1 as its scaling function. Moreover,

$$\widehat{\varphi_1}(x) = e^{-\pi x i} \frac{\sin \pi x}{\pi x}$$

Let $\widehat{h}_1(t) := \widehat{h}_2(t) := \psi_H(t)$. Since

$$\widehat{\psi_H}(x) = i e^{-\pi x i} \frac{\sin^2(\pi x/2)}{\pi x/2}, \tag{5.10}$$

we see that the hypotheses of Theorem 5.8 are satisfied. Thus, if

$$\widehat{h}(x) := \widehat{h_1}(x) \widehat{h_2}(x) \mu(x), \tag{5.11}$$

where $\mu(x)$ is given by (5.7), we know that h is an MRA orthonormal wavelet.

If we require more information, we may apply Theorem 5.6 or Theorem 5.7 instead:

If $\varphi := \varphi_1 * \varphi_1$, then

$$\widehat{\varphi}(x) := \widehat{\varphi_1}(x)\widehat{\varphi_1}(x) = \widehat{N_2}(x) = e^{-2\pi x i} \left(\frac{\sin \pi x}{\pi x} \right)^2.$$

Since

$$\Phi(x) = \sum_{k \in \mathbb{Z}} |\widehat{N_2}(x + k)|^2,$$

from, e. g., [**15**, (4.2.10)], we conclude that

$$[r(x)]^2 = \Phi(x) = 1/3 + (2/3)\cos^2 \pi x > 0.$$

This implies that the conditions of Theorem 5.6 are satisfied, and therefore that the function h defined by (5.11) is an MRA orthonormal wavelet.

Let us now find an explicit representation for $\widehat{h}$.

By direct computation,

$$\mu(x) = \sqrt{3} e^{-\pi x i} \left[\frac{1 + 2\sin^2 \pi x/2}{(1 + 2\cos^2 \pi x/2)(1 + 2\cos^2 \pi x)} \right]^{1/2}.$$

Thus, applying (5.10) we have:

$$\widehat{h}(x) = [\widehat{\psi_H}(x)]^2 \mu(x) =$$

$$- \sqrt{3} e^{-3\pi x i} \frac{\sin^4(\pi x/2)}{(\pi x/2)^2} \left[\frac{1 + 2\sin^2 \pi x/2}{(1 + 2\cos^2 \pi x/2)(1 + 2\cos^2 \pi x)} \right]^{1/2}.$$

We can also verify our conclusions by a third method: The function

$$\widehat{\psi}(x) := \sqrt{3} e^{-\pi x i} \frac{\sin^4(\pi x/2)}{(\pi x/2)^2} \left[\frac{1 + 2\sin^2 \pi x/2}{(1 + 2\cos^2 \pi x/2)(1 + 2\cos^2 \pi x)} \right]^{1/2}$$

is a Battle–Lemarié wavelet, which is known to be an MRA orthonormal wavelet (cf. [**17**, p. 148]). Since $\widehat{h}(x)$ is the product of $\psi(x)$ and a unimodular 1–periodic function, applying Theorem 4.7 we deduce that h is an MRA orthonormal wavelet.

References

1. H. A. Aimar, A. L. Bernardis, and O. P. Gorosito, Perturbations of the Haar wavelet by convolution, *Proc. Amer. Math. Soc.* **129** (2001), 3619–3621. **MR** 2002g:42043.
2. P. Auscher, Toute base d'ondelettes régulières de $L^2(\mathbb{R})$ est issue d'une analyse multi–reésolution régulière, *C. R. Acad. Sci. Paris*, Sér. 1 **315** (1992), 1227–1230.
3. P. Auscher, Solution of two problems on wavelets, *J. Geom. Anal.* **5** (1995), 181–236.
4. A. Aldroubi and M. Unser, Families of multiresolution and wavelet spaces with optimal properties, *Numer. Funct. Anal. and Optimiz.* **14** (1993), 417–446. **MR** 94j:42045.
5. J. J. Benedetto and S. Li, MRA frames with applications, *ICASSP'93*, Minneapolis, III (April 1992), 304–307.
6. J. J. Benedetto and O. M. Treiber, Wavelet frames: multiresolution analysis and extension principles, in "Wavelet Transforms and Time–Frequency Signal Analysis" (L. Debnath, Ed.), pp. 3–36, Birkhäuser, Boston, 2001.
7. J. J. Benedetto and D. F. Walnut, Gabor frames for L^2 and related spaces, in "Wavelets: Mathematics and Applications" (J. J. Benedetto and M. W. Frazier, Eds.), pp. 97–162, CRC Press, Boca Raton, FL, 1994. **MR** 94i:42040.
8. M. Bownik, Riesz wavelets and generalized multiresolution analyses *Appl. Comput. Harm. Anal.* **14** No. 3 (2003), 181–194.
9. E. O. Brigham, "The Fast Fourier Transform", Prentice–Hall, Englewood Cliffs, 1974.
10. P. Casazza, O. Christensen and N. Kalton, Frames of translates, *Collect. Math.* **52** (2001), 35–59. **MR** 2002f:42002.
11. Di–R. Chen, On the splitting trick snd wavelet frame packets, *SIAM J. Math. Anal.* **31** (2000), 726–739.
12. O. Christensen, Frames and the projection method, *Appl. Comput. Harm. Analysis* **1** (1993), 50-53. **MR** 95a:42041.
13. O. Christensen, Frame perturbations, *Proc. American Math. Soc.* **123** (1995), 1217–1220. **MR** 95e:46024.
14. O. Christensen, "An Introduction to Frames and Riesz Bases", Birkhäuser, Boston, 2003.
15. C. K. Chui, "An Introduction to Wavelets", Academic Press, San Diego, 1992. **MR** 93f:42055.
16. C. K. Chui and X. Shi, Littlewood–Paley inequalities for frames and wavelets, *SIAM J. Math. Anal.* **24** (1993), 263–277. **MR** 94d:42039.
17. I. Daubechies, "Ten Lectures on Wavelets," SIAM, Philadelphia, 1992. **MR** 93e:42045.
18. M. Frazier, B. Jawerth and G. Weiss, "Littlewood–Paley Theory and the Study of Function Spaces", CBMS–AMS Regional Conference Series, Vol. **79**, AMS, 1991.
19. M. Frazier, G. Garrigós, K. Wang, and G. Weiss, A characterization of functions that generate wavelet amd related expansions, J. Fourier Anal. and Applic. **4** (1997), 884–906.
20. S. J. Favier and R. A. Zalik, On the stability of frames and Riesz bases, *Appl. Comput. Harm. Anal.* **2** (1995), 160–173. **MR** 96e:42030
21. N. K. Govil and R. A. Zalik, Perturbations of the Haar Wavelet, *Proc. American Math. Soc.* **125** (1997), 3363–3370. **MR** 97m:42025.
22. K. Gröchenig, "Foundations of Time–Frequency Analysis", Birkhäuser, Boston, 2001.
23. K. Gröchenig, Acceleration of the frame algorithm, *IEEE Trans. Signal Proc.* **41** (1993), 3331-3340.
24. G. Gripenberg, A necessary and sufficient condition for the existence of a father wavelet, *Studia Math.* **114** (1995), 207–226. **MR** 96d:42049.
25. H. Helson, "Lectures on Invariant Subspaces", Academic Press, New York, 1964.
26. E. Hernández, D. Labate and G. Weiss, A Unified Characterization of Reproducing Systems Generated by a Finite Family, II, *J. Geom. Analysis* **12** (2002), 615–662.
27. E. Hernández, X. Wang and G. Weiss, Characterization of wavelets, scaling functions and wavelets associated with multiresolution analyses, in "Function spaces, Interpolation Spaces, and Related Topics. Proceedings of the workshop, Haifa, Israel, June 7-13, 1995". (Y.

Brudnyi et al., eds.) Ramat Gan: Bar-Ilan University/ distr. by the American Mathematical Society, Isr. Math. Conf. Proc. 13, 51-87 (1999). **MR** 2000j:42046.

28. E. Hernández and G. Weiss, "A First Course on Wavelets", CRC Press, Boca Raton, FL, 1996. **MR** 97i:42015.
29. Y. Katznelsom, "An introduction to Harmonic Analysis", 2nd. corrected edition, Dover, New York, 1976.
30. H. O. Kim, R. Y. Kim and J. K. Lim, Characterizations of biorthogonal wavelets which are associated with biorthogonal multiresolution analyses, *Appl. Comput. Harm. Anal.* **11** (2001), 263–272.
31. H. O. Kim, R. Y. Kim and J. K. Lim, Quasi–biorthogonal frame multiresolution analyses and wavelets, *Adv. Comput. Math.* **18** (2003), 269–296.
32. H. O. Kim, R. Y. Kim, Y. H. Lee and J. K. Lim, On Riesz wavelets associated with multiresolution analyses, *Appl. Comput. Harm. Anal.* **13** (2002), 138–150.
33. D. Larson, W.–S. Tang, and E. Weber, Riesz wavelets and multiresolution structures, *SPIE Proceedings* **4478**, *Wavelet Applications in Signal and Image Processing IX* (2001), 254-262.
34. S. Mallat, "A Wavelet Tour of Signal Processing", 2nd ed., Academic Press, San Diego, 1999.
35. C. Micchelli and H. Prautzsch, Refinement and subdivision for spaces of integer translates of a compactly supported function, in "Numerical Analysis 1987" (D. F. Griffiths and G. A. Watson, Eds.), 192–222 Pitman Res. Notes in Math. **170** (1987).
36. M. Paluszyński, H. Šikić, G. Weiss and S. Xiao, Generalized low pass filters and MRA frame wavelets, *J. Geom. Anal.* **11** (2001), 311–342. **MR** 2002g:42054.
37. M. Paluszyński, H. Šikić, G. Weiss and S. Xiao, Tight frame wavelets, their dimension functions, MRA tight frame wavelets and connectivity properties. *Adv. Comput. Math.* **18** (2003), 297–327.
38. M. Papadakis, H. Šikić and G. Weiss, The characterization of low pass filters and some basic properties of wavelets, scaling functions and related concepts, *J. Fourier Anal. Appl.* **5** (1999), 495–521. **MR** 2001d:42026.
39. S. Schuler, "Theory And Applications Of The Sine–Gabor Wavelet Frame", Ph. D. Dissertation, Department of Electrical Engineering, University of Florida, 1997.
40. G. Strang and T. Nguyen, "Wavelets and Filter Banks", Wellesley–Cambridge Press, Wellesley, Massachussetts, 1996. **MR** 98b:94003.
41. X. Wang, "The Study Of Wavelets From The Properties Of Their Fourier Transforms", Ph. D. dissertation, Washington University in St. Louis, 1995.
42. The Wutam Consortium, Basic properties of wavelets, *J. Fourier Anal. Appl.* **4** (1998), 575–594. **MR** 99i:42056.
43. R. M. Young, "An Introduction to Nonharmonic Fourier Series", Revised 1st ed., Academic Press, San Diego, 2002. **MR** 2002b:42001.
44. R. A. Zalik, A Class of Quasi–orthogonal Wavelet Bases, in "Wavelets, Multiwavelets and their Applications" (A. Aldroubi and E. B. Lin, Eds.), 81–94, Contemporary Mathematics, Vol. 216, American Mathematical Society, Providence, RI, 1998. **MR** 99e:42053.
45. R. A. Zalik, Riesz bases and multiresolution analyses, *Appl. Comput. Harm. Analysis* **7** (1999), 315–351. **MR** 2000i:42024.
46. A. Zygmund, "Trigonometric Series", Vol. I and II, 2nd. ed., Cambridge University Press, Cambridge, repr. 1968.

DEPARTMENTO DE MATÉMATICA, UNIVERSIDAD NACIONAL DE MAR DEL PLATA, ARGENTINA
E-mail address: `algonzal@mdp.edu.ar`

DEPARTMENT OF MATHEMATICS, AUBURN UNIVERSITY, AL 36849–5310
E-mail address: `zalik@auburn.edu`

Contemporary Mathematics
Volume **345**, 2004

The existence of Gabor bases and frames

Deguang Han and Yang Wang

ABSTRACT. For an arbitrary full rank lattice Λ in $\mathbb{R}^{2d}$ and a function $g \in L^2(\mathbb{R}^d)$ the Gabor (or Weyl-Heisenberg) system is $\mathbf{G}(\Lambda, g) := \{e^{2\pi i\langle \ell, x\rangle} g(x-\kappa) \mid (\kappa,\ \ell) \in \Lambda\}$. It is well-known that a necessary condition for $\mathbf{G}(\Lambda, g)$ to be an orthonormal basis for $L^2(\mathbb{R}^d)$ is that the density of Λ satisfies $D(\Lambda) = 1$. However, except for symplectic lattices it remains an unsolved question whether $D(\Lambda) = 1$ is sufficient for the existence of a $g \in L^2(\mathbb{R}^d)$ such that $\mathbf{G}(\Lambda, g)$ is an orthonormal basis. We investigate this problem and prove that this is true for some of the important cases. In particular we show that this is true for $\Lambda = M\mathbb{Z}^d$ where M is either a block triangular matrix or any rational matrix with $|\det M| = 1$. Moreover, if M is rational we prove that there exists a compactly supported g such that $\mathbf{G}(\Lambda, g)$ is an orthonormal basis. We also obtain similar results for Gabor frames when $D(\Lambda) \geq 1$.

1. Introduction

Let Λ be a full rank lattice in $\mathbb{R}^{2d}$ and $g \in L^2(\mathbb{R}^d)$. The *Gabor (or Weyl-Heisenberg) system associated with* Λ *and* g is the following family of functions in $L^2(\mathbb{R}^d)$:

$$\mathbf{G}(\Lambda, g) := \left\{ e^{2\pi i\langle \ell, x\rangle} g(x-\kappa) \,\middle|\, (\kappa,\ \ell) \in \Lambda \right\}. \tag{1.1}$$

Gabor systems were introduced for the purpose of signal processing, and they are closely related to the representation of the Heisenberg group. In this paper we consider Gabor orthonormal bases and Gabor frames. Recall that a family of functions $\{f_j\}$ in $L^2(\mathbb{R}^d)$ is a frame if there exist constants $C_1, C_2 > 0$ such that

$$C_1\|f\|_2^2 \leq \sum_j |\langle f, f_j\rangle|^2 \leq C_2\|f\|_2^2 \tag{1.2}$$

for all $f \in L^2(\mathbb{R}^d)$. The constants C_1, C_2 are called the *frame bounds* for the frame. A frame $\{f_j\}$ is called a *tight frame* if $C_1 = C_2$, and a *Parseval tight frame* if $C_1 = C_2 = 1$. A sequence $\{f_j\}$ is called *Bessel* if it satisfies only the upper bound inequality (1.2). Recall that a sequence $\{f_j\}$ in $L^2(\mathbb{R}^d)$ is called a *Riesz basis* for

The second author is supported in part by the National Science Foundation, grants DMS-0070586 and DMS-0139261.

$L^2(\mathbb{R}^d)$ if it is complete in $L^2(\mathbb{R}^d)$ and there exist positive constants C_1 and C_2 such that

$$C_1 \sum_j |c_j|^2 \leq \Big|\sum_j c_j f_j\Big|^2 \leq C_2 \sum_j |c_j|^2 \tag{1.3}$$

for all $\{c_j\} \in l^2$. Frames are natural generalizations of Riesz bases by allowing redundancies. A fundamental question in the study of Gabor bases and frames is to find conditions on g and Λ such that $\mathbf{G}(\Lambda, g)$ is a basis or frame. Well known is the following theorem, often referred to as the Density Theorem for Gabor systems:

THE DENSITY THEOREM. *Let Λ be a full rank lattice in $\mathbb{R}^{2d}$.*

(A) *If there exists a $g \in L^2(\mathbb{R}^d)$ such that $\mathbf{G}(\Lambda, g)$ is complete in $L^2(\mathbb{R}^d)$, then $D(\Lambda) \geq 1$.*

(B) *If there exists a $g \in L^2(\mathbb{R}^d)$ such that $\mathbf{G}(\Lambda, g)$ is a Riesz basis for $L^2(\mathbb{R}^d)$, then $D(\Lambda) = 1$.*

Part (A) of the above theorem was proved in dimension $d = 1$ for separable lattices $\Lambda = \alpha\mathbb{Z} \times \beta\mathbb{Z}$ by Daubechies ([Da1], [Da2]) under the additional assumption that $\alpha\beta$ is rational. She in fact gave a constructive proof in that setting. As a corollary of a result about von Neumann algebras associated with lattices, M. Rieffel indeed proved the theorem earlier in [**Rie**] for any separable lattice $\Lambda = \alpha\mathbb{Z} \times \beta\mathbb{Z}$ with any α and β. In higher dimensions the Density Theorem was proved by Ramanathan and Steger [**RSt**], who also proved that under the assumption that $\mathbf{G}(\Lambda, g)$ is a frame we have $D^-(\Lambda) \geq 1$, where $D^-(\cdot)$ denotes the lower Beurling density, even when the lattice condition on Λ is relaxed. Part(B) of the theorem was also proved in [**RSt**], and without the lattice condition on Λ. There have been a great deal of research related to the density of Λ in a Gabor system in various contexts using various different approaches, see e.g. [**RSh**], [**CDH**], [**FS1**], [**GH1**], [**Gro**], [**Wa**] and the references therein.

This paper concerns the converse of the Density Theorem. We consider the following questions on the existence of Gabor bases and frames: Let Λ be a full rank lattice in $\mathbb{R}^{2d}$. If $D(\Lambda) = 1$ can we always find a $g \in L^2(\mathbb{R}^d)$ such that $\mathbf{G}(\Lambda, g)$ is an orthonormal basis or a Riesz basis for $L^2(\mathbb{R}^d)$? If $D(\Lambda) \geq 1$ can we always find a $g \in L^2(\mathbb{R}^d)$ such that $\mathbf{G}(\Lambda, g)$ is a frame or even a tight frame for $L^2(\mathbb{R}^d)$?

The existence questions were answered for separable lattices $\Lambda = \mathcal{K} \times \mathcal{L}$ in which both $\mathcal{K}$ and $\mathcal{L}$ are full rank lattices in $\mathbb{R}^d$. For $d = 1$ the answer is trivial. With $\Lambda = \alpha\mathbb{Z} \times \beta\mathbb{Z}$ we may simply choose $g = \frac{1}{\sqrt{|\alpha|}}\chi_{[0,|\alpha|)}$. Then it is an easy exercise to check that $\mathbf{G}(\Lambda, g)$ is a tight frame for $L^2(\mathbb{R})$ if $D(\Lambda) \geq 1$ and an orthonormal basis if $D(\Lambda) = 1$. The same results also hold in higher dimensions, although the problem became highly nontrivial. If $D(\Lambda) = 1$ then there exists a $g \in L^2(\mathbb{R}^d)$ such that $\mathbf{G}(\Lambda, g)$ is an orthonormal basis for $L^2(\mathbb{R}^d)$. If $D(\Lambda) \geq 1$ then there exists a $g \in L^2(\mathbb{R}^d)$ such that $\mathbf{G}(\Lambda, g)$ is a tight frame for $L^2(\mathbb{R}^d)$, see Han and Wang [**HW**]. They proved the existence results by studying the existence of domains that tile simultaneously by two different lattices.

However, for non-separable lattices Λ the existence questions remain unsolved for $d > 1$. For $d = 1$ the existence results hold, and can be proved easily by transforming the lattice Λ into a separable lattice using symplectic matrices, see Gröchenig [**Gro**]. We shall give a short review of the subject in Section 3. Key to this result is the fact that every lattice in $\mathbb{R}^2$ is symplectic. This is no longer true for $\mathbb{R}^{2d}$ with $d > 1$. While the existence results hold for symplectic lattices in any dimension, they appear to be difficult questions for non-symplectic lattices. In this paper we prove the existence results for a large class of non-symplectic lattices. A useful fact is that the existence of bases and frames are "almost" equivalent to the existence of complete Gabor systems.

THEOREM 1.1. *Let Λ be a full rank lattice in $\mathbb{R}^{2d}$ with $D(\Lambda) = 1$. The following are equivalent:*

(A) *There exists a function $g \in L^2(\mathbb{R}^d)$ such that $\mathbf{G}(\Lambda, g)$ is complete in $L^2(\mathbb{R}^d)$, and the set $\{h : \mathbf{G}(\Lambda, h)$ is Bessel$\}$ is dense in $L^2(\mathbb{R}^d)$.*
(B) *There exists a function $g \in L^2(\mathbb{R}^d)$ such that $\mathbf{G}(\Lambda, g)$ is a Riesz basis for $L^2(\mathbb{R}^d)$.*
(C) *There exists a function $g \in L^2(\mathbb{R}^d)$ such that $\mathbf{G}(\Lambda, g)$ is an orthonormal basis for $L^2(\mathbb{R}^d)$.*

A similar result holds for the existence of frames.

THEOREM 1.2. *Let Λ be a full rank lattice in $\mathbb{R}^{2d}$ with $D(\Lambda) \geq 1$. The following are equivalent:*

(A) *There exists a function $g \in L^2(\mathbb{R}^d)$ such that $\mathbf{G}(\Lambda, g)$ is complete in $L^2(\mathbb{R}^d)$, and the set $\{h : \mathbf{G}(\Lambda, h)$ is Bessel$\}$ is dense in $L^2(\mathbb{R}^d)$*
(B) *There exists a function $g \in L^2(\mathbb{R}^d)$ such that $\mathbf{G}(\Lambda, g)$ is a frame for $L^2(\mathbb{R}^d)$.*
(C) *There exists a function $g \in L^2(\mathbb{R}^d)$ such that $\mathbf{G}(\Lambda, g)$ is a tight frame for $L^2(\mathbb{R}^d)$.*

We conjecture that the condition that $\{h : \mathbf{G}(\Lambda, h)$ is Bessel$\}$ is dense in $L^2(\mathbb{R}^d)$ stated in (A) is automatically satisfied.

Note that any full rank lattice Λ in $\mathbb{R}^{2d}$ can be written as $\Lambda = R\mathbb{Z}^{2d}$ where R is a non-singular $2d \times 2d$ matrix. The density of Λ is then $D(\Lambda) = |\det(R)|^{-1}$. We shall call $|\det(R)|$ the *volume* of Λ and denote it by $\mathrm{v}(\Lambda)$. We prove the existence of bases and frames for certain non-symplectic lattices:

THEOREM 1.3. *Let Λ be a full rank lattice in $\mathbb{R}^{2d}$, $\Lambda = R\mathbb{Z}^{2d}$. Suppose that $R = TM$ where T is symplectic and $M = \begin{bmatrix} A & B \\ C & D \end{bmatrix}$ with either $B = 0$ or $C = 0$, where A, B, C, D are all $d \times d$ matrices. Then the following statements hold.*

(A) *There exists a function $g \in L^2(\mathbb{R}^d)$ such that $\mathbf{G}(\Lambda, g)$ is an orthonormal basis for $L^2(\mathbb{R}^d)$ if and only if $D(\Lambda) = 1$.*
(B) *There exists a function $g \in L^2(\mathbb{R}^d)$ such that $\mathbf{G}(\Lambda, g)$ is a tight frame for $L^2(\mathbb{R}^d)$ if and only if $D(\Lambda) \geq 1$.*

A particularly interesting case is when Λ is rational, i.e. $\Lambda = R\mathbb{Z}^{2d}$ and R has rational entries. We obtain a stronger conclusion with the following theorem:

THEOREM 1.4. *Let $\Lambda = R\mathbb{Z}^{2d}$ for a rational matrix R. If $D(\Lambda) = 1$ then there exists a compactly supported $g \in L^2(\mathbb{R}^d)$ such that $\mathbf{G}(\Lambda, g)$ is an orthonormal basis for $L^2(\mathbb{R}^d)$. If $D(\Lambda) \geq 1$ then there exists a compactly supported $g \in L^2(\mathbb{R}^d)$ such that $\mathbf{G}(\Lambda, g)$ is a tight frame for $L^2(\mathbb{R}^d)$.*

We point out that the existence of a compactly supported $g \in L^2(\mathbb{R}^d)$ such that $\mathbf{G}(\Lambda, g)$ is a basis (frame) is not solved even for separable lattices for $d > 1$, or for non-separable lattices in dimension $d = 1$. We conjecture that for some lattices Λ with $D(\Lambda) = 1$ there exists no compactly supported g such that $\mathbf{G}(\Lambda, g)$ is a basis even in dimension $d = 1$. In particular, we conjecture that for $\Lambda = R\mathbb{Z}^2$ with $R = \begin{bmatrix} 1 & b \\ 0 & 1 \end{bmatrix}$ and $b \notin \mathbb{Q}$ there exists no compactly supported g such that $\mathbf{G}(\Lambda, g)$ is an orthonormal basis for $L^2(\mathbb{R})$.

The rest of this paper is organized as follows: In Section 2 we briefly review our lattice tiling results and discuss how they can be used in the separable lattice case. These results are also needed in Section 4. Section 3 is a short review on symplectic lattices and the Stone-von Neumann theorem. Both sections are provided for the convenience of the readers. Section 4 is devoted to proving Theorems 1.1-1.4.

The authors wish to thank J. P. Gabardo and Chris Heil for helpful comments.

2. Separable Lattices

This section is devoted to reviewing some of the results in Han and Wang [**HW**] on the existence of Gabor bases and frames for separable lattices. We shall be using these results to prove the new results stated in Section 1. Let Λ be a full rank separable lattice, $\Lambda = A\mathbb{Z}^d \times B\mathbb{Z}^d$. Motivated by the one dimensional case, one simple idea to prove the existence of a Gabor basis $\mathbf{G}(\Lambda, g)$ is to test whether g can be chosen as a characteristic function. This naturally leads to a tiling problem for two different lattices. To state the problem and the results we need to recall some notations.

Let Ω be a measurable set in $\mathbb{R}^d$ (not necessarily bounded), and let $\mathcal{L}$ be a full rank lattice in $\mathbb{R}^d$. We say Ω *tiles* $\mathbb{R}^d$ by $\mathcal{L}$, or Ω is a *fundamental domain* of $\mathcal{L}$, if

(i) $\bigcup_{\ell \in \mathcal{L}} (\Omega + \ell) = \mathbb{R}^d$ a.e.;

(ii) $(\Omega + \ell) \cap (\Omega + \ell')$ has Lebesgue measure 0 for any $\ell \neq \ell'$ in $\mathcal{L}$.

We say that Ω *packs* $\mathbb{R}^d$ by $\mathcal{L}$ if only (ii) holds. Equivalently, Ω tiles $\mathbb{R}^d$ by $\mathcal{L}$ if and only if

$$\sum_{\ell \in \mathcal{L}} \chi_\Omega(x - \ell) = 1 \text{ for a.e. } x \in \mathbb{R}^d, \tag{2.1}$$

and Ω packs $\mathbb{R}^d$ by $\mathcal{L}$ if and only if

$$\sum_{\ell \in \mathcal{L}} \chi_\Omega(x - \ell) \leq 1 \text{ for a.e. } x \in \mathbb{R}^d. \tag{2.2}$$

Clearly, $\mu(\Omega) = \mathrm{v}(\mathcal{L})$ if Ω tiles by $\mathcal{L}$, and $\mu(\Omega) \leq \mathrm{v}(\mathcal{L})$ if Ω packs by $\mathcal{L}$, where μ denotes the Lebesgue measure. Furthermore, if Ω packs $\mathbb{R}^d$ by $\mathcal{L}$ and $\mu(\Omega) = \mathrm{v}(\mathcal{L})$, then Ω necessarily tiles $\mathbb{R}^d$ by $\mathcal{L}$.

Now let $\Lambda = \mathcal{K} \times \mathcal{L}$ be a full rank lattice in $\mathbb{R}^{2d}$ with $\mathrm{v}(\Lambda) = \mathrm{v}(\mathcal{K}) \cdot \mathrm{v}(\mathcal{L}) = 1$. Write $\mathcal{K} = A\mathbb{Z}^d$ and $\mathcal{L} = B\mathbb{Z}^d$. Then $|\det A| = |\det(B^T)^{-1}|$. It not hard to check that the question of whether there exists a characteristic function which generates an orthonormal Gabor basis is equivalent to the question of whether the two lattices $A\mathbb{Z}^d$ and $(B^T)^{-1}\mathbb{Z}^d$ have a common fundamental domain. Similarly, if $\mathrm{v}(\Lambda) \leq 1$, then the question of whether there exists a characteristic function which generates a tight Gabor frame is equivalent to the question of whether there exists a measurable set Ω such that it tiles $\mathbb{R}^d$ by the lattice $A\mathbb{Z}^d$ and packs by the lattice $(B^T)^{-1}\mathbb{Z}^d$. The tiling question has been studied in the area of tiling theory for many years (cf. [Ko]) and is closely related to a well-known open problem of Steinhaus' which asks whether there exists a common fundamental domain for all the lattices $R_\theta\mathbb{Z}^2$, where R_θ is the rotation matrix by angle θ. The following result was proved in Han and Wang [**HW**]:

THEOREM 2.1 ([**HW**]). *Let $\mathcal{L}, \mathcal{K}$ be two full rank lattices in $\mathbb{R}^d$ such that* $\mathrm{v}(\mathcal{L}) \geq \mathrm{v}(\mathcal{K})$. *Then there exists a measurable set Ω in $\mathbb{R}^d$ such that Ω tiles $\mathbb{R}^d$ by $\mathcal{K}$ and packs $\mathbb{R}^d$ by $\mathcal{L}$. In particular, when $\mathrm{v}(\mathcal{L}) = \mathrm{v}(\mathcal{K})$, then there exists a measurable set Ω in $\mathbb{R}^d$ such that Ω tiles $\mathbb{R}^d$ by both $\mathcal{K}$ and $\mathcal{L}$.*

If in addition both $\mathcal{K}$ and $\mathcal{L}$ are rational then Ω can be chosen to be compact, see [**HW**], Corollary 2.4. This yields the following:

THEOREM 2.2 ([**HW**]). *Let $\Lambda = A\mathbb{Z}^d \times B\mathbb{Z}^d$ be a full rank lattice in $\mathbb{R}^{2d}$ with* $\mathrm{v}(\Lambda) \leq 1$. *Suppose that A and B are in $\mathcal{M}_n(\mathbb{Q})$. Then there exists a bounded set Ω such that $\mathbf{G}(\Lambda, g)$ is a tight Gabor frame for $L^2(\mathbb{R}^d)$ when $g = \chi_\Omega$.*

These two results yield the following existence result for Gabor bases and frames:

THEOREM 2.3 ([**HW**]). *Let $\mathcal{L}$, $\mathcal{K}$ be two full rank lattices in $\mathbb{R}^d$ and let $\Lambda = \mathcal{K} \times \mathcal{L}$.*

(i) *If $\mathrm{v}(\mathcal{L})\mathrm{v}(\mathcal{K}) = 1$ then there exists a function $g \in L^2(\mathbb{R}^d)$ such that $\mathbf{G}(\Lambda, g)$ is an orthonormal basis for $L^2(\mathbb{R}^d)$.*
(ii) *If $\mathrm{v}(\mathcal{L})\mathrm{v}(\mathcal{K}) \leq 1$ then there exists a function $g \in L^2(\mathbb{R}^d)$ such that $\mathbf{G}(\Lambda, g)$ is a tight frame for $L^2(\mathbb{R}^d)$.*

3. Symplectic Lattices

This section is devoted to a short review of symplectic lattices and the Stone-von Neumann Theorem. A matrix $M \in \mathcal{M}_{2d}(\mathbb{R})$ is *symplectic* if $M^TJM = J$ where $J = \begin{bmatrix} 0 & -I_d \\ I_d & 0 \end{bmatrix}$. If we write $M = \begin{bmatrix} A & B \\ C & D \end{bmatrix}$ then M is symplectic if and only if $AC^T = A^TC$, $BD^T = B^TD$ and $A^TD - C^TB = I_d$. We shall use $\mathrm{Sp}(d, \mathbb{R})$ to denote the set of all symplectic matrices in $\mathcal{M}_{2d}(\mathbb{R})$. A lattice Λ in $\mathbb{R}^{2d}$ is *symplectic* if $\Lambda = \alpha M\mathbb{Z}^{2d}$ for some $\alpha \in \mathbb{R} \setminus \{0\}$ and $M \in \mathrm{Sp}(d, \mathbb{R})$.

Symplectic matrices arise from the study of the Heisenberg group $\mathbb{H}_d$ and the Schrödinger representation on $\mathbb{H}_d$. The Heisenberg group is $\mathbb{H}_d = \mathbb{R}^d \times \mathbb{R}^d \times \mathbb{T}$ with the group multiplication given by

$$(x, y, \tau)(u, v, \eta) = (x + u, y + v, \tau\eta e^{i\pi(\langle x,v\rangle - \langle y,u\rangle)})$$

for $x, y \in \mathbb{R}^d$ and $\tau, \eta \in \mathbb{T} := \{\lambda \in \mathbb{C} : |\lambda| = 1\}$. The Schrödinger (irreducible) unitary representation π of $\mathbb{H}_d$ on $L^2(\mathbb{R})$ is given by

$$\pi(x, y, \tau)f(t) = \tau e^{\pi ixy} e^{2\pi iyt} f(t + x), \qquad f \in L^2(\mathbb{R}).$$

STONE-VON NEUMANN THEOREM. *Let ρ be any irreducible unitary representation of $\mathbb{H}_d$ on a Hilbert space $\mathcal{H}$ such that*

$$\rho(0, 0, \tau)h = \tau h, \qquad \tau \in \mathbb{T}, \quad h \in \mathcal{H}.$$

Then ρ is unitarily equivalent to the Schrödinger representation π, i.e., there is a unitary operator $U\colon \mathcal{H} \to L^2(\mathbb{R}^d)$ such that

$$\rho(x, y, \tau) = U^*\pi(x, y, \tau)U, \qquad (x, y, \tau) \in \mathbb{H}_d.$$

Note that any $M \in \mathrm{Sp}(d, \mathbb{R})$ preserves the symplectic form $[Mz, Mw] = [z, w]$ for all $z, w \in \mathbb{R}^{2d}$, where $[z, w] = \langle x, v\rangle - \langle u, y\rangle$ with $z = [x, y]^T$ and $w = [u, v]^T$. Hence $\rho(x, y, \tau) = \pi(M[x, y]^T, \tau)$ defines an irreducible unitary representation for $\mathbb{H}_d$ and satisfies the condition

$$\rho(0, 0, \tau)h = \pi(0, 0, \tau)h = \tau h, \qquad h \in L^2(\mathbb{R}^d),\ \tau \in \mathbb{T}.$$

It follows from the Stone-von Neumann Theorem that there exists a unitary operator, say $\sigma(M)$, on $L^2(\mathbb{R}^d)$ such that

$$\pi(M[x, y]^T, \tau) = \sigma(M)\rho(x, y, \tau)\sigma(M)^{-1}, \qquad x,\ y \in \mathbb{R}^d,\ \tau \in \mathbb{T}.$$

This leads to the following:

LEMMA 3.1. *Let Λ be a full rank lattice in $\mathbb{R}^{2d}$ and let $M \in \mathrm{Sp}(d, \mathbb{R})$. Then there is a function $g \in L^2(\mathbb{R}^d)$ such that $\mathbf{G}(\Lambda, g)$ is an orthonormal basis (resp. Riesz basis, frame, tight frame) for $L^2(\mathbb{R}^d)$ if and only if there is a function $h \in L^2(\mathbb{R}^d)$ such that $\mathbf{G}(M\Lambda, h)$ is an orthonormal basis (resp. Riesz basis, frame, tight frame) for $L^2(\mathbb{R}^d)$.*

In general it is hard to find σ. However, for certain M, $\sigma(M)$ are familiar operators. For example, if $M = \begin{bmatrix} 0 & I \\ -I & 0 \end{bmatrix}$ then $\sigma(M) = i^{1/2}\mathcal{F}^{-1}$, where $\mathcal{F}$ denotes the Fourier transform; if $M = \begin{bmatrix} I & 0 \\ C & I \end{bmatrix}$ with $C = C^T$ then $(\sigma(M)f)(t) = e^{-\pi i\langle t, Ct\rangle} f(t)$. For more details about Stone-von Neumann Theorem and representations of $\mathbb{H}_d$ we refer to [**Fol**] or [**Gro**].

For any 2×2 real matrix M, it is obvious that $M \in \mathrm{Sp}(1, \mathbb{R})$ if and only if $\det(M) = 1$. Now any full rank lattice Λ in $\mathbb{R}^2$ can be expressed as $\Lambda = M(\alpha\mathbb{Z}^2)$ with $\alpha \in \mathbb{R}$ and $\det(M) = 1$. Since the existence results hold for the lattice $\alpha\mathbb{Z}^2$, Lemma 3.1 now yields the following theorem, see [**Gro**]:

THEOREM 3.2. *Let Λ be a full rank lattice in $\mathbb{R}^2$.*

(A) *Suppose that $D(\Lambda) = 1$ then there exists a function $g \in L^2(\mathbb{R})$ such that $\mathbf{G}(\Lambda, g)$ is an orthonormal basis for $L^2(\mathbb{R}^d)$.*
(B) *Suppose that $D(\Lambda) \geq 1$ then there exists a function $g \in L^2(\mathbb{R})$ such that $\mathbf{G}(\Lambda, g)$ is a tight frame for $L^2(\mathbb{R}^d)$.*

The factorization of a matrix into a product of a scalar and a symplectic matrix can no longer be done for all matrices in higher dimensions. This is the main obstacle in proving the existence of Gabor bases and frames for a given lattice Λ.

4. Proof of Theorems

We first prove Theorem 1.3, which we divide up as two lemmas. Note that the phase τ in $\pi(x, y, \tau) \in \mathbb{H}_d$ does not affect the basis or frame property of a Gabor system. We will simply let $\tau = 1$ and denote $\pi(x, y, 1)$ by $\pi(x, y)$.

LEMMA 4.1. *Let Λ be a full rank lattice in $\mathbb{R}^{2d}$ such that $\Lambda = M\mathbb{Z}^{2d}$ with*

$$M = \begin{bmatrix} A & 0 \\ B & D \end{bmatrix}.$$

If $\mathrm{v}(\Lambda) = 1$ (resp. $\mathrm{v}(\Lambda) \leq 1$), then there exists a function $g \in L^2(\mathbb{R}^d)$ such that $\mathbf{G}(\Lambda, g)$ is an orthonormal basis (resp. tight frame) for $L^2(\mathbb{R}^d)$. Moreover g can be chosen such that $|g(t)|$ is the scalar multiple of a characteristic function.

PROOF. We will apply a matrix T (which is not necessarily a symplectic matrix, and hence the Stone-von Neumann Theorem does not apply here) to Λ such that $T\Lambda$ is a separable lattice. We then apply Theorem 2.2 to obtain g. Let $C = -BA^{-1}$ and

$$T = \begin{bmatrix} I & 0 \\ C & I \end{bmatrix}.$$

Then

$$TM = \begin{bmatrix} A & 0 \\ 0 & D \end{bmatrix}.$$

A simple calculation shows that for any $x, y \in \mathbb{R}^d$ and $f \in L^2(\mathbb{R}^d)$,

$$\begin{aligned}
\pi(T[x,y]^T)f(t) &= \pi(x, Cx+y)f(t) \\
&= e^{2\pi i\langle Cx+y,t\rangle+\pi i\langle x,Cx+y\rangle} f(t+x) \\
&= e^{\pi i\langle x,(C^T-C)t\rangle} e^{-\pi i\langle t,Ct\rangle} e^{2\pi i\langle y,t\rangle+\pi i\langle x,y\rangle} e^{\pi i\langle t+x,C(t+x)\rangle} f(t+x) \\
&= e^{\pi i\langle x,(C^T-C)t\rangle} U\pi(x,y)U^{-1}f(t),
\end{aligned}$$

where $(Uf)(t) = e^{-\pi i\langle t,Ct\rangle}f(t)$.

We first consider the case $D(\Lambda) \geq 1$, which is equivalent to $\mathrm{v}(\Lambda) = |\det(AD)| \leq 1$. Hence $|\det A| \leq |\det(D^T)^{-1}|$. By Theorem 2.1 there exists a measurable set Ω in $\mathbb{R}^d$ such that Ω tiles $\mathbb{R}^d$ by $A\mathbb{Z}^d$ and packs $\mathbb{R}^d$ by $(D^T)^{-1}\mathbb{Z}^d$. Thus for any $f \in L^2(\mathbb{R}^d)$ we have

$$\|f\|^2 = \sum_{m\in\mathbb{Z}^d} \|P_m f\|^2$$

and

$$\|P_m f\|^2 = \sum_{n\in\mathbb{Z}^d} |\langle f, \pi(Am, Dn)h(t)\rangle|^2$$

where $h(t) = \frac{1}{\sqrt{|\det A|}}\chi_\Omega$ and P_m is the orthogonal projection onto $L^2(\Omega + Am)$. Now we define $g(t) = U^{-1}h(t)$. We claim that $\{\pi(x,y)g : (x,y) \in \Lambda\}$ is a tight frame for $L^2(\mathbb{R}^d)$ (and hence $\mathbf{G}(\Lambda, g)$ is a tight frame). Indeed, for any $f \in L^2(\mathbb{R}^d)$ we have

$$\begin{aligned}
\sum_{(x,y)\in\Lambda} |\langle f, \pi(x,y)g\rangle|^2 &= \sum_{(x,y)\in\Lambda} |\langle f, e^{\pi i\langle x,(C-C^T)t\rangle} U^{-1}\pi(T(x,y)^T)Ug\rangle|^2 \\
&= \sum_{(m,n)\in\mathbb{Z}^d\times\mathbb{Z}^d} |\langle e^{-\pi i\langle Am,(C-C^T)t\rangle} Uf, \pi(Am,\tilde{D}n)^T)h\rangle|^2 \\
&= \sum_{(x,y)\in\Lambda} |\langle e^{-\pi i\langle x,(C-C^T)t\rangle} Uf, \pi(T(x,y)^T)h\rangle|^2 \\
&= \sum_{m\in\mathbb{Z}^d}\sum_{n\in\mathbb{Z}^d} |\langle e^{-\pi i\langle Am,(C-C^T)t\rangle} Uf, \pi(Am,Dn)^T)h\rangle|^2 \\
&= \sum_{m\in\mathbb{Z}^d} \|P_m e^{-\pi i\langle Am,(C-C^T)t\rangle} Uf\|^2 \\
&= \sum_{m\in\mathbb{Z}^d} \|P_m f\|^2 = \|f\|^2.
\end{aligned}$$

Thus $\{\pi(x,y)g : (x,y) \in \Lambda\}$ is a tight frame as claimed. In fact it is a Parseval tight frame. This shows that $\mathbf{G}(\Lambda, g)$ is a tight frame for $L^2(\mathbb{R}^d)$.

In the case of $D(\Lambda) = 1$ we follow the exact same procedure. However in this case Ω tiles by both $A\mathbb{Z}^d$ and $(D^T)^{-1}\mathbb{Z}^d$. This yields a Gabor orthonormal basis $\mathbf{G}(\Lambda, g)$ for $L^2(\mathbb{R}^d)$, see also [**HW**].

We remark that in both cases $|g| = \frac{1}{\sqrt{|\det(A)|}}\chi_\Omega$. □

LEMMA 4.2. *Let Λ be a full rank lattice in $\mathbb{R}^{2d}$ such that $\Lambda = M\mathbb{Z}^{2d}$ with*

$$M = \begin{bmatrix} A & B \\ 0 & D \end{bmatrix}.$$

If $\mathrm{v}(\Lambda) = 1$ (resp. $\mathrm{v}(\Lambda) \leq 1$), then there exists a function $g \in L^2(\mathbb{R}^d)$ such that $\mathbf{G}(\Lambda, g)$ is an orthonormal basis (resp. tight frame) for $L^2(\mathbb{R}^d)$. Moreover g can be chosen such that $|\widehat{g}(t)|$ is the scalar multiple of a characteristic function.

PROOF. Note that the Fourier transform of $g_{\kappa,\ell} := e^{2\pi\langle \ell, x\rangle} g(x-\kappa)$ is $e^{2\pi\langle \ell,\kappa\rangle}\widehat{g}_{\ell,-\kappa}$. Hence $\mathbf{G}(\Lambda, g)$ is an orthonormal basis (resp. tight frame) if and only if $\mathbf{G}(\tilde{\Lambda}, \widehat{g})$ is, where $\tilde{\Lambda} = \tilde{M}\mathbb{Z}^{2d}$ with

$$\tilde{M} = \begin{bmatrix} D & 0 \\ -B & -A \end{bmatrix}.$$

The lemma now follows from Lemma 4.1. □

PROOF OF THEOREM 1.3. Let $\Lambda = R\mathbb{Z}^{2d}$ with $R = TM$ where T is symplectic and M is block triangular. Let $\Gamma = M\mathbb{Z}^{2d}$. We have already established the existence of an orthonormal Gabor basis (resp. tight Gabor frame) if $D(\Gamma) = 1$ (resp. $D(\Gamma) \geq 1$). The theorem now follows directly from Lemma 3.1. □

COROLLARY 4.3. *Let Λ be a full rank lattice in $\mathbb{R}^{2d}$ such that $\Lambda = M\mathbb{Z}^{2d}$ with*

$$M = \begin{bmatrix} A & B \\ C & D \end{bmatrix}.$$

and assume that $\mathrm{v}(\Lambda) = 1$ (resp. $\mathrm{v}(\Lambda) \leq 1$). If either CA^{-1} or BD^{-1} is symmetric, then there exists a function $g \in L^2(\mathbb{R}^d)$ such that $\mathbf{G}(\Lambda, g)$ is an orthonormal basis (resp. tight frame) for $L^2(\mathbb{R}^d)$.

PROOF. If CA^{-1} is symmetric, we define $T = \begin{bmatrix} I & 0 \\ -CA^{-1} & I \end{bmatrix}$; whereas if BD^{-1} is symmetric we let $T = \begin{bmatrix} I & -BD^{-1} \\ 0 & I \end{bmatrix}$. The matrix T is in $\mathrm{Sp}(d, \mathbb{R})$. Furthermore let $\Gamma = TM\mathbb{Z}^{2d}$ then there exists an $h \in L^2(\mathbb{R}^d)$ such that $\mathbf{G}(\Gamma, h)$ is an orthonormal basis (resp. a tight frame) because TM is block triangular and $\mathrm{v}(\Gamma) = \mathrm{v}(\Lambda) = 1$ (resp. $\mathrm{v}(\Gamma) = \mathrm{v}(\Lambda) \leq 1$). Since $\Lambda = T^{-1}\Gamma$ and T^{-1} is symplectic, the corollary now follows directly from Lemma 3.1. □

We now prove Theorem 1.4. To do so we need to establish the following lemma.

LEMMA 4.4. *Let $M \in \mathcal{M}_n(\mathbb{Z})$. Then there exists a unimodular $P \in \mathcal{M}_n(\mathbb{Z})$ such that MP is lower triangular.*

PROOF. Let $[a_{11}, a_{12}, \cdots, a_{1n}]$ be the first row of M with $\gcd(a_{1j}) = d_1$. It follows from elementary number theory that

$$x_1 a_{11} + x_2 a_{12} + \cdots + x_n a_{1n} = d_1$$

for some integers $x_1, x_2, \ldots, x_n$ with $\gcd(x_j) = 1$. Let P_1 be the unimodular integer matrix with $[x_1, x_2, \cdots, x_n]^T$ as its first column. Such a matrix is well known to exist, see Newman [**Ne**], Theorem II.1, page 13. Observe that

$$M_1 := MP_1 = \begin{bmatrix} d_1 & b_{12} & \cdots & b_{1n} \\ b_{21} & b_{22} & \cdots & b_{2n} \\ \vdots & \vdots & \ddots & \vdots \\ b_{n1} & b_{n2} & \cdots & b_{nn} \end{bmatrix}.$$

Since each b_{1j} is a linear combination of a_{1j}'s, $d_1 \mid b_{1j}$ for $j = 2, ..., n$. Therefore we may use column Gaussian elimination to reduce M_1 to

$$M_2 = \begin{bmatrix} d_1 & 0 & \cdots & 0 \\ c_{21} & c_{22} & \cdots & c_{2n} \\ \vdots & \vdots & \ddots & \vdots \\ c_{n1} & c_{n2} & \cdots & c_{nn} \end{bmatrix}.$$

Note that $M_2 = M_1 P_2$ with a unimodular $P_2 \in \mathcal{M}_n(\mathbb{Z})$. Now the lemma is proved by induction on n. □

PROOF OF THEOREM 1.4. We have $\Lambda = R\mathbb{Z}^{2d}$ with $R \in \mathcal{M}_{2d}(\mathbb{Q})$. Write $R = \frac{1}{q}\tilde{R}$ such that $\tilde{R} \in \mathcal{M}_{2d}(\mathbb{Z})$ and $q \in \mathbb{Z}$. It follows from Lemma 4.4 that $\tilde{R} = \tilde{T}P$ where $\tilde{T} \in \mathcal{M}_{2d}(\mathbb{Z})$ is a lower triangular integral matrix and P is unimodular integral matrix. Thus $\tilde{R}\mathbb{Z}^{2d} = \tilde{T}P\mathbb{Z}^{2d} = \tilde{T}\mathbb{Z}^{2d}$. Set $T = \frac{1}{q}\tilde{T}$ and thus $\Lambda = T\mathbb{Z}^{2d}$.

Now T is lower triangular, and write $T = \begin{bmatrix} A & 0 \\ B & D \end{bmatrix}$. Therefore in the case $D(\Lambda) = 1$ it follows from Theorem 1.3 and its proof that there exists a $g \in L^2(\mathbb{R}^d)$ such that $\mathbf{G}(\Lambda, g)$ is an orthonormal basis, with $|g| = \frac{1}{\sqrt{|\det(A)|}}\chi_\Omega$ for any domain Ω that tiles by both $A\mathbb{Z}^d$ and $(D^T)^{-1}\mathbb{Z}^d$. Since both A and D are rational, we may choose a bounded domain Ω ([**HW**], Corollary 2.4). Hence we may choose a compactly supported g. In the case $D(\Lambda) \geq 1$ the same argument shows that we may choose a g with $|g| = \frac{1}{\sqrt{|\det(A)|}}\chi_\Omega$, where Ω can be any domain that tiles by $A\mathbb{Z}^d$ and packs by $(D^T)^{-1}\mathbb{Z}^d$. Again a bounded such Ω exists since both A and D are rational ([**HW**]). The theorem is now proved. □

We conclude the paper by proving Theorems 1.1 and 1.2.

PROOF OF THEOREM 1.1 AND 1.2. Clearly, (B) imples (A), and it is also well-known that (B) and (C) are equivalent by using the fact that the "frame operator" commutes with both translations and modulations. Thus it is sufficient to prove that (A) implies (C). Let T_κ and M_ℓ be the unitary operators defined by

$$T_\kappa f(x) = f(x - \kappa)$$

and

$$M_\ell f(x) = e^{2\pi i <\ell, x>} f(x)$$

for $f \in L^2(\mathbb{R}^d)$. Then

$$\mathcal{U}_\Lambda := \{M_\ell T_\kappa \mid (\kappa, \ell) \in \Lambda\}$$

is a group-like unitary system in the terminology of [GH1] and [GH2]. Now assume that there exists a $g \in L^2(\mathbb{R}^d)$ such that $\mathbf{G}(\Lambda, g)$ is complete in $L^2(\mathbb{R}^d)$, and the set $\{h : \mathbf{G}(\Lambda, h) \text{ is Bessel}\}$ is dense in $L^2(\mathbb{R}^d)$. The group-like unitary system $\mathcal{U}_\Lambda$

satisfies the condition required in Theorem 2 in [GH2]. Thus there exists a function f such that $\mathbf{G}(\Lambda, f)$ is a normalized tight frame (orthonormal basis in the case of $D(\Lambda) = 1$) for $L^2(\mathbb{R}^d)$. □

References

[CDH] O. Christensen, B. Deng and C. Heil, Density of Gabor frames, *Appl. Comput. Harm. Anal* **7**(1999), 292–304.

[Da1] I. Daubechies, The wavelet transformation, time-frequency localization and s ignal processing, *IEEE Trans. Inform. Theory*, **36** (1990), 961-1005.

[Da2] I. Daubechies, *Ten Lectures on Wavelets*, CBMS-NSF regional conference series in Applied Math., Philadelphia (1992)

[DLL] I. Daubechies, H. Landau and Z. Landau, Gabor time frequency lattices and the Wexler-Raz identity *J. Fourier. Anal. Appl.* **1** (1995), 437-478.

[DS] R. Duffin and A.Shaeffer, A class of non-harmonic Fourier series, *Trans. Amer. Math. Soc.* 72 (1952) 341-366

[FS1] H. G. Feichtinger and T. Strohmer (eds), *Gabor Analysis and Algorithms: Theory and Applications,* Applied and Numerical Harmonic Analysis, Birkhäuser, 1998.

[FS2] H. G. Feichtinger and T. Strohmer (eds), *Advances in Gabor Analysis,* Applied and Numerical Harmonic Analysis, Birkhäuser, 2002.

[Fol] G. B. Folland, *Harmonic Analysis in Phase Space,* Annals of Math. Studies, Princeton Univ. Press, Princeton (NJ), 1989.

[GH1] J. P. Gabardo and D. Han, Frame representations for group-like unitary operator systems, *J. Operator Theory,* to appear.

[GH2] J. P. Gabardo and D. Han, Subspace Weyl-Heisenberg frames, *J. Fourier Analysis and Appl.*, **7** (2001), 419–433.

[GH3] J. P. Gabardo and D. Han, Aspects of Gabor analysis and operator algebras, *Advances in Gabor Analysis,* Eds: H. G. Feichtinger and T. Strohmer, Birkhäuser, 2002, 129-150.

[Ga] D. Gabor, Theory of Communication, *J. Inst. Elec. Eng.* (London) **93** (1946), 429–457.

[Gro] K. H. Gröchenig, *Foundations of Time-Frequency Analysis,* Applied and Numerical Harmonic Analysis, Birkhäuser, 2001.

[HL] D. Han and D. Larson, Frames, bases and group parametrizations, *Memoirs Amer. Math. Soc.,* **697** (2000).

[HW] D. Han and Y. Wang, Lattice tiling and Gabor frames, *Geometric and Functional Analysis,* 11(2001), 742–758.

[Ko] M. Kolountzakis, Multi-lattice tiles, *Internat. Math. Res. Notices* **19** (1997), 937-952.

[Ne] M. Newman, *Integral Matrices*, Academic Press, 1972.

[Rie] M. A. Rieffel, von Neumann algebras associated with pairs of lattices in Lie groups, *Math. Ann.* **257**(1981), 403–413.

[RSh] A. Ron and Z. Shen, Weyl-Heisenberg frames and Riesz bases in $L^2(\mathbb{R}^d)$, *Duke Mathematical Journal* **89** (1997), 237–282.

[RSt] J. Ramanathan and T. Steger, Incompleteness of sparse coherent states, *Appl. Comp. Harm. Anal.* **2** (1995), 148–153.

[Wa] Y. Wang, Sparse complete Gabor systems on a lattice, preprint.

Department of Mathematics, University of Central Florida, Orlando, FL 32816, USA

E-mail address: `dhan@pegasus.cc.ucf.edu`

School of Mathematics, Georgia Institute of Technology, Atlanta, Georgia 30332, USA.

E-mail address: `wang@math.gatech.edu`

Contemporary Mathematics
Volume **345**, 2004

Co-affine systems in $\mathbb{R}^d$

Brody Dylan Johnson

ABSTRACT. The proof of non-existence for co-affine frames is extended from the one-dimensional setting [**GLWW**] to the case of expansive dilation matrices in $\mathbb{R}^d$. The problem of identifying subspaces on which co-affine systems may admit frame-type inequalities is then considered. In the context of multiresolution analysis it is shown that frame-type inequalities may hold on certain fundamental subspaces of an MRA. Finally, necessary conditions are given for a general co-affine system to admit frame-type inequalities on band-limited subspaces. In the case of the co-affine system generated by the Shannon wavelet these results dictate that Parseval's identity holds on the band-limited subspace having bandwidth 2, but cannot hold with any larger bandwidth.

1. Introduction

Let $A \in GL_d(\mathbb{R})$ be expansive, i.e., each eigenvalue of A is strictly larger than 1 in modulus. Let D be the dilation operator induced by A, given by $Df(x) = \sqrt{|\det A|}f(Ax)$. Let T_y, $y \in \mathbb{R}^d$, be the translation operator, defined by $T_y f(x) = f(x-y)$. Finally, Fourier transform, $\hat{f}$, of $f \in L^1 \cap L^2(\mathbb{R}^d)$ shall be taken to be

$$\hat{f}(\xi) := \int_{\mathbb{R}^d} f(x) e^{-2\pi i \xi \cdot x} dx.$$

Together, the dilation and translation operators form the backbone of wavelet theory. The classic notion of a *dyadic orthonormal wavelet* is a function $\psi \in L^2(\mathbb{R})$ for which the collection $\{2^{\frac{j}{2}}\psi(2^j x - k) : j, k \in \mathbb{Z}\}$ is an orthonormal basis for $L^2(\mathbb{R})$ [**D, HW**]. More generally, wavelet theory involves the study of collections composed of certain dilates and translates of a finite number of generating functions. Given a finite collection $\Psi = \{\psi_1, \ldots, \psi_L\} \subset L^2(\mathbb{R}^d)$, the *affine system generated by* Ψ is defined to be the collection

$$X(\Psi) = \{D^j T_k \psi_\ell : 1 \leq \ell \leq L, j \in \mathbb{Z}, k \in \mathbb{Z}^d\}.$$

As in the dyadic orthonormal case, one might be interested merely in those $\Psi \subset L^2(\mathbb{R}^d)$ for which $X(\Psi)$ is an orthonormal basis for $L^2(\mathbb{R}^d)$, but in many instances this restriction is unnecessary and one may instead consider the more general class of affine systems which comprise frames for $L^2(\mathbb{R}^d)$.

2000 *Mathematics Subject Classification.* Primary 42C40.
Key words and phrases. co-affine, quasi-affine, multiresolution analysis.

In a separable Hilbert space, $\mathbb{H}$, a collection $\{h_j\}_{j\in J} \subset \mathbb{H}$ is a *frame* if there exist constants $0 < B_1 \leq B_2 < \infty$ such that for all $f \in \mathbb{H}$

$$B_1\|f\|_{\mathbb{H}}^2 \leq \sum_{j\in J} |\langle f, h_j\rangle_{\mathbb{H}}|^2 \leq B_2\|f\|_{\mathbb{H}}^2. \tag{1.1}$$

The constants B_1 and B_2 are referred to as the lower and upper frame bounds, respectively, and in the case that one may choose $B_1 = B_2$ the frame is said to be *tight.* In particular, when one may choose $B_1 = B_2 = 1$ the frame is said to be *Parseval.* Any collection possessing an upper frame bound is called a *Bessel sequence.*

The term *wavelet* will be used here to refer to a generator or set of generators Ψ (as above) for which $X(\Psi)$ is a frame for $L^2(\mathbb{R}^d)$. When $X(\Psi)$ is additionally an orthonormal collection the term *orthonormal wavelet* will also be used. The literature in wavelet theory has been dominated by the study of affine systems, but there are alternative systems that have also played important roles in the theory. The most notable alternative systems are the *quasi-affine* systems, which were introduced by Ron and Shen as a tool for characterizing affine frames using the theory of shift-invariant spaces [**RS**]. With Ψ as above, the *quasi-affine system generated by* Ψ is the collection

$$\begin{aligned} X^q(\Psi) =& \{D^j T_k \psi_\ell : 1 \leq \ell \leq L, j \geq 0, k \in \mathbb{Z}^d\} \\ & \cup \{|\det A|^{\frac{j}{2}} T_k D^j \psi_\ell : 1 \leq \ell \leq L, j < 0, k \in \mathbb{Z}^d\}. \end{aligned}$$

Notice that $X^q(\Psi)$ is $\mathbb{Z}^d$-shift invariant, a property stemming from the reversal of the dilation and translation operators for the scales $j < 0$. The principle result relating affine and quasi-affine systems says that when A is an integral, expansive matrix then $X(\Psi)$ is a frame for $L^2(\mathbb{R}^d)$ if and only if $X^q(\Psi)$ is a frame for $L^2(\mathbb{R}^d)$ and, moreover, that in either case the frame bounds of the two systems are identical [**RS, CSS**]. The result also holds for Bessel sequences. Despite the prevalence of affine systems in the literature, there have been a number of other interesting works dealing with or related to quasi-affine systems [**Bo1, GLWW, HKMT, HLW, HLWW, La, MZ**]. In particular, the works [**HKMT**] and [**MZ**] deal with the à trous algorithm, which is a shift-invariant version of the discrete wavelet transform with strong ties to quasi-affine systems [**Jo**].

It is natural to wonder what would result from the reversal of the dilation and translation operators at *each* scale $j \in \mathbb{Z}$, as opposed to just the negative scales as in the quasi-affine systems.

DEFINITION 1.1 ([**GLWW**]). Let $\Psi = \{\psi_\ell\}_{\ell=1}^L \subset L^2(\mathbb{R}^d)$ and let $c := \{c_{\ell;j}\}$, $1 \leq \ell \leq L$, $j \in \mathbb{Z}$, be any numerical sequence. The *weighted co-affine system generated by* Ψ *and* c, denoted $X^*(\Psi, c)$, is the collection

$$X^*(\Psi, c) = \{\psi^*_{\ell;j,k} := c_{\ell;j} T_k D^j \psi_\ell : 1 \leq \ell \leq L, j \in \mathbb{Z}, k \in \mathbb{Z}^d\}.$$

Gressman *et al.* showed that when $d = 1$, $A = a > 1$, and $\Psi = \{\psi\}$, the weighted co-affine system $X^*(\Psi, c)$ can never comprise a frame for $L^2(\mathbb{R})$ [**GLWW**]. This suggests that finitely generated co-affine systems should always fail to constitute frames for $L^2(\mathbb{R}^d)$ (at least for expansive dilations), but as unions of shift-invariant systems one would still expect that co-affine systems should be able to provide stable, frame-like reconstruction formulas on proper subspaces of $L^2(\mathbb{R}^d)$.

The intent of this paper is threefold. First, the non-existence result of [**GLWW**] will be extended to the case of finitely generated co-affine systems associated with expansive dilation matrices in $\mathbb{R}^d$. Second, the framework of multiresolution analysis shall be used to develop frame-like reconstruction formulas for certain fundamental subspaces of an MRA. Finally, the technique of the non-existence result will be used to derive necessary conditions for a co-affine system to admit a frame-type inequality on a given band-limited subspace.

2. Non-existence of co-affine frames

The following lemma generalizes a useful calculation from [**GLWW**] to $\mathbb{R}^d$.

LEMMA 2.1. *If $X^*(\Psi, c)$ is a Bessel system for $L^2(\mathbb{R}^d)$ then for each $f \in L^2(\mathbb{R}^d)$,*

$$\int_{\mathcal{D}} w(x)dx = \int_{\mathbb{R}^d} \Big(\sum_{\ell=1}^{L} \sum_{j\in\mathbb{Z}} |c_{\ell;j}|^2 |\det A|^{-j} |\hat{\psi}(A^{-j}\xi)|^2 \Big) |\hat{f}(\xi)|^2 d\xi, \tag{2.1}$$

where $w(x)$ is the $\mathbb{Z}^d$-periodic function defined by

$$w(x) = \sum_{\ell=1}^{L} \sum_{j\in\mathbb{Z}} \sum_{k\in\mathbb{Z}^d} \left|\langle T_x f, \psi^*_{\ell;j,k}\rangle\right|^2,$$

and $\mathcal{D}$ represents the unit cube in $\mathbb{R}^d$.

PROOF. The fact that the system is Bessel implies that the integral in (2.1) is finite. Moreover, the integrand is non-negative, allowing one to interchange the order of the sums and the integral. Observe that

$$\begin{aligned}
\int_{\mathcal{D}} w(x)dx &= \int_{\mathcal{D}} \sum_{\ell=1}^{L} \sum_{j\in\mathbb{Z}} \sum_{k\in\mathbb{Z}} \left|\langle T_x f, \psi^*_{\ell;j,k}\rangle\right|^2 dx \\
&= \sum_{\ell=1}^{L} \sum_{j\in\mathbb{Z}} |c_{\ell;j}|^2 \int_{\mathcal{D}} \sum_{k\in\mathbb{Z}^d} \left|\langle f, T_{-x+k} D^j \psi_\ell\rangle\right|^2 dx \\
&= \sum_{\ell=1}^{L} \sum_{j\in\mathbb{Z}} |c_{\ell;j}|^2 \int_{\mathbb{R}^d} \left|\langle f, T_x D^j \psi_\ell\rangle\right|^2 dx \\
&= \sum_{\ell=1}^{L} \sum_{j\in\mathbb{Z}} |c_{\ell;j}|^2 \int_{\mathbb{R}^d} \left|\langle \hat{f}, |\det A|^{\frac{-j}{2}} \hat{\psi}_\ell((A^T)^{-j}\,\cdot\,) e^{-2\pi i \langle x, \cdot\rangle}\rangle\right|^2 dx \\
&= \sum_{\ell=1}^{L} \sum_{j\in\mathbb{Z}} |c_{\ell;j}|^2\, |\det A|^{-j} \int_{\mathbb{R}^d} \left|\big(\hat{f}\overline{\hat{\psi}_\ell((A^T)^{-j}\,\cdot\,)}\big)\check{}(x)\right|^2 dx \\
&= \int_{\mathbb{R}^d} \Big(\sum_{\ell-1}^{L} \sum_{j\in\mathbb{Z}} |c_{\ell;j}|^2\, |\det A|^{-j}\, \left|\hat{\psi}_\ell((A^T)^{-j}\xi)\right|^2 \Big) |\hat{f}(\xi)|^2 dx.
\end{aligned}$$

□

The next lemma describes a well-known fact about matrices that plays an important role in the handling of expansive dilations. One can find a proof of the lemma in [**HLW**].

LEMMA 2.2. *Suppose $A \in GL_d(\mathbb{R})$ and that there exist $\alpha, \beta \in \mathbb{R}$ such that $0 < \alpha < |\lambda| < \beta < \infty$ for each eigenvalue λ of A. Then there exists $C = C(A, \alpha, \beta) \geq 1$ such that*

$$C^{-1}\alpha^j |x| \leq |A^j x| \leq C\beta^j |x|, \tag{2.2}$$

for all $x \in \mathbb{R}^d$ and each positive integer j.

THEOREM 2.3. *There exist no weighted co-affine frames for $L^2(\mathbb{R}^d)$.*

PROOF. Let $\Psi = \{\psi_1, \dots, \psi_L\} \subset L^2(\mathbb{R})$ and $c = \{c_{\ell;j}\}_{1 \leq \ell \leq L, j \in \mathbb{Z}}$ be a fixed numerical sequence, $c \neq 0$. Proceeding by contradiction, suppose that $X^*(\Psi, c)$ is a frame with bounds $0 < B_1 \leq B_2 < \infty$. Notice that for any fixed $j_0 \in \mathbb{Z}$, $k_0 \in \mathbb{Z}^d$, and $1 \leq \ell_0 \leq L$ one has

$$|c_{\ell_0;j_0}|^4 \|\psi^*_{\ell_0,j_0,k_0}\|^4 \leq \sum_{\ell=1}^{L} \sum_{j \in \mathbb{Z}} \sum_{k \in \mathbb{Z}^d} \left|\langle \psi^*_{\ell_0;j_0,k_0}, \psi^*_{\ell;j,k} \rangle\right|^2 \leq B_2 |c_{\ell_0;j_0}|^2 \|\psi_{\ell_0}\|^2,$$

which implies

$$|c_{\ell;j}|^2 \leq B_2 \|\psi_\ell\|^{-2}. \tag{2.3}$$

It follows from Lemma 2.1 that for a.e. $\xi \in \mathbb{R}^d$

$$B_1 \leq \sum_{\ell=1}^{L} \sum_{j \in \mathbb{Z}} |c_{\ell;j}|^2 |\det A|^{-j} \left|\hat{\psi}_\ell\big((A^T)^{-j}\xi\big)\right|^2 \leq B_2. \tag{2.4}$$

It is by means of (2.4) that a contradiction will be derived.

Let $1 < \lambda \leq \Lambda < \infty$, respectively, be strict lower and upper bounds of the eigenvalues of A. Let $E_{a,b}$ be the set defined by

$$E = E_{a,b} = \{x \in \mathbb{R}^d : a < |x| < b\},$$

where $0 < a < b < \infty$. It is important to understand the overlap of $(A^T)^{j_1} E_{a,b}$ and $(A^T)^{j_2} E_{a,b}$ for $j_1, j_2 \in \mathbb{Z}$. Since $A \in GL_n(\mathbb{R})$ it is sufficient to consider $j_1 \geq 0$ and $j_2 = 0$. Observe that if $x \in E_{a,b}$ and $j > 0$ then

$$C^{-1}\lambda^j a < |(A^T)^j x| < C\Lambda^j b, \tag{2.5}$$

with C as in Lemma 2.2. Thus, in order for $(A^T)^j E$ to be disjoint from E one must have

$$\lambda^j \geq C\frac{b}{a}.$$

Fixing $a = 1$ and $b = \lambda$ there exists $J \geq 1$ (finite) such the last inequality is satisfied for all $j \geq J$. Let $E := E_{1,\lambda}$. The upshot of these observations is that the sets $\{(A^T)^j E\}_{j \in \mathbb{Z}}$ have finite overlap, i.e., $A^j E \cap E = \emptyset$ for sufficiently large integers j. This property is crucial for the following argument. In fact, if $x \in (A^T)^{j_0} E$, $j_0 \in \mathbb{Z}$, then x belongs to at most $2J - 1$ sets in the collection $\{(A^T)^j E\}_{j \in \mathbb{Z}}$. Let $n \in \mathbb{Z}$ and observe by (2.4) that

$$\begin{aligned}\int_{(A^T)^{-n}E} B_1 d\xi &\leq \int_{(A^T)^{-n}E} \Big(\sum_{\ell=1}^{L} \sum_{j \in \mathbb{Z}} |c_{\ell;j}|^2 \, |\det A|^{-j} \, \left|\hat{\psi}_\ell\big((A^T)^{-j}\xi\big)\right|^2\Big) d\xi \\ &= \int_E \Big(\sum_{\ell=1}^{L} \sum_{j \in \mathbb{Z}} |c_{\ell;j}|^2 \, |\det A|^{-(j+n)} \, \left|\hat{\psi}_\ell\big((A^T)^{-(j+n)}\xi\big)\right|^2\Big) d\xi\end{aligned}$$

$$= \int_E \Big(\sum_{\ell=1}^{L} \sum_{m\in\mathbb{Z}} |c_{\ell;-(m+n)}|^2 \, |\det A|^m \, \big|\hat{\psi}_\ell\big((A^T)^m\xi\big)\big|^2 \Big) d\xi$$

$$\leq \sum_{\ell=1}^{L} B_2 \|\psi_\ell\|^{-2} \sum_{m\in\mathbb{Z}} \int_{(A^T)^m E} |\hat{\psi}_\ell(\xi)|^2 d\xi$$

$$\leq (2J-1) \sum_{\ell=1}^{L} B_2 \|\psi_\ell\|^{-2} \int_{\mathbb{R}^d} |\hat{\psi}_\ell(\xi)|^2 d\xi$$

$$\leq (2J-1)\, B_2\, L.$$

Letting μ denote the volume of the unit sphere in $\mathbb{R}^d$, the left-hand side of the last chain of inequalities becomes

$$\int_{A^{-n}E} B_1 d\xi = |\det A|^{-n} \mu(\lambda^d - 1) B_1,$$

from which it follows that

$$\mu(\lambda^d - 1) B_1 \leq |\det A|^n \, (2J-1)\, B_2\, L \tag{2.6}$$

for each $n \in \mathbb{Z}$. By choosing $-n$ large enough one obtains the desired contradiction. □

3. Multiresolution analysis and co-affine systems

In the last section it was seen that co-affine systems can never constitute frames for $L^2(\mathbb{R}^d)$. This leads naturally to questions about which subspaces admit frame-like reconstructions in terms of co-affine systems. One setting in which this will be possible is that of the familiar multiresolution analysis, a tool that has fueled much research in wavelet theory. Accordingly, it will be assumed throughout this section not only that A is expansive, but also that A is integral, i.e., $A\mathbb{Z}^d \subset \mathbb{Z}^d$.

Recall that a *multiresolution analysis* (MRA) is a collection of closed subspaces $\{V_j\}_{j\in\mathbb{Z}} \subseteq L^2(\mathbb{R}^d)$ such that:

(i) $V_j \subseteq V_{j+1}$, $j \in \mathbb{Z}$;
(ii) $f \in V_j$ if and only if $Df \in V_{j+1}$, $j \in \mathbb{Z}$;
(iii) $\bigcap_{j\in\mathbb{Z}} V_j = \{0\}$;
(iv) $\overline{\bigcup_{j\in\mathbb{Z}} V_j} = L^2(\mathbb{R}^d)$;
(v) There exists $\varphi \in V_0$ (a *scaling function*) such that $\{T_k\varphi\}_{k\in\mathbb{Z}^d}$ is an orthonormal basis for V_0.

The subspace V_0 is often referred to as the *core subspace* of the MRA. Notice that V_0 is invariant under integral translations.

The importance of multiresolution analysis in wavelet theory is immense. The early works of Mallat [**Ma**] and Daubechies [**D**] offered constructions of dyadic orthonormal wavelets relying on the structure of MRAs and since that time there have been generalizations and extensions too numerous to mention here in any detail. One example of a multi-dimensional construction of MRA wavelets (tight affine frames) may be found in [**Bo1**] and a nice treatment of the one-dimensional theory of dyadic orthonormal wavelets may be found in [**HW**].

EXAMPLE 3.1. Consider the Shannon wavelet,

$$\psi(x) = -2\frac{\sin(2\pi x) + \cos(\pi x)}{\pi(2x+1)},$$

and the associated scaling function

$$\varphi(x) = \frac{\sin(\pi x)}{\pi x}.$$

Notice that $\hat{\varphi}(\xi) = \chi_{[-2^{-1},2^{-1}]}(\xi)$ and thus the principal shift-invariant space V_0 generated by φ consists precisely of those $L^2(\mathbb{R})$ functions whose Fourier transforms are supported in $[-2^{-1}, 2^{-1}]$. The collection $\{T_k\varphi\}_{k\in\mathbb{Z}}$ forms an orthonormal basis for V_0. Defining $V_j = \{f \in L^2(\mathbb{R}) : \operatorname{supp}(\hat{f}) \subseteq [-2^{j-1}, 2^{j-1}]\}$ it is easy to see that $\{V_j\}_{j\in\mathbb{Z}}$ is a dyadic MRA. The subspaces V_j are merely dyadic band-limited subspaces with bandwidth 2^j.

Given a scaling function φ for an MRA, it is an elementary fact that there exists $m_0 \in L^\infty(\mathbb{T}^d)$ (the *low-pass filter* associated to φ) such that $\hat{\varphi}(A^T\xi) = m_0(\xi)\,\hat{\varphi}(\xi)$. The $\mathbb{Z}^d$-periodic low-pass filter must also satisfy

$$\sum_{s=0}^{|\det A|-1} |m_0(\xi + (A^T)^{-1}\vartheta_s)|^2 = 1, \quad \text{a.e.}\xi \in \mathbb{T}^d,$$

where $\{\vartheta_s\}_{s=0}^{|\det A|-1}$ is a complete set of distinct coset representatives of $\mathbb{Z}^d/A^T\mathbb{Z}^d$ with $\vartheta_0 = 0$. Given $m_1, \ldots, m_L \in L^\infty(\mathbb{T}^d)$ (*high-pass filters*) one can construct $\Psi = \{\psi_\ell\}_{1\le\ell\le L}$ via the refinement equations

$$\hat{\psi}_\ell(A^T\xi) = m_\ell(\xi)\,\hat{\varphi}(\xi), \quad 1 \le \ell \le L.$$

Given a low-pass filter, m_0, there are two conditions on the high-pass filters that will be useful below, namely,

$$\sum_{\ell=0}^{L} |m_\ell(\xi)|^2 = 1, \quad \text{a.e. } \xi \in \mathbb{T}^d, \tag{3.1}$$

and

$$\sum_{\ell=0}^{L} |m_\ell(\xi + (A^T)^{-1}\vartheta_s)|^2 = \delta_{0,s}, \quad \text{a.e.}\xi \in \mathbb{T}^d. \tag{3.2}$$

Observe that (3.2) implies (3.1). It will be seen below how these assumptions on the high-pass filters will lead to two different reconstruction formulas for fundamental subspaces of a given MRA in terms of appropriate co-affine systems.

Associated with an MRA $\{V_j\}_{j\in\mathbb{Z}}$, the multiresolution approximation operators P_j are defined by

$$P_j f = \sum_{k\in\mathbb{Z}^d} \langle f, D^j T_k\varphi\rangle D^j T_k\varphi \tag{3.3}$$

for $f \in L^2(\mathbb{R}^d)$ and $j \in \mathbb{Z}$. The operator P_j is actually the orthogonal projection onto V_j, since φ is a scaling function for the MRA.

One may also construct multiresolution operators using the co-affine structure. Let P_j^* and Q_j^*, respectively, be the co-affine approximation and detail operators at the scale $j \in \mathbb{Z}$. For $f \in L^2(\mathbb{R}^d)$ these operators are given by

$$(3.4) \qquad P_j^* f := \sum_{k \in \mathbb{Z}^d} \langle f, \varphi_{j,k}^* \rangle \varphi_{j,k}^* \text{ and } Q_j^* f := \sum_{\ell=1}^{L} \sum_{k \in \mathbb{Z}^d} \langle f, \psi_{\ell;j,k}^* \rangle \psi_{\ell;j,k}^*,$$

where $\psi_{\ell;j,k}^* := c_{\ell;j} T_k D^j \psi_\ell$ and $\varphi_{j,k}^* = T_k D^j \varphi$. For $j \leq 0$ these operators resemble the analogous multiresolution operators for the refinable quasi-affine systems as defined in [**Jo**] and, through this association, it is easy to see that P_j^* and Q_j^* are bounded when φ is a scaling function for an MRA, see, e.g., Proposition 6 of [**Jo**]. The following lemma from [**Jo**] summarizes the key properties of the co-affine multiresolution operators.

LEMMA 3.2. [**Jo**] *Let φ and Ψ as above and suppose that the coefficient sequence associated to $X^*(\Psi, c)$ is such that $c_{\ell;j} = |\det A|^{\frac{j}{2}}$, $1 \leq \ell \leq L$, $j < 0$.*

(a) *For each $f \in L^2(\mathbb{R}^d)$, $\|P_j^* f\| \to 0$ as $j \to -\infty$.*
(b) *If the filters $m_0, \ldots, m_L$ satisfy (3.1), then $P_j^* + Q_j^* = P_{j+1}^*$, $j < 0$.*
(c) *If the filters $m_0, \ldots, m_L$ satisfy (3.2) and $c_{\ell;0} = 1$, $1 \leq \ell \leq L$, then $P_0^* + Q_0^* = P_1$, where, as above, P_1 is the orthogonal projection onto V_1.*

REMARK 3.3. The conditions on $\{c_{\ell;j}\}_{\ell,j}$ in Lemma 3.2 owe some explanation. Notice that Lemma 3.2 (b) depends only on $\{c_{\ell;j}\}$, $j < 0$, while Lemma 3.2 (c) additionally employs the coefficients $\{c_{\ell;0}\}$. These normalizations are borrowed from the quasi-affine structure and, moreover, the properties presented in Lemma 3.2 are essentially a restatement of the analogous properties of quasi-affine systems (see [**Jo**]).

The next proposition shows how one may decompose V_0 using a *truncated* co-affine system under the fairly weak filter condition (3.1).

PROPOSITION 3.4. *Let φ and Ψ as above and suppose that $c = \{c_{\ell;j}\}_{j \in \mathbb{Z}}$ is defined by*

$$c_{\ell;j} = \begin{cases} 0, & j \geq 0 \\ |\det A|^{\frac{j}{2}}, & j < 0. \end{cases}$$

If the filters $m_1, \ldots, m_L$ satisfy (3.1), then the collection $P_0\big(X^*(\Psi, c)\big)$ is a Parseval frame for V_0, i.e., for each $f \in V_0$,

$$(3.5) \qquad \sum_{\ell=1}^{L} \sum_{j \in \mathbb{Z}} \sum_{k \in \mathbb{Z}^d} |\langle f, \psi_{\ell;j,k}^* \rangle|^2 = \|f\|^2.$$

PROOF. Applying Lemma 3.2 (b) one finds that for $f \in V_0$,

$$f = P_0 f = \lim_{J \to \infty} \sum_{j=-J}^{-1} \sum_{\ell=1}^{L} \sum_{k \in \mathbb{Z}^d} \langle f, \psi_{\ell;j,k}^* \rangle \psi_{\ell;j,k}^*,$$

where the limit converges in $L^2(\mathbb{R}^d)$. By taking inner products with f on each side of this equation it follows that

$$\|f\|^2 = \lim_{J \to \infty} \sum_{\ell=1}^{L} \sum_{j=-J}^{-1} \sum_{k \in \mathbb{Z}^d} |\langle f, \psi_{\ell;j,k}^* \rangle|^2 = \sum_{j<0} \sum_{\ell=1}^{L} \sum_{k \in \mathbb{Z}^d} |\langle f, \psi_{\ell;j,k}^* \rangle|^2.$$

Because $c_{\ell;j} = 0$ whenever $j \geq 0$, (3.5) now follows. Finally, that (3.5) is equivalent to $P_0\big(X^*(\Psi, c)\big)$ being a Parseval frame for V_0 follows from the fact that $P_0 f = f$ for $f \in V_0$. □

REMARK 3.5. One must include the projection P_0 in the statement of Proposition 3.4 because, even after truncation, the collection $X^*(\Psi, c)$ is not necessarily contained in V_0.

By additionally imposing the filter conditions (3.2) one may marginally increase the utility of co-affine systems by truncating the system at $j = 1$ rather than $j = 0$.

PROPOSITION 3.6. *Let φ and Ψ as above and suppose that $c = \{c_{\ell;j}\}_{j\in\mathbb{Z}}$ is defined by*

$$c_{\ell;j} = \begin{cases} 0, & j > 0 \\ |\det A|^{\frac{j}{2}}, & j \leq 0. \end{cases}$$

If the filters $m_1, \ldots, m_L$ satisfy (3.2)*, then the collection $P_1\big(X^*(\Psi, c)\big)$ is a Parseval frame for V_1.*

PROOF. The proof follows that of Proposition 3.4, except in this case one may also appeal to Lemma 3.2 (c). □

REMARK 3.7. For the sake of simplicity this discussion has been limited to the context of MRAs, but one could prove results similar to Propositions 3.4 and 3.6 for properly defined *dual* systems in which separate filters were used for analysis and synthesis, e.g., biorthogonal wavelets.

EXAMPLE 3.8. Again consider the Shannon wavelet of Example 3.1. By defining $m_0(\xi)$ by

$$m_0(\xi) = \begin{cases} 1, & 0 \leq |\xi| \leq \frac{1}{4} \\ 0, & \frac{1}{4} < |\xi| < \frac{1}{2}. \end{cases}$$

and $m_1(\xi)$ by

$$m_1(\xi) = \begin{cases} 0, & 0 \leq |\xi| \leq \frac{1}{4} \\ e^{2\pi i \xi}, & \frac{1}{4} < |\xi| < \frac{1}{2}. \end{cases}$$

one can verify the refinement equations $\hat{\varphi}(2\xi) = m_0(\xi)\hat{\varphi}(\xi)$ and $\hat{\psi}(2\xi) = m_1(\xi)\hat{\varphi}(\xi)$. Moreover, the filter equations (3.2) are satisfied and, thus, Proposition 3.6 applies. In this case the collection $\{2^j\psi(2^j(x-k))\}_{j\leq 0, k\in\mathbb{Z}^d}$ is itself a Parseval frame for $V_1 = \big\{f \in L^2(\mathbb{R}) : \operatorname{supp}(f) \subseteq [-1,1]\big\}$. The projection P_1 is unnecessary because of the fact that V_1 is band-limited and, hence, shift-invariant under all $\mathbb{R}^d$ translations.

Propositions 3.4 and 3.6 indicate that truncated co-affine systems may provide stable reconstruction formulas on useful subspaces of $L^2(\mathbb{R}^d)$, admitting frame-like inequalities on core subspaces of a given MRA. One deficit of these results is that the elements of $X^*(\Psi, c)$ for scales $j \geq 1$ are ignored. In the next section, this problem will be approached from another angle, where necessary conditions will be considered for a co-affine system to admit a frame-type inequality on band-limited subspaces of $L^2(\mathbb{R}^d)$. The goal is to quantify the contribution of the positive scales.

4. Band-limited subspaces spanned by co-affine systems

Example 3.8 provides motivation for understanding necessary conditions under which frame inequalities (or equalities) of the sort (3.5) may hold for a given co-affine system on a band-limited subspace of $L^2(\mathbb{R}^d)$. For $\Omega \in \mathbb{R}_+$, let $\mathcal{B}_\Omega$ be the band-limited subspace

$$\mathcal{B}_\Omega = \{f \in L^2(\mathbb{R}^d) : \operatorname{supp}(\hat{f}) \subseteq [-\Omega, \Omega]^d\}.$$

Let $\Psi = \{\psi_1, \ldots, \psi_L\} \subset L^2(\mathbb{R}^d)$ and $c = \{c_{\ell;j}\}_{1 \le \ell \le L, j \in \mathbb{Z}}$ be such that $X^*(\Psi, c)$ is a Bessel system with bound B_2 relative to an expanding dilation matrix $A \in GL_d(\mathbb{R})$. Suppose further that for all $f \in \mathcal{B}_\Omega$ one has

$$B_1 \|f\|^2 \le \sum_{\ell=1}^{L} \sum_{j,k \in \mathbb{Z}} \left|\langle f, \psi^*_{\ell;j,k}\rangle\right|^2.$$

It follows from Lemma 2.1 that

$$\sum_{j \in \mathbb{Z}} |c_{\ell;j}|^2 \, |\det A|^j \left|\hat{\psi}\big((A^T)^j \xi\big)\right|^2 \le B_2, \quad \text{a.e. } \xi \in \mathbb{R}^d \tag{4.1}$$

and

$$\sum_{j \in \mathbb{Z}} |c_{\ell;j}|^2 \, |\det A|^j \left|\hat{\psi}\big((A^T)^j \xi\big)\right|^2 \ge B_1, \quad \text{a.e. } \xi \in [-\Omega, \Omega]^d. \tag{4.2}$$

The main idea behind Theorem 2.3 was to contradict (4.2) by integrating the inequality over a carefully chosen set E. In particular, the argument hinged on the fact that the dilates of E have finite overlap, i.e., $A^j E \cap E = \emptyset$ for sufficiently large integers j. In order to apply this argument to the current problem one must additionally assume that E be contained in $[-\Omega, \Omega]$. The result of these observations is the following proposition.

PROPOSITION 4.1. *Suppose that $X^*(\Psi, c)$ is a Bessel system with bound $B_2 > 0$ relative to an expanding dilation matrix $A \in GL_d(\mathbb{R})$. Suppose further that there exists $0 < B_1 < B_2$ such that for each $f \in \mathcal{B}_\Omega$,*

$$B_1 \|f\|^2 \le \sum_{\ell=1}^{L} \sum_{j,k \in \mathbb{Z}} \left|\langle f, \psi^*_{\ell;j,k}\rangle\right|^2.$$

If $E \subseteq [-\Omega, \Omega]$ and $A^j E \cap E = \emptyset$ for each integer $j \ge J$, then, necessarily,

$$|E| B_1 \le (2J - 1) B_2 \, L, \tag{4.3}$$

where $|E|$ is the Lebesgue measure of $E \subset \mathbb{R}^d$.

As an application of Proposition 4.1 the situation of the Shannon wavelet will be reexamined under the umbrella of all dyadic orthonormal wavelets for $L^2(\mathbb{R})$.

COROLLARY 4.2. *Let $\psi \in L^2(\mathbb{R})$ be a dyadic orthonormal wavelet and fix $c = \{c_j\}_{j \in \mathbb{Z}}$ so that $c_j = 1$ if $j > 0$ and $c_j = 2^{\frac{j}{2}}$ if $j \le 0$. Then $X^*(\psi, c)$ does not satisfy Parseval's identity on any subspace $\mathcal{B}_\Omega$, $\Omega > 1$.*

PROOF. In the terminology of Proposition 4.1, $L = 1$ and $B_1 = B_2 = 1$. Two separate cases will be considered: $\Omega \ge 2$ and $1 < \Omega < 2$. Case 1: $\Omega \ge 2$. Letting $E = [-2, -1] \cup [1, 2]$, one has that $E \subset [-\Omega, \Omega]$ as well as the fact that $2^j E \cap E = \emptyset$ whenever $j \ge 1$ (j an integer). Hence, $J = 1$ and $|E| = 2$. The necessary condition (4.3) then becomes $2 \le 1$, a contradiction.

Case 2: $1 < \Omega < 2$. Let $\Omega = 1 + t$, where $0 < t < 1$. Suppose by way of contradiction that $X^*(\psi, c)$ does yield Parseval's identity on $\mathcal{B}_\Omega$. Let $E = [-(1+t), -\frac{1+t}{2}] \cup [\frac{1+t}{2}, 1+t]$. The sets $2^j E$, $j \in \mathbb{Z}$, are pairwise disjoint (J=1) and, thus, Proposition 4.1 implies

$$|E| = 2\frac{t+1}{2} = t + 1 \leq 1,$$

which is a contradiction since $0 < t < 1$. □

In the case of the co-affine system generated by the Shannon wavelet, Corollary 4.2 implies that Parseval's identity cannot hold on any band-limited subspace $\mathcal{B}_\Omega$, $\Omega > 1$. Thus, the result of Proposition 3.6 is in some sense sharp.

Acknowledgments

The author would like to thank the authors of [**GLWW**] for their inspiration as well as Chris Heil and Keri Kornelson for many helpful comments and suggestions.

References

[Bo1] M. Bownik, *Tight frames of multidimensional wavelets,* J. Fourier Analysis and Appl., **3** (1997), 525-542.

[Bo2] M. Bownik *Quasi-affine systems and the Calderón condition*, Proceedings of the Conference on Harmonic Analysis, Mt. Holyoke, MA, Contemp. Math., (2001), to appear.

[CSS] C. Chui, X. Shi, and J. Stöckler, *Affine Frames, Quasi-Affine Frames and Their Duals*, Adv. Comp. Math., **8** (1998), 1-17.

[D] I. Daubechies, *Ten lectures on wavelets,* CBMS-NSF Regional Conference Series in Applied Mathematics, Vol. 61, Society for Industrial and Applied Mathematics, Philadelphia, PA, (1992).

[GLWW] P. Gressman, D. Labate, G. Weiss, and E. Wilson, *Affine, quasi-affine, and co-affine wavelets*, Beyond Wavelets, Academic Press, (2001).

[HLW] E. Hernández, D. Labate, and G. Weiss, *A unified characterization of reproducing systems generated by a finite family, II*, J. Geom. Anal., **12**(4), (2002), 615-662.

[HLWW] E. Hernández, D. Labate, G. Weiss, and E. Wilson, *Oversampling, quasi-affine frames, and wave packets*, preprint, (2003).

[HKMT] M. Holschneider, R. Kronland-Martinet, J. Morlet, and P. Tchamitchian, "A Real-Time Algorithm for Signal Analysis with the Help of the Wavelet Transform," *Wavelets, Time-Frequency Methods and Phase Space*, Springer-Verlag, Berlin, (1989).

[HW] E. Hernández and G. Weiss, *A first course on wavelets*, CRC Press, Boca Raton, FL, (1996).

[Jo] B. Johnson, *On the relationship between quasi-affine systems and the à trous algorithm*, Collect. Math., **53**(2), (2002), 187-210.

[La] R. Laugesen, *Translational averaging for completeness, characterization, and oversampling of wavelets*, Collect. Math., **53**(3) (2002), 211-249.

[Ma] S. Mallat, *A theory of multiresolution signal decomposition: the wavelet representation*, IEEE Trans. Pattern Anal. Machine Intell., **11** (1989), 674-693.

[MZ] S. Mallat and S. Zhong, *Characterization of signals from multiscale edges,* IEEE Trans. Patt. Anal. and Mach. Intell., **14**(7), (1992) 710-732.

[RS] A. Ron and Z. Shen, *Affine systems in $L_2(\mathbb{R}^d)$: The analysis of the analysis operator*, J. Funct. Anal., **148**, (1997), 408-447.

School of Mathematics, Georgia Institute of Technology, Atlanta, Georgia 30332-0160

Current address: Department of Mathematics and Mathematical Computer Science, Saint Louis University, Saint Louis, Missouri 63103

E-mail address: brody@slu.edu

Contemporary Mathematics
Volume **345**, 2004

Rank-one decomposition of operators and construction of frames

Keri A. Kornelson and David R. Larson

ABSTRACT. The construction of frames for a Hilbert space $\mathcal{H}$ can be equated to the decomposition of the frame operator as a sum of positive operators having rank one. This realization provides a different approach to questions regarding frames with particular properties and motivates our results. We find a necessary and sufficient condition under which any positive finite-rank operator B can be expressed as a sum of rank-one operators with norms specified by a sequence of positive numbers $\{c_i\}$. Equivalently, this result proves the existence of a frame with B as it's frame operator and with vector norms $\{\sqrt{c_i}\}$. We further prove that, given a non-compact positive operator B on an infinite dimensional separable real or complex Hilbert space, and given an infinite sequence $\{c_i\}$ of positive real numbers which has infinite sum and which has supremum strictly less than the essential norm of B, there is a sequence of rank-one positive operators, with norms given by $\{c_i\}$, which sum to B in the strong operator topology.

These results generalize results by Casazza, Kovačević, Leon, and Tremain, in which the operator is a scalar multiple of the identity operator (or equivalently the frame is a tight frame), and also results by Dykema, Freeman, Kornelson, Larson, Ordower, and Weber in which $\{c_i\}$ is a constant sequence.

Introduction

The existence and characterization of frames with a variety of additional properties is an area of active research which has resulted in recent papers including [**1**], [**2**], and [**6**]. These questions can also be phrased in terms of the expression of positive operators as sums of rank-one positive operators. Therefore, although motivated by applications in signal processing, the results have independent interest to operator theory.

Throughout this paper, $\mathcal{H}$ will be a real or complex separable Hilbert space. For $\mathbb{J}$ a real or countably infinite index set, the collection of vectors $\{x_j\}_{j\in\mathbb{J}} \subset \mathcal{H}$

2000 *Mathematics Subject Classification.* Primary 42C15, 47N40; Secondary 47C05, 46B28.

Key words and phrases. frames, tight frames, rank-one operators, positive operators, rank-one decomposition.

The research of the first author was supported in part by an NSF VIGRE postdoctoral fellowship. The research of the second author was supported in part by a grant from the NSF.

is a *frame* if there exist positive constants $D \geq C > 0$ such that for every $x \in \mathcal{H}$:

$$C\|x\|^2 \leq \sum_{j \in \mathbb{J}} |\langle x, x_j \rangle|^2 \leq D\|x\|^2$$

A frame is called a *tight frame* if $C = D$, and a *Parseval frame* if $C = D = 1$.

We use elementary tensor notation for a rank-one operator on $\mathcal{H}$. Given $u, v, x \in \mathcal{H}$, the operator $u \otimes v$ is defined by $(u \otimes v)x = \langle x, v \rangle u$ for $x \in \mathcal{H}$. The operator $u \otimes u$ is a projection if and only if $\|u\| = 1$.

Given a frame $\{x_i\}$ for $\mathcal{H}$, the frame operator is defined to be the map $S : \mathcal{H} \to \mathcal{H}$ taking $x \to \sum_i \langle x, x_i \rangle x_i$. We can write S in the form:

$$S = \sum_i x_i \otimes x_i \tag{1}$$

where the convergence is in the strong operator topology (SOT). It is a well-known result that $\{x_i\}$ is a tight frame with frame bound λ if and only if $S = \lambda I$, where I is the identity operator on $\mathcal{H}$. (see [**8**]) Therefore, finding a tight frame is equivalent to decomposing a scalar multiple of the identity as a sum of rank-one operators.

Given $\mathcal{H}$ a finite dimensional Hilbert space, and $\{c_i\}$ a sequence of positive real numbers, Casazza, Kovačević, Leon, and Tremain found in [**2**] a necessary and sufficient condition (called the Fundamental Frame Inequality) for the existence of a tight frame for $\mathcal{H}$ with the norms of the vectors given by $\{c_i\}$. As stated above, this is equivalent to decomposing a scalar multiple of the identity as a sum of rank-one positive operators with prescribed norms. In Section 1, we generalize this result to the case where the scalar is replaced with an arbitrary positive invertible operator on $\mathcal{H}$ (in fact, we state the result for positive operators which are not necessarily invertible), thus obtaining the condition for the existence of non-tight frames with specified norms.

In Section 2, $\mathcal{H}$ is an infinite dimensional separable real or complex Hilbert space. Recall that the essential norm of an operator on $\mathcal{H}$ is defined to be $\|B\|_{ess} = \inf\{\|B - K\|\}$ where K is a compact operator on $\mathcal{H}$. We prove that, given B a non-compact positive operator on $\mathcal{H}$ and given $\{c_i\}$ a sequence of positive real numbers which sum to infinity and for which $\sup_i\{c_i\} < \|B\|_{ess}$, B can be expressed as a sum of rank-one operators having norms given by $\{c_i\}$, with convergence in the SOT. This implies that every positive invertible operator with essential norm strictly greater than $\sup_i\{c_i\}$ is the frame operator for a frame with prescribed norms $\{\sqrt{c_i}\}$. This result is the generalization of the infinite-dimensional result in [**2**], which restricts to tight frames. It also generalizes the results in [**6**] regarding ellipsoidal tight frames, in which the sequence $\{c_i\}$ is a constant sequence.

The authors would like to thank Eric Ricard for his helpful comments during the Linear Analysis Workshop at Texas A&M University in August, 2003.

1. Finite dimensions

The following lemma was shown to us by Marc Ordower. It was a key step in the proof of Proposition 6 in [**6**].

LEMMA 1. *Let B be a positive operator with rank n and nonzero eigenvalues $b_1 \geq b_2 \geq \cdots \geq b_n > 0$ and with a corresponding orthonormal set of eigenvectors $\{e_i\}_{i=1}^n$. If c is a positive number with $b_j \geq c \geq b_{j+1}$ for some $1 \leq j \leq n-1$, then* *there is a unit vector x in span$\{e_j, e_{j+1}\}$ such that $B - c(x \otimes x)$ is positive and has rank $n-1$.*

PROOF. If A is a finite-rank self-adjoint operator and $c \geq \|A\|$ is a constant, then the least eigenvalue of A is given by $\|(A+cI)^{-1}\|^{-1} - c$. It follows that the function giving the least eigenvalue is continuous on the set of finite-rank self-adjoint operators. Let $A_x = B - c(x \otimes x)$ for x an arbitrary unit vector in span$\{e_j, e_{j+1}\}$. The least eigenvalue of A_x is nonnegative for $x = e_j$ and nonpositive for $x = e_{j+1}$, hence span$\{e_j, e_{j+1}\}$ contains a unit vector x for which the least eigenvalue of A_x is zero. Therefore, A_x is positive of rank $n-1$.

□

REMARK 1. Since the trace of A_x is $(\sum_1^n b_i) - c$, and eigenvalues other than b_j and b_{j+1} are unchanged, the remaining nonzero eigenvalue must be

$$\tilde{b} = b_j + b_{j+1} - c$$

Moreover, the new eigenvalue maintains the jth position in the ordering of all the eigenvalues:

$$b_1 \geq \cdots \geq b_{j-1} \geq \tilde{b} \geq b_{j+2} \geq \cdots \geq b_n$$

THEOREM 2. *Let B be a positive operator with rank n and nonzero eigenvalues $b_1 \geq b_2 \geq \cdots \geq b_n > 0$. Let $\{c_i\}_{i=1}^k$; $k \geq n$ be a nonincreasing sequence of positive numbers such that $\sum_{i=1}^k c_i = \sum_{j=1}^n b_j$. There exist unit vectors $\{x_i\}_{i=1}^k$ such that*

$$B = \sum_{i=1}^k c_i (x_i \otimes x_i)$$

if and only if for each p, $1 \leq p \leq n-1$,

$$\sum_{i=1}^p c_i \leq \sum_{i=1}^p b_i \tag{2}$$

Note: When a positive invertible operator B can be written in the form of this theorem, we say that B admits a *rank-one decomposition corresponding to* $\{c_i\}$.

REMARK 2. The hypotheses in Theorem 2 reduces to exactly the Fundamental Frame Inequality from [**2**] when $b_1 = b_2 = \cdots = b_n$. Thus, an immediate corollary of Theorem 2 is the result from [**2**] that the Fundamental Frame Inequality is a necessary and sufficient condition for the existence of a tight frame for $\mathbb{R}^n$ consisting of vectors with norms $\sqrt{c_i}$; $i = 1, \ldots, k$. We include this as Corollary 5.

PROOF OF THEOREM 2. Assuming the property (2) holds, we use strong induction on the rank n of the operator. The case $n = 1$ when B is already a rank-one operator is clear. Let $\{e_i\}_1^n$ be an orthonormal set of eigenvectors for $b_1, b_2, \ldots, b_n$, respectively.
Case 1: If $b_1 \geq c_1 \geq b_n$, there exists an index l ; $1 \leq l \leq n-1$ such that $b_l \geq c_1 \geq b_{l+1}$. By Lemma 1, there exists a unit vector x in the span of e_l and e_{l+1} such that $B - c_1(x \otimes x)$ has rank $n-1$ and eigenvalues in nonincreasing order $b_1, \ldots, b_{l-1}, \tilde{b}, b_{l+2}, \ldots, b_n$, where $\tilde{b} = b_l + b_{l+1} - c_1$. Let $x_1 = x$. It remains to check that the inequalities (2) hold for the operator $B - c_1(x_1 \otimes x_1)$ and the sequence $\{c_2, \cdots, c_k\}$. Clearly, for $1 \leq r \leq l$, we have

$$c_2 + \cdots + c_r \leq c_1 + \cdots + c_{r-1} \leq b_1 + \cdots b_{r-1}$$

For $l+1 \le r \le n-1$,

$$\begin{aligned} c_1 + \cdots + c_l + c_{l+1} + \cdots + c_r &\le b_1 + \cdots + b_l + b_{l+1} + \cdots + b_r \\ \Rightarrow c_2 + \cdots + c_l + c_{l+1} + \cdots + c_r &\le b_1 + \cdots + (b_l + b_{l+1} - c_1) + \cdots + b_r \\ \Rightarrow c_2 + \cdots + c_l + c_{l+1} + \cdots + c_r &\le b_1 + \cdots + \tilde{b} + \cdots + b_r \end{aligned}$$

Since we have reduced to an operator of rank $n-1$ which satisfies the hypotheses of the theorem, induction gives the remaining elements of the rank-one decomposition of B corresponding to the sequence $\{c_i\}_1^k$.

Case 2: If $c_1 < b_n$, then we are unable to use Lemma 1. We select rank-one operators to subtract from B which preserve the rank and decrease the smallest eigenvalue until the Case 1 property is attained.

Let p be the largest integer such that $c_1 + c_2 + \cdots + c_p < b_n$. Note that Case 2 can only occur for $k > n$. Since the sums of each sequence are equal, $c_{p+1} + \cdots + c_k > b_1 + \cdots + b_{n-1}$, but because $c_{p+1} \le c_1 \le b_n \le b_{n-1}$, the sum on the left must have more terms. Therefore, $k - p > n-1$, or alternatively, $1 \le p \le k-n$. Let $x_i = e_n$ for each $i = 1, 2, \ldots, p$. The operator $B - \sum_{i=1}^p c_i(x_i \otimes x_i)$ still has rank n and the eigenvalues in decreasing order are $b_1, b_2, \ldots, b_{n-1}, \tilde{b}$, where $\tilde{b} = b_n - \sum_{i=1}^p c_i$. By the selection of p, c_{p+1} exceeds or equals $\tilde{b}$, the smallest eigenvalue of $B - \sum_{i=1}^p c_i(x_i \otimes x_i)$. The method of Case 1 can now be applied to the operator $B - \sum_{i=1}^p c_i(x_i \otimes x_i)$ and the remaining sequence $c_{p+1}, \ldots, c_k$.

Conversely, assume B has a rank-one decomposition corresponding to $\{c_i\}$. Given a fixed j with $1 \le j \le k$, define P to be the orthogonal projection onto the span of $\{x_i\}_{i=1}^j$. Clearly, rank $P \le j$. For each i, define $P_i = x_i \otimes x_i$. We then have:

$$PBP = \sum_{i=1}^k c_i P P_i P \ge \sum_{i=1}^j c_i P P_i P = \sum_{i=1}^j c_i P_i$$

and therefore $\operatorname{trace}(PBP) \ge \operatorname{trace}\left(\sum_{i=1}^j c_i P_i\right) = c_1 + c_2 + \cdots c_j$. We next show that $\operatorname{trace}(PBP) \le b_1 + b_2 + \cdots + b_n$, which will complete the proof.

Let $\{e_i\}_{i=1}^n$ be as above, and for notational purposes, set $b_{n+1} = 0$. For $1 \le i \le n$, define $Q_i = e_1 \otimes e_1 + \cdots + e_i \otimes e_i$. A calculation verifies that B can be written

$$B = \sum_{i=1}^n (b_i - b_{i+1}) Q_i$$

Then, $PBP = \sum_{i=1}^n (b_i - b_{i+1}) P Q_i P$. Observe that $\operatorname{trace}(PQ_iP) = \operatorname{trace}(PQ_i) \le \min\{\operatorname{rank} P, \operatorname{rank} Q_i\} \le \min\{j, i\}$. Therefore,

$$\begin{aligned} \operatorname{trace}(PBP) &= \sum_{i=1}^n (b_i - b_{i+1}) \operatorname{trace}(PQ_iP) \\ &\le \sum_{i=1}^j (b_i - b_{i+1}) i + j \sum_{i=j+1}^n (b_i - b_{i+1}) \\ &= (b_1 + b_2 + \cdots + b_j - j b_{j+1}) + j(b_{j+1} - b_{n+1}) \\ &= b_1 + b_2 + \cdots + b_j \end{aligned}$$

□

Theorem 2 will be used in the proof of Proposition 7. To demonstrate the theorem, we have included the following example for the case where $\mathcal{H} = \mathbb{R}^2$.

EXAMPLE 1. Let $B = \begin{bmatrix} 5 & 0 \\ 0 & 4 \end{bmatrix}$. We find the rank-one decomposition of B corresponding to $\{c_i\}_{i=1}^4 = 3, 3, 2, 1$.

Since $c_1 < b_2$ (Case 2), we take $x_1 = \begin{bmatrix} 1 \\ 0 \end{bmatrix}$. This leaves $B - 3(x_1 \otimes x_1) = \begin{bmatrix} 5 & 0 \\ 0 & 1 \end{bmatrix}$ to decompose. We now have c_2 between the two eigenvalues 5 and 1 (Case 1). Let $x(t) = \sqrt{1-t}\begin{bmatrix} 1 \\ 0 \end{bmatrix} + \sqrt{t}\begin{bmatrix} 0 \\ 1 \end{bmatrix} = \begin{bmatrix} \sqrt{1-t} \\ \sqrt{t} \end{bmatrix}$. We wish to find t such that $\begin{bmatrix} 5 & 0 \\ 0 & 1 \end{bmatrix} - 3(x(t) \otimes x(t))$ has rank one, i.e. the determinant is zero. Straightforward calculation shows the solution $t = \frac{1}{6}$. Take $x_2 = x(\frac{1}{6}) = \begin{bmatrix} \sqrt{\frac{5}{6}} \\ \sqrt{\frac{1}{6}} \end{bmatrix}$. The remainder $\begin{bmatrix} 5 & 0 \\ 0 & 1 \end{bmatrix} - 3(x_2 \otimes x_2) = \begin{bmatrix} \frac{5}{2} & -\frac{\sqrt{5}}{2} \\ -\frac{\sqrt{5}}{2} & \frac{1}{2} \end{bmatrix}$ has rank one and it's range is spanned by the unit vector $z = \begin{bmatrix} \sqrt{\frac{5}{6}} \\ -\sqrt{\frac{1}{6}} \end{bmatrix}$. We then must have $x_3 = x_4 = z$, which completes the decomposition.

$$3\begin{bmatrix} 0 & 0 \\ 0 & 1 \end{bmatrix} + 3\begin{bmatrix} \frac{5}{6} & \frac{\sqrt{5}}{6} \\ \frac{\sqrt{5}}{6} & \frac{1}{6} \end{bmatrix} + (2+1)\begin{bmatrix} \frac{5}{6} & -\frac{\sqrt{5}}{6} \\ -\frac{\sqrt{5}}{6} & \frac{1}{6} \end{bmatrix} = \begin{bmatrix} 5 & 0 \\ 0 & 4 \end{bmatrix} = B$$

The next example demonstrates that a rank-one decomposition is impossible when property (2) does not hold.

EXAMPLE 2. The operator $B = \begin{bmatrix} 5 & 0 & 0 \\ 0 & 2 & 0 \\ 0 & 0 & 2 \end{bmatrix}$ does not have a rank-one decomposition corresponding to the sequence $\{c_i\}_{i=1}^3 = \{4, 4, 1\}$. To see this, assume that we can find unit vectors x_1, x_2, x_3 such that $B = 4(x_1 \otimes x_1) + 4(x_2 \otimes x_2) + (x_3 \otimes x_3)$. Let P be the projection onto the span of $\{x_1, x_2\}$. Then $\operatorname{trace}(PBP) = \sum_{i=1}^3 c_i \operatorname{trace} P(x_i \otimes x_i)P \geq 4 + 4 = 8$, because $P(x_i \otimes x_i)P = x_i \otimes x_i$, which has trace 1.

Given P any projection with rank two, using the argument from the last paragraph of Theorem 2, we have $\operatorname{trace}(PBP)$ less than or equal to 7, i.e. the sum of the largest two eigenvalues. The contradiction implies B does not have a rank-one decomposition corresponding to $\{4, 4, 1\}$.

The following proposition lends insight into the underlying geometry of tight frames. We include it for independent interest.

PROPOSITION 3 (Alternate version of $\mathbb{R}^2$ case). *Let B be a positive invertible operator on $\mathbb{R}^2$, and express B as* $\operatorname{diag}(b_1, b_2)$, *where $b_1 \geq b_2 > 0$. Let $\{c_i\}_{i=1}^k$ be* *positive constants such that* $c_1 \geq c_2 \geq \cdots \geq c_k$ *and*

$$\sum_{i=1}^{k} c_i = b_1 + b_2$$

If $c_1 \leq b_1$, *then there exist unit vectors* $\{x_i\}_{i=1}^{k}$ *such that:*

$$B = \sum_{i=1}^{k} c_i \, (x_i \otimes x_i)$$

PROOF. Let

$$x_i = \begin{bmatrix} \cos\theta_i \\ \sin\theta_i \end{bmatrix} \quad \text{so} \quad x_i \otimes x_i = \begin{bmatrix} \cos^2\theta_i & \cos\theta_i \sin\theta_i \\ \cos\theta_i \sin\theta_i & \sin^2\theta_i \end{bmatrix}$$

Then let $B = \operatorname{diag}(b_1, b_2)$ and $\{c_i\}_{i=1}^{k}$ a sequence of positive numbers such that $\sum_1^k c_i = b_1 + b_2$ and $c_1 \geq c_2 \geq \cdots \geq c_k$. Then $B = \sum_{i=1}^{k} c_i \, (x_i \otimes x_i)$ if and only if the following equations hold:

$$\begin{aligned} c_1 \cos^2\theta_1 + c_2 \cos^2\theta_2 + \cdots + c_k \cos^2\theta_k &= b_1 \\ c_1 \sin^2\theta_1 + c_2 \sin^2\theta_2 + \cdots + c_k \sin^2\theta_k &= b_2 \\ c_1 \cos\theta_1 \sin\theta_1 + c_2 \cos\theta_2 \sin\theta_2 + \cdots + c_k \cos\theta_k \sin\theta_k &= 0 \end{aligned} \tag{3}$$

These equations, under the condition that $\sum_1^k c_i = b_1 + b_2$, are equivalent to:

$$\begin{aligned} \sum_1^k c_i \cos(2\theta_i) &= b_1 - b_2 \\ \sum_1^k c_i \sin(2\theta_i) &= 0 \end{aligned} \tag{4}$$

By defining $y_i = c_i \begin{bmatrix} \cos 2\theta_i \\ \sin 2\theta_i \end{bmatrix}$, the above equations become the vector equation:

$$\sum_{i=1}^{k} y_i - \begin{bmatrix} b_1 - b_2 \\ 0 \end{bmatrix} = \begin{bmatrix} 0 \\ 0 \end{bmatrix} \tag{5}$$

Using the "tip-to-tail" method of adding vectors, we can solve this equation by finding a polygon with $k+1$ sides having lengths $c_1, c_2, \ldots, c_k, b_1 - b_2$. We must place the side with length $b_1 - b_2$ along the x-axis, but all of the angles between sides are variable. Such a polygon always exists for $k \geq 2$ provided the greatest side length is less than or equal to the sum of the lengths of the remaining sides. Since the longest vector has either length c_1 or $b_1 - b_2$, there is a solution when $c_1 \leq b_1$:

$$\begin{aligned} c_1 &\leq b_1 \\ 2c_1 &\leq (b_1 - b_2) + (b_1 + b_2) \\ &= (b_1 - b_2) + \sum_{i=1}^{k} c_i \\ c_1 &\leq (b_1 - b_2) + (c_2 + c_3 + \cdots + c_k) \\ \text{and} \quad b_1 - b_2 &\leq b_1 + b_2 \\ &= c_1 + c_2 + \cdots + c_k \end{aligned}$$

□

The following is the frame version of Theorem 2.

COROLLARY 4. *Let $\mathcal{H}$ be a Hilbert space with finite dimension n. Let B be a positive invertible operator on $\mathcal{H}$ with eigenvalues $b_1 \geq b_2 \geq \cdots \geq b_n > 0$, and for $k \geq n$ let $\{c_i\}_{i=1}^k$ be a sequence of positive numbers with $c_1 \geq c_2 \geq \cdots \geq c_k$ such that $\sum_{i=1}^k c_i = \sum_{j=1}^n b_j$. There exists a frame for $\mathcal{H}$ with k vectors $\{x_i\}_{i=1}^k$ having frame operator B and such that $\|x_i\| = \sqrt{c_i}, 1 \leq i \leq k$, if and only if condition (2) from Theorem 2 is satisfied.*

PROOF. B has a rank-one decomposition corresponding to $\{c_i\}_{i=1}^k$ if and only if there exist unit vectors $\{y_i\}_{i=1}^k$ such that

$$B = \sum_{i=1}^{k} c_i(y_i \otimes y_i) = \sum_{i=1}^{k} \sqrt{c_i}y_i \otimes \sqrt{c_i}y_i$$

From the expression (1) of the frame operator, $\{x_i\}_{i=1}^k = \{\sqrt{c_i}y_i\}_{i=1}^k$ is a frame with frame operator B and $\|x_i\| = \sqrt{c_i}$. □

An immediate corollary is Theorem 5.1 from [**2**], which finds the Fundamental Frame Inequality as the necessary and sufficient condition for the existence of a tight frame with lengths of the vectors prescribed by the sequence. We state this result here.

COROLLARY 5. [**2**] *Let $\mathcal{H}$ be a Hilbert space with finite dimension n, and for some $k \geq n$, let $\{a_i\}_{i=1}^k$ be a sequence of positive numbers such that $a_1 \geq a_2 \geq \cdots \geq a_k > 0$. If*

$$a_1^2 \leq \frac{1}{n} \sum_{i=1}^{k} a_i^2 \qquad \textit{(Fund. Frame Inequality)}$$

then there exist unit vectors $\{x_i\}_{i=1}^k$ such that the vectors

$$y_i = a_i x_i$$

form a tight frame for $\mathcal{H}$. The frame bound will be

$$\lambda = \frac{1}{n} \sum_{i=1}^{k} a_i^2$$

PROOF. This case satisfies the hypotheses of Corollary 4 for $B = \lambda I$ and $c_i = a_i^2$, $i = 1, \ldots, k$ since each $b_i = \lambda$, and $a_1^2 \leq \lambda$ implies the remaining inequalities. □

2. Infinite dimensions

The main result, Theorem 6, of this section (or more particularly Corollary 12) is a generalization of Theorem 5.4 from [**2**], which we list here as Corollary 8. We will use Proposition 7, the operator-theoretic formulation of Corollary 8, in the proof of Theorem 6.

Corollary 12 is also a generalization of Theorem 2 from [**6**]. This theorem states that, given B a positive operator with $\|B\|_{ess} > 1$, there exists an ellipsoidal tight frame for B with frame bound 1. This is equivalent to the existence of a frame of unit vectors with frame operator B^{-2}, which is the special case of Corollary 12 with $\{c_i\}$ the constant sequence in which each $c_i = 1$.

THEOREM 6. *Let B be a positive non-compact operator in $\mathcal{B}(\mathcal{H})$ for $\mathcal{H}$ a real or complex Hilbert space with infinite dimension. If $\{c_i\}_{i=1}^{\infty}$ is a bounded sequence of positive numbers with $\sup_i c_i < \|B\|_{ess}$ and $\sum_i c_i = \infty$, then there is a sequence of rank-one projections $\{P_i\}_1^{\infty} \subset \mathcal{B}(\mathcal{H})$ such that*

$$B = \sum_{i=1}^{\infty} c_i P_i$$

where convergence is in the strong operator topology.

EXAMPLE 3. The condition that $\sup_i\{c_i\}$ be strictly less than $\|B\|_{ess}$ cannot be dropped in general. For instance, if $B = I$ the identity, and

$$\{c_i\}_{i=1}^{\infty} = \{\frac{2}{3}, 1, 1, 1, \cdots\}$$

then there is no series $\{P_i\}$ of projections with $\sum_i c_i P_i = I$. Equivalently, there is no sequence of vectors forming a Parseval frame in which the norms of the vectors $\{x_i\}$ are each $\|x_i\| = \sqrt{c_i}$. If, in fact, such a frame did exist, the vectors $\{x_2, x_2, \ldots\}$ would necessarily form an orthonormal set. The assumption that the frame is Parseval then implies that x_1 is a unit vector, which is a contradiction.

We will prove Theorem 6 in a series of steps, using the finite dimensional Theorem 2. The first step is Proposition 7, which, as we previously stated, is a reformulation of Theorem 5.4 from [**2**]. The result is theirs, but our operator-theoretic proof is different from the one given in [**2**].

PROPOSITION 7. *Let $\mathcal{H}$ be an infinite-dimensional separable Hilbert space. Let $\{c_i\}_{i=1}^{\infty}$ be a sequence of numbers with $0 < c_i \leq 1$, and suppose $\sum_i c_i = \infty$ and $\sum_i (1 - c_i) = \infty$. Then there is a sequence of rank-one projections $\{P_i\}_{i=1}^{\infty}$ such that*

$$I = \sum_{i=1}^{\infty} c_i P_i$$

with the sum converging in the strong operator topology.

An immediate corollary is the precise statement of Theorem 5.4 from [**2**], which gives the existence of a Parseval frame for $\mathcal{H}$ with norms of the frame elements coming from the sequence.

COROLLARY 8. [**2**] *If $\mathcal{H}$ is an infinite-dimensional real or complex Hilbert space, and if $\{a_i\}$ is a sequence of real numbers with $0 < a_i \leq 1$ such that $\sum_i a_i^2 = \infty$ and $\sum_i (1 - a_i^2) = \infty$, then there is a Parseval frame $\{x_i\}$ for $\mathcal{H}$ such that $\|x_i\| = a_i$ for all i.*

PROOF OF PROPOSITION 7. For each $n \in \mathbb{N}$, let $s(n)$ denote the smallest integer such that

$$c_1 + c_2 + \cdots + c_{s(n)} > n$$

(Such a value exists for every n since $\sum_i c_i = \infty$.) Denote what we will call the *residual* by

$$r(n) = c_1 + c_2 + \cdots + c_{s(n)} - n$$

and denote the *integer gap* by

$$g(n) = s(n) - n$$

Then $0 < r(n) \leq 1$, and clearly $g(n) \leq g(n+1)$ for all $n \in \mathbb{N}$. Denote $\delta_i = 1 - c_i$, and let $d(n) = \delta_1 + \delta_2 + \cdots + \delta_n$. We have

$$\begin{aligned} g(n) = s(n) - n &> s(n) - (c_1 + \cdots c_{s(n)}) \\ &= d(s(n)) \\ &\geq d(n) \end{aligned}$$

By hypothesis, $\sum_i \delta_i = \infty$, so $d(n) \to \infty$ and thus $g(n) \to \infty$. We can therefore inductively choose an increasing sequence $\{n_i\}_{i=1}^{\infty}$ of natural numbers such that for all $i \geq 2$,

$$g(n_i) \geq g(n_{i-1}) + 1 \tag{6}$$

and

$$d\left(s(n_{i-1}) + n_i - n_{i-1} + 1\right) - d\left(s(n_{i-1})\right) > 2 \tag{7}$$

Note that condition (6) implies that $s(n_i) - s(n_{i-1}) \geq n_i - n_{i-1} + 1$, and condition (7) implies that:

$$c_{(s(n_{i-1}+1))} + \cdots + c_{(s(n_{i-1})+n_i-n_{i-1}+1)} < n_i - n_{i-1} - 1 \tag{8}$$

Let $\{e_i\}_1^{\infty}$ be an orthonormal basis for $\mathcal{H}$, and let $E_i = e_i \otimes e_i$ for all $i \in \mathbb{N}$. Let

$$B_1 = E_1 + E_2 + \cdots + E_{n_1} + r(n_1)E_{n_1+1}$$

and for $i \geq 2$, let

$$B_i = ((1 - r(n_{i-1}))\, E_{n_{i-1}+1} + \left[E_{n_{i-1}+2} + \cdots + E_{n_i}\right] + r(n_i)E_{n_i+1}$$

This gives

$$\operatorname{trace}(B_1) = c_1 + \cdots + c_{s(n_1)}$$

and for $i \geq 2$,

$$\begin{aligned} \operatorname{trace}(B_i) &= [1 - r(n_{i-1})] + [n_i - n_{i-1} - 1] + r(n_i) \\ &= [r(n_i) + n_i] - [r(n_{i-1}) + n_{i-1}] \\ &= \sum_{j=1}^{s(n_i)} c_j - \sum_{j=1}^{s(n_{i-1})} c_j \\ &= \sum_{j=s(n_{i-1})+1}^{s(n_i)} c_j \end{aligned}$$

We have $\operatorname{rank}(B_1) = n_1 + 1$, and for $i \geq 2$, we have $\operatorname{rank}(B_i) \leq n_i - n_{i-1} + 1$. When $i \geq 2$, at least the first $n_i - n_{i-1} - 1$ (in decreasing order) of the eigenvalues of B_i are 1. (That is, all but the last two eigenvalues of B_i are 1.) This, together with the condition (8), implies that B_i and the sequence $\{c_{s(n_{i-1})+1}, \ldots, c_{s(n_i)}\}$ satisfy the hypotheses of Theorem 2. To see this, for $1 \leq p \leq n_i - n_{i-1} - 1$, the inequalities (2) in Theorem 2 are trivially satisfied. For $p = n_i - n_{i-1}$, the inequality (8) implies that the sum of the first $n_i - n_{i-1}$ of the terms in $\{c_{s(n_i-1)+1}, \ldots, c_{s(n_i)}\}$ is less than $n_i - n_{i-1} - 1$, which is less than or equal to the sum of the first $n_i - n_{i-1}$ eigenvalues of B_i.

Therefore, for $i \geq 2$, there exist rank-one projections $\{P_{s(n_{i-1})+1}, \ldots, P_{s(n_i)}\}$ in the range of B_i such that

$$B_i = c_{s(n_{i-1})+1} P_{s(n_{i-1})+1} + \cdots + c_{s(n_i)} P_{s(n_i)}$$

For the special case B_1, since there is at most one eigenvalue different from 1, the hypotheses of Theorem 2 are clearly satisfied, and so there exist rank-one projections $P_1, \ldots P_{s(n_1)}$ such that

$$B_1 = \sum_{j=1}^{s(n_1)} c_j P_j$$

For all $i \in \mathbb{N}$,

$$B_1 + \cdots + B_i = E_1 + \cdots + E_{n_i} + r(n_i) E_{n_i+1}$$

Since $\{r(n_i)\}$ is bounded, the sum $\sum_{i=1}^{\infty} B_i$ converges strongly to the identity I. It follows that $\sum_{i=1}^{\infty} c_i P_i$ converges strongly to I. □

COROLLARY 9. *Let E be an infinite rank projection in $\mathcal{B}(\mathcal{H})$, and let $\{c_i\}$ be a sequence of real numbers with $0 < c_i \leq 1$ such that $\sum_i c_i = \infty$ and $\sum_i (1 - c_i) = \infty$. Then E has a rank-one decomposition corresponding to $\{c_i\}$:*

$$E = \sum_{i=1}^{\infty} c_i P_i$$

PROOF. Apply Proposition 7 to the subspace $E\mathcal{H}$ and the result follows immediately. □

In [**6**], it was proven (Theorem 2) that if B is a positive operator with essential norm strictly greater that 1, then B has a projection decomposition. That is, $B = \sum_i P_i$, where $\{P_i\}$ are self-adjoint projections and convergence is in the strong operator topology. Clearly $\{P_i\}$ can be taken to be rank-one. We need here a stronger result, that each of the projections can be required to have infinite rank.

PROPOSITION 10. *Let B be a positive operator in $\mathcal{B}(\mathcal{H})$ for $\mathcal{H}$ a real or complex separable Hilbert space with infinite dimension, and suppose $\|B\|_{ess} > 1$. Then B has an infinite-rank projection decomposition.*

PROOF. The hypothesis that $\|B\|_{ess} > 1$ implies there exists a sequence $\{Q_i\}_{i=1}^{\infty}$ of mutually orthogonal infinite-rank projections commuting with B such that $\|B|_{Q_i\mathcal{H}}\|_{ess} > 1$ for all i, and with $\sum_i Q_i = I$. Let $A_i = B|_{Q_i\mathcal{H}}$. By Theorem 2 in [**6**], each A_i has a projection decomposition:

$$A_i = \sum_{j=1}^{\infty} E_{ij}$$

Here we can assume without loss of generality that for each i, $\{E_{i,j}\}_{j=1}^{\infty}$ is an infinite sequence of nonzero projections because A_i is not finite rank. Note that for each j, the projections $\{E_{1,j}, E_{2,j}, \dots,\}$ are mutually orthogonal. Let

$$E_j = E_{1,j} + E_{2,j} + \cdots$$

with the sum converging strongly. Then E_j is an infinite-rank projection, and we have $\sum_{j=1}^{\infty} E_j = B$, as required.

□

REMARK 3. If $\mathcal{E}$ is a countably infinite set of positive numbers, then it is well-known that $\sum_{\lambda \in \mathcal{E}} \lambda$ makes sense as an extended real number, independent of enumeration. It is also clear that if $\sum_{\lambda \in \mathcal{E}} \lambda = \infty$, then there exists a partition $\{\mathcal{E}_i\}_{i=1}^{\infty}$ of $\mathcal{E}$ into infinitely many subsets such that for each i, $\sum_{\lambda \in \mathcal{E}_i} \lambda = \infty$. Similarly, if $\{c_i\}_{i=1}^{\infty}$ is a sequence of positive numbers such that $\sum_i c_i = \infty$, there exists a partition of the sequence into infinite subsequences $\Lambda_i = \{c_{i,j}\}_{j=1}^{\infty}$ such that $\sum_{j=1}^{\infty} a_{i,j} = \infty$ for each i. We will use this partition in the proof of the next theorem.

The reader will note that Theorem 11 is an alternate restatement of Theorem 6. We include it because it has a simpler form and has independent interest. We prove it first and then obtain Theorem 6 by a simple scaling argument.

THEOREM 11. *Let B be a positive operator in $\mathcal{B}(\mathcal{H})$ for $\mathcal{H}$ with $\|B\|_{ess} > 1$. Let $\{c_i\}_{i=1}^{\infty}$ be any sequence of numbers with $0 < c_i \leq 1$ such that $\sum_i c_i = \infty$. Then there exists a sequence of rank-one projections $\{P_i\}_{i=1}^{\infty}$ such that*

$$B = \sum_{i=1}^{\infty} c_i P_i$$

PROOF. Choose α, $0 < \alpha < 1$ such that $\|\alpha B\|_{ess} > 1$. Using Proposition 10, write $\alpha B = \sum_n Q_n$, a projection decomposition with each Q_n having infinite rank. Next, let $\gamma_i = \alpha c_i$ for all i, and partition the sequence $\{\gamma_i\}$ into subsequences Λ_n as described in Remark 3. Note that for each n, $\sup_j\{\gamma_{n,j}\} < 1$. Apply Corollary 9 to each operator Q_n with the subsequence Λ_n, obtaining rank-one projections $\{F_{n,j}\}_{j=1}^{\infty}$ with

$$Q_n = \sum_{j=1}^{\infty} \gamma_{n,j} F_{n,j}$$

Scaling by α, we have the rank-one decomposition of B:

$$B = \frac{1}{\alpha} \sum_{n=1}^{\infty} \sum_{j=1}^{\infty} \gamma_{n,j} F_{n,j} = \sum_{n=1}^{\infty} \sum_{j=1}^{\infty} c_{n,j} F_{n,j}$$

□

PROOF OF THEOREM 6. We apply Theorem 11. Given B non-compact, we have $m := \|B\|_{ess} > 0$ and $\sup_i c_i < m$. Take $\alpha = \frac{1}{m-\epsilon}$ for some $\epsilon > 0$, so that we have $\|\alpha B\|_{ess} > 1$, and let $\gamma_i = \alpha c_i$, which gives $\sup_i \gamma_i < 1$. By Theorem 11, there is a rank-one decomposition for αB corresponding to the sequence $\{\gamma_i\}$:

$$\alpha B = \sum_{i=1}^{\infty} \gamma_i P_i$$

Scaling the above by $m - \epsilon$ gives the final result. □

The following is the frame-theoretic version of Theorem 6. It is now clear that Theorem 5.4 from [**2**] is the special case of Corollary 12 in which B is the identity operator, and Theorem 2 from [**6**] is the case where $\{c_i\}$ is a constant sequence.

COROLLARY 12. *Let $\mathcal{H}$ be a real or complex separable Hilbert space of infinite dimension, and let B be any bounded positive invertible operator on $\mathcal{H}$. Let $\{a_i\}$ be an arbitrary sequence of positive numbers with infinite sum and with* $\sup_i\{a_i^2\} < \|B\|_{ess}$. *Then there is a frame $\{z_i\}$ for $\mathcal{H}$ with frame operator B such that $\|z_i\| = a_i$ for all i.*

PROOF. Let $c_i = a_i^2$, and let $\alpha = \sup_i\{a_i^2\} = \sup_i\{c_i\}$. Then $\alpha < \|B\|_{ess}$, so by Theorem 6, B has a rank-one decomposition corresponding to $\{c_i\}$. Let $\{x_i\}$ be unit vectors in the range of the rank-one operators of this decomposition, and let $z_i = \sqrt{c_i}x_i$ for all i.

$$B = \sum_i c_i(x_i \otimes x_i) = \sum_i z_i \otimes z_i$$

By the expression (1) of the frame operator, $\{z_i\}$ is the desired frame. □

References

[1] John J. Benedetto and Matthew Fickus. Finite normalized tight frames. *Adv. Comput. Math.*, 18(2-4):357–385, February 2003.

[2] Peter G. Casazza, Jelena Kovačević, Manuel Leon, and Janet C. Tremain. Custom built tight frames. Preprint, 2002.

[3] Ingrid Daubechies. *Ten Lectures on Wavelets.* CBMS-NSF Regional Conference Series in Applied Mathematics. Society for Industrial and Applied Mathematics (SIAM), Philadelphia, PA, 1992.

[4] Ingrid Daubechies, A. Grossmann, and Y. Meyer. Painless nonorthogonal expansions. *J. Math. Phys.*, 27(5):1271–1283, May 1986.

[5] R.J. Duffin and A.C. Schaeffer. A class of nonharmonic Fourier series. *Trans. Amer. Math. Soc.*, 72(2):341–366, March 1952.

[6] Ken Dykema, Dan Freeman, Keri Kornelson, Dave Larson, Marc Ordower, and Eric Weber. Ellipsoidal tight frames and projection decomposition of operators. Preprint, 2003.

[7] Vivek K. Goyal, Jonathan A. Kelner, and Jelena Kovačević. Quantized frame expansions with erasures. *Appl. Comput. Harmon. Anal.*, 10:203–233, 2001.

[8] Deguang Han and David R. Larson. Frames, bases, and group representations. *Mem. Amer. Math Soc.*, 147(697):1–94, September 2000.

[9] Eugenio Hernández and Guido Weiss. *A First Course in Wavelets.* CRC Press, LLC, Boca Raton, FL, 1996.

[10] R. Horn and C. Johnson. *Topics in Matrix Analysis.* Cambridge University Press, 1991.

[11] Kadison and Ringrose. *Fundamentals of the Theory of Operator Algebras, Volume 1,* volume 15 of *Graduate Studies in Mathematics.* American Mathematical Society, Providence, RI, 1997.

TEXAS A&M UNIVERSITY
E-mail address: keri@math.tamu.edu

TEXAS A&M UNIVERSITY
E-mail address: larson@math.tamu.edu

Contemporary Mathematics
Volume **345**, 2004

An approach to the study of wave packet systems

Demetrio Labate, Guido Weiss, and Edward Wilson

ABSTRACT. The traditional study of reproducing systems involves the Gabor systems, that are generated by the action of translations and modulations on a single or finite family of functions in $L^2(\mathbb{R}^n)$, and the affine systems, where the dilations are used rather than the modulations. In this paper, we show that the interplay of all three operators yield a wide variate of reproducing systems, and we employ the term *wave packet systems*, which has been used by other authors, to describe those function systems generated by the combined action of a class of translations, modulations and dilations on a finite family of functions. We will examine in detail both the continuous and discrete versions of these systems. We shall show that these systems can be studied by using a unified approach that the authors have developed in some of their previous work.

1. Preliminaries

Before embarking in this study, it will be useful to introduce some notation and definitions. We consider three fundamental operators on $L^2(\mathbb{R}^n)$: the *translations* $T_y : (T_y f)(x) = f(x-y)$, where $y \in \mathbb{R}^n$; the *dilations* $D_a : (D_a f)(x) = |\det a|^{1/2} f(ax)$, where $a \in GL_n(\mathbb{R})$; and the *modulations* $M_\nu : (M_\nu f)(x) = e^{2\pi i \nu \cdot x} f(x)$, where $\nu \in \mathbb{R}^n$.

The Fourier transform is defined as

$$\hat{f}(\xi) = (\mathcal{F}f)(\xi) = \int_{\mathbb{R}^n} f(x)\, e^{-2\pi i \xi \cdot x}\, dx,$$

and the inverse Fourier transform is

$$\check{f}(x) = (\mathcal{F}^{-1}f)(x) = \int_{\mathbb{R}^n} f(\xi)\, e^{2\pi i x \cdot \xi}\, d\xi.$$

The following proposition, which is easily verified, states some basic properties of the translation, modulation and dilation operators.

PROPOSITION 1.1. *Let*

$$G = \{U = c\, D_a\, M_\nu\, T_y : c \in \mathbb{C},\ |c| = 1,\ (a, \nu, y) \in GL_n(\mathbb{R}) \times \mathbb{R}^n \times \mathbb{R}^n\}. \tag{1.1}$$

2000 *Mathematics Subject Classification.* 42C15, 42C40.
Key words and phrases. Affine systems, Gabor systems, frames, wavelets, wave packets.
The authors were supported in part by a grant from Southwestern Bell.

G is a subgroup of the group of unitary operators on $L^2(\mathbb{R}^n)$ which is preserved by the action of the operator $U \mapsto \widehat{U}$, where $\widehat{U}\,\hat{f} = (Uf)^\wedge$. In particular, we have:

(i) $T_y\, M_\nu = e^{-2\pi i \nu \cdot y}\, M_\nu\, T_y$;
(ii) $D_a\, T_y = T_{a^{-1}y}\, D_a$;
(iii) $D_a\, M_\nu = M_{a^T\nu}\, D_a$;
(iv) *for* $U = c\, D_a\, M_\nu\, T_y$, *then* $\widehat{U} = c\, D_{a^I}\, T_\nu\, M_{-y}$, *where* $a^I = (a^T)^{-1}$.

2. Wave packet systems

In [**CF**], Córdoba and Fefferman introduced "wave packets" as those families of functions obtained by applying certain collections of dilations, modulations and translations to the Gaussian function. In this paper, we will adopt the same expression to describe, more generally, any collections of functions which are obtained by applying a combination of dilations, modulations and translations to a finite family of functions in $L^2(\mathbb{R}^n)$. Unlike the original wave packets of Córdoba and Fefferman, the systems that we will consider will not always be "well localized" in time and frequency.

DEFINITION 2.1. Let $\Psi = \{\psi^\ell : 1 \le \ell \le L\} \subset L^2(\mathbb{R}^n)$, where L is a finite integer, and let $S \subseteq GL_n(\mathbb{R}) \times \mathbb{R}^n$. The *continuous wave packet system* relative to S generated by Ψ is the collection:

$$\mathcal{W}_S(\Psi) = \big\{D_a\, M_\nu\, T_y\, \psi^\ell : (a,\nu) \in S, y \in \mathbb{R}^n, 1 \le \ell \le L\big\}. \tag{2.1}$$

Let us make several remarks about this definition.

Note that the map

$$(a,\nu,y) \mapsto U^{(0)}_{(a,\nu,y)} = D_a\, M_\nu\, T_y$$

is a one-to-one function from $S \times \mathbb{R}^n$ into G, where G is the group introduced by (1.1). By changing the oder of the operators, we can also define the following one-to-one functions from $S \times \mathbb{R}^n$ into G:

$$U^{(1)}_{(a,\nu,y)} = D_a\, T_y\, M_\nu$$
$$U^{(2)}_{(a,\nu,y)} = T_y\, D_a\, M_\nu$$
$$U^{(3)}_{(a,\nu,y)} = M_\nu\, D_a\, T_y$$
$$U^{(4)}_{(a,\nu,y)} = T_y\, M_\nu\, D_a$$
$$U^{(5)}_{(a,\nu,y)} = M_\nu\, T_y\, D_a.$$

In view of Proposition 1.1, we can generate alternate continuous wave packet systems $\mathcal{W}^{(i)}_S(\Psi)$, with $1 \le i \le 5$, by replacing the operator $U^{(0)}_{(a,\nu,y)} = D_a\, T_y\, M_\nu$ in Definition 2.1 with the operators $U^{(i)}_{(a,\nu,y)}$, $1 \le i \le 5$. The systems $\mathcal{W}^{(0)}_S(\Psi) = \mathcal{W}_S(\Psi)$ and $\mathcal{W}^{(1)}_S(\Psi)$ are equivalent since they only differ by a unimodular scalar. The same is true for the systems $\mathcal{W}^{(4)}_S(\Psi)$ and $\mathcal{W}^{(5)}_S(\Psi)$. The remaining systems, on the other hand, have substantial differences that we will explicitly discuss later.

One could also define more general wave packet systems of the form:

$$\{U_p\, \psi^\ell : p \in \mathcal{P}, 1 \le \ell \le L\},$$

where L is a finite integer, $\mathcal{P}$ is a parameter set and the map $p \mapsto U_p$ is one-to-one from $\mathcal{P}$ into G, where G is given by (1.1). We will consider some specific examples of this type in the following.

Finally, observe that we will also consider discrete analogs of the wave packet systems given by Definition 2.1. These systems are obtained by letting S in Definition 2.1 be countable and by sampling the translations over a lattice. The reader is probably already familiar with special cases of these systems, including Gabor systems, wavelets and their close variants. Some more general examples will be discussed in Section 3.

A very important class of wave packet systems are those associated with a reproducing formula. More precisely, we introduce the following definition.

DEFINITION 2.2. Let $S \subseteq GL_n(\mathbb{R}) \times \mathbb{R}^n$ and λ be a measure on S. Fix i, where $0 \le i \le 5$. Then the system $\mathcal{W}_S^{(i)}(\Psi)$ is a *continuous Parseval frame wave packet system* relative to (S,λ) for $L^2(\mathbb{R}^n)$, provided that the functions

$$(a,\nu,y) \longmapsto \langle f, U^{(i)}_{(a,\nu,y)}\,\psi\rangle$$

are $\lambda-$measurable for all $f,\, \psi \in L^2(\mathbb{R}^n)$, and

$$\|f\|^2 = \sum_{\ell=1}^{L} \int_{S\times\mathbb{R}^n} |\langle f, U^{(i)}_{(a,\nu,y)}\,\psi^\ell\rangle|^2\, d\lambda(a,\nu)\, dy$$

for all $f \in L^2(\mathbb{R}^n)$.

The following theorem gives a characterization of all those families $\Psi \subset L^2(\mathbb{R}^n)$ such that the wave packet system $\mathcal{W}_S(\Psi) = \mathcal{W}_S^{(0)}(\Psi)$ is a continuous Parseval frame wave packet system for $L^2(\mathbb{R}^n)$.

THEOREM 2.1. *Let $\Psi = \{\psi^\ell : 1 \le \ell \le L\} \subset L^2(\mathbb{R}^n)$. The system $\mathcal{W}_S(\Psi)$, given by (2.1), is a continuous Parseval frame wave packet system relative to (S,λ) for $L^2(\mathbb{R}^n)$ if and only if*

$$\Delta_\Psi(\xi) = \sum_{\ell=1}^{L} \int_S |\hat\psi^\ell(a^I\xi - z)|^2\, d\lambda(a,z) = 1 \qquad \textit{for a.e. } \xi \in \mathbb{R}^n. \tag{2.2}$$

PROOF. Observe first that, for each $\psi^\ell \in L^2(\mathbb{R}^n)$, $\ell = 1,\dots,L$, using Proposition 1.1 we obtain that $D_a\, M_\nu\, T_y\, \psi^\ell = e^{2\pi i y\cdot\nu}\, T_{a^{-1}y}\, D_a\, M_\nu\, \psi^\ell$. Hence, using again Proposition 1.1, we have that $\big(D_a\, M_\nu\, T_y\, \psi^\ell\big)^\wedge = e^{2\pi i y\cdot\nu}\, M_{-a^{-1}y}\, D_{a^I}\, T_\nu\, \hat\psi^\ell$, and, thus,

$$|\langle f, D_a\, M_\nu\, T_y\, \psi^\ell\rangle| = |\langle \hat f, M_{-a^{-1}y}\, D_{a^I}\, T_\nu\, \hat\psi^\ell\rangle|. \tag{2.3}$$

Let $u = a^{-1}y$, so that $|\det a|\, du = dy$, and, then, using (2.3) and the Plancherel Theorem, we have that for every $f \in L^2(\mathbb{R}^n)$:

$$\begin{aligned}
\int_{\mathbb{R}^n} |\langle f, D_a\, M_\nu\, T_y\, \psi^\ell\rangle|^2\, dy &= \int_{\mathbb{R}^n} |\langle \hat{f}, M_{-u}\, D_{a^I}\, T_\nu\, \hat{\psi}^\ell\rangle|^2\, |\det a|\, du \\
&= \int_{\mathbb{R}^n} \Big| \int_{\mathbb{R}^n} \hat{f}(\xi)\, \overline{\hat{\psi}^\ell(a^I\xi - \nu)}\, e^{2\pi i u\cdot\xi} \Big|^2\, du \\
&= \int_{\mathbb{R}^n} \Big| \{\hat{f}\, \overline{\hat{\psi}^\ell(a^I \cdot -\nu)}\}^\vee(u)\, d\xi \Big|^2\, du \\
&= \int_{\mathbb{R}^n} |\hat{f}(\xi)|^2\, |\hat{\psi}^\ell(a^I\xi - \nu)|^2\, d\xi. \qquad (2.4)
\end{aligned}$$

Thus, using (2.4) and Fubini's theorem, we see that the system $\mathcal{W}_S(\Psi)$ is a continuous Parseval frame if and only if

$$\begin{aligned}
\|f\|^2 &= \sum_{\ell=1}^{L} \int_S \int_{\mathbb{R}^n} |\langle f, D_a\, M_\nu\, T_y\, \psi^\ell\rangle|^2\, dy\, d\lambda(a,\nu) \\
&= \sum_{\ell=1}^{L} \int_S \int_{\mathbb{R}^n} |\hat{f}(\xi)|^2\, |\hat{\psi}^\ell(a^I\xi - \nu)|^2\, d\xi\, d\lambda(a,\nu) \\
&= \int_{\mathbb{R}^n} |\hat{f}(\xi)|^2\, \Delta_\Psi(\xi)\, d\xi,
\end{aligned}$$

for all $f \in L^2(\mathbb{R}^n)$, and (2.2) follows easily. □

Notice that the main step in the proof of Theorem 2.1 is the application of the Plancherel Theorem in (2.4). This computation shows the essential role played by the translation variable y (or $u = a^{-1}y$) in the proof of Theorem 2.1 and, also, explains the reason for using *all* translations T_y, $y \in \mathbb{R}^n$, in Definition 2.1. Later we will consider a similar situation in the case of discrete wave packet systems, which will involve *all* the translations arising from a lattice. Some authors (e.g., [**ACM**], [**CDH**], [**CKS**], [**HK**], [**W**]) have considered the Beurling density and several variants of this notion in order to understand the restrictions on the dispersion of elements in situations where a Parseval frame is obtained from an irregular translation set, as well as irregular dilation and/or modulation sets. To our knowledge, however, there are no results to date on irregular translation sets with an abstract measure replacing the Lebesgue measure. In this paper, we restrict ourselves to lattices ($\mathbb{Z}^n$) or continuous ($\mathbb{R}^n$) translations.

It is clear that, by choosing a different order of the operators, one obtains other versions of Theorem 2.1 for the systems $\mathcal{W}_S^{(i)}(\Psi)$, $1 \le i \le 5$. Indeed, using arguments similar to the ones for Theorem 2.1, we obtain the following characterizing equalities.

COROLLARY 2.2. *Let* $\Psi = \{\psi^\ell : 1 \le \ell \le L\} \subset L^2(\mathbb{R}^n)$. *For* $1 \le i \le 5$, *the system* $\mathcal{W}_S^{(i)}(\Psi)$ *is a continuous Parseval frame wave packet system relative to* (S, λ) *for* $L^2(\mathbb{R}^n)$ *if and only if*

$$\Delta_\Psi^{(i)}(\xi) = 1 \qquad \text{for a.e. } \xi \in \mathbb{R}^n, \qquad (2.5)$$

where

$$\Delta_\Psi^{(1)}(\xi) = \Delta_\Psi(\xi),$$

$$\Delta_\Psi^{(2)}(\xi) = \sum_{\ell=1}^{L} \int_S |\hat\psi^\ell(a^I\xi - \nu)|^2 \, |\det a|^{-1} \, d\lambda(a,\nu),$$

$$\Delta_\Psi^{(3)}(\xi) = \sum_{\ell=1}^{L} \int_S |\hat\psi^\ell(a^I(\xi - \nu))|^2 \, d\lambda(a,\nu),$$

$$\Delta_\Psi^{(4)}(\xi) = \Delta_\Psi^{(5)}(\xi) = \sum_{\ell=1}^{L} \int_S |\hat\psi^\ell(a^I(\xi - \nu))|^2 \, |\det a|^{-1} \, d\lambda(a,\nu).$$

Thus we have seen that the changes in the order in which the operators are taken give us four different characterizations corresponding to the four functions $\Delta_\Psi = \Delta_\Psi^{(1)}, \Delta_\Psi^{(2)}, \Delta_\Psi^{(3)}, \Delta_\Psi^{(4)} = \Delta_\Psi^{(5)}$.

2.1. Examples. We will describe a number of examples of continuous wave packet systems, according to Definition 2.1, for different choices of the set S.

Example (a). Let $S = \{0\} \times \mathbb{R}^n \simeq \mathbb{R}^n$ and $\psi \in L^2(\mathbb{R}^n)$, then

$$\mathcal{W}_S(\psi) = \mathcal{G}_S(\psi) = \{M_\nu \, T_y \, \psi : \nu \in S, y \in \mathbb{R}^n\}$$

is the *continuous Gabor system* (also known as the *Weyl–Heisenberg system*) associated with S. Furthermore, if one chooses λ to be Lebesgue measure, then Theorem 2.1 shows that $\mathcal{W}_S(\psi)$ is a continuous Parseval frame relative to $(S, d\nu)$ for any function $\psi \in L^2(\mathbb{R}^n)$ with $\|\psi\| = 1$ (this property of Gabor systems is well known; see, for example, [**G**, Ch. 3]).

Example (b). Let $S = \{0\} \times \mathbb{Z}^n \simeq \mathbb{Z}^n$ and $\psi \in L^2(\mathbb{R}^n)$, then

$$\mathcal{W}_S(\psi) = \mathcal{G}_S(\psi) = \{M_k \, T_y \, \psi : k \in S, y \in \mathbb{R}^n\}$$

is the continuous Gabor system associated with S. If one chooses λ to be the counting measure, then it follows from Theorem 2.1 that $\mathcal{W}_S(\psi)$ is a continuous Parseval frame relative to (S, λ) iff

$$\sum_{k \in \mathbb{Z}^n} |\hat\psi(\xi - k)|^2 = 1 \quad \text{for a.e. } \xi \in \mathbb{R}^n. \tag{2.6}$$

Observe that equation (2.6) is equivalent to the property that $\{T_k \, \psi : k \in \mathbb{Z}\}$ is an orthonormal system (cf. [**HW**, Prop. 2.1.11]).

Example (c). Let $S \subseteq GL_n(\mathbb{R}) \times \{0\} \simeq GL_n(\mathbb{R})$, then

$$\mathcal{W}_S(\Psi) = \mathcal{A}_S(\Psi) = \{D_a \, T_y \, \psi^\ell : a \in S, y \in \mathbb{R}^n, 1 \le \ell \le L\}$$

is the *continuous affine system* associated with S. If S is a closed subgroup of $GL_n(\mathbb{R})$ or, more generally, a topological group continuously embedded in $GL_n(\mathbb{R})$, then it is natural to take λ to be the right Haar measure on S. In this case, the function $\Delta_\Psi(\xi)$ in (2.2) is S-invariant, in the sense that $\Delta_\Psi(a^I \, \xi) = \Delta_\Psi(\xi)$ for all $a \in S$, $\xi \in \mathbb{R}^n$, and equation (2.2) is the classical Calderón condition (see [**C**], where this equation was originally obtained, or [**WW**], for the generalizations). In [**LWWW**], it is shown that the affine group $G = S \ltimes \mathbb{R}^n$ must be non–unimodular if there exists a continuous Parseval frame $\mathcal{A}_S(\Psi)$. Conversely, if G is non–unimodular and almost every $\xi \in \mathbb{R}^n$ has a compact ε–stabilizer $K_\xi^\varepsilon = \{a \in S : |a^I \, \xi - \xi| \le \varepsilon\}$, then one can find $\psi \in L^2(\mathbb{R}^n)$ for which $\mathcal{A}_S(\Psi)$ is a continuous Parseval frame relative to (S,λ). In [**LWWW**], the authors have considered other examples involving either integration over a coset space of a group or over products of subgroups. Also observe that the continuous Gabor systems in (a) may be viewed as obtained by integration over a coset space of the Weyl–Heisenberg group (see, for example, [**F**] or [**G**] for more details on this approach to the continuous Gabor systems).

The following two examples, unlike the previous ones, are not continuous Parseval frames. However, they can still be described according to Definition 2.1.

Example (d). Let $S = \{((1+\nu^2)^{1/4} I_n, \nu) : \nu \in \mathbb{R}^n\} \subset GL_n(\mathbb{R}) \times \mathbb{R}^n$, where I_n is the $n \times n$ identity matrix, and $\phi(x) = 2^{n/4}\, e^{-\pi x^2}$, $x \in \mathbb{R}^n$. Then the functions

$$\mathcal{W}_S(\phi) = \{T_y\, M_\nu\, D_a\, \phi : \, (a,\nu) \in S, y \in \mathbb{R}^n\}$$

are the wave packets introduced by Còrdoba and Fefferman in [**CF**]. In the same paper, Còrdoba and Fefferman show that their wave packets form, "approximately", a continuous Parseval frame, in the sense that, for each $f \in \mathcal{S}(\mathbb{R}^n)$, the system $\mathcal{W}_S(\phi)$ is reproducing up to a small error that is controlled by a Sobolev norm of f.

Example (e). The following example, which is a generalization of the wave packets of Còrdoba and Fefferman, is due to Hogan and Lakey [**HL**]. Let $S = \{(|\nu|^{1/2}, \nu) : \nu \in \mathbb{R}\} \subset \mathbb{R} \times \mathbb{R}$ and define the collections

$$\mathcal{W}_S(\psi) = \{\psi_{(y,\nu)} = T_y\, M_\nu\, D_{a(\nu)}\, \psi : \, (a(\nu),\nu) \in S, y \in \mathbb{R}\}.$$

In [**HL**] it is shown that, if $\psi \in L^1 \bigcap L^2(\mathbb{R})$ is real and if $|\xi|\, |\hat{\psi}(\xi)| \in L^1(\mathbb{R})$, then there are constants $0 < A < B < \infty$ such that

$$A\, \|f\|^2 \le \iint_{\mathbb{R}^2} \left|\langle f, \psi_{(y,\nu)}\rangle\right|^2 dy\, d\nu \le B\, \|f\|^2,$$

for all $f \in L^2(\mathbb{R})$.

3. Discrete wave packet systems

In this section, we will examine the case of discrete wave packet systems. These systems are discrete analogs of the continuous wave packets considered in the previous section, and are obtained by restricting the ordered pairs of dilations and modulations to a countable set and by sampling the translations over a regular lattice (while, in the continuous case, the translations involve all $y \in \mathbb{R}^n$).

More precisely, let $\Psi = \{\psi^1, \dots, \psi^L\} \subset L^2(\mathbb{R}^n)$, $\mathcal{P}$ be a countable set and, corresponding to each $p \in \mathcal{P}$, let $a_p \in GL_n(\mathbb{R})$ and $\nu_p \in \mathbb{R}^n$. We define the (discrete) *wave packet systems* generated by Ψ relative to $\mathcal{P}$ as the collections

$$\mathcal{W}_{\mathcal{P}}(\Psi) = \{D_{a_p}\, M_{\nu_p}\, T_k\, \psi^\ell : k \in \mathbb{Z}^n, p \in \mathcal{P}, \ell = 1, \dots, L\}. \tag{3.1}$$

Special cases of the set $\mathcal{W}_{\mathcal{P}}(\Psi)$ are the classical (discrete) Gabor and affine systems. Indeed, if $\mathcal{P} = \mathbb{Z}^n$, $\nu_p = b\,p$, where $b \in GL_n(\mathbb{R})$, $p \in \mathcal{P}$, and $a_p = I_n$, for each $p \in \mathcal{P}$, where I_n is the $n \times n$ identity matrix, then

$$\mathcal{W}_{\mathcal{P}}(\Psi) = \mathcal{G}_b(\Psi) = \{M_{bp}\, T_k\, \psi^\ell : p, k \in \mathbb{Z}^n, \ell = 1, \dots, L\}$$

is the Gabor system generated by Ψ. Alternatively, if $\mathcal{P} = \mathbb{Z}$, $a_p = a^p$, where $a \in GL_n(\mathbb{R})$, and $\nu_p = 0$, for each $p \in \mathcal{P}$, then

$$\mathcal{W}_{\mathcal{P}}(\Psi) = \mathcal{A}_a(\Psi) = \{D_a^p\, T_k\, \psi^\ell : p \in \mathbb{Z}, k \in \mathbb{Z}^n, \ell = 1, \dots, L\}$$

is the affine system generated by Ψ and the integer powers of the dilation matrix $a \in GL_n(\mathbb{R})$.

We will also consider variants of the (discrete) wave packet systems obtained by using a larger collection of translations and an appropriate renormalization of the generators. More precisely, we define the (discrete) *oversampled wave packet systems* generated by Ψ relative to the set $\mathcal{P}$ and the matrices $\{\gamma_p\}_{p\in\mathcal{P}} \subset GL_n(\mathbb{R})$ as the collections

$$(3.2)\quad \mathcal{W}_{\mathcal{P}}^{\gamma_p}(\Psi) = \{D_{a_p}\, M_{\nu_p}\, T_{\gamma_p^{-1} k}\Big(\frac{1}{\sqrt{|\det \gamma_p|}}\,\psi^\ell\Big) : k \in \mathbb{Z}^n,\, p \in \mathcal{P},\, \ell = 1, \ldots, L\}.$$

The notion of oversampling plays a very important role, for example, in the case of the affine systems, and Definition (3.2) includes, as special cases, the oversampled affine systems of Chui and Shi [**CS**] and the quasi affine systems of Ron and Shen [**RSb**] (cf. [**HLWW**] for a more detailed study of the notion of oversampling in relation to affine systems and other reproducing systems). For example, let $\mathcal{P} = \mathbb{Z}$, $a_p = a^p$, where $a \in GL_n(\mathbb{R})$ and $p \in \mathcal{P}$, $\nu_p = 0$, as in the case of affine systems, and, in addition, let $\gamma_p = I_n$, if $p \geq 0$, $\gamma_p = a^{-p}$, if $p < 0$. Under these assumptions, the collections $\mathcal{W}_{\mathcal{P}}(\psi)$ are the quasi affine systems

$$\mathcal{W}_{\mathcal{P}}(\psi) = \mathcal{A}_a^q(\psi) = \{\psi_{p,k}^q : p \in \mathbb{Z}, k \in \mathbb{Z}^n\},$$

where

$$\psi_{p,k}^q = \begin{cases} \sqrt{|\det a^p|}\, D_a^p\, T_{a^p k}\, \psi, & k \in \mathbb{Z}^n,\, p < 0 \\ D_a^p\, T_k\, \psi, & k \in \mathbb{Z}^n,\, p \geq 0. \end{cases}$$

This example also illustrates the role of the normalization factor $\frac{1}{\sqrt{|\det \gamma_p|}}$ in Definition 3.2.

More general examples of oversampled wave packet systems will be considered in Section 3.2.

3.1. Unified theory of discrete reproducing systems. In order to describe the properties of the discrete wave packet systems that we will consider in this paper, it is useful to recall the notion of a frame.

A countable family $\{e_j : j \in \mathcal{J}\}$ of elements in a separable Hilbert space $\mathcal{H}$ is a *frame* if there exist constants $0 < A \leq B < \infty$ satisfying

$$A\,\|v\|^2 \leq \sum_{j\in\mathcal{J}} |\langle v, e_j\rangle|^2 \leq B\,\|v\|^2$$

for all $v \in \mathcal{H}$. The constants A and B are called the *lower* and *upper frame bounds*, respectively. If only the right hand side inequality holds, we say that $\{e_j : j \in \mathcal{J}\}$ is a *Bessel system* with constant B. A frame is a *tight frame* if A and B can be chosen so that $A = B$, and is a *Parseval frame* (also called *normalized tight frame*) if $A = B = 1$. Thus, if $\{e_j : j \in \mathcal{J}\}$ is a Parseval frame in $\mathcal{H}$, then

$$(3.3)\qquad \|v\|^2 = \sum_{j\in\mathcal{J}} |\langle v, e_j\rangle|^2$$

for each $v \in \mathcal{H}$. This is equivalent to the reproducing formula

$$(3.4)\qquad v = \sum_{j\in\mathcal{J}} \langle v, e_j\rangle\, e_j$$

for all $v \in \mathcal{H}$, where the series in (3.4) converges in the norm of $\mathcal{H}$ (we refer the reader to [**HW**, Ch. 8] for the basic properties of frames).

One main problem in the study of wavelets, Gabor systems and related systems is to determine conditions on the families $\Psi \subset L^2(\mathbb{R}^n)$ such that these systems form a basis or a Parseval frame or, more generally, a frame. In this section, we will examine this problem in relation to the wave packet systems.

The following simple observation, which is adapted from [**HLWW**, Prop. 2.1] shows that in order for the system $\mathcal{W}_{\mathcal{P}}^{\gamma_p}(\Psi)$, given by (3.2), to be a frame (or even a Bessel system), there are some restrictions on the choice of the oversampling matrices $\{\gamma_p\}_{p\in\mathcal{P}}$.

PROPOSITION 3.1. *If the oversampled wave packet system $\mathcal{W}_{\mathcal{P}}^{\gamma_p}(\Psi)$, given by (3.2), is a Bessel system with constant $B>0$, then, for each $\ell = 1,\dots,L$,*

$$|\det \gamma_p| \geq \frac{1}{B}\,\|\psi^\ell\|^2 \qquad \textit{for each } p \in \mathcal{P}.$$

PROOF. Since $\mathcal{W}_{\mathcal{P}}^{\gamma_p}(\Psi)$ is a Bessel system with constant β, then

$$\sum_{\ell=1}^{L}\sum_{p\in\mathcal{P}}\sum_{k\in\mathbb{Z}^n} |\langle f, \psi_{p,k}^\ell\rangle|^2 \leq \beta\,\|f\|^2 \tag{3.5}$$

for all $f \in L^2(\mathbb{R}^n)$, where $\psi_{p,k}^\ell = |\det\gamma_p|^{-1/2}\, D_{a_p}\, M_{\nu_p}\, T_{\gamma_p^{-1}k}\,\psi^\ell$. Equation (3.5) implies that, for any $p_0 \in \mathcal{P}$, $k_0 \in \mathbb{Z}^n$, $1 \leq \ell_0 \leq L$, we have:

$$|\langle \psi_{p_0,k_0}^{\ell_0}, \psi_{p_0,k_0}^{\ell_0}\rangle|^2 \leq \beta\,\|\psi_{p_0,k_0}^{\ell_0}\|^2. \tag{3.6}$$

Since $\|\psi_{p_0,k_0}^{\ell_0}\|^2 = |\det\gamma_{p_0}|^{-1}\,\|\psi^{\ell_0}\|^2$, from (3.6) we deduce:

$$|\det\gamma_{p_0}|^{-2}\,\|\psi^{\ell_0}\|^4 \leq \beta\,|\det\gamma_{p_0}|^{-1}\,\|\psi^{\ell_0}\|^2,$$

and, thus, $|\det\gamma_{p_0}| \geq \beta^{-1}\,\|\psi^{\ell_0}\|^2$, for all $p_0 \in \mathcal{P}$, $1 \leq \ell_0 \leq L$. □

In [**HLW**] and [**HLWW**], the authors and their collaborators have developed a general approach to the study of reproducing systems generated by a finite family $\Psi \subset L^2(\mathbb{R}^n)$. After recalling some basic results from these papers, and making some additional observations, we will apply this approach to the study of the discrete wave packet systems $\mathcal{W}_{\mathcal{P}}(\Psi)$, given by (3.1).

Let $\mathcal{P}$ be a countable collection of indices, $\{g_p : p \in \mathcal{P}\}$ a family of functions in $L^2(\mathbb{R}^n)$ and $\{C_p : p \in \mathcal{P}\}$ a corresponding collection of matrices in $GL_n(\mathbb{R})$ and consider the families of the form

$$\Phi_{\{C_p\}}^{\{g_p\}} = \{T_{C_p k}\, g_p : k \in \mathbb{Z}^n, p \in \mathcal{P}\}. \tag{3.7}$$

We obtained a characterization of all those $\{g_p\}_{p\in\mathcal{P}}$ such that $\Phi_{\{C_p\}}^{\{g_p\}}$ is a Parseval frames for $L^2(\mathbb{R}^n)$. In order to state the main result, we need to introduce the following notation. Define:

$$\Lambda = \bigcup_{p\in\mathcal{P}} C_p^I\,(\mathbb{Z}^n)\,, \tag{3.8}$$

where $C_p^I = (C_p^T)^{-1}$, and, for $\alpha \in \Lambda$, let

$$\mathcal{P}_\alpha = \{p \in \mathcal{P} : \alpha \in C_p^I\,\mathbb{Z}^n\}. \tag{3.9}$$

If $\alpha = 0 \in \Lambda$, then $\mathcal{P}_0 = \mathcal{P}$ (since $C_p^T\,0 = 0$ for all $p \in \mathcal{P}$); otherwise the best we can say is that $\mathcal{P}_\alpha \subset \mathcal{P}$. Also, let

$$\mathcal{D} = \mathcal{D}_E = \{f \in L^2(\mathbb{R}^n) \colon \hat f \in L^\infty(\mathbb{R}^n) \text{ and } \operatorname{supp}\hat f \text{ is compact in } \mathbb{R}^n \setminus E\}, \tag{3.10}$$

where E is a subspace of $\mathbb{R}^n$ of dimension smaller than n to be specified later, and appropriately chosen for each of the systems defined by (3.7) (for example, we use $E = \{0\}$ in Proposition 3.8 and a nontrivial set E in Section 3.3). It is clear that $\mathcal{D}$ is a dense subspace of $L^2(\mathbb{R}^n)$. We have the following result.

THEOREM 3.2 ([**HLW**]). *Let $\{g_p\}_{p\in\mathcal{P}}$ be a collection of functions in $L^2(\mathbb{R}^n)$, where $\mathcal{P}$ is a countable indexing set, and $\{C_p\}_{p\in\mathcal{P}} \subset GL_n(\mathbb{R})$. Assume the local integrability condition (L.I.C.):*

$$L(f) = \sum_{p\in\mathcal{P}} \sum_{m\in\mathbb{Z}^n} \int_{supp\,\hat{f}} |\hat{f}(\xi + C_p^I m)|^2 \frac{1}{|\det C_p|} |\hat{g}_p(\xi)|^2 \, d\xi < \infty \tag{3.11}$$

for all $f \in \mathcal{D}$, where $\mathcal{D}$ is given by (3.10) and $C_p^I = (C_p^T)^{-1}$. Then the system $\Phi_{\{C_p\}}^{\{g_p\}}$, given by (3.7), is a Parseval frame for $L^2(\mathbb{R}^n)$ if and only if

$$\sum_{p\in\mathcal{P}_\alpha} \frac{1}{|\det C_p|} \hat{g}_p(\xi) \, \overline{\hat{g}_p(\xi + \alpha)} = \delta_{\alpha,0} \quad \textit{for a.e. } \xi \in \mathbb{R}^n, \tag{3.12}$$

for each $\alpha \in \Lambda$, where δ is the Kronecker delta for $\mathbb{R}^n$.

In the same paper, we also obtained a similar theorem for the characterization of dual systems (cf. [**HLW**, Thm.9.1]), as well as the following result about Bessel families (cf. [**HLW**, Prop.4.1]) which will be useful in the next section.

PROPOSITION 3.3 ([**HLW**]). *Let $\{g_p\}_{p\in\mathcal{P}} \subset L^2(\mathbb{R}^n)$, where $\mathcal{P}$ is a countable set, and $\{C_p\}_{p\in\mathcal{P}} \subset GL_n(\mathbb{R})$. If the system $\Phi_{\{C_p\}}^{\{g_p\}}$, given by (3.7), is Bessel with constant $B > 0$, then*

$$\sum_{p\in\mathcal{P}} \frac{1}{|\det C_p|} |\hat{g}_p(\xi)|^2 \le B \quad \textit{for a.e. } \xi \in \mathbb{R}^n. \tag{3.13}$$

In the case of general frames (not just Parseval frames) no characterization results in the spirit of Theorem 3.2 are known for the collections $\Phi_{\{C_p\}}^{\{g_p\}}$. Nevertheless, the following theorem, which uses an idea from [**D**, Sec. 3.3], gives sufficient conditions for $\Phi_{\{C_p\}}^{\{g_p\}}$ to be a frame for $L^2(\mathbb{R}^n)$.

THEOREM 3.4. *Let $\{g_p\}_{p\in\mathcal{P}} \subset L^2(\mathbb{R}^n)$, where $\mathcal{P}$ is countable, and $\{C_p\}_{p\in\mathcal{P}} \subset GL_n(\mathbb{R})$. Suppose that there are $0 < \alpha \le \beta < \infty$ such that*

$$\alpha \le \sum_{p\in\mathcal{P}} \frac{1}{|\det C_p|} |\hat{g}_p(\xi)|^2 \le \beta, \qquad \textit{for all } \xi \in \mathbb{R}^n \tag{3.14}$$

and that

$$\sum_{m\neq 0} \sup_{\xi\in\mathbb{R}^n} \sum_{p\in\mathcal{P}} \frac{1}{|\det C_p|} |\overline{\hat{g}_p(\xi)} \, \hat{g}_p(\xi + C_p^I m)| < \alpha. \tag{3.15}$$

Then the family $\Phi_{\{C_p\}}^{\{g_p\}}$, given by (3.7), is a frame for $L^2(\mathbb{R}^n)$ with lower and upper frame bounds

$$A = \alpha - \sum_{m\neq 0} \sup_{\xi\in\mathbb{R}^n} \sum_{p\in\mathcal{P}} \frac{1}{|\det C_p|} |\overline{\hat{g}_p(\xi)} \, \hat{g}_p(\xi + C_p^I m)|$$

and

$$B = \beta + \sum_{m\neq 0} \sup_{\xi\in\mathbb{R}^n} \sum_{p\in\mathcal{P}} \frac{1}{|\det C_p|} |\overline{\hat{g}_p(\xi)}\, \hat{g}_p(\xi + C_p^I m)|,$$

respectively.

PROOF. It suffices to prove the theorem for $f \in \mathcal{D}$, where $\mathcal{D}$ is given by (3.10). We fix $p \in \mathcal{P}$ and estimate first the quantity $\sum_{k\in\mathbb{Z}^n} |\langle f, T_{C_p k}\, g_p\rangle|^2$. Since $(T_{C_p k}\, g_p)^\wedge(\xi) = e^{-2\pi i C_p k\cdot\xi}\, \hat{g}(\xi)$, it follows from the Plancherel theorem that

$$\sum_{k\in\mathbb{Z}^n} |\langle f, T_{C_p k}\, g_p\rangle|^2 = \sum_{k\in\mathbb{Z}^n} \left| \int_{\mathbb{R}^n} \hat{f}(\xi)\, \overline{\hat{g}_p(\xi)}\, e^{2\pi i C_p k\cdot\xi}\, d\xi \right|^2 . \tag{3.16}$$

Let $\mathbb{T}^n = [0,1)^n$. Since $\mathbb{R}^n = \bigcup_{l\in\mathbb{Z}^n} \{C^I(\mathbb{T}^n - l)\}$ is a disjoint union, the integral in (3.16) can be written in the form

$$\sum_{l\in\mathbb{Z}^n} \int_{C_p^I(\mathbb{T}^n)} \hat{f}(\xi - C_p^I l)\, \overline{\hat{g}_p(\xi - C_p^I l)}\, e^{2\pi i C_p k\cdot\xi}\, d\xi =$$

$$= \int_{C_p^I(\mathbb{T}^n)} \Big(\sum_{l\in\mathbb{Z}^n} \hat{f}(\xi - C_p^I l)\, \overline{\hat{g}_p(\xi - C_p^I l)} \Big)\, e^{2\pi i C_p k\cdot\xi}\, d\xi.$$

Since the function $\sum_{l\in\mathbb{Z}^n} \hat{f}(\xi - C_p^I l)\, \overline{\hat{g}_p(\xi - C_p^I l)}$ is $C_p^I\, \mathbb{Z}^n$-periodic and belongs to $L^2(C_p^I \mathbb{T}^n)$ (recall that $f \in \mathcal{D}$), it follows that the right hand side of (3.16) is, up to a constant, the square of the l^2-norm of the Fourier coefficients of this $C_p^I\, \mathbb{Z}^n$-periodic function, with respect to the orthonormal basis

$$\{\sqrt{|\det C_p|}\, e^{2\pi i C_p k\cdot\xi} : k \in \mathbb{Z}^n\}$$

of $L^2(C_p^I \mathbb{T}^n)$. Therefore, from this observation we obtain:

$$\sum_{k\in\mathbb{Z}^n} |\langle f, T_{C_p k}\, g_p\rangle|^2 = \frac{1}{|\det C_p|} \int_{C_p^I \mathbb{T}^n} |\sum_{l\in\mathbb{Z}^n} \hat{f}(\xi - C_p^I l)\, \overline{\hat{g}_p(\xi - C_p^I l)}|^2\, d\xi, \tag{3.17}$$

From equation (3.17), making the change of indices $u = l + m$ we obtain:

$$|\det C_p| \sum_{k\in\mathbb{Z}^n} |\langle f, T_{C_p k}\, g_p\rangle|^2 =$$

$$= \int_{C_p^I \mathbb{T}^n} \sum_{l,u\in\mathbb{Z}^n} \hat{f}(\xi + C_p^I l)\, \overline{\hat{g}_p(\xi + C_p^I l)}\, \overline{\hat{f}(\xi + C_p^I u)}\, \hat{g}_p(\xi + C_p^I u)\, d\xi$$

$$= \int_{C_p^I \mathbb{T}^n} \sum_{l,m\in\mathbb{Z}^n} \hat{f}(\xi + C_p^I l)\, \overline{\hat{g}_p(\xi + C_p^I l)}\, \overline{\hat{f}(\xi + C_p^I (l+m))}\, \hat{g}_p(\xi + C_p^I (l+m))\, d\xi$$

$$= \sum_{m\in\mathbb{Z}^n} \int_{\mathbb{R}^n} \hat{f}(\xi)\, \overline{\hat{g}_p(\xi)}\, \overline{\hat{f}(\xi + C_p^I m)}\, \hat{g}_p(\xi + C_p^I m)\, d\xi.$$

From this expression, summing over $p \in \mathcal{P}$ and splitting the sum in m into the case $m = 0$ and $m \neq 0$ we obtain:

$$\sum_{p\in\mathcal{P}} \sum_{k\in\mathbb{Z}^n} |\langle f, T_{C_p k}\, g_p\rangle|^2 = \int_{\mathbb{R}^n} |\hat{f}(\xi)|^2 \sum_{p\in\mathcal{P}} \frac{1}{|\det C_p|} |\hat{g}_p(\xi)|^2\, d\xi + R(f), \tag{3.18}$$

where

$$R(f) = \sum_{p\in\mathcal{P}} \sum_{m\neq 0} \frac{1}{|\det C_p|} \int_{\mathbb{R}^n} \hat{f}(\xi)\, \overline{\hat{g}_p(\xi)}\, \overline{\hat{f}(\xi + C_p^I m)}\, \hat{g}_p(\xi + C_p^I m)\, d\xi. \tag{3.19}$$

Using the Cauchy-Schwarz inequality and the change of variables $\eta = \xi + C_p^I m$, for each $p \in \mathcal{P}$ we have:

$$\begin{aligned}
&\int_{\mathbb{R}^n} \hat{f}(\xi)\, \overline{\hat{g}_p(\xi)}\, \overline{\hat{f}(\xi + C_p^I m)}\, \hat{g}_p(\xi + C_p^I m)\, d\xi \leq \\
&\quad \leq \Big(\int_{\mathbb{R}^n} |\hat{f}(\xi)|^2\, |\overline{\hat{g}_p(\xi)}\, \hat{g}_p(\xi + C_p^I m)|\, d\xi\Big)^{1/2} \Big(\int_{\mathbb{R}^n} |\hat{f}(\xi + C_p^I m)|^2 \times \\
&\quad \times |\overline{\hat{g}_p(\xi)}\, \hat{g}_p(\xi + C_p^I m)|\, d\xi\Big)^{1/2} \\
&\quad = \Big(\int_{\mathbb{R}^n} |\hat{f}(\xi)|^2\, |\overline{\hat{g}_p(\xi)}\, \hat{g}_p(\xi + C_p^I m)|\, d\xi\Big)^{1/2} \Big(\int_{\mathbb{R}^n} |\hat{f}(\eta)|^2\, |\overline{\hat{g}_p(\eta - C_p^I m)}\, \hat{g}_p(\eta)|\, d\eta\Big)^{1/2}.
\end{aligned}$$

Thus, using the inequality $2\,ab \leq a^2 + b^2$, and summing over m, from the above expression we obtain that

$$\begin{aligned}
\sum_{m\neq 0} \int_{\mathbb{R}^n} \hat{f}(\xi)\, \overline{\hat{g}_p(\xi)}\, \overline{\hat{f}(\xi + C_p^I m)}\, \hat{g}_p(\xi + C_p^I m)\, d\xi &\leq \\
\leq \sum_{m\neq 0} \int_{\mathbb{R}^n} |\hat{f}(\xi)|^2\, |\overline{\hat{g}_p(\xi)}\, \hat{g}_p(\xi + C_p^I m)|\, d\xi.
\end{aligned}$$

Using this observation into (3.19) we obtain:

$$\begin{aligned}
R(f) &\leq \sum_{m\neq 0} \int_{\mathbb{R}^n} |\hat{f}(\xi)|^2 \sum_{p\in\mathcal{P}} \frac{1}{|\det C_p|}\, |\overline{\hat{g}_p(\xi)}\, \hat{g}_p(\xi + C_p^I m)|\, d\xi \\
&\leq \|f\|^2 \sum_{m\neq 0} \sup_{\xi\in\mathbb{R}^n} \sum_{p\in\mathcal{P}} \frac{1}{|\det C_p|}\, |\overline{\hat{g}_p(\xi)}\, \hat{g}_p(\xi + C_p^I m)|. \qquad (3.20)
\end{aligned}$$

The proof now follows from (3.18) and (3.20), and by making appropriate choices of $f \in \mathcal{D}$. □

3.2. Discrete wave packet Parseval frames. We will now apply the unified theory presented in the previous section to the study of the discrete wave packet systems. As a first application, we will use Theorem 3.2 to deduce a characterization of the functions that generate a discrete wave packet system Parseval frame, in the same spirit as the results which are known in the case of affine and Gabor systems (cf. [**HLW**] and the references there).

We obtain the following characterization of all families $\Psi = \{\psi^1, \ldots, \psi^L\} \subset L^2(\mathbb{R}^n)$, for which the wave packet systems $\mathcal{W}_{\mathcal{P}}^{\gamma_p}(\Psi)$, given by (3.2) form a Parseval frame for $L^2(\mathbb{R}^n)$, provided the L.I.C. is satisfied. It is clear that the corresponding characterization for the systems $\mathcal{W}_{\mathcal{P}}(\Psi)$, given by (3.1), follows by considering the case $\gamma_p = I_n$, for each $p \in \mathcal{P}$.

THEOREM 3.5. *Let $\Psi = \{\psi^1, \ldots, \psi^L\} \subset L^2(\mathbb{R}^n)$, $\mathcal{P}$ be countable and, corresponding to each $p \in \mathcal{P}$, let $a_p, \gamma_p \in GL_n(\mathbb{R})$ and $\nu_p \in \mathbb{R}^n$. Assume the L.I.C.:*

$$L(f) = \sum_{p\in\mathcal{P}} \sum_{m\in\mathbb{Z}^n} \int_{supp\, \hat{f}} |\hat{f}(\xi + a_p^T \gamma_p^T m)|^2\, |\hat{\psi}(a_p^I \xi - \nu_p)|^2\, d\xi < \infty, \tag{3.21}$$

for all $f \in \mathcal{D}$, where $\mathcal{D}$ is given by (3.10). Then the system $\mathcal{W}_{\mathcal{P}}(\Psi)$, given by (3.1), is a Parseval frame for $L^2(\mathbb{R}^n)$ if and only if

$$\sum_{p\in\mathcal{P}_\alpha} \hat{\psi}(a_p^I\,\xi - \nu_p)\,\overline{\hat{\psi}(a_p^I(\xi+\alpha) - \nu_p)} = \delta_{\alpha,0} \quad \text{for a.e. } \xi \in \mathbb{R}^n, \tag{3.22}$$

for each $\alpha \in \Lambda = \bigcup_{p\in\mathcal{P}} a_p^T\,\gamma_p^T\,\mathbb{Z}^n$, where $\mathcal{P}_\alpha = \{p \in \mathcal{P} : \alpha \in a_p^T\,\gamma_p^T\,\mathbb{Z}^n\}$.

PROOF. Since

$$D_{a_p}\,M_{\nu_p}\,T_{\gamma_p^{-1}k}\,\psi = e^{2\pi i \nu_p\cdot\gamma_p^{-1}k}\,T_{a_p^{-1}\gamma_p^{-1}k}\,D_{a_p}\,M_{\nu_p}\,\psi,$$

using the notation introduced before Theorem 3.2 we can write equation (3.1) as

$$\mathcal{W}_{\mathcal{P}}(\psi) = \{T_{C_p\,k}\,g_p : k \in \mathbb{Z},\, p \in \mathcal{P}\},$$

where

$$C_p = a_p^{-1}\gamma_p^{-1} \quad \text{and} \quad g_p = \frac{e^{2\pi i \nu_p\cdot\gamma_p^{-1}k}}{\sqrt{|\det \gamma_p|}}\,D_{a_p}\,M_{\nu_p}\,\psi.$$

Thus, we have:

$$\Lambda = \bigcup_{p\in\mathcal{P}} a_p^T\,\gamma_p^T\,\mathbb{Z}^n, \quad \text{and, for } \alpha \in \Lambda, \quad \mathcal{P}_\alpha = \{p \in \mathcal{P} : \alpha \in a_p^T\,\gamma_p^T\,\mathbb{Z}\}.$$

With the assumptions that we made for $\mathcal{P}$ and g_p, the functional $L(f)$, given by the left hand side of (3.11), takes the form

$$L(f) = \sum_{p\in\mathcal{P}}\sum_{m\in\mathbb{Z}^n}\int_{\operatorname{supp}\hat{f}} |\hat{f}(\xi + a_p^T\,\gamma_p^T\,m)|^2\,|\hat{\psi}(a_p^I\,\xi - \nu_p)|^2\,d\xi,$$

and, by (3.21), we have that $L(f) < \infty$ for all $f \in \mathcal{D}$. It now follows from Theorem 3.2 that the system $\mathcal{W}_{\mathcal{P}}(\Psi)$ is a Parseval frame if and only if (3.22) holds. □

As in the case of continuous wave packet systems, we have different versions of Theorem 3.5 if we change the order of the operators in Definition 3.1. Let

$$\begin{aligned}
U^{(1)}_{(p,k)} &= D_{a_p}\,T_{\gamma_p^{-1}k}\,M_{\nu_p},\\
U^{(2)}_{(p,k)} &= T_{\gamma_p^{-1}k}\,D_{a_p}\,M_{\nu_p},\\
U^{(3)}_{(p,k)} &= M_{\nu_p}\,D_{a_p}\,T_{\gamma_p^{-1}k},\\
U^{(4)}_{(p,k)} &= T_{\gamma_p^{-1}k}\,M_{\nu_p}\,D_{a_p},\\
U^{(5)}_{(p,k)} &= M_{\nu_p}\,T_{\gamma_p^{-1}k}\,D_{a_p}.
\end{aligned}$$

We obtain alternate versions of the discrete wave packet systems $\mathcal{W}^{(i)}_{\mathcal{P}}(\Psi)$, with $1 \le i \le 5$, by replacing the operator $U^{(0)}_{(p,k)} = D_{a_p}\,T_{\gamma_p^{-1}k}\,M_{\nu_p}$ in Definition 3.1 with the operators $U^{(i)}_{(a,\nu,y)}$, $1 \le i \le 5$. We have the following characterization result.

COROLLARY 3.6. *Let $\Psi = \{\psi^1, \ldots, \psi^L\} \subset L^2(\mathbb{R}^n)$, $\mathcal{P}$ be countable and, corresponding to each $p \in \mathcal{P}$, let $a_p, \gamma_p \in GL_n(\mathbb{R})$ and $\nu_p \in \mathbb{R}^n$. Assume $L^{(i)}(f) < \infty$,*

$1 \le i \le 5$, *for all* $f \in \mathcal{D}$, *where* $\mathcal{D}$ *is given by* (3.10) *and*

$$\begin{aligned} L^{(1)}(f) &= L(f) = \sum_{p\in\mathcal{P}} \sum_{m\in\mathbb{Z}^n} \int_{supp\,\hat f} |\hat f(\xi + a_p^T \gamma_p^T m)|^2 \, |\hat\psi(a_p^I \xi - \nu_p)|^2 \, d\xi, \\ L^{(2)}(f) &= \sum_{p\in\mathcal{P}} \sum_{m\in\mathbb{Z}^n} \int_{supp\,\hat f} |\hat f(\xi + \gamma_p^T m)|^2 \, |\hat\psi(a_p^I \xi - \nu_p)|^2 \, d\xi, \\ L^{(3)}(f) &= \sum_{p\in\mathcal{P}} \sum_{m\in\mathbb{Z}^n} \int_{supp\,\hat f} |\hat f(\xi + a_p^T \gamma_p^T m)|^2 \, |\hat\psi(a_p^I(\xi - \nu_p))|^2 \, d\xi, \\ L^{(4)}(f) &= L^{(5)}(f) = \sum_{p\in\mathcal{P}} \sum_{m\in\mathbb{Z}^n} \int_{supp\,\hat f} |\hat f(\xi + \gamma_p^T m)|^2 \, |\hat\psi(a_p^I(\xi - \nu_p))|^2 \, d\xi. \end{aligned}$$

Then the system $\mathcal{W}_{\mathcal{P}}^{(i)}(\Psi)$, $1 \le i \le 5$, *is a Parseval frame for* $L^2(\mathbb{R}^n)$ *if and only if* $E_\alpha^{(i)}(\xi) = \delta_{\alpha,0}$, *for a.e.* $\xi \in \mathbb{R}^n$, *for each* $\alpha \in \Lambda^{(i)}$, *where*

$$\begin{aligned} E_\alpha^{(1)}(\xi) &= E_\alpha^{(2)}(\xi) = E_\alpha(\xi) = \sum_{p\in\mathcal{P}_\alpha} \hat\psi(a_p^I \xi - \nu_p)\, \overline{\hat\psi(a_p^I(\xi+\alpha) - \nu_p)}, \\ E_\alpha^{(3)}(\xi) &= E_\alpha^{(4)}(\xi) = E_\alpha^{(5)}(\xi) = \sum_{p\in\mathcal{P}_\alpha} \hat\psi(a_p^I(\xi - \nu_p))\, \overline{\hat\psi(a_p^I(\xi+\alpha - \nu_p))}, \end{aligned}$$

and $\Lambda^{(1)} = \Lambda^{(3)} = \Lambda = \bigcup_{p\in\mathcal{P}} a_p^T \gamma_p^T \mathbb{Z}^n$, $\Lambda^{(2)} = \Lambda^{(4)} = \Lambda^{(5)} = \bigcup_{p\in\mathcal{P}} \gamma_p^T \mathbb{Z}^n$, $\mathcal{P}_\alpha^{(1)} = \mathcal{P}_\alpha^{(3)} = \mathcal{P}_\alpha = \{p \in \mathcal{P} : \alpha \in a_p^T \gamma_p^T \mathbb{Z}^n\}$, $\mathcal{P}_\alpha^{(2)} = \mathcal{P}_\alpha^{(4)} = \mathcal{P}_\alpha^{(5)} = \{p \in \mathcal{P} : \alpha \in \gamma_p^T \mathbb{Z}^n\}$.

There are special classes of wave packet systems $\mathcal{W}_{\mathcal{P}}(\Psi)$ for which the L.I.C., given by (3.21), is satisfied, and, as a consequence, one obtains versions of Theorem 3.5 which do not contain this hypothesis. For simplicity, in the following, we will restrict our attention to one dimensional systems with one generator. However, the same ideas can be extended to higher dimensions.

The following theorem deals with the case of (one-dimensional) wave packet systems where the dilations are generated by the integer powers of a real number. We have the following simple characterization result.

THEOREM 3.7. *Let* $\psi \in L^2(\mathbb{R})$, $\mathcal{P}$ *be countable and*

$$\mathcal{W}_{\mathcal{P}}(\psi) = \{D_{a^{j_p}} M_{\nu_p} T_k \psi : k \in \mathbb{Z}, p \in \mathcal{P}\}, \tag{3.23}$$

where $j_p \in \mathbb{Z}$, $\nu_p, a \in \mathbb{R}$, $a > 1$, *and, for each* $j \in \mathbb{Z}$, *there is a finite number* K_j *of indices* $p \in \mathcal{P}$ *for which* $j_p = j$. *Then the system* $\mathcal{W}_{\mathcal{P}}(\psi)$, *given by* (3.23), *is a Parseval frame for* $L^2(\mathbb{R})$ *if and only if*

$$\sum_{p\in\mathcal{P}_\alpha} \hat\psi(a^{-j_p}\xi - \nu_p)\, \overline{\hat\psi(a^{-j_p}(\xi+\alpha) - \nu_p)} = \delta_{\alpha,0} \quad \text{for a.e. } \xi \in \mathbb{R}, \tag{3.24}$$

for each $\alpha \in \Lambda = \bigcup_{p\in\mathcal{P}} a^{j_p}\mathbb{Z}$, *where* $\mathcal{P}_\alpha = \{p \in \mathcal{P} : \alpha \in a^{j_p}\mathbb{Z}\}$.

In order to prove this theorem, we need the following result which uses an idea from [**HLW**, Prop.5.2] (where a similar result is proved in the case of affine systems).

PROPOSITION 3.8. *Let $\psi \in L^2(\mathbb{R})$, $\mathcal{P}$ be countable, ν_p, $a \in \mathbb{R}$, $a > 1$, and, for each $j \in \mathbb{Z}$, assume that there is a finite number K_j of indices $p \in \mathcal{P}$ for which $j_p = j$. Also assume that*

$$\sum_{p\in\mathcal{P}} |\hat{\psi}(a^{-j_p}\xi - \nu_p)|^2 \leq \beta \quad \text{for a.e. } \xi \in \mathbb{R}^n, \tag{3.25}$$

for some $\beta > 0$. Then

$$L(f) = \sum_{p\in\mathcal{P}} \sum_{m\in\mathbb{Z}} \int_{\text{supp}\, \hat{f}} |\hat{f}(\xi + a^{j_p}m)|^2 \, |\hat{\psi}(a^{-j_p}\xi - \nu_p)|^2 \, d\xi < \infty, \tag{3.26}$$

for all $f \in \mathcal{D}$, where $\mathcal{D}$ is given by (3.10) with $E = \{0\}$.

To prove this proposition, we need the following two lemmas that are special cases of Lemma 5.10 and Lemma 5.11 from [**HLW**] (it suffices to set $E = \{0\}$). For $r \in \mathbb{R}$, $r > 0$, define:

$$\mathcal{Q}(r) = \{x \in \mathbb{R} : \frac{1}{r} < |x| < r\}.$$

LEMMA 3.9. *Let $a \in \mathbb{R}$, $a > 1$, and $r \in \mathbb{R}$. There exists $N = N(a, r) \in \mathbb{N}$ such that*

$$\#\{j \in \mathbb{Z} : a^j \, \eta \in Q(r)\} \leq N$$

for all $\eta \in \mathbb{R}^n$.

LEMMA 3.10. *Let $a \in \mathbb{R}$, $a > 1$, and $r \in \mathbb{R}$. There exists $C = C(a, r) \in \mathbb{R}$ such that*

$$\#\{m \in \mathbb{Z}^n \setminus \{0\} : a^j \, m \in Q(r)\} \leq C \, a^{-j}$$

for all $j \in \mathbb{Z}$.

We can now prove the Proposition 3.8.

PROOF OF PROPOSITION 3.8.

Without loss of generality, we can assume that, for each $f \in \mathcal{D}$, there exists $r \in \mathbb{R}$ such that $\text{supp}\, \hat{f} \subset \mathcal{Q}(r)$.

Write $L_0(f)$ for the sum of the terms in (3.26) for which $m = 0$ and $L_1(f)$ for the sum of the terms in (3.26) for which $m \neq 0$. Then $L(f) = L_0(f) + L_1(f)$.

We first estimate $L_0(f)$. If $m = 0$, then, using (3.25), we have:

$$L_0(f) = \sum_{p\in\mathcal{P}} \int_{\mathcal{Q}(r)} |\hat{f}(\xi)|^2 \, |\hat{\psi}(a^{-j_p}\xi - \nu_p)|^2 \, d\xi \, \leq \, \beta \, \|f\|^2 \, < \infty. \tag{3.27}$$

We now estimate $L_1(f)$. Using the change of variable $a^{-j_p}\xi = \eta$, from (3.26) we have:

$$L_1(f) = \sum_{p\in\mathcal{P}} \sum_{m\neq 0} \int_{a^{j_p}\eta\in\text{supp}\, \hat{f}} |\hat{f}(a^{j_p}(\eta + m)|^2 \, |\hat{\psi}(\eta - \nu_p)|^2 \, a^{j_p} \, d\eta. \tag{3.28}$$

Suppose $m \in \mathbb{Z} \setminus \{0\}$, $a^{j_p} \eta \in Q(r)$ and $a^{j_p}(\eta + m) \in Q(r)$; then, for $j_p \in \mathbb{Z}$, we have:

$$|a^{j_p} m| \leq |a^{j_p}(\eta + m)| + |a^{j_p}\eta| < r + r = 2r.$$

Thus, with the notation we introduced before Lemma 3.9, we have that

$$\{m \in \mathbb{Z}\setminus\{0\} : a^{j_p}\eta \in Q(r) \text{ and } a^{j_p}(\eta+m) \in Q(r)\} \subset \{m \in \mathbb{Z}\setminus\{0\} : a^{j_p}m \in Q(2r)\},$$

for every $j_p \in \mathbb{Z}$. By Lemma 3.10, the number of elements in the last set does not exceed $C(a,2r)\,a^{-j_p}$, for each $j_p \in \mathbb{Z}$. Thus, it follows that

$$L_1(f) \le C(a,2r)\|\hat{f}\|_\infty^2 \sum_{p\in\mathcal{P}} \int_{a^{j_p}\eta\in Q(r)} |\hat{\psi}(\eta-\nu_p)|^2\,d\eta.$$

Next, observe that, by Lemma 3.9, the number of distinct $j_p \in \mathbb{Z}$ such that $a^{j_p}\eta \in Q(r)$ does not exceed a fixed number $N(a,r)$, independently of $\eta \in \mathbb{R}$. Furthermore, by hypothesis, for any (of these finitely many distinct) j_p there is a finite number K_{j_p} indices $p' \in \mathcal{P}$ such that $j_p = j_{p'}$. Hence, letting $K(a,r)$ to be the maximum of the (finitely many) K_{j_p}, we have:

$$L_1(f) \le K(a,r)\,C(a,2r)\,N(a,r)\,\|\hat{f}\|_\infty^2\,\|\hat{\psi}\|_2^2 < \infty. \tag{3.29}$$

Finally, from (3.27) and (3.29), we deduce that, if $f \in \mathcal{D}$, then $L(f) < \infty$. □

We can now prove Theorem 3.7.

PROOF OF THEOREM 3.7.
We apply Theorem 3.5 with $a_p = a^{j_p}$ and $\gamma_p = 1$, for each $p \in \mathcal{P}$. Under these assumptions, equation (3.22) becomes equation (3.24) and, thus, in order to complete the proof, we only need to show that the L.I.C. (3.21) is satisfied. In order to show that this is the case, we apply Proposition 3.8. Indeed, if the system $\mathcal{W}_\mathcal{P}(\psi)$, given by (3.23), is a Parseval frame, then, by Proposition 3.3 applied to $\mathcal{W}_\mathcal{P}(\psi)$, we obtain (3.25). Conversely, if the equalities (3.24) are satisfied, then also in this case (3.25) is true (it suffices to take $\alpha = 0$ in (3.24)). In both cases, we can apply Proposition 3.8 and, thus, we obtain condition (3.26) (which is exactly the L.I.C. (3.21) under our assumptions for a_p, γ_p). □

As we mentioned at the beginning of Section 3, the oversampled wave packet systems $\mathcal{W}_\mathcal{P}^{\gamma_p}(\psi)$, given by (3.2), are obtained from the corresponding wave packet systems by using a larger collection of translations and an appropriate renormalization of the generators. For example, corresponding to the wave packet system $\mathcal{W}_\mathcal{P}(\psi)$, given by (3.23), we define the oversampled wave packet system generated by $\psi \in L^2(\mathbb{R})$, relative to the oversampling coefficient $\gamma \in \mathbb{Z}\setminus\{0\}$, as the collections of the form

$$\mathcal{W}_\mathcal{P}^{\gamma}(\psi) = \{\gamma^{-1/2}\,D_{a^{j_p}}\,M_{\nu_p}\,T_{\gamma^{-1}k}\,\psi : k \in \mathbb{Z},\, p \in \mathcal{P}\}, \tag{3.30}$$

where $j_p \in \mathbb{Z}$, ν_p, $a \in \mathbb{R}$. Observe that these systems include as special cases the classical oversampled affine systems introduced by Chui and Shi [**CS**], where $\nu_p = 0$ for each $p \in \mathcal{P}$.

Another interesting class of oversampled systems is obtained by considering the following collection of oversampling coefficients γ_p, $p \in \mathcal{P}$:

$$\gamma_p = \begin{cases} a^{-j_p} & \text{if } j_p < 0, \\ 1 & \text{if } j_p \ge 0. \end{cases} \tag{3.31}$$

Corresponding to the wave packet system $\mathcal{W}_\mathcal{P}(\psi)$, given by (3.23), we then define the *quasi affine wave packet system* generated by $\psi \in L^2(\mathbb{R})$ as the collections of the form

$$\mathcal{W}_\mathcal{P}^{q}(\psi) = \{\gamma_p^{-1/2}\,D_{a^{j_p}}\,M_{\nu_p}\,T_{\gamma_p^{-1}k}\,\psi : k \in \mathbb{Z},\, p \in \mathcal{P}\}, \tag{3.32}$$

where $j_p \in \mathbb{Z}$, ν_p, $a \in \mathbb{R}$ and γ_p is given by (3.31). The reason for the terminology is that these systems include, as special cases, the quasi affine systems of Ron and Shen [**RSb**], where $\nu_p = 0$ for each $p \in \mathcal{P}$. As we will later show, these systems also inherit the basic fundamental properties of the quasi affine systems.

In particular, it is simple to verify that the quasi affine wave packet systems $\mathcal{W}^q_{\mathcal{P}}(\psi)$, unlike the corresponding wave packet systems $\mathcal{W}_{\mathcal{P}}(\psi)$, are *shift-invariant*, that is, they are invariant with respect to integer translations, when $a \in \mathbb{Z}$. Indeed, to see that this is the case, let $\psi^q_{p,k} = \gamma_p^{-1/2} D_{a^{j_p}} M_{\nu_p} T_{\gamma_p^{-1}k}\, \psi$, where γ_p is given by (3.31). It is clear that, using this notation, we can write $\mathcal{W}^q(\psi) = \{\psi^q_{p,k} : p \in \mathcal{P}\}$. Next consider the effect of a shifts T_ℓ, $\ell \in \mathbb{Z}$, on the system $\mathcal{W}^q(\psi)$. If $j < 0$, by changing the order of the operators, we have:

$$T_\ell\, \psi^q_{p,k} = a^{j_p/2}\, T_\ell\, D_a^{j_p}\, M_{\nu_p}\, T_{a^{j_p}k}\, \psi = a^{j_p/2}\, e^{-2\pi i \nu_p \cdot a^{j_p}\ell}\, D_a^{j_p}\, M_{\nu_p}\, T_{a^{j_p}(k+\ell)}\, \psi = \\ = e^{-2\pi i \nu_p \cdot a^{j_p}\ell}\, \psi^q_{p,k+\ell}.$$

Similarly, if $j \geq 0$, by changing again the order of the operators we have:

$$T_\ell\, \psi^q_{p,k} = T_\ell\, D_a^{j_p}\, M_{\nu_p}\, T_k\, \psi = e^{-2\pi i \nu_p \cdot a^{j_p}\ell}\, D_a^{j_p}\, M_{\nu_p}\, T_{k+a^{j_p}\ell}\, \psi = e^{-2\pi i \nu_p \cdot a^{j_p}\ell}\, \psi^q_{p,k+a^{j_p}\ell},$$

and $k + a^{j_p}\ell \in \mathbb{Z}$, since $a \in \mathbb{Z}$ and $j_p \geq 0$. Observe that the scalars $e^{-2\pi i \nu_p \cdot a^{j_p}\ell}$ play no role in the wave packets frame expansions.

Another basic property of the oversampled wave packet systems that we have just described is that they preserve the tight frame property of the corresponding wave packet system. More precisely, we have the following result.

THEOREM 3.11. *Let $\psi \in L^2(\mathbb{R})$ and assume that the wave packet system $\mathcal{W}_{\mathcal{P}}(\psi)$, given by (3.23), where $a \in \mathbb{Z} \setminus \{0\}$, is a Parseval frame for $L^2(\mathbb{R})$. Also assume that, in (3.23), for each $j \in \mathbb{Z}$, there is only a finite number K_j of indices $p \in \mathcal{P}$ for which $j_p = j$. We then have the following:*

(a) *If $\gamma \in \mathbb{Z} \setminus a\mathbb{Z}$, then the corresponding oversampled wave packet system $\mathcal{W}^\gamma_{\mathcal{P}}(\psi)$, given by (3.30), is also a Parseval frame for $L^2(\mathbb{R})$.*
(b) *The corresponding quasi affine wave packet system $\mathcal{W}^q_{\mathcal{P}}(\psi)$, given by (3.32), is also a Parseval frame for $L^2(\mathbb{R})$.*

This theorem extends similar results for the affine and Gabor systems (cf. [**HLWW**] and the references in there). In particular, if $\mathcal{P} = \mathbb{Z}$ and $\nu_p = 0$ for each $p \in \mathcal{P}$, then, as we observed before, $\mathcal{W}_{\mathcal{P}}(\psi) = \mathcal{A}_a(\psi)$ is the affine system, $\mathcal{W}^\gamma_{\mathcal{P}}(\psi) = \mathcal{A}^\gamma_a(\psi)$ is the corresponding oversampled affine system, and, in this case, Theorem 3.11 gives the Second Oversampling Theorem originally discovered by Chui and Shi [**CS**].

Before proving Theorem 3.11, we will make a few observations about the proof, which uses the same main idea in both cases (a) and (b) of the theorem. In our proof, we will apply Theorem 3.2 to the systems $\mathcal{W}^\gamma_{\mathcal{P}}(\psi)$ and $\mathcal{W}^q_{\mathcal{P}}(\psi)$ and obtain the corresponding characterization equations, as we did for the system $\mathcal{W}_{\mathcal{P}}(\psi)$. Then we will observe that every characterization equation of the systems $\mathcal{W}^\gamma_{\mathcal{P}}(\psi)$ and $\mathcal{W}^q_{\mathcal{P}}(\psi)$ is also a characterization equation of the corresponding wave packet system $\mathcal{W}_{\mathcal{P}}(\psi)$. The proof of Theorem 3.11 uses similar arguments to those of Theorem 3.7. Here, in order to avoid undue repetitions, we will only sketch the main steps of the proof.

PROOF OF THEOREM 3.11.

If the wave packet system $\mathcal{W}_{\mathcal{P}}(\psi)$ is a Parseval frame for $L^2(\mathbb{R})$, then, by Theorem 3.7, equations (3.24) are satisfied and, as a consequence, the inequality (3.26) is also satisfied. Furthermore, since $a \in \mathbb{Z}$, for each $\alpha_0 \in \Lambda = \bigcup_{p\in\mathcal{P}} a^{j_p}\,\mathbb{Z}$, there are unique $j_0 \in \mathbb{Z}$ and $q_0 \in \mathbb{Z} \setminus a\,\mathbb{Z}$ such that $\alpha_0 = a^{j_0}\,q_0$. Thus,

$$\mathcal{P}_{\alpha_0} = \{p \in \mathcal{P} : a^{-j_p}\,\alpha_0 \in \mathbb{Z}\} = \{p \in \mathcal{P} : a^{-j_p+j_0}\,q_0 \in \mathbb{Z}\} = \{p \in \mathcal{P} : -j_p + j_0 \geq 0\},$$

and, using the change of variable $\eta = a^{-j_0}\xi$, from (3.24) we obtain

$$\sum_{p\in\mathcal{P}_{\alpha_0}} \hat{\psi}(a^{-j_p+j_0}\eta - \nu_p)\,\overline{\hat{\psi}(a^{-j_p+j_0}(\eta + q_0) - \nu_p)} = \delta_{q_0,0} \quad \text{for a.e. } \eta \in \mathbb{R}. \tag{3.33}$$

Case (a). We apply Theorem 3.5 to the system $\mathcal{W}^{\gamma}_{\mathcal{P}}(\psi)$. This gives us that, if the L.I.C.

$$L^{\gamma}(f) = \sum_{p\in\mathcal{P}}\sum_{m\in\mathbb{Z}} \int_{\operatorname{supp}\hat{f}} |\hat{f}(\xi + \gamma\,a^{j_p}m)|^2\,|\hat{\psi}(a^{-j_p}\xi - \nu_p)|^2\,d\xi < \infty \tag{3.34}$$

is satisfied for all $f \in \mathcal{D}$, where $\mathcal{D}$ is given by (3.10) with $E = \{0\}$, then the system $\mathcal{W}^{\gamma}_{\mathcal{P}}(\psi)$ is a Parseval frame for $L^2(\mathbb{R})$ if and only if

$$\sum_{p\in\mathcal{P}^{\gamma}_{\alpha}} \hat{\psi}(a^{-j_p}\xi - \nu_p)\,\overline{\hat{\psi}(a^{-j_p}(\xi + \alpha) - \nu_p)} = \delta_{\alpha,0} \quad \text{for a.e. } \xi \in \mathbb{R}, \tag{3.35}$$

for each $\alpha \in \Lambda^{\gamma} = \bigcup_{p\in\mathcal{P}} \gamma\,a^{j_p}\,\mathbb{Z}$, where $\mathcal{P}^{\gamma}_{\alpha} = \{p \in \mathcal{P} : \alpha \in \gamma\,a^{j_p}\,\mathbb{Z}\}$. Observe that, since $\gamma \in \mathbb{Z}$, (3.26) implies (3.34). Also observe that, since $a \in \mathbb{Z}$ and $\gamma \in \mathbb{Z}\setminus a\,\mathbb{Z}$, for each $\alpha_1 \in \Lambda^{\gamma}$, there are are unique $j_1 \in \mathbb{Z}$ and $q_1 \in \mathbb{Z} \setminus a\,\mathbb{Z}$ such that $\alpha_1 = a^{j_1}\,\gamma\,q_1$. Thus, we have that

$$\begin{aligned}\mathcal{P}^{\gamma}_{\alpha_1} &= \{p \in \mathcal{P} : \gamma^{-1}\,a^{-j_p}\,\alpha_1 \in \mathbb{Z}\} = \{p \in \mathcal{P} : a^{-j_p+j_1}\,q_1 \in \mathbb{Z}\} = \\ &= \{p \in \mathcal{P} : -j_p + j_1 \geq 0\},\end{aligned}$$

and, using the change of variable $\eta = a^{-j_1}\xi$, from (3.35) we obtain

$$\sum_{p\in\mathcal{P}^{\gamma}_{\alpha_1}} \hat{\psi}(a^{-j_p+j_1}\eta - \nu_p)\,\overline{\hat{\psi}(a^{-j_p+j_1}(\eta + \gamma\,q_1) - \nu_p)} = \delta_{q_1,0} \quad \text{for a.e. } \eta \in \mathbb{R}. \tag{3.36}$$

Since $\gamma \in \mathbb{Z}\setminus a\,\mathbb{Z}$, then also $\gamma\,q_1 \in \mathbb{Z}\setminus a\,\mathbb{Z}$, and, consequently, (3.36) holds whenever (3.33) holds. Thus, $\mathcal{W}^{\gamma}_{\mathcal{P}}(\psi)$ is a Parseval frame for $L^2(\mathbb{R})$.

Case (b). Under the assumptions that, for each $p \in \mathcal{P}$,

$$C_p = \begin{cases} 1, & j_p < 0 \\ a^{-j_p} & j_p \geq 0 \end{cases} \qquad g_p = \begin{cases} e^{2\pi i \nu_p \cdot a^{j_p} k}\,a^{j_p/2}\,D_a^{j_p}\,M_{\nu_p}\,\psi, & j_p < 0 \\ e^{2\pi i \nu_p \cdot k}\,D_a^{j_p}\,M_{\nu_p}\,\psi, & j_p \geq 0, \end{cases}$$

the family $\{T_{C_p k}\,g_p : p \in \mathcal{P}\}$ is exactly the system $\mathcal{W}^{q}_{\mathcal{P}}(\psi)$, and so we can apply again Theorem 3.5. We obtain that, if the L.I.C.

$$\begin{aligned} L^{q}(f) = &\sum_{\{p\in\mathcal{P}:j_p<0\}}\sum_{m\in\mathbb{Z}} \int_{\operatorname{supp}\hat{f}} |\hat{f}(\xi + m)|^2\,|\hat{\psi}(a^{-j_p}\xi - \nu_p)|^2\,d\xi + \\ &+ \sum_{\{p\in\mathcal{P}:j_p\geq 0\}}\sum_{m\in\mathbb{Z}} \int_{\operatorname{supp}\hat{f}} |\hat{f}(\xi + a^{j_p}m)|^2\,|\hat{\psi}(a^{-j_p}\xi - \nu_p)|^2\,d\xi < \infty \end{aligned} \tag{3.37}$$

is satisfied for all $f \in \mathcal{D}$, where $\mathcal{D}$ is given by (3.10) with $E = \{0\}$, then the system $\mathcal{W}^q_{\mathcal{P}}(\psi)$ is a Parseval frame for $L^2(\mathbb{R})$ if and only if

$$\sum_{p \in \mathcal{P}^q_m} \hat{\psi}(a^{-j_p}\xi - \nu_p)\,\overline{\hat{\psi}(a^{-j_p}(\xi + m) - \nu_p)} = \delta_{m,0} \quad \text{for a.e. } \xi \in \mathbb{R}, \tag{3.38}$$

for each $m \in \Lambda^q = \mathbb{Z}$, where $\mathcal{P}^q_m = \{p \in \mathcal{P} : m \in a^{j_p}\,\mathbb{Z}\}$. Observe that, as in part (a), (3.26) implies (3.37). Also observe that, for each $m_1 \in \mathbb{Z}$, there are are unique $j_1 \in \mathbb{Z}$, $j_1 \geq 0$, and $q_1 \in \mathbb{Z} \setminus a\,\mathbb{Z}$ such that $m_1 = a^{j_1}\,q_1$. Thus,

$$\mathcal{P}^q_{m_1} = \{p \in \mathcal{P} : a^{-j_p + j_1}\,q_1 \in \mathbb{Z}\} = \{p \in \mathcal{P} : -j_p + j_1 \geq 0\},$$

and, using the change of variable $\eta = a^{-j_1}\xi$, from (3.38) we obtain

$$\sum_{p \in \mathcal{P}^q_{m_1}} \hat{\psi}(a^{-j_p + j_1}\eta - \nu_p)\,\overline{\hat{\psi}(a^{-j_p + j_1}(\eta + q_1) - \nu_p)} = \delta_{q_1,0} \quad \text{for a.e. } \eta \in \mathbb{R}. \tag{3.39}$$

It is now clear that (3.39) holds whenever (3.33) holds (observe that $j_1 \geq 0$ while, in (3.33), $j_0 \in \mathbb{Z}$ and, thus, the converse implication does not hold). Thus, $\mathcal{W}^q_{\mathcal{P}}(\psi)$ is a Parseval frame for $L^2(\mathbb{R})$. □

Using Theorem 3.11 we can show that the continuous wave packets $\mathcal{W}_S(\Psi)$, given by (2.1), are, in a certain sense, the "ultimate oversampling" over the discrete wave packets $\mathcal{W}_{\mathcal{P}}(\Psi)$, given by (3.1) (a similar observation for the affine systems can be found in [**WW**]) . This also sheds some light into the discretization process that leads from the continuous to the discrete systems. For simplicity, we consider the one-dimensional case only.

Let $\psi \in L^2(\mathbb{R})$ and consider the (discrete) wave packet system $\mathcal{W}_{\mathcal{P}}(\psi)$, given by (3.23), with $a \in \mathbb{Z} \setminus \{0\}$. Assume that $\mathcal{W}_{\mathcal{P}}(\Psi)$ is a Parseval frame for $L^2(\mathbb{R})$. Now consider the corresponding oversampled wave packet systems $\mathcal{W}^N_{\mathcal{P}}(\psi)$, given by(3.30), where the $N \in \mathbb{Z} \setminus a\mathbb{Z}$. By Theorem 3.11, the systems $\mathcal{W}^N_{\mathcal{P}}(\psi)$ are also a Parseval frame for $L^2(\mathbb{R})$. Thus, for each $f \in L^2(\mathbb{R})$, we have

$$f = \frac{1}{N} \sum_{p \in \mathcal{P}} \sum_{k \in \mathbb{Z}} \langle f, D_a^{j_p}\, M_{\nu_p}\, T_{N^{-1}k}\, \psi\rangle\, D_a^{j_p}\, M_{\nu_p}\, T_{N^{-1}k}\, \psi,$$

with convergence in $L^2(\mathbb{R})$, for each $N \in \mathbb{Z} \setminus a\,\mathbb{Z}$. We can interpret this sum as a Riemann sum. Thus, choosing a sequence of $N \in \mathbb{Z} \setminus a\,\mathbb{Z}$, and taking the limit for N tending to ∞, we obtain that, for each $f \in L^2(\mathbb{R})$

$$f = \sum_{p \in \mathcal{P}} \int_{\mathbb{R}} \langle f, D_a^{j_p}\, M_{\nu_p}\, T_y\, \psi\rangle\, D_a^{j_p}\, M_{\nu_p}\, T_y\, \psi\, dy,$$

with convergence in $L^2(\mathbb{R})$. This shows that the system $\mathcal{W}_S(\Psi)$, given by (2.1) where $S = \{(a^{j_p}, \nu_p) : p \in \mathcal{P}\}$, is a continuous Parseval frame wave packet system relative to S for $L^2(\mathbb{R})$.

3.3. A very general example. In [**HLWW**], we gave an example of a Parseval frame wave packet system associated with a general disjoint covering of the real line by intervals. In this section, we extend this example to dimension $n = 2$, where the triangles assume the role played by the intervals. This construction presents a very general example of wave packet systems: in fact, the dilations do not have to be powers of a matrix and the modulations do not have to be associated with a lattice.

Consider a triangularization of $\mathbb{R}^2$. More precisely, let τ_1 be the triangle of vertices $(0,0)$, $(1,0)$, $(0,1)$, and consider a tiling of $\mathbb{R}^2$ given by the union of countably many non-degenerate triangles $\{\tau_j\}_{j\in\mathbb{N}}$:

$$\mathbb{R}^2 = \bigcup_{j\in\mathbb{N}} \tau_j, \tag{3.40}$$

where the triangles τ_j have disjoint interiors. For each triangle τ_j, $j \in \mathbb{N}$, with vertices $u_j = (u_j^{(1)}, u_j^{(2)})$, $v_j = (v_j^{(1)}, v_j^{(2)})$ and $w_j = (w_j^{(1)}, w_j^{(2)})$, let the matrix a_j be determined by

$$a_j^T = \begin{pmatrix} v_j^{(1)} - u_j^{(1)} & w_j^{(1)} - u_j^{(1)} \\ v_j^{(2)} - u_j^{(2)} & w_j^{(2)} - u_j^{(2)}. \end{pmatrix}$$

It is clear that, for each $j \in \mathbb{N}$, $\tau_j = a_j^T \tau_1 + u_j$, and this shows that, by choosing an appropriate ordering of the vertices, to each triangle τ_j there is a uniquely associated dilation $a_j \in GL_2(\mathbb{R})$ and translation $u_j \in \mathbb{R}^2$ mapping τ_1 into τ_j.

Consider the wave packet systems

$$\mathcal{W}(\psi) = \{D_{a_j} M_{\nu_j} T_k \psi : k \in \mathbb{Z}^2, j \in \mathbb{N}\}, \tag{3.41}$$

with $\nu_j = (a_j^T)^{-1} u_j = a_j^I u_j$ and $\hat{\psi}(\xi) = \chi_{\tau_1}(\xi)$, $\xi \in \mathbb{R}^2$. We will now apply Theorem 3.5 to show that $\mathcal{W}(\psi)$ is a Parseval frame for $L^2(\mathbb{R}^2)$.

Since $|\hat{\psi}(a_j^I \xi - \nu_j)| = |\hat{\psi}(a_j^I(\xi - u_j))| = \chi_{\tau_j}(\xi)$, the left hand side of the L.I.C., given by (3.21), takes the form:

$$\begin{aligned} L(f) &= \sum_{j\in\mathbb{N}} \sum_{m\in\mathbb{Z}} \int_{\operatorname{supp} \hat{f}} |\hat{f}(\xi + a_j^T m)|^2 \, |\hat{\psi}((a_j^T)^{-1}\xi - \nu_j))|^2 \, d\xi \\ &= \sum_{j\in\mathbb{N}} \sum_{m\in\mathbb{Z}} \int_{\operatorname{supp} \hat{f} \cap \tau_j} |\hat{f}(\xi + a_j^T m)|^2 \, d\xi. \end{aligned} \tag{3.42}$$

We need to show that $L(f) < \infty$ for all $f \in \mathcal{D}$, where $\mathcal{D}$ is a dense subspace of $L^2(\mathbb{R}^2)$. Take $\mathcal{D} = \{f \in L^2(\mathbb{R}^2) : \|\hat{f}\|_\infty < \infty, \operatorname{supp} \hat{f}$ compact and $\operatorname{supp} \hat{f} \subseteq$ finite union of interiors of $\tau_j\}$. This clearly implies that the intersection $\operatorname{supp} \hat{f} \bigcap \tau_j$ is nonzero only for finitely many j. Furthermore, since $a_j \in GL_2(\mathbb{R})$, for each $j \geq 1$ the norm $\|a_j^T m\|$ tends to infinity as $\|m\|$ goes to infinity and, as a consequence, there are only finitely many $m \in \mathbb{Z}$ such that the integral in (3.42) is nonzero. Combining these two observations, it follows that $L(f) < \infty$ for all $f \in \mathcal{D}$.

By Theorem 3.5, in order to show that the wave packet system $\mathcal{W}(\psi)$ is a Parseval frame it only remains to show that ψ satisfies the characterizing equations (3.22) which, in this case, are the two equations:

$$\sum_{j\in\mathbb{N}} |\hat{\psi}(a_j^I \xi - \nu_j)|^2 = 1 \quad \text{for a.e. } \xi \in \mathbb{R}^2, \tag{3.43}$$

$$\sum_{j\in\mathcal{P}_\alpha} \hat{\psi}(a_j^I \xi - \nu_j) \, \overline{\hat{\psi}(a_j^I(\xi + \alpha) - \nu_j)} = 0, \quad \text{for a.e. } \xi \in \mathbb{R}^2, \text{ if } \alpha \neq 0, \tag{3.44}$$

where $\mathcal{P}_\alpha = \{j \in \mathcal{P} : a_j^I \alpha \in \mathbb{Z}^2\}$ and $\alpha \in \bigcup_{j\geq 1} a_j^T \mathbb{Z}^2$. As we observed before, we have that $\hat{\psi}(a_j^I \xi - \nu_j) = \chi_{\tau_j}(\xi)$, and so (3.40) implies equation (3.43). Next, consider equation (3.44) with $\alpha \neq 0$. Fix $j \in \mathcal{P}_\alpha$ and let $\eta = a_j^I \xi - \nu_j$. Under these assumptions, we have

$$\hat{\psi}(a_j^I\xi - \nu_j)\,\overline{\hat{\psi}(a_j^I(\xi+\alpha) - \nu_j)} = \chi_{\tau_1}(\eta)\,\chi_{\tau_1}(\eta + (a_j^T)^{-1}\alpha). \tag{3.45}$$

Since $j \in \mathcal{P}_\alpha$, $\alpha \neq 0$, then $a_j^I\,\alpha \in \mathbb{Z}^2 \setminus \{0\}$ and, because the interiors of the triangles are pairwise disjoint, the expression (3.45) vanishes almost everywhere. This implies that equation (3.44) is also satisfied and, thus, our system is a Parseval frame for $L^2(\mathbb{R}^2)$.

Observe that we can also show directly, without using Theorem 3.5, that the system $\mathcal{W}(\psi)$ is a Parseval frame for $L^2(\mathbb{R}^2)$. In order to do this, observe first that the collection $\{M_k\,\chi_{\tau_1} : k \in \mathbb{Z}^2\}$ is a Parseval frame for $L^2(\tau_1)$. As we observed before, we have the following tiling of $\mathbb{R}^2$:

$$\bigcup_{j\in\mathbb{N}} D_{a_j^I}\,T_{\nu_j}\,\tau_1 = \mathbb{R}^2,$$

where the union is disjoint. This implies that the collection

$$\{D_{a_j^I}\,T_{\nu_j}\,M_{-k}\,\chi_{\tau_1} : k \in \mathbb{Z}^2, j \in \mathbb{N}\} = \{\big(D_{a_j}\,M_{\nu_j}\,T_k\,(\chi_{\tau_1})^\vee\big)^\wedge : k \in \mathbb{Z}^2, j \in \mathbb{N}\}$$

is a Parseval frame for $L^2(\mathbb{R}^2)$.

Also observe that the choice of the triangle τ_1 plays no special role in this example; in fact, the construction can easily be modified by choosing a different initial triangle. Finally, observe that this construction generalizes to higher dimensions. Indeed, if the triangles are replaced by n-dimensional simplices, then all the computations carry over in a similar way.

3.4. Discrete wave packet frames. By applying Theorem 3.4 to the wave packet systems $\mathcal{W}_\mathcal{P}(\Psi)$, given in (3.23), one obtains results very much in the spirit of those found in [**D**, Ch. 3] for Gabor and affine systems, or in [**HL**], where one can find examples of wave packet systems that are a frame for $L^2(\mathbb{R})$. In particular, we are able to recover the results in [**D**] and [**HL**]. For the sake of brevity, we will not examine this situation further in this paper.

References

[ACM] A. Aldroubi, C. Cabrelli, and U. Molter, *Wavelets on irregular grids with arbitrary dilation matrices and frame atoms for $L^2(\mathbb{R}^n)$*, preprint (2003).

[C] A. Calderón *Intermediate spaces and interpolation, the complex method*, Stud. Math. **24** (1964), 113–190.

[CCMW] C. Chui, W. Czaja, M. Maggioni, and G. Weiss *Characterization of general tight wavelets frames with matrix dilations and tightness preserving oversampling*, J. Fourier Anal. Appl. **8** (2002), 173–200.

[CDH] O. Christensen, B. Deng, and C. Heil, *Density of Gabor frames*, Appl. Comput. Harmon. Anal. **7** (1999), 33–47.

[CS] C. Chui, and X. Shi, *$n\times$oversampling preserves any tight affine frame for odd n*, Proc. Am. Math. Soc. **121(2)** (1994), 511-517.

[CF] A. Córdoba and C. Fefferman, *Wave packets and Fourier integral operators*, Comm. Partial Diff. Eq. **3** (1978), 979-1005.

[D] I. Daubechies, *Ten Lectures on Wavelets*, CBMS–NSF Regional Conference Series in Applied mathematics, Vol.61, SIAM, Philadelphia, 1992.

[CKS] W. Czaja, G. Kutyniok, and D. Speegle, *Density of wave packet systems*, preprint (2003).

[F] G. B. Folland, *Harmonic Analysis on Phase Space*, Princeton University Press, Princeton, NJ, 1989.

[G] K. Gröchenig, *Foundations of Time–Frequency Analysis*, Birkäuser, Boston, 2001.

[HK] C. Heil and G. Kutyniok, *Density of weighted wavelet frames*, J. Geometric Analysis **13** (2003), 479–493.

[HLW] E. Hernández, D. Labate, and G. Weiss, *A unified characterization of reproducing systems generated by a finite family*, J. Geom. Anal. **12(4)** (2002), 615–662.

[HLWW] E. Hernández, D. Labate, G. Weiss, and E. Wilson *Oversampling, quasi affine frames and wave packets*, Appl. Comput. Harmon. Anal. (to appear) (2003) .

[HW] E. Hernández, G. Weiss, *A First Course on Wavelets*, CRC Press, Boca Raton FL, 1996.

[HL] J. A. Hogan and J. D. Lakey, *Extensions of the Heisenberg group by dilations and frames*, Appl. Comput. Harmon. Anal. **2** (1995), 174–199.

[LWWW] R. S. Laugesen, N. Weaver, G. Weiss, and E. Wilson, *A characterization of the higher dimensional groups associated with continuous wavelets*, J. Geom. Anal. **12(1)** (2001), 89–102.

[RSb] A. Ron, and Z. Shen, *Affine systems in* $L_2(\mathbb{R}^d)$*: the analysis of the analysis operator*, J. Funct. Anal. **148** (1997), 408-447.

[W] Y. Wang, *Wavelets, tilings, and spectral sets*, Duke Math J. **114** (2002), 43–57.

[WW] G. Weiss, and E. Wilson, *The mathematical theory of wavelets*, Proceeding of the NATO–ASI Meeting. Harmonic Analysis 2000 – A Celebration. Kluwer Publisher, 2001.

DEPARTMENT OF MATHEMATICS, WASHINGTON UNIVERSITY, SAINT LOUIS, MISSOURI 63130
E-mail address: dlabate@math.wustl.edu

DEPARTMENT OF MATHEMATICS, WASHINGTON UNIVERSITY, SAINT LOUIS, MISSOURI 63130
E-mail address: guido@math.wustl.edu

DEPARTMENT OF MATHEMATICS, WASHINGTON UNIVERSITY, SAINT LOUIS, MISSOURI 63130
E-mail address: enwilson@math.wustl.edu

Contemporary Mathematics
Volume **345**, 2004

Convolution for Gabor systems and Newton's method

M. C. Lammers

ABSTRACT. We use the operators associated with Gabor systems to create a convolution-like multiplication for the Gabor systems that have finite upper frame bounds. In the critically sampled case this yields a C^* algebra while in the oversampled case we create a non-commutative Banach algebra. In both cases we are able to generate duals and canonical tight frames by finding inverses with respect to this algebraic structure. Using this convolution we produce a Newton's method algorithm for the computation of dual windows and canonical tight windows.

1. Introduction

In 1952 Duffin and Schaeffer [**3**] defined frames:

DEFINITION 1.1. A sequence $\{f_n\}_{n\in\mathbb{Z}}$ of elements of a Hilbert space H is called a *frame* if there are positive constants A (the lower frame bound) and B (the upper frame bound) such that

$$A\|f\|^2 \leq \sum_{n\in\mathbb{Z}} |\langle f, f_n\rangle|^2 \leq B\|f\|^2, \quad \text{for all} \quad f \in H.$$

Of interest in many applications are frames that are generated by applying unitary operators to a single function in $L^2(\mathbb{R})$. Gabor systems are sequences generated by using translations and modulations of a single function. For a complete reference we refer the reader to the book of K. Gröchenig [**5**]. Given $f \in L^2(\mathbb{R})$ we let these unitary operators be denoted by

$$\text{Translation: } T_a f(x) = f(x-a),\ a \in \mathbb{R} \text{ and}$$
$$\text{Modulation: } M_b f(x) = e^{2\pi i b x} f(x),\ b \in \mathbb{R}.$$

DEFINITION 1.2. If $a, b \in \mathbb{R}^+$ and $g \in L^2(\mathbb{R})$ we call $\{M_{mb}T_{na}g\}_{m,n\in\mathbb{Z}}$ a *Gabor system* (also known as a *Weyl-Heisenberg system*) and denote it by $\mathcal{G}(g, a, b)$. We call the function g the *window* for the system and say that the parameters a, b determine the lattice of the system.

In this work we fix a window and develop an algebraic structure for Gabor systems that is dependent on this fixed window. Fixing a function in the study of Gabor systems is quite common. For example, one of the most important areas of research involving Gabor systems is the study of modulation spaces. The norm of a modulation space is defined by $\|f\|_{M^{p,q}} = \|\langle f, M_w T_x g\rangle\|_{L^{p,q}(\mathbb{R}^2)}$ for some fixed window g. Given that the norm is dependent on a fixed window it is likely that the algebraic structure developed here using this same window will have applications to theses spaces as well as $L^2(\mathbb{R})$. We refer to this multiplicative structure as a convolution because in the the critically sampled case $a = b = 1$ the Zak transform turns this Banach algebra multiplication into pointwise multiplication.

Gabor systems are used in much the same way as a basis to represent a function in $L^2(\mathbb{R})$. That is, one would like to represent a function f as $f = \sum c_{mn} M_{mb} T_{na} g$. Unlike a basis the sequence c_{mn} need not be unique. At first, the fact that this sequence c_{mn} is not unique may seem to be a liability but in fact it has proved to be an asset in many applications. In order to generate such representations it is necessary to find a dual system associated with a given Gabor system.

Associated to a Gabor system there exists a system which is self-dual called the canonical tight frame. In [**10**], Janssen and Strohmer produce an algorithm based on Newton's method to compute this canonical tight frame. We use the algebraic structure mentioned above to produce a Newton's method-like algorithm for computing duals and canonical tight frames. The method in this paper may be preferable to the method above because it does not require inverting the frame operator of a new system at each step of the algorithm. Independently, Janssen has recently assembled an impressive arsenal of fast algorithms for computing canonical duals and tight frames [**9**]. Although the algorithms are similar in spirit there are several distinct differences. One difference is that in the method presented here there is no condition on the ratio of the frame bounds for the algorithms to converge. Another difference is that the algebra itself in this work is dependent on a fixed frame so the recursions may yield different approximations of the dual windows for different fixed frames. Since it is still unknown if the so-called canonical dual is the best dual for all settings, the fact that this algorithm also allows us to compute different dual windows which are dependent on different tight frames holds promise for future applications.

2. Preliminaries

There are several operators that arise naturally in the study Gabor systems $\mathcal{G}(g, a, b)$. Using the notation from [**5**], we define the *coefficient operator* or *analysis operator* $C_g : L^2(\mathbb{R}) \to \ell^2(\mathbb{Z}^2)$ and the *synthesis operator* or *reconstruction operator* $D_g : \ell^2(\mathbb{Z}^2) \to L^2(\mathbb{R})$ associated with $\mathcal{G}(g, a, b)$ by

$$C_g(f) = \{\langle f, M_{mb} T_{na} g\rangle\}_{m,n\in\mathbb{Z}} \text{ and } D_g(c) = \sum_{m,n\in\mathbb{Z}} c_{m,n} M_{mb} T_{na} g.$$

We point out that the operator D_g is the adjoint of C_g and since

$$\|C_g(f)\|^2 = \sum_{m,n\in\mathbb{Z}} |\langle f, M_{mb} T_{na} g\rangle|^2 \text{ for all } f \in L^2(\mathbb{R}),$$

it follows that the coefficient operator and the synthesis operators are bounded if and only if the system $\mathcal{G}(g, a, b)$ has a finite upper frame bound B. When a system has a finite upper frame bound but not necessarily a lower frame bound we say that it is a *Bessel system*. We will be interested in composing analysis and synthesis operators from different systems $\mathcal{G}(g,a,b)$ and $\mathcal{G}(h,a,b)$ and we will denote these by

$$S_{g,h}(f) = D_h C_g(f) = \sum_{m,n\in\mathbb{Z}} \langle f, M_{mb}T_{na}g\rangle M_{mb}T_{na}h.$$

If we compose the synthesis and analysis operators from the same system we define this to be *frame operator associated with* $\mathcal{G}(g,a,b)$ and denote it by

$$S_g(f) = D_g C_g(f) = \sum_{m,n\in\mathbb{Z}} \langle f, M_{mb}T_{na}g\rangle M_{mb}T_{na}g.$$

When we do this we will always be considering systems generated by the same parameters a,b. A very important fact about operators of the form $S_{g,h}$ is that they commute with the translation and modulation operators determined by the parameters of those systems. This is shown by a straight forward change of variable argument and since it is crucial to a number of the computations that follow we include it here. For all $j,k\in\mathbb{Z}$

$$\begin{aligned}(2.1)\qquad S_{g,h}(M_{jb}T_{ka}f) &= \sum_{m,n\in\mathbb{Z}} \langle M_{jb}T_{ka}f, M_{mb}T_{na}g\rangle M_{mb}T_{na}h\\ &= \sum_{m,n\in\mathbb{Z}} \langle f, T_{-ka}M_{-jb}M_{mb}T_{na}g\rangle M_{mb}T_{na}h\\ &= \sum_{m,n\in\mathbb{Z}} \langle f, e^{-2\pi i(m-j)na}M_{(m-j)b}T_{(k-n)a}g\rangle\\ &\qquad\cdot e^{-2\pi i(m-j)na}M_{jb}T_{ka}M_{(m-j)b}T_{(k-n)a}h\\ &= M_{jb}T_{ka}\sum_{m,n\in\mathbb{Z}} \langle f, M_{(m-j)b}T_{(k-n)a}g\rangle M_{(m-j)b}T_{(k-n)a}h\\ &= M_{jb}T_{ka}S_{g,h}(f).\end{aligned}$$

It is easy to show that if $\mathcal{G}(g,a,b)$ is a frame then S is a positive invertible operator and there exist $\gamma\in L^2(R)$ so that the system $\mathcal{G}(\gamma,a,b)$ together with $\mathcal{G}(g,a,b)$ yield the following representations of any $f\in L^2(\mathbb{R})$:

$$f = D_g C_\gamma(f) = \sum_{m,n\in\mathbb{Z}} \langle f, M_{mb}T_{na}\gamma\rangle M_{mb}T_{na}g \text{ (equivalently } D_gC_\gamma = I)$$

and

$$f = D_\gamma C_g(f) = \sum_{m,n\in\mathbb{Z}} \langle f, M_{mb}T_{na}g\rangle M_{mb}T_{na}\gamma \text{ (equivalently } D_\gamma C_g = I).$$

In this case we refer to $\mathcal{G}(\gamma,a,b)$ as a *dual frame* of $\mathcal{G}(g,a,b)$ and we call frames which are their own duals *1-tight frames* or *Parseval frames.* Dual frames need not be unique but there is one of particular interest. Because S_g and S_g^{-1} are self adjoint and commute with the translation and modulation operators of the system we have

$$\begin{aligned} f = S_g S_g^{-1}(f) &= \sum_{m,n\in\mathbb{Z}} \langle f, S_g^{-1}(M_{mb}T_{na}g)\rangle M_{mb}T_{na}g\\ &= \sum_{m,n\in\mathbb{Z}} \langle f, M_{mb}T_{na}S_g^{-1}(g)\rangle M_{mb}T_{na}g.\end{aligned}$$

This motivates us to define the *canonical dual* of a Gabor frame $\mathcal{G}(g,a,b)$ to be $\mathcal{G}(g^d,a,b)$ where $g^d = S_g^{-1}(g)$. Similarly, we define the *canonical tight frame* of a Gabor frame $\mathcal{G}(g,a,b)$ to be $\mathcal{G}(g^t,a,b)$ where $g^t = S_g^{-\frac{1}{2}}(g)$ because

$$f = S_g S_g^{-1}(f) = \sum_{m,n\in\mathbb{Z}} \langle f, M_{mb}T_{na}S_g^{-\frac{1}{2}}(g)\rangle M_{mb}T_{na}S_g^{-\frac{1}{2}}(g).$$

One of the fundamental tools for analyzing Gabor systems is the Zak transform. Originally introduced by Ge'lfand [**4**], Zak[**12**] rediscovered it while studying problems in solid state physics. The work of Janssen [**7, 8**] established this transform as an indispensable part of time frequency analysis and Gabor systems.

DEFINITION 2.1. We define the *Zak transform* of a function $f(x) \in L^2(\mathbb{R})$ to be

$$U_Z(f)(t,w) = \sum_{k\in\mathbb{Z}} f(t-k)e^{2\pi ikw} = \sum_{j,k\in\mathbb{Z}} \langle f, e^{2\pi ijt}\mathbf{1}_{[k,k+1)}\rangle e^{2\pi ijt}e^{2\pi ikw},\ t,w\in\mathbb{R}$$

$$\text{where } \mathbf{1}_{[k,k+1)}(x) = \begin{cases} 1 \text{ if } x\in[k,k+1) \\ 0 \text{ else} \end{cases}.$$

We may interpret the Zak transform as a unitary operator from $L^2(\mathbb{R})$ to $L^2[0,1]^2$.

Much like the Fourier transform, the Zak transform has the two important properties regarding translation and modulation of a function:

$$U_Z(e^{2\pi ijt}f) = e^{2\pi ijt}U_Z(f) \text{ and } U_Z(T_kf) = e^{2\pi ikw}U_Z(f). \tag{2.2}$$

Applying the two properties in 2.2 to frame operators gives

$$U_Z(S_{g,h}(f)) = U_Z(g)\overline{U_Z(h)}U_Z(f) \tag{2.3}$$

and this provides the essential steps of the proof for characterizing $\mathcal{G}(g,1,1)$ systems that produce frames.

THEOREM 2.2. [**7**] *The system $\mathcal{G}(g,1,1)$ yields a Gabor frame with bounds A and B iff*

$$A \le |U_Z(g)(t,w)| \le B \textit{ for a.e. } (t,w)\in[0,1]^2.$$

This result has been generalized by Zibulski-Zeevi [**13**] in the case $ab = \frac{p}{q}$ is rational. Their extension uses the Zak matrices, the elements of which are defined by

$$\mathbf{\Phi}^f(t,w) = \frac{1}{\sqrt{p}}\left(U_Z(f)(t-l\frac{p}{q}, w+\frac{k}{p})\right)_{k=0\ldots p-1, l=0\ldots q-1}.$$

THEOREM 2.3. *For $ab = \frac{p}{q} < 1$ the system $\mathcal{G}(g,a,b)$ is a frame with bounds A and B iff*

$$AI_{p\times p} \le \mathbf{\Phi}^g(t,w)\mathbf{\Phi}^g(t,w)^* \le BI_{p\times p}.$$

Finally we recall an inequality from operator theory. If S is a positive operator and $AI \le S \le BI$ then

$$\|I - \frac{S}{B}\|_{op} \le \frac{B-A}{B} < 1 \tag{2.4}$$

where $\|\cdot\|_{op}$ is the operator norm of a bounded operator on $L^2(\mathbb{R})$. In particular, if $\mathcal{G}(g,a,b)$ is a frame then $AI \le S_g \le BI$ and

$$\|I - \frac{S_g}{B}\|_{op} \le \frac{B-A}{B} < 1. \tag{2.5}$$

3. The critically sampled case a=b=1

We present the simple case here for it motivates the oversampled case. When the parameters $a = b = 1$, or even $ab = 1$, Gabor systems that are frames are Riesz bases and the operators C_g and D_g are invertible so things are very straightforward in this case. We designate the set of functions whose Gabor system of the form $\mathcal{G}(g,1,1)$ has a finite upper frame bound as the space Z. In light of Theorem 2.2, another way to characterize this space is as the subspace generated by applying the inverse Zak transform to the subspace $L^\infty[0,1]^2$ of $L^2[0,1]^2$, i.e.

$$Z = U_Z^{-1}(L^\infty[0,1]^2) \text{ where } \|g\|_Z = \operatorname*{ess\ sup}_{t,w\in[0,1]^2} |U_Z(g)(t,w)|.$$

We create a convolution and involution on the space Z by picking a Gabor system that is an orthonormal basis. A natural choice is $\mathcal{G}(\mathbf{e},1,1)$, with $\mathbf{e} = \mathbf{1}_{[0,1)}(x)$. We chose to label this function $\mathbf{e}$ for it corresponds to the multiplicative identity for the algebra and it should not be confused with the constant $e \equiv 2.71828\cdots$. Now with $\mathbf{e}$ fixed we define

$$g \odot f = D_g C_{\mathbf{e}}(f) = \sum_{m,n\in\mathbb{Z}} \langle f, M_m T_n \mathbf{e}\rangle M_m T_n g,$$
$$\text{and } f^* = D_{\mathbf{e}} C_f(\mathbf{e}) = \sum_{m,n\in\mathbb{Z}} \langle \mathbf{e}, M_m T_n f\rangle M_m T_n \mathbf{e}.$$

Applying the commutation relation in (2.1) and the fact that D_g and C_g are adjoints we have

$$\begin{aligned}
(f^*)^* &= \sum_{m,n\in\mathbb{Z}} \langle \mathbf{e}, M_m T_n f^*\rangle M_m T_n \mathbf{e} \\
&= \sum_{m,n\in\mathbb{Z}} \langle \mathbf{e}, M_m T_n D_{\mathbf{e}} C_f(\mathbf{e})\rangle M_m T_n \mathbf{e} \\
&= \sum_{m,n\in\mathbb{Z}} \langle \mathbf{e}, D_{\mathbf{e}} C_f(M_m T_n \mathbf{e})\rangle M_m T_n \mathbf{e} \\
&= \sum_{m,n\in\mathbb{Z}} \langle D_f C_{\mathbf{e}}(\mathbf{e}), M_m T_n \mathbf{e}\rangle M_m T_n \\
&= \sum_{m,n\in\mathbb{Z}} \langle f, M_m T_n \mathbf{e}\rangle M_m T_n \mathbf{e} = f
\end{aligned}$$

as we would expect for an involution. Our motivation for referring to the multiplication $\odot$ as a convolution is that the Zak transform of this convolution becomes a pointwise multiplication in the Zak transform domain. That is,

$$U_Z(g \odot f) = U_Z(f)\overline{U_z(\mathbf{e})}U_Z(g) = U_Z(f)U_Z(g) \tag{3.1}$$

since $\overline{U_Z(\mathbf{1}_{[0,1)}(x))} = \overline{\mathbf{1}_{[0,1)}(\mathbf{t})\mathbf{1}_{[0,1)}(w)} = \mathbf{1}_{[0,1)}(t)\mathbf{1}_{[0,1)}(w)$.

PROPOSITION 3.1. *The Zak transform acts as a *-isomorphism from Z to $L^\infty[0,1]^2$. Moreover, $\mathbf{e}$ acts as an identity for this algebra.*

PROOF. The fact that U_Z preserves the algebraic structure follows from (3.1). Computing

$$U_Z(f^*) = U_Z(\mathbf{e})\overline{U_Z(f)}U_Z(\mathbf{e}) = \overline{U_Z(f)}$$

shows that U_Z respects the involution.

Finally since $\mathcal{G}(\mathbf{e}, 1, 1)$ is an orthonormal basis, $\mathbf{e} \odot f = D_{\mathbf{e}}C_{\mathbf{e}}(f) = f = D_f C_{\mathbf{e}}(\mathbf{e}) = f \odot \mathbf{e}$ and $\mathbf{e}$ acts as the identity for the convolution. □

In view of Theorem 2.2 it is easy to see that Gabor frames in this critically sampled case correspond exactly to the invertible elements of the the C^*-algebra $L^\infty[0,1]^2$. One quick result from looking at the Gabor systems which are Bessel as the C^*-algebra is that we have a way to compute dual frames from standard techniques in Banach algebras.

PROPOSITION 3.2. *If $g \in Z$ and $\|g - \mathbf{e}\|_Z < 1$ then g is invertible. Moreover,*

$$\begin{aligned}(g)^{-1} &= \mathbf{e} + \sum_{k=1}^{\infty}(\mathbf{e} - g)^k \\ &= \mathbf{e} + (\mathbf{e} - g) + (\mathbf{e} - g) \odot (\mathbf{e} - g) + (\mathbf{e} - g) \odot (\mathbf{e} - g) \odot (\mathbf{e} - g) + \cdots \\ &= \mathbf{e} + (\mathbf{e} - g) + D_{\mathbf{e}-g}C_{\mathbf{e}}(\mathbf{e} - g) + D_{\mathbf{e}-g}C_{\mathbf{e}}\left(D_{\mathbf{e}-g}C_{\mathbf{e}}(\mathbf{e} - g)\right) + \cdots\end{aligned}$$

and $\mathcal{G}((g^{-1})^*, 1, 1)$ is the dual frame of $\mathcal{G}(g, 1, 1)$.

PROOF. The invertibility of g and the series representation of g^{-1} above follow directly from techniques of Banach Algebras. We will present these details in the more complicated oversampled case but since they are so similar we omit them here. So it remains to show that $(g^{-1})^*$ is the window for the dual frame generated by the window g. By the definition of inverse we have $g \odot g^{-1} = \mathbf{e}$ and $g \odot g^{-1} \odot f = f$. Translating this into operator notation we get

$$D_g C_{\mathbf{e}} D_{g^{-1}} C_{\mathbf{e}} = I.$$

Utilizing the commuting relations from 2.1 and recalling $(g^{-1})^* = D_{\mathbf{e}}C_{g^{-1}}(\mathbf{e})$ we have

$$\begin{aligned}C_{\mathbf{e}}D_{g^{-1}}C_{\mathbf{e}}(f) &= \{\langle D_{g^{-1}}C_{\mathbf{e}}(f), M_m T_n \mathbf{e}\rangle\}_{m,n\in\mathbb{Z}} \\ &= \{\langle f, D_{\mathbf{e}}C_{g^{-1}}M_m T_n \mathbf{e}\rangle\}_{m,n\in\mathbb{Z}} \\ &= \{\langle f, M_m T_n D_{\mathbf{e}}C_{g^{-1}}(\mathbf{e})\rangle\}_{m,n\in\mathbb{Z}} \\ &= C_{D_{\mathbf{e}}C_{g^{-1}(\mathbf{e})}} = C_{(g^{-1})^*}(f).\end{aligned} \tag{3.2}$$

Hence, $D_g C_{(g^{-1})^*} = I$ which implies $\mathcal{G}((g^{-1})^*, 1, 1)$ is the dual frame of $\mathcal{G}(g, 1, 1)$. □

4. Oversampled case ab<1

In the critically sampled case the way we created our convolution was to pick a Gabor system which was an orthonormal basis, but in the oversampled case no such orthonormal basis exists. This fact, known as the Balian-Low Theorem, was discovered by Balian [**1**] and separately by Low [**11**] and is one of prime tenets of a branch of harmonic analysis known as time-frequency analysis. Given this restriction, instead of picking a Gabor system which is an orthonormal basis we shall chose one which is a Parseval frame. That is, for a fixed $ab < 1$, chose $\alpha \in L^2(\mathbb{R})$ so $D_\alpha C_\alpha = I$. In reality we don't even need it to be a Parseval frame for any frame will do but choosing a Parseval frame will simplify the notation. We create the multiplication for Bessel systems of the form $\mathcal{G}(g,a,b)$ in the same way as before, i.e.,

$$f \odot_\alpha g = D_f C_\alpha(g) = \sum_{m,n\in\mathbb{Z}} \langle g, M_{mb}T_{na}\alpha\rangle M_{mb}T_{na}f.$$

As we pointed out in the introduction, $\mathcal{G}(g,a,b)$ is a Bessel system iff $\|C_g\|_{op} = \|D_g\|_{op} < \infty$. So we will define the space $Z_{ab} \subset L^2(\mathbb{R})$ with the norm $\|g\|_{Z_{ab}} = \|C_g\|_{op} = \|D_g\|_{op}$.

THEOREM 4.1. *Fix $ab < 1$, ab rational, and let $\mathcal{G}(\alpha,a,b)$ be any Parseval frame. Then the space of functions so that $\mathcal{G}(g,a,b)$ is a Bessel system, Z_{ab}, is a Banach algebra under $\odot_\alpha$ multiplication.*

PROOF. We recall the fundamental property of the composition of the operators which makes things work:

$$D_g C_f(M_{mb}T_{na}h) = M_{mb}T_{na}D_g C_f(h).$$

So an adjoint version of (3.2) above give us for all $g,h,k \in Z_{ab}$ and $c \in \ell^2(\mathbb{Z}^2)$ that

$$\begin{aligned} D_g C_h D_k(c) &= D_g C_h \sum_{m,n\in\mathbb{Z}} c_{m,n} E_{mb} T_{na} k \\ &= \sum_{m,n\in\mathbb{Z}} c_{m,n} D_g C_h E_{mb} T_{na} k \\ &= \sum_{m,n\in\mathbb{Z}} c_{m,n} E_{mb} T_{na} D_g C_h(k) = D_{D_g C_h(k)}(c). \end{aligned} \tag{4.1}$$

Since α is a Parseval frame, $\|\alpha\|_{Z_{ab}} = 1$, and

$$\begin{aligned} \|f \odot_\alpha g\|_{Z_{ab}} = \|D_{D_f C_\alpha(g)}\|_{op} &= \|D_f C_\alpha D_g\|_{op} \\ &\le \|f\|_{Z_{ab}}\|\alpha\|_{Z_{ab}}\|g\|_{Z_{ab}} = \|f\|_{Z_{ab}}\|g\|_{Z_{ab}}. \end{aligned}$$

The following calculations, which follow easily form the linearity of the operators and the identity in (4.1),

$$(f+g)\odot_\alpha h = D_{f+g}C_\alpha(h) = D_f C_\alpha(h) + D_g C_\alpha(h) = f\odot_\alpha h + g \odot_\alpha h$$

$$\begin{aligned} \text{and } (f\odot_\alpha g)\odot_\alpha h &= D_{(f\odot_\alpha g)}C_\alpha(h) \\ &= D_{D_f C_\alpha(g)} C_\alpha(h) \\ &= D_f C_\alpha D_g C_\alpha(h) = f \odot_\alpha (g \odot_\alpha h) \end{aligned}$$

verify that the multiplication is both distributive and associative.

Completeness in the rationally oversampled case follows from the $L^\infty[0,1]^2$ norm we used in the $a = b = 1$ case. We illustrate this with the $a = 1$, $b = \frac{1}{2}$ case. In this case $\mathbf{\Phi}^g(t,w)\mathbf{\Phi}^g(t,w)^* = |U_Z(g)(t,w)|^2 + |U_Z(g)(t-\dfrac{1}{2},w)|^2$ and

$$\begin{aligned} \|g\|_{Z_{1,\frac{1}{2}}} &= \operatorname*{ess\,sup}_{t,w\in[0,1]^2} (|U_Z(g)(t,w)|^2 + |U_Z(g)(t-\frac{1}{2},w)|^2)^{\frac{1}{2}} \\ &= \|(|U_Z(g)(t,w)|^2 + |U_Z(g)(t-\frac{1}{2},w)|^2)^{\frac{1}{2}}\|_{L^\infty[0,1]^2}. \end{aligned}$$

Now consider a sequence $\{g_n\}$ that is Cauchy in the $\|\cdot\|_{Z_{1,\frac{1}{2}}}$ norm. The definition of the norm above and gives us $\operatorname{ess\,sup}_{t,w\in[0,1]^2} |U_Z(g_n)(t,w)| < \infty$ for each n and

$$\operatorname*{ess\,sup}_{t,w\in[0,1]^2} |U_Z(g_n - g_m)(t,w)| \to 0.$$

Now by the linearity of U_Z we have

$$\operatorname*{ess\,sup}_{t,w\in[0,1]^2} |U_Z(g_n)(t,w) - U_Z(g_m)(t,w)| \to 0.$$

This implies that $\{U_Z(g_n)(t,w)\}$ is a Cauchy sequence in $L^\infty[0,1]^2$ and the rest follows.

For the more general case ab rational case one must take the supremum over the eigenvalues of the matrix $\mathbf{\Phi}^g(t,w)\mathbf{\Phi}^g(t,w)^*$. This will also reduce to properties of Zak transform of g and the completeness of $L^\infty[0,1]^2$. □

Similar algebraic structures, such as the twisted convolutions, have been used effectively by Groechnig and Leinert [**6**] on a subalgebra of the rotation algebra to produce fundamental results about invertibility of the frame operator in the modulation spaces. This twisted convolution, however, is not dependent on any frame and therefore yields results of a different nature than those below. The function α acts as a *left identity* (but not right) for the algebra since $\alpha \odot_\alpha f = D_\alpha C_\alpha(f) = f$. In fact any dual window of $\mathcal{G}(\alpha,a,b)$ will act as a left identity. This being the case, it is easy to see that we will not have a C^*-algebra in the oversampled case since a left identity must also be a right identity in a C^*-algebra. While the multiplication $\odot_\alpha$ is associative and distributive it is not commutative. This can be seen by taking any dual window β of α with $\beta \neq \alpha$. For in this case, $\alpha \odot_\alpha \beta = \beta \neq \alpha = \beta \odot_\alpha \alpha$. Even though α is only a left identity, for simplification we will still refer to g as a *right inverse* of f if the equation $f \odot_\alpha g = \alpha$ is satisfied. At this point we will abuse notation a little and define $f^* = D_\alpha C_f(\alpha)$. Since α is not a right identity this is not an involution in the traditional sense because $(f^*)^* \neq f$. However, $(f^*)^* = f \odot_\alpha \alpha$ and the notation helps tie the oversampled case to the critically sampled one. Now we are able to make the connection between inverses of the algebra and frames in the rationally oversampled case.

PROPOSITION 4.2. *Fix $ab < 1$, ab rational, and let $\mathcal{G}(\alpha,a,b)$ be any Parseval frame. If h is a right inverse for g, with respect to the multiplication $\odot_\alpha$, then $\mathcal{G}(g,a,b)$ is a frame and $\mathcal{G}(h^*,a,b)$ is a dual frame.*

PROOF. If $h \in Z_{ab}$ and α fixed we have

$$C_\alpha D_h C_\alpha(f) = C_{D_\alpha C_h(\alpha)}(f) = C_{h^*}(f). \tag{4.2}$$

This is the oversampled version of (3.2) and again relies on the commuting relations from (2.1).

Now let h be a right inverse for g with respect to $\odot_\alpha$. Then $g \odot_\alpha h = \alpha$ and $g \odot_\alpha h \odot_\alpha f = f$. In operator form this becomes

$$I = D_g C_\alpha D_h C_\alpha = D_g C_{D_\alpha C_h(\alpha)} = D_g C_{h^*}.$$

This implies that $\mathcal{G}(g,a,b)$ is a frame and $\mathcal{G}(h^*,a,b)$ is a dual. □

Duals for Gabor systems are not unique and the dual produced in Proposition 4.2 need not be the canonical dual because it is highly dependent on α. In Theorem 4.7 we will show that we can produce the canonical dual in a similar way but first, given the relationship between duals and inverses, we present a geometric sum representation of a right inverse of an element from the algebra.

THEOREM 4.3. *Fix $ab < 1$, ab rational, and let $\mathcal{G}(\alpha, a, b)$ be any Parseval frame. Now consider the Banach algebra Z_{ab} with multiplication defined by $\odot_\alpha$. If $g \in Z_{ab}$ and $\|\alpha - g\|_{Z_{ab}} < r < 1$ then g has a right inverse*

$$h = \alpha + \sum_{k=1}^{\infty} (\alpha - g \odot_\alpha \alpha)^k.$$

That is, $g \odot_\alpha h = \alpha$.

PROOF. We only have a left identity and our Banach algebra is not commutative but our construction is essentially the same as the argument for a commutative Banach algebra with identity. We actually find a right inverse for $g \odot_\alpha \alpha$ and show that it is also a right inverse for g. In order to use the geometric series here we point out that

$$\|\alpha - g \odot_\alpha \alpha\|_{Z_{ab}} = \|(\alpha - g) \odot_\alpha \alpha\|_{Z_{ab}} \leq \|\alpha - g\|_{Z_{ab}} \|\alpha\|_{Z_{ab}} = \|\alpha - g\|_{Z_{ab}}.$$

This implies $\|h\|_{Z_{ab}} \leq \|\alpha\|_{Z_{ab}} + \sum_{k=1}^{\infty} \|\alpha - g\|_{Z_{ab}} < \infty$ and so the series converges.

Let $y = \alpha - g \odot_\alpha \alpha$ and $h_n = \alpha + \sum_{k=1}^{n} y^k$. Now we compute $g \odot_\alpha h_n$ using the facts that $g \odot_\alpha h_n = (g \odot_\alpha \alpha) \odot_\alpha h_n$ and $y = y \odot_\alpha \alpha$:

$$\begin{aligned} g \odot_\alpha h_n = (\alpha - y) \odot_\alpha h_n &= (\alpha - y) \odot_\alpha \left(\alpha + y + (y)^2 ... (y)^n\right) \\ &= \left(\alpha + y + (y)^2 ... (y)^n\right) - \left(y + (y)^2 ... (y)^{n+1}\right) \\ &= \alpha - (y)^{n+1}. \end{aligned}$$

We observe that $\|(y)^{n+1}\|_{Z_{ab}} \leq r^n$ so $(y)^{n+1} \to 0$, and the completeness of the algebra gives us that there exist an h so

$$g \odot_\alpha h = \lim_{n \to \infty} g \odot_\alpha h_n = \alpha.$$

We point out that that h is a right inverse of $g \odot_\alpha \alpha$ as well. □

This gives us the immediate corollary.

COROLLARY 4.4. *Consider $\mathcal{G}(g, a, b)$, $g \in Z_{ab}$ with ab rational and suppose there exists a normalized tight frame $\mathcal{G}(\alpha, a, b)$ such that $\|\alpha - g\|_{Z_{ab}} < r < 1$. Then $\mathcal{G}(g, a, b)$ is a frame.*

We can repeat this argument again for any left identity, i.e., any dual window of α. That is, let β be any dual window of α and let h_1 satisfy the equation $g \odot_\alpha h_1 = \beta$. By the same argument we get $g_1 = D_\alpha C_{h_1}(\alpha)$ is a dual of g as well. However, it not necessarily different from the one obtained by finding a right inverse with respect to α. This is easily seen by letting $g = \alpha$ for all duals produced in this manner will be α. Since Corollary 4.4 is a perturbation type result we would like to compare it with the more general perturbation theorem of Christensen and Heil.

THEOREM 4.5. [**2**] *Let H be a Hilbert space, $\{f_i\}_{i=1}^{\infty}$ be a frame with bounds A, B and let $\{g_i\}_{i=1}^{\infty} \subset H$. If there exists $R < A$ so that*

$$\sum_{i=1}^{\infty} |\langle h, f_i - g_i\rangle_{\mathcal{H}}|^2 \leq R\,\|h\|^2 \text{ for all } h \in H,$$

then $\{g_i\}_{i=1}^{\infty}$ is a frame with bounds $A(1-\sqrt{\frac{R}{A}})^2$ and $B(1-\sqrt{\frac{R}{A}})^2$.

The next corollary states the perturbation result of Christensen and Heil in this setting and gives a relationship between a dual for the perturbation and the dual of the original frame.

COROLLARY 4.6. *Fix a normalized tight frame $\mathcal{G}(\alpha, a, b)$ with ab rational and let g have a right inverse, g^{-1}, with respect to the algebra determined by $\odot_\alpha$. That is $g \odot_\alpha g^{-1} = \alpha$. If $\|f-g\|_{Z_{ab}} < \frac{1}{\|g^{-1}\|_{Z_{ab}}}$, then f has a right inverse and $\mathcal{G}(g, a, b)$ is a frame. Moreover, $f \odot_\alpha g^{-1}$ has a right inverse h and $\mathcal{G}((g^{-1} \odot_\alpha h)^*, a, b)$ is a dual frame of of $\mathcal{G}(f, a, b)$.*

PROOF. We first need to bound the norm to use the theorem above:

$$\begin{aligned}
\|\alpha - f \odot_\alpha g^{-1}\|_{Z_{ab}} &= \|g \odot_\alpha g^{-1} - f \odot_\alpha g^{-1}\|_{Z_{ab}} \\
&= \|(g-f) \odot_\alpha g^{-1}\|_{Z_{ab}} \leq \|g-f\|_{Z_{ab}} \|g^{-1}\|_{Z_{ab}} < 1.
\end{aligned}$$

By Theorem 4.3 above $f \odot_\alpha g^{-1}$ has a right inverse h. This implies $f \odot_\alpha g^{-1} \odot_\alpha h = \alpha$ and by the associativity of the algebra $g^{-1} \odot_\alpha h$ is a right inverse of f. The rest follows from Proposition 4.2. □

Now we show that finding a right inverse of $S_g(\alpha)$ is enough to invert the frame operator and produce the canonical dual window g^d.

THEOREM 4.7. *If $\mathcal{G}(g, a, b)$ is a frame with ab rational, $\mathcal{G}(\alpha, a, b)$ is a Parseval frame and h is the right inverse of $S_g(\alpha)$ then*

$$S_g^{-1}(f) = \sum_{m,n\in\mathbb{Z}} \langle f, M_{mb}T_{na}\alpha\rangle M_{mb}T_{na}h$$

and the canonical dual window $g^d = S_g^{-1}(g) = h \odot_\alpha g$.

PROOF. First we observe

$$S_g(f) = S_g D_\alpha C_\alpha(f) = D_{S_g(\alpha)} C_\alpha(f) = S_g(\alpha) \odot_\alpha f. \tag{4.3}$$

Hence, if h is a right inverse of $S_g(\alpha)$ under $\odot_\alpha$ multiplication we have

$$S_g(h) = \alpha \text{ which implies } S_g^{-1}(\alpha) = h.$$

Since $\mathcal{G}(\alpha, a, b)$ is a Parseval frame and S_g^{-1} commutes with M_{mb} and T_{na} we get

$$S_g^{-1}(f) = \sum_{m,n\in\mathbb{Z}} \langle f, M_{mb}T_{na}\alpha\rangle S_g^{-1}(M_{mb}T_{na}\alpha) = \sum_{m,n\in\mathbb{Z}} \langle f, M_{mb}T_{na}\alpha\rangle M_{mb}T_{na}h.$$

Now from 4.3 we have $g^d = S_g^{-1}(g) = S_g^{-1}(\alpha) \odot_\alpha g = h \odot_\alpha g$ □

Before we continue, let us state the more general result in the language of the operators associated with frames. By an argument similar to that above, in order to find the window for the canonical dual of a Gabor frame $\mathcal{G}(g,a,b)$ it is enough to find any Gabor frame $\mathcal{G}(h,a,b)$ with dual $\mathcal{G}(k,a,b)$ so that $S_g(m) = h$. Then

$$g^d = S_g^{-1}(g) = S_g^{-1}(D_h C_k(g)) = D_{S_g^{-1}(h)} C_k(g) = D_m C_k(g).$$

If we show $\|\frac{S_g(\alpha)}{B} - \alpha\|_{Z_{ab}} < 1$, then combining this with Theorems 4.3 and 4.7 we may use these methods to compute the canonical dual for any frame. Because it will be needed later we actually show something a little more general.

PROPOSITION 4.8. *If $\mathcal{G}(\alpha,a,b)$ is a Parseval frame then for any $\mathcal{G}(g,a,b)$ we have* $\|\frac{S_g^n(\alpha)}{B^n} - \alpha\|_{Z_{ab}} < 1$.

PROOF. If A and B are the frame bounds of $\mathcal{G}(g,a,b)$ we have $AI \le S_g \le BI$. So this implies $A^n I \le S_g^n \le B^n I$ and $\|\frac{S_g^n}{B^n} - I\|_{op} < 1$, where the norm here is the usual operator norm. Without loss of generality assume $B = 1$, for we may always consider $g' = \frac{g}{\sqrt{B}}$. Next we use 4.1 to observe that

$$S_g D_\alpha = D_{S_g(\alpha)},\ S_g^2 D_\alpha = D_{S_g^2(\alpha)},\ \cdots,\ S_g^n D_\alpha = D_{S_g^n(\alpha)},\ \cdots.$$

Using this and the fact that $\|D_\alpha\|_{op} = 1$ we have

$$\begin{aligned}
\|S_g^n(\alpha) - \alpha\|_{Z_{ab}} &= \|D_{S_g^n(\alpha)} - D_\alpha\|_{op} \\
&= \|S_g^n D_\alpha - D_\alpha\|_{op} \\
&= \|(S_g^n - I) D_\alpha\|_{op} \\
&\le \|(S_g^n - I)\|_{op} \|D_\alpha\|_{op} < 1.
\end{aligned}$$

□

The canonical tight frame window is represented by $g^t = S_g^{-\frac{1}{2}}(g)$. We show that this may be computed with a binomial series. Since we will get much better convergence in the next section and because the proof is not very different in nature than the proof of Theorem 4.3 we only give a sketch of it here.

THEOREM 4.9. *Let $\mathcal{G}(\alpha,a,b)$ produce a normalized tight frame with ab rational. Now consider the Banach algebra Z_{ab} with multiplication defined by $\odot_\alpha$. If $g \in Z_{ab}$ and $\|\alpha - g\|_{Z_{ab}} < r < 1$ then $g \odot_\alpha \alpha$ has a "inverse square root", i.e.,*

$$k = \alpha + \sum_{k=1}^{\infty} \binom{-\frac{1}{2}}{k} (g \odot_\alpha \alpha - \alpha)^k \text{ that is } g \odot_\alpha k \odot_\alpha k = \alpha.$$

PROOF. Similar to above we let $y = g \odot_\alpha \alpha - \alpha$ and $k_n = \alpha + \sum_{k=1}^n y^k$. To illustrate the convergence we will compute $g \odot_\alpha k_2 \odot_\alpha k_2 - \alpha$ then we generalize.

Once again it is important to point out that $y = y \odot_\alpha \alpha$.

$$\begin{aligned}
&\|g \odot_\alpha k_2 \odot_\alpha k_2 - \alpha\| \\
&\qquad = \|g \odot_\alpha \alpha \odot_\alpha k_2 \odot_\alpha k_2 - \alpha\| \\
&\qquad = \|(y+\alpha) \odot_\alpha k_2 \odot_\alpha k_2 - \alpha\| \\
&\qquad = \left\|(y+\alpha) \odot_\alpha \left(\alpha - \frac{1}{2}y + \frac{3}{8}y \odot_\alpha y\right) \odot_\alpha \left(\alpha - \frac{1}{2}y + \frac{3}{8}y \odot_\alpha y\right) - \alpha\right\| \\
&\qquad = \left\|-\frac{5}{8}y^3 - \frac{15}{64}y^5 + \frac{9}{64}y^5\right\| < 3r^3.
\end{aligned}$$

For k_n we point out that the first n terms of the series cancel and we are left with the powers $y^{n+1}, y^{n+2}, \cdots, y^{2n+1}$. By the triangle inequality we have

$$\|g \odot_\alpha \alpha - k_n \odot_\alpha k_n\| \le (n+1)r^{n+1},$$

so $\|g \odot_\alpha \alpha - k_n \odot_\alpha k_n\| \to 0$. □

Finally we see that this allows us to find $S_g^{-\frac{1}{2}}$. Given $\mathcal{G}(g,a,b)$ is a frame by Proposition 4.8 we can know $\|S_{g'}(\alpha) - \alpha\|_{Z_{ab}} < 1$. Next find k such that $S_{g'}(\alpha) \odot_\alpha k \odot_\alpha k = \alpha$ so we get

$$f = S_{g'}(\alpha) \odot_\alpha k \odot_\alpha k \odot_\alpha f = S_{g'}(D_k C_\alpha D_k C_\alpha)(f)$$

Hence, $S_{g'}^{-\frac{1}{2}} = D_k C_\alpha$ and $g^t = D_k C_\alpha(g) = k \odot_\alpha g$.

5. Newton's method

The sections above give representations for dual windows and canonical tight frame windows associated with a Gabor system in terms of a sum which converges geometrically in norm. While the methods above produce a power series representation, the convergence is somewhat slow. Inspired by Newton's method, we give other algorithms for finding dual windows dependent on a fixed Parseval frame, the canonical dual frame window and the canonical tight frame window for a Gabor system .

Oversampled systems, duals and g^d.

Here we apply a Newton's method to the oversampled case to find duals. One way of applying Newton's method to find an inverse for $c \in \mathbb{R}$ is to consider the function

$$f(x) = x^{-1} - c \text{ and the sequence } x_{n+1} = x_n - \frac{f(x_n)}{f'(x_n)} = 2x_n - x_n^2 c.$$

If the algebra were commutative and had an identity this approach would be fine but since this is not the case we compensate for this with some minor adjustments. We recall, by Proposition 4.2, that to find a dual for a system $\mathcal{G}(g,a,b)$ it is enough to find a right inverse h and then $\mathcal{G}(h^*,a,b)$ is a dual frame. So first we will create a sequence h_n that converge quadratically to the right inverse h and show that this implies the associated sequence h_n^* converges quadratically to a dual window h^*.

Now we point out that if $g \odot_\alpha h = \alpha$ then $g \odot_\alpha h \odot_\alpha \alpha = \alpha$ so consider any $h_0 = h_0 \odot_\alpha \alpha$ so that $\|g \odot_\alpha h_0 - \alpha\|_{Z_{ab}} = r < 1$. We will let h_{n+1} be defined by

$$h_{n+1} = 2h_n - h_n \odot_\alpha g \odot_\alpha h_n = 2h_n - D_{h_n} C_\alpha \left(D_g C_\alpha(h_n)\right).$$

Note that since $h_0 = h_0 \odot_\alpha \alpha$ for each n we have $h_n = h_n \odot_\alpha \alpha$ so computing the square yields

$$\begin{aligned}(g \odot_\alpha h_n - \alpha)^2 &= (g \odot_\alpha h_n \odot_\alpha \alpha - \alpha)^2 \\ &= (g \odot_\alpha h_n \odot_\alpha \alpha - \alpha) \odot_\alpha (g \odot_\alpha h_n \odot_\alpha \alpha - \alpha) \\ &= (g \odot_\alpha h_n) \odot_\alpha (g \odot_\alpha h_n) \odot_\alpha \alpha - 2(g \odot_\alpha h_n \odot_\alpha \alpha) - \alpha.\end{aligned}$$

Now we may use the identity above to get the following inequality :

$$\begin{aligned}\|g \odot_\alpha h_{n+1} - \alpha\|_{Z_{ab}} &= \|g \odot_\alpha h_{n+1} \odot_\alpha \alpha - \alpha\|_{Z_{ab}} \\ &= \|g \odot_\alpha (2h_n \odot_\alpha \alpha - h_n \odot_\alpha g \odot_\alpha h_n \odot_\alpha \alpha) \odot_\alpha \alpha - \alpha\|_{Z_{ab}} \\ &= \|(g \odot_\alpha h_n) \odot_\alpha (g \odot_\alpha h_n) \odot_\alpha \alpha - 2(g \odot_\alpha h_n \odot_\alpha \alpha) - \alpha\|_{Z_{ab}} \\ &\leq \|g \odot_\alpha h_n \odot_\alpha \alpha - \alpha\|_{Z_{ab}}^2 = \|g \odot_\alpha h_n - \alpha\|_{Z_{ab}}^2.\end{aligned}$$

Hence, we have h_n converges quadratically in the Z_{ab} norm to the right inverse h. Now consider the sequence h_n^*. We will show that $D_g C_{h_n^*}$ converges quadratically to the identity in the operator norm and therefore, h_n^* converges to a dual window of the frames $\mathcal{G}(g, a, b)$. Remembering that $I = D_\alpha C_\alpha$, the same techniques as in 4.1 and 4.2 yield

$$\begin{aligned}\|D_g C_{h_n^*} - I\|_{op} &= \|D_g C_{D_\alpha C_{h_n}(\alpha)} - D_\alpha C_\alpha\|_{op} \\ &= \|D_g C_\alpha D_{h_n} C_\alpha - D_\alpha C_\alpha\|_{op} \\ &= \|(D_g C_\alpha D_{h_n} C - D_\alpha) C_\alpha\|_{op} \\ &\leq \|D_g C_\alpha D_{h_n} C - D_\alpha\|_{op} \\ &= \|D_{D_g C_\alpha(h_n)} - D_\alpha\|_{op} \\ &= \|g \odot_\alpha h_n - \alpha\|_{Z_{ab}} < r^{2^n}.\end{aligned} \tag{5.1}$$

The above inequality gives a nice bound for how close the composition of analysis and synthesis operators from the two systems are at each step of the iteration.

Because of Theorem 4.7, in order to find the canonical dual window of g it is enough to find the right inverse h of $S_g(\alpha)$ and then $g^d = h \odot_\alpha g$. In order to apply the algorithm we need to show that there exist some h_0 so that $\|S_g(\alpha) \odot_\alpha h_0 - \alpha\| < 1$. Using Proposition 4.8 we see we can let $h_0 = \dfrac{S_g(\alpha)}{B^2}$. Then to find approximations of the canonical dual window we let $g_n^d = h_n \odot_\alpha g$.

The critically sampled system, $S_g^{-\frac{1}{2}}$ and g^t.

As is normally the case, the critically sampled systems are not as interesting as the oversampled ones. However, we will compute the canonical tight dual here first to try and help motivated the algorithm in the oversampled case. Recall, $\mathbf{e} = \mathbf{1}_{[0,1)}$ is an identity for this algebra.

Let k be a positive invertible element in the Z Algebra, i.e., there exist $A, B > 0$ such that $A < Z(k) < B$. Then adapting Newton's method for the real function $f(x) = \frac{1}{x^2} - k$ gives

$$\gamma_{n+1} = \frac{3\gamma_n - k \odot (\gamma_n)^3}{2} = \frac{3\gamma_n - D_k C_{\mathbf{e}} D_{\gamma_n} C_{\mathbf{e}} D_{\gamma_n} C_{\mathbf{e}}(\gamma_n)}{2}.$$

For proper choice of γ_0 we get $\gamma_n \to \gamma$ and $D_k C_{\mathbf{e}} D_\gamma C_{\mathbf{e}} D_\gamma C_{\mathbf{e}} = I$. Indeed, if $\|k \odot (\gamma_0)^2 - \mathbf{e}\| = \|D_k C_{\mathbf{e}} D_{C_{\mathbf{e}}\gamma_0}(\gamma_0) - \mathbf{e}\|_Z < r < 1$ then

$$\left\|k \odot (\gamma_{n+1})^2 - \mathbf{e}\right\|_Z = \left\|k \odot \frac{3\gamma_n - k \odot (\gamma_n)^3}{2} \odot \frac{3\gamma_n - k \odot (\gamma_n)^3}{2} - \mathbf{e}\right\|_Z$$

$$= \left\|\tfrac{1}{4}(k \odot (\gamma_n)^2 - 4\mathbf{e}) \odot (k \odot (\gamma_n)^2 - \mathbf{e})^2\right\|_Z \tag{5.2}$$

$$= \left\|\tfrac{1}{4}(D_k C_{\mathbf{e}} D_{\gamma_n} C_{\mathbf{e}}(\gamma_n) - 4\mathbf{e}) \odot (D_k C_{\mathbf{e}} D_{\gamma_n} C_{\mathbf{e}}(\gamma_n) - \mathbf{e})^2\right\|_Z$$

$$\leq \left\|\tfrac{1}{4}(D_k C_{\mathbf{e}} D_{\gamma_n} C_{\mathbf{e}}(\gamma_n) - 4\mathbf{e})\right\|_Z \left\|(D_k C_{\mathbf{e}} D_{\gamma_n} C_{\mathbf{e}}(\gamma_n) - \mathbf{e})^2\right\|_Z$$

$$\leq \frac{1}{4}\left(\|D_k C_{\mathbf{e}} D_{\gamma_n} C_{\mathbf{e}}(\gamma_n) - \mathbf{e} - 3\mathbf{e}\|_Z\right) r^2$$

$$\leq \frac{1}{4}\left(\|D_k C_{\mathbf{e}} D_{\gamma_n} C_{\mathbf{e}}(\gamma_n) - \mathbf{e}\|_Z + \|3\mathbf{e}\|_Z\right) r^2$$

$$\leq \frac{1}{4}\left(\|D_k C_{\mathbf{e}} D_{\gamma_n} C_{\mathbf{e}}(\gamma_n) - \mathbf{e}\|_Z + 3\right) r^2 < \frac{r+3}{4} r^2 < r^2. \tag{5.3}$$

Where the equality in 5.2 follows from factoring in a commutative Banach algebra and the inequality in 5.3 follow from the fact that $r < 1$. So $k \odot (\gamma)^2 = e$ and we have an algorithm that converges quadratically to the inverse square root of an element from the algebra. We again observe $S_g = D_{S_g(\mathbf{e})} C_{\mathbf{e}}$. Hence, if $k = S_g(\mathbf{e})$ then $S_g D_\gamma C_e D_\gamma C_e = I$ and $S_g^{-\frac{1}{2}} = D_\gamma C_e$.

Oversampled systems, $S_g^{-\frac{1}{2}}$ and g^t.

Finally, we produce a Newton's Method to find $S_g^{-\frac{1}{2}}$ in the oversampled case. Things become complicated because we do not have a commutative algebra so in this case we will specify the starting point of the algorithm and get around this problem.

By Proposition 4.8 we know $\|\frac{S_g^3(\alpha)}{B^3} - \alpha\| < 1$. So if we let $k = S_g(\alpha)$ as we did above and $\gamma_0 = \frac{S_g(\alpha)}{B^3} = \frac{k}{B^3}$ the condition $\|k \odot_\alpha (\gamma_0)^2 - \alpha\|_{Z_{ab}} = \|k \odot_\alpha \gamma_0 \odot_\alpha \gamma_0 - \alpha\|_{Z_{ab}} < r < 1$ is satisfied. Now we preceded the same as we did above letting the iteration be defined by:

$$\gamma_{n+1} = \frac{3\gamma_n - \gamma_0 \odot_\alpha \gamma_n^3}{2}.$$

We have delicately maneuvered around the issue of non-commutativity because in this case the iterations above are just polynomials in k. This and the fact that

$$k = S_g(\alpha) = S_g(\alpha) \odot_\alpha \alpha = k \odot_\alpha \alpha$$

allows us to proceed exactly as above and get

$$\begin{aligned}\left\|(\gamma_0 \odot_\alpha (\gamma_{n+1})^2 - \alpha)\right\|_{Z_{ab}} &= \left\|\tfrac{1}{4}(\gamma_0 \odot_\alpha (\gamma_n)^2 - 4\alpha)(\gamma_0 \odot_\alpha (\gamma_n)^2 - \alpha)^2\right\|_{Z_{ab}} \\ &\leq \frac{1}{4}\left(\left\|\gamma_0 \odot_\alpha (\gamma_n)^2 - \alpha\right\|_{Z_{ab}} + 3\right) r^2 < r^2.\end{aligned}$$

In operator notation this becomes

$$S_g(D_\gamma C_\alpha D_\gamma C_\alpha)(f) = D_{S_g(\alpha)} C_\alpha (D_\gamma C_\alpha D_\gamma C_\alpha)(f) = \gamma_0 \odot_\alpha (\gamma)^2 \odot_\alpha f = f$$

and therefore $S_g^{-\frac{1}{2}} = D_\gamma C_\alpha$. If we wish to have an algorithm that computes g^t it is enough to consider $g_n^t = \gamma_n \odot_\alpha g$. This follows form

$$g^t = S_g^{-\frac{1}{2}}(g) = D_\gamma C_\alpha(g) = \gamma \odot_\alpha g$$

and the convergence follows from

$$\|g_n^t - g^t\|_{Z_{ab}} = \|(\gamma_n - \gamma) \odot_\alpha g\|_{Z_{ab}} \leq \|(\gamma_n - \gamma)\|_{Z_{ab}} \|g\|_{Z_{ab}}.$$

References

[1] R. Balian. Un principe d'incertitude fort en théorie du signal ou en mécanique quantique. *C. R. Acad. Sci. Paris Sér. II Méc. Phys. Chim. Sci. Univers Sci. Terre*, 292(20):1357–1362, 1981.

[2] O. Christensen and C. Heil. Perturbations of Banach frames and atomic decompositions. *Math. Nachr.*, 185:33–47, 1997.

[3] R. J. Duffin and A. C. Schaeffer. A class of nonharmonic Fourier series. *Trans. Amer. Math. Soc.*, 72:341–366, 1952.

[4] I. M. Gel'fand. Eigenfunction expansions for equations with periodic coefficients. *Dokl. Akad. Nauk. SSR*, 41(8):2679–2687, 1959.

[5] K. Gröchenig. *Foundations of time-frequency analysis*. Applied and Numerical Harmonic Analysis. Birkhäuser Boston Inc., Boston, MA, 2001.

[6] K. Gröchenig and M. Leinert. Wiener's lemma for twisted convolution and Gabor frames. *Preprint*, 2003.

[7] A. J. E. M. Janssen. Bargmann transform, Zak transform, and coherent states. *J. Math. Phys.*, 23(5):720–731, 1982.

[8] A. J. E. M. Janssen. The Zak transform: a signal transform for sampled time-continuous signals. *Philips J. Res.*, 43(1):23–69, 1988.

[9] A. J. E. M. Janssen. Analysis of some fast algorithms to compute canonical windows for Gabor frames. *Preprint*, 2003.

[10] A. J. E. M. Janssen and T. Strohmer. Characterization and computation of canonical tight windows for Gabor frames. *J. Fourier Anal. Appl.*, 8(1):1–28, 2002.

[11] F. Low. Complete sets of wave packets. In *A Passion for Physics-Essay in Honor of Geoffery Chew*, pages 17–22. World Scientific, Singapore, 1985.

[12] J. Zak. Finite translations in solid state physics. *Phys. Rev. Lett.*, 19:1385–1397, 1967.

[13] M. Zibulski and Y. Y. Zeevi. Oversampling in the Gabor scheme. *IEEE Trans. Signal Process.*, 41(8):2679–2687, 1993.

Department of Mathematics, University of North Carolina at Wilmington, Wilmington, North Carolina 28403, USA

E-mail address: `lammersm@uncw.edu`

Contemporary Mathematics
Volume **345**, 2004

Wavelets, wavelet sets, and linear actions on $\mathbb{R}^n$

Gestur Ólafsson and Darrin Speegle

Abstract. Wavelet and frames have become a widely used tool in mathematics, physics, and applied science during the last decade. This article gives an overview over some well known results about the continuous and discrete wavelet transforms and groups acting on $\mathbb{R}^n$. We also show how this action can give rise to wavelets, and in particular, MSF wavelets in $L^2(\mathbb{R}^n)$.

Introduction

The classical wavelet system consists of a single function $\psi \in L^2(\mathbb{R})$ such that $\{ 2^{j/2}\psi(2^j x + k) \mid j, k \in \mathbb{Z} \}$ is an orthonormal basis for $L^2(\mathbb{R})$. There has been quite a bit of recent interest in relaxing various aspects of the definition of wavelets, in particular in higher dimensions. For example, one can allow multiple functions $\psi^i, \dots, \psi^L$, an arbitrary matrix of dilations, and an arbitrary lattice of translations. One could relax even further to allow a group of dilations, or perhaps even just a set of dilations and translations. A first question one would ask, then, is: for which collections of dilations and translations do there exist wavelets? We will begin by reviewing some well-known results concerning this central question. Then, we will show that there is a fundamental connection between the papers of Dai, Diao, Gu and Han [**15**], Fabec and Ólafsson [**21**], Laugesen, Weaver, Weiss and Wilson [**46**], and Wang [**57**]. One argument that a survey paper such as this one is useful is that, even though these eleven authors are active in the field, there is only one cross-reference of the above papers in the references of the other papers.

We now describe briefly the connection between the papers listed above. All four papers are concerned with constructing reproducing systems consisting of dilations and translations of a function. That is, they consider triples $(\mathcal{D}, \mathcal{T}, M)$, where $\mathcal{D}$ is some collection of invertible matrices, $\mathcal{T}$ is some collection of points in $\mathbb{R}^n$, and M a non-trivial closed subspace of $L^2(\mathbb{R}^n)$. Then, they ask whether there is a function ψ such that $\{ \psi_{a,k} = |\det a|^{1/2} \psi(a(x) + k) \mid a \in \mathcal{D}, k \in \mathcal{T} \}$ is a frame, normalized tight frame, or even an orthonormal basis for M.

2000 *Mathematics Subject Classification.* 42C40, 43A85.

Key words and phrases. Wavelet transform, frames, Lie groups, square integrable representations, reductive groups.

Research supported by NSF grants DMS-0070607 and DMS-0139783.

In [**15**], it is assumed that $\mathcal{D} = \{ a^j \mid j \in \mathbb{Z} \}$ for some expansive matrix a, that $\mathcal{T} = \mathbb{Z}^n$, and that M is an a-invariant subspace of $L^2(\mathbb{R}^n)$. In [**21**], the assumptions are that, $\mathcal{D}$ is constructed as a subset of a particular type of group H, that $\mathcal{T}$ is a full rank lattice depending on H, and finally that M is of the form $M = \{ f \in L^2(\mathbb{R}^n) \mid \mathrm{Supp}(\hat{f}) \subseteq \overline{\mathcal{O}} \}$, where $\mathcal{O} \subset \mathbb{R}^n$ is an open H-orbit. In [**46**], it is assumed that $\mathcal{D}$ is a group, $\mathcal{T} = \mathbb{Z}^n$, and $M = L^2(\mathbb{R}^n)$. In [**57**], it is assumed that $\mathcal{D}$ and $\mathcal{T}$ satisfy non-algebraic conditions relating to the existence of fundamental regions (see Section 1 for details) and $M = L^2(\mathbb{R}^n)$. Moreover, all four papers—either explicitly or implicitly—are concerned primarily with the existence of functions of the form $\hat{\psi} = \chi_\Omega$.

When put in this general framework, it becomes clear that the four papers are related in spirit and scope. What we will show below is that they are also related in that results in [**15**] can be used to remove technical assumptions from results in [**57**]. The improved results in [**57**] can then be used to improve the results in [**46**] and [**21**]. We will improve the results in [**21**] by removing the dependence of the lattice on the group, and by constructing an orthonormal basis where a normalized tight frame was constructed before. The proof of the main Theorem in [**21**] will also be simplified. Finally, we improve the results in [**46**] by replacing normalized tight frame system with a wavelet system.

We will attempt to make these technical improvements to the theorems in these papers with a minimal amount of technical work. In particular, where possible, we will apply theorem quoting proofs. The primary exception to this is Theorem 1.18, where we essentially need to check that the details of an argument in [**18**] go through in a slightly more general setting.

1. Wavelet sets

We start this section by recalling some simple definitions and facts about wavelets, wavelet sets, and tilings. For a measurable set $\Omega \subseteq \mathbb{R}^n$ we denote by χ_Ω the indicator function of the set Ω and by $|\Omega| = \int \chi_\Omega(x)\, dx$ the measure of Ω.

DEFINITION 1.1. Let (M, μ) be a measure space. A countable collection $\{\Omega_j\}$ of subsets of M is a (measurable) tiling of M if $\mu(M \setminus \bigcup_j \Omega_j) = 0$, and $\mu(\Omega_i \cap \Omega_j) = 0$ for $i \neq j$.

DEFINITION 1.2. Let $\mathcal{T} \subset \mathbb{R}^n$ and $\mathcal{D} \subset \mathrm{GL}(n, \mathbb{R})$. We say that $\mathcal{D}$ is a *multiplicative tiling set* of $\mathbb{R}^n$ if there exists a set $\Omega \subset \mathbb{R}^n$ of positive measure such that $\{ d\Omega \mid d \in \mathcal{D} \}$ is a tiling of $\mathbb{R}^n$. The set Ω is said to be a *multiplicative $\mathcal{D}$-tile.* We say $\mathcal{D}$ is a *bounded multiplicative tiling set* of $\mathbb{R}^n$ if there is a multiplicative $\mathcal{D}$-tile Ω which is bounded and such that $0 \notin \overline{\Omega}$.

Similarly, we say that $\mathcal{T}$ is an *additive tiling set* of $\mathbb{R}^n$ if there exists a set $\Omega \subset \mathbb{R}^n$ such that $\{ \Omega + x \mid x \in \mathcal{T} \}$ is a tiling of $\mathbb{R}^n$. The set Ω is said to be an *additive $\mathcal{T}$-tile.* Again, we add the word bounded if Ω can be chosen to be a bounded set (with no restriction on being bounded away from 0).

A set Ω is a *$(\mathcal{D}, \mathcal{T})$-tiling set* if it is a $\mathcal{D}$ multiplicative tiling set and a $\mathcal{T}$ additive tiling set.

Note that this definition does not coincide with the definition of Wang [**57**]. Wang defines a multiplicative tiling set to be what we have defined to be a bounded multiplicative tiling set. We feel that boundedness properties of $\mathcal{D}$-tiles are interesting properties, but they should not be part of a definition of tiling.

Multiplicative and additive tilings of $\mathbb{R}^n$ show up in wavelet theory and other branches of analysis in a natural way.

DEFINITION 1.3. A function $\varphi \in L^2(\mathbb{R}^n)$ is called a *wavelet* if there exists a subset $\mathcal{D} \subset \mathrm{GL}(n, \mathbb{R})$ and a subset $\mathcal{T} \subset \mathbb{R}^n$ such that

$$\mathcal{W}(\varphi; \mathcal{D}, \mathcal{T}) := \{ \, |\det d|^{1/2} \varphi(dx + k) \mid d \in \mathcal{D}, \, k \in \mathcal{T} \, \}$$

forms an orthonormal basis for $L^2(\mathbb{R}^n)$. The set $\mathcal{D}$ is called the *dilation set* for φ, the set $\mathcal{T}$ is called the *translation set* for φ, and we say that φ is a $(\mathcal{D}, \mathcal{T})$*-wavelet.*

Normalize the Fourier transform by

$$\mathcal{F}(f)(\lambda) = \hat{f}(\lambda) = \int_{\mathbb{R}^n} f(x) e^{-2\pi i (\lambda, x)} \, dx \, .$$

We set $f^{\vee}(x) = \hat{f}(-x)$. Then $f = (\hat{f})^{\vee}$. For simplicity we set $e_\lambda(x) = e^{2\pi i (\lambda, x)}$.

DEFINITION 1.4. Let $\Omega \subset \mathbb{R}^n$ be measurable with positive, but finite measure. We say that Ω is *a wavelet set* if there exists a pair $(\mathcal{D}, \mathcal{T})$, with $\mathcal{D} \subset \mathrm{GL}(n, \mathbb{R})$ and $\mathcal{T} \subset \mathbb{R}^n$ such that $\chi_\Omega^\vee$ is a $(\mathcal{D}, \mathcal{T})$-wavelet. If $\chi_\Omega^\vee$ is a $(\mathcal{D}, \mathcal{T})$-wavelet, then we say that Ω is a $(\mathcal{D}, \mathcal{T})$*-wavelet set.*

DEFINITION 1.5. A measurable set $\Omega \subset \mathbb{R}^n$ with finite positive measure is called *a spectral set* if there exists a set $\mathcal{T} \subset \mathbb{R}^n$ such that the sequence of functions $\{e_\lambda\}_{\lambda \in \mathcal{T}}$ forms an orthogonal basis for $L^2(\Omega)$. If this is the case we say that $\mathcal{T}$ is the *spectrum of* Ω, and say that $(\Omega, \mathcal{T})$ is a spectral pair

Now, after given this list of definitions, let us recall some results, questions, and conjectures on how these concepts are tied together. A first result, which has appeared in several places [**16, 36, 40**] is

THEOREM 1.6. *A measurable set $\Omega \subset \mathbb{R}$ is a wavelet set for the pair $\mathcal{D} = \{ 2^n \mid n \in \mathbb{Z} \}$ and $\mathcal{T} = \mathbb{Z}$ if and only if Ω is a $(\mathcal{D}, \mathcal{T})$-tiling set.*

The proof of Theorem 1.6 given in [**16**], for example, works to prove the following

THEOREM 1.7. *Let a be an invertible matrix. A measurable set $\Omega \subset \mathbb{R}^n$ is a wavelet set for the pair $\mathcal{D} = \{ a^n \mid n \in \mathbb{Z} \}$ and the full rank lattice $\mathcal{T}$ if and only if Ω is a $(\mathcal{D}, \mathcal{T})$-tiling set.*

For the general case we have now the following two related questions:

QUESTION 1 (Wang, [**57**]). For which sets $\mathcal{D} \subset \mathrm{GL}(n, \mathbb{R})$, $\mathcal{T} \subset \mathbb{R}^n$ do there exist $(\mathcal{D}, \mathcal{T})$-wavelets?

QUESTION 2. For which sets $\mathcal{D} \subset \mathrm{GL}(n, \mathbb{R})$, $\mathcal{T} \subset \mathbb{R}^n$ do there exist $(\mathcal{D}, \mathcal{T})$-wavelet sets?

Clearly, if there exists a $(\mathcal{D}, \mathcal{T})$-wavelet set, then there exists a $(\mathcal{D}, \mathcal{T})$-wavelet, but, what is interesting, is that the converse may also be true. In particular, there are currently no examples known of sets $(\mathcal{D}, \mathcal{T})$ for which there exist wavelets, but for which there do not exist $(\mathcal{D}, \mathcal{T})$-wavelet sets. Therefore we can state the third natural question:

QUESTION 3. Is it true that if there exists a $(\mathcal{D}, \mathcal{T})$-wavelet, then there exists a $(\mathcal{D}, \mathcal{T})$-wavelet set?

So far, all evidence points to a positive answer for question 3. (Though, we should point out that question 3 has mostly been thought about in the case that $\mathcal{D}$ is a singly generated group and $\mathcal{T}$ is a full rank lattice, so it is possible that there is a relatively easy counterexample to the question posed in this generality.) When $\mathcal{D}$ is generated by a single matrix a and $\mathcal{T}$ is a full rank lattice, it is known [**17**] that if a is expansive, then there exist $(\mathcal{D}, \mathcal{T})$-wavelet sets. Moreover, it is also known [**11, 13, 14**] in the expansive case that there exist $(\mathcal{D}, \mathcal{T})$-wavelets that do not come from a wavelet set if and only if there is a $j \neq 0$ such that $(a^T)^j(\mathcal{T}^*) \cap \mathcal{T}^* \neq \{0\}$, where $\mathcal{T}^*$ is the dual lattice defined by $\mathcal{T}^* = \{\alpha \in \mathbb{R}^n \mid \forall \beta \in \mathcal{T} : (\alpha, \beta) \in \mathbb{Z}\}$. In particular, for *most* pairs of this type, the *only* wavelets that exist come from wavelet sets. When $\mathcal{D}$ is generated by a not necessarily expansive matrix a and $\mathcal{T}$ is a lattice, then the handful of $(\mathcal{D}, \mathcal{T})$-wavelets known all come from $(\mathcal{D}, \mathcal{T})$-wavelet sets.

There is also a stronger version of question 3 due to Larson [**45**] in the one dimensional case.

QUESTION 4 (Larson, [**45**]). Is it true that if ψ is a $(\mathcal{D}, \mathcal{T})$-wavelet, then there is a $(\mathcal{D}, \mathcal{T})$-wavelet set $E \subseteq \operatorname{supp}(\hat{\psi})$?

This problem is open even for the "classical" case of dimension 1 with dilations by powers of 2 and translations by integers. We name two partial answers. The first is given by Rzeszotnik in his PhD Thesis, and the second is due to Rzeszotnik and the second author of this paper.

THEOREM 1.8 (Rzeszotnik, [**52**] Corollary 3.10). *Every multiresolution analysis (MRA) $(2^j, \mathbb{Z})$-wavelet contains in the support of its Fourier transform an MRA $(\mathcal{D}, \mathcal{T})$-wavelet set.*

THEOREM 1.9. [**53**] *If ψ is a classical wavelet and the set $K = \operatorname{supp}(\hat{\psi})$ satisfies*
(1) $\sum_{k\in\mathbb{Z}} \chi_K(\xi + k) \leq 2$ *a.e.;*
(2) $\sum_{k\in\mathbb{Z}} \chi_K(2^j\xi) \leq 2$ *a.e.*
Then K contains a wavelet set.

Qing Gu has an unpublished example which shows that the techniques in [**53**] do not extend to the case that $\sum_{k\in\mathbb{Z}} \chi_K(\xi + k) \leq 3$ a.e. and $\sum_{j\in\mathbb{Z}} \chi_K(2^j\xi) \leq 3$ a.e.

Tilings and spectral sets are related by the *Fuglede conjecture* [**27**]:

CONJECTURE 1 (Fuglede). *A measurable set Ω, with positive and finite measure is a spectral set if and only if Ω is an additive $\mathcal{T}$-tile for some set $\mathcal{T}$.*

The conjecture, in general, still remains unsolved, even if several partial results have been obtained [**41, 44, 42, 43, 57**]. In June 2003 it was shown by Tao [**56**] that the conjecture in false in dimension 5 and higher. We will not discuss those articles, but concentrate on the important paper [**57**] by Wang, which also made the first serious attempt at studying $(\mathcal{D}, \mathcal{T})$-wavelet sets when $\mathcal{D}$ is not even a subgroup of $\mathrm{GL}(n, \mathbb{R})$, and $\mathcal{T}$ is not a lattice. We need two more definitions before we state some of Wang's results. Let $a \in \mathrm{GL}(n, \mathbb{R})$. A set $\mathcal{D} \subseteq \mathrm{GL}(n, \mathbb{R})$ is said to be a invariant if $\mathcal{D}a = \mathcal{D}$. The multiplicative tiling set $\mathcal{D}$ is said to satisfy the *interior condition* if there exists a multiplicative $\mathcal{D}$-tile Ω such that $\Omega^o \neq \emptyset$. Similarly the spectrum $\mathcal{T} \subset \mathbb{R}^n$ satisfies the *interior condition* if there exists a measurable set

$\Omega \subset \mathbb{R}^n$ such that $\Omega^o \neq \emptyset$ and $(\Omega, \mathcal{T})$ is a spectral pair. With these definitions we can state two of Wang's main results:

THEOREM 1.10 (Wang,[**57**]). *Let $\mathcal{D} \subset \mathrm{GL}(n, \mathbb{R})$ and $\mathcal{T} \subset \mathbb{R}^n$. Let $\Omega \subset \mathbb{R}^n$ be measurable, with positive and finite measure. If Ω is a multiplicative $\mathcal{D}^T$-tile and $(\Omega, \mathcal{T})$ is a spectral pair, then Ω is a $(\mathcal{D}, \mathcal{T})$-wavelet set. Conversely, if Ω is a $(\mathcal{D}, \mathcal{T})$-wavelet set and $0 \in \mathcal{T}$, then Ω is a multiplicative $\mathcal{D}^T$-tile and $(\Omega, \mathcal{T})$ is a spectral pair.*

THEOREM 1.11 (Wang,[**57**]). *Let $\mathcal{D} \subset \mathrm{GL}(n, \mathbb{R})$ such that $\mathcal{D}^T := \{ d^T \mid d \in \mathcal{D} \}$ is a bounded multiplicative tiling set, and let $\mathcal{T} \subset \mathbb{R}^n$ be a spectrum, with both $\mathcal{D}^T$ and $\mathcal{T}$ satisfying the interior condition. Suppose that $\mathcal{D}^T$ is a-invariant for some expanding matrix a and $\mathcal{T} - \mathcal{T} \subset \mathcal{L}$ for some full rank lattice $\mathcal{L}$ of $\mathbb{R}^n$. Then, there exists a wavelet set Ω with respect to $\mathcal{D}$ and $\mathcal{T}$.*

In his paper, Wang states "The assumption that $\mathcal{D}^T$ ... have the interior condition is most likely unnecessary. All known examples of multiplicative tiling sets admit a tile having nonempty interior." In this section, we will in fact show that the assumption that $\mathcal{D}^T$ satisfies the interior condition is indeed unnecessary, but not by proving that every multiplicative tiling set admits a tile having nonempty interior. Instead, we will use a Lebesgue density argument as in [**15, 18**]. Moreover, the assumption of multiplicative tiling sets having prototiles that are bounded and bounded away from the origin is not a "wavelet" assumption, but rather it is motivated from the point of view of tiling questions and the relation between translation and dilation tilings of the line. From the point of view of wavelets, by Theorem 1.10, one does not always wish to restrict to bounded multiplicative tiling sets. There are, however, some benefits of obtaining wavelet sets that are bounded and bounded away from the origin - especially if they also satisfy some additional properties. For example, if the sets are the finite union of intervals, one can use these wavelets to show that theorems about the poor decay of wavelets in $L^2(\mathbb{R}^n)$ for "bad" dilations are optimal. Along these lines, Bownik [**12**] showed that if a is irrational and $\psi_1, \ldots, \psi_L$ is an $(a, \mathbb{Z})$-multiwavelet, then there is an i such that for each $\delta > 0$, $\limsup_{|x| \to \infty} |\psi_i| \, |x|^{1+\delta} = \infty$. He also showed that this result is sharp by finding wavelet sets for each of these dilations that are the union of at most three intervals. Another possibility is to use wavelet sets that are the finite union of intervals (and satisfy several extra conditions) as a start point for the smoothing techniques in [**16, 40**]. However, these two advantages come from having wavelet sets that are not only bounded and bounded away from the origin, but also the finite union of intervals. In the construction considered in [**57**], it is not clear at all whether the end wavelet sets can be chosen to be the finite union of nice sets. In fact, the construction by Benedetto and Leon was used originally exactly to construct fractal-like wavelet sets.

Since the general question of existence of wavelet sets is phrased not in terms of sets bounded and bounded away from the origin, but arbitrary measurable sets, we will also show that the assumption that there exist a multiplicative tiling set that is bounded and bounded away from the origin is unnecessary. This will be done by showing that whenever there is a set that tiles $\mathbb{R}^n$ by $\mathcal{D}$ dilations, where $\mathcal{D}$ is invariant under an expansive matrix, then there exists a bounded multiplicative tiling set for $\mathcal{D}$.

We begin with some easy observations that were also in [**57**]. We say that sets U and V in $\mathbb{R}^n$ are a-dilation equivalent if there is a partition $\{ U_k \mid k \in \mathbb{Z} \}$ of U such that $\{ a^k U_k \mid k \in \mathbb{Z} \}$ is a partition of V.

LEMMA 1.12. *Let $\mathcal{D} \subset \mathrm{GL}(n, \mathbb{R})$ be invariant under an invertible matrix a. If Ω is a multiplicative $\mathcal{D}$-tile and Ω_0 is a-dilation equivalent to Ω, then Ω_0 is a multiplicative $\mathcal{D}$-tile.*

PROOF. Let S_k be a partition of Ω such that $\Omega_0 = \bigcup_{k\in\mathbb{Z}} a^k S_k$. Then,

$$\begin{aligned}\bigcup_{d\in\mathcal{D}} d\Omega_0 &= \bigcup_{d\in\mathcal{D}} \bigcup_{j\in\mathbb{Z}} da^j S_j \\ &= \bigcup_{j\in\mathbb{Z}} \bigcup_{d\in\mathcal{D}} da^j S_j \\ &= \bigcup_{j\in\mathbb{Z}} \bigcup_{d\in\mathcal{D}} dS_j = \mathbb{R}^n.\end{aligned}$$

Similarly, one shows that $|d_1\Omega_0 \cap d_2\Omega_0| = 0$ for all $d_1 \neq d_2$ in $\mathcal{D}$. □

LEMMA 1.13. *Let a be an expansive matrix and Ω_0, Ω_1 be such that for $i = 1, 2$ we have $|a^j\Omega_i \cap a^k\Omega_i| = 0$ for all $j \neq k$. Then, Ω_0 is a equivalent to Ω_1 if and only if $\bigcup_{j\in\mathbb{Z}} a^j\Omega_0 = \bigcup_{j\in\mathbb{Z}} a^j\Omega_1$ a.e.*

PROPOSITION 1.14. *Let $\mathcal{D} \subset \mathrm{GL}(n, \mathbb{R})$ be invariant under an expansive matrix a. If $\mathcal{D}$ is a multiplicative tiling set, then there is a set Ω_0 bounded and bounded away from the origin such that $\{ d\Omega_0 \mid d \in \mathcal{D} \}$ tiles $\mathbb{R}^n$. In particular, $\mathcal{D}$ is a bounded multiplicative tiling set.*

PROOF. It is widely known that a is expansive if and only if there is an ellipsoid $\mathcal{E}$ such that $\overline{\mathcal{E}} \subset a\mathcal{E}^\circ$. In this case, it is easy to check that $\Omega_1 = a\mathcal{E} \setminus \mathcal{E}$ is a bounded multiplicative tiling set for $\{ a^j \mid j \in \mathbb{Z} \}$; that is, Ω_1 is bounded and bounded away from the origin, and $\{ a^j\Omega_1 \mid j \in \mathbb{Z} \}$ tiles $\mathbb{R}^n$. Let $S_j = a^j\Omega_1 \cap \Omega$, and $\Omega_0 = \bigcup_{j\in\mathbb{Z}} a^{-j} S_j$. It is clear that $\Omega_0 \subset \Omega_1$, so it is bounded and bounded away from the origin. Moreover, since $\{ a^j\Omega_1 \mid j \in \mathbb{Z} \}$ is a tiling of $\mathbb{R}^n$, it follows that $\{ S_j \mid j \in \mathbb{Z} \}$ is a partition of Ω; hence, Ω_0 is a-dilation equivalent to Ω. Therefore, by lemma 1.14, Ω_0 is a multiplicative tiling set. □

Next, we turn to showing that the assumption of a multiplicative tile with non-empty interior is unnecessary. We have (combining Lemma 2 and Theorem 1 of [**15**]):

THEOREM 1.15 ([**15**]). *Let M be a measurable subset of $\mathbb{R}^n$ with positive measure satisfying $aM = M$ for some expansive matrix a. Then, there exists a set $E \subset M$ such that $\{ E + k \mid k \in \mathbb{Z}^n \}$ tiles $\mathbb{R}^n$ and $\{ a^jE \mid j \in \mathbb{Z} \}$ tiles M.*

Suppose that we are considering classes of $(\mathcal{D}, \mathcal{T})$-wavelets, where $\mathcal{D} = \{ a^j \mid j \in \mathbb{Z} \}$ and $\mathcal{T}$ is a full rank lattice. It is a general principle that one can either assume that a is in (real) Jordan form, in which case one must deal with arbitrary full rank lattices, or one can assume that the lattice $\mathcal{T} = \mathbb{Z}^n$, in which case one needs to consider all matrices of the form bab^{-1}. In particular, if one is working with expansive matrices, it is almost always permissible to restrict attention to translations by $\mathbb{Z}^n$. While this is clear to experts in the field, it is likely that researchers new to this field are not aware that the above theorem is really a theorem about arbitrary full rank lattices.

Indeed, let M be a measurable subset of $\mathbb{R}^n$ with positive measure satisfying $aM = M$, for some expansive matrix a. Let $\mathcal{L}$ be a full rank lattice in $\mathbb{R}^n$. Then, there is an invertible matrix b such that $b\mathcal{L} = \mathbb{Z}^n$. The set bM is bab^{-1} invariant, and bab^{-1} is an expansive matrix, so there is a set F such that $\{ F+k \mid k \in \mathbb{Z}^n \}$ tiles $\mathbb{R}^n$ and $\{ ba^jb^{-1}F \mid j \in \mathbb{Z} \}$ tiles bM. We claim that for $E = b^{-1}F$, $\{ E+k \mid k \in \mathcal{L} \}$ tiles $\mathbb{R}^n$ and $\{ a^jE \mid j \in \mathbb{Z} \}$ tiles M. Indeed,

$$\begin{aligned}
\bigcup_{k\in\mathcal{L}} E+k &= \bigcup_{k\in\mathcal{L}} b^{-1}F+k \\
&= \bigcup_{k\in\mathcal{L}} b^{-1}(F+bk) \\
&= \bigcup_{k\in\mathbb{Z}^n} b^{-1}(F+k) \\
&= b^{-1}\Big(\bigcup_{k\in\mathbb{Z}^n} (F+k)\Big) \\
&= \mathbb{R}^n .
\end{aligned}$$

One can similarly show the disjointness of translates by $\mathcal{L}$. To see that $\{ a^jE \mid j \in \mathbb{Z} \}$ tiles M, note that

$$\begin{aligned}
\bigcup_{j\in\mathbb{Z}} a^jE &= \bigcup_{j\in\mathbb{Z}} a^jb^{-1}F \\
&= b^{-1}\Big(\bigcup_{j\in\mathbb{Z}} ba^jb^{-1}F\Big) \\
&= b^{-1}bM = M.
\end{aligned}$$

Again, disjointness of the dilates is immediate. Thus, we have proved the following theorem, that seems to be well known:

THEOREM 1.16 ([**15**]). *Let M be a measurable subset of $\mathbb{R}^n$ with positive measure satisfying $aM = M$ for some expansive matrix a, and let $\mathcal{T}$ be a full rank lattice in $\mathbb{R}^n$. Then, there exists a set $E \subset M$ such that $\{ E+k \mid k \in \mathcal{T} \}$ tiles $\mathbb{R}^n$ and $\{ a^jE \mid j \in \mathbb{Z} \}$ tiles M.*

Theorem 1.16 can be used to give an easy proof of Theorem 1.11 removing three of the assumptions, but adding the assumption that the translation set is a full rank lattice.

THEOREM 1.17. *Let $\mathcal{D} \subset \mathrm{GL}(n, \mathbb{R})$ be such that $\mathcal{D}^T$ is a multiplicative tiling set. Suppose also that $\mathcal{D}^T$ is a invariant for some expansive matrix a. Let $\mathcal{T} \subset \mathbb{R}^n$ be a full rank lattice. Then, there exists a $(\mathcal{D}, \mathcal{T})$-wavelet set E.*

PROOF. By assumption, there exists a set Ω such that $\mathcal{D}^T(\Omega)$ is a tiling of $\mathbb{R}^n$. Consider the set $M = \bigcup_{j\in\mathbb{Z}} a^j\Omega$. The set M is clearly a invariant, and $\{ a^j(\Omega) \mid j \in \mathbb{Z} \}$ is a measurable partition of M so by 1.16, there exists a set E such that $\{ a^j(E) \mid j \in \mathbb{Z} \}$ tiles M and $\{ E+k \mid k \in \mathcal{L}^* \}$ tiles $\mathbb{R}^n$. By Lemmas 1.12 and 1.13, since E is a-equivalent to Ω, $\{ d^TE \mid d \in \mathcal{D} \}$ tiles $\mathbb{R}^n$. That is, E is a $(\mathcal{D}, \mathcal{T})$-wavelet set, as desired. □

We have exhibited above the essential nature of the argument in [**57**]. That is, what is desired is a general criterion for the following question:

QUESTION 5. Given an expansive matrix a, a full rank lattice $\mathcal{L}$ and two sets Ω_1 and Ω_2, when does there exist a set Ω that is a-equivalent to Ω_1 and $\mathcal{L}$ equivalent to Ω_2?

In the above case, we were forced to restrict to the case that Ω_2 is a fundamental region for the full rank lattice $\mathcal{L}$, since that is what was shown in [**15**]. As a final generalization in this section, we show that what is really necessary is that Ω_2 contain a neighborhood of the origin. The reader should compare the theorem below with the statement and proofs of the theorems in [**17**] and [**18**].

THEOREM 1.18. *Let a be an expansive matrix and $\Omega_1 \subset \mathbb{R}^n$ a set of positive measure such that $|a^j\Omega_1 \cap a^k\Omega_1| = 0$ whenever $j \neq k$. Let $M = \bigcup_{j\in\mathbb{Z}} a^j\Omega_1$. Let $\mathcal{L} \subset \mathbb{R}^n$ be a full rank lattice and $\Omega_2 \subset \mathbb{R}^n$ such that $|\Omega_2 + k_1 \cap \Omega_2 + k_2| = 0$ for $k_1 \neq k_2 \in \mathcal{L}$ and such that there exists $\epsilon > 0$ such that $M \cap B_\epsilon(0) \subset \Omega_2 \cap B_\epsilon(0)$. Then, there exists a set Ω such that Ω is a equivalent to Ω_1 and $\mathcal{L}$ equivalent to Ω_2.*

Before proving Theorem 1.18, we state and prove its main corollary, which is Theorem 2.1 of [**57**] with all but one technical assumption removed.

COROLLARY 1.19. *Let $\mathcal{D} \subset \mathrm{GL}(n,\mathbb{R})$ be such that $\mathcal{D}^T$ is a multiplicative tiling set. Let $\mathcal{T}$ be a spectrum with interior such that there exists a full rank lattice such that $\mathcal{T} - \mathcal{T} \subset \mathcal{L}$. Then, if $\mathcal{D}^T$ is a-invariant for some expansive matrix a, there exists a $(\mathcal{D}, \mathcal{T})$-wavelet set.*

PROOF. Since translations of spectral sets are again spectral sets, we may assume without loss of generality that Ω_2 contains 0 as an interior point. By Lemma 3.1 of [**57**], $(\Omega_2 + k_1) \cap (\Omega_2 + k_2)$ has measure 0 whenever $k_1 \neq k_2 \in \mathcal{L}^*$. So, by 1.18, there is a set Ω that is a equivalent to Ω_1 and $\mathcal{L}^*$ equivalent to Ω_2. By Lemma 1.12, $\{\, d^T\Omega \mid d \in \mathcal{D}\,\}$ tiles $\mathbb{R}^n$, and by Lemma 3.2 in [**57**], $(\Omega, \mathcal{T})$ is a spectral set. Therefore, Ω is a $(\mathcal{D}, \mathcal{T})$-wavelet set. □

We turn now to proving Theorem 1.18. We begin by noting that arguing as in the proof of Theorem 1.16, one can restrict to the case $\mathcal{L} = \mathbb{Z}^n$. Next, we need to extract the following lemma from the proof of Corollary 1 in [**15**], then we will follow very closely the proof in [**18**].

LEMMA 1.20. *Let a be an expansive matrix in $\mathrm{GL}(n,\mathbb{R})$. Let F_0 be a set of positive measure such that $|a^jF_0 \cap a^kF_0| = 0$ whenever $j \neq k$. Let $E = [-1/2, 1/2]^n$. Then, for every $\epsilon > 0$, there exists $k_0 \in \mathbb{Z}$ and $\ell_0 \in \mathbb{Z}^n$ such that $|a^{k_0}F_0 \cap (E + \ell_0)| > 1 - \epsilon$.*

The proof of Lemma 1.20 is a clever use of a Lebesgue density argument, which we will not repeat here.

PROOF OF THEOREM 1.18. First, note that as in the case of Theorem 1.16, Lemma 1.20 is really a lemma about arbitrary full rank lattices $\mathcal{L}$. Moreover, one can replace $E = [-1/2, 1/2]^n$ by any subset E of a fundamental region of $\mathcal{L}$ to get the following formally stronger lemma.

LEMMA 1.21. *Let a be an expansive matrix in $\mathrm{GL}(n,\mathbb{R})$. Let F_0 be a set of positive measure such that $|a^jF_0 \cap a^kF_0| = 0$ whenever $j \neq k$. Let $\mathcal{L} \subset \mathbb{R}^n$ be a full rank lattice with fundamental region Ω. Then, for every set $E \subset \Omega$ and every $\epsilon > 0$, there exists $k_0 \in \mathbb{Z}$ and $\ell_0 \in \mathcal{L}$ such that $|a^{k_0}F_0 \cap (E + \ell_0)| \geq (1 - \epsilon)|E|$.*

Turning to the proof of 1.18, note that by Theorem 1.14, we may assume without loss of generality that Ω_1 is bounded and bounded away from the origin. We may also assume that Ω_2 is contained in a convex, centrally symmetric fundamental region of $\mathcal{L}$. The rest of the proof follows very closely the proof of Theorem 3.7 in [**18**], with Lemma 1.21 playing the role of Proposition 3.5 in [**18**]. We will construct a family $\{ G_{ij} \mid i \in \mathbb{N}, j \in \{1,2\} \}$ of measurable sets whose a-dilates form a measurable partition of Ω_1 and whose translates by vectors in $\mathcal{L}$ form a measurable partition of Ω_2. Then

$$\Omega := \bigcup G_{ij} \tag{1.1}$$

is the set desired in Theorem 1.18. Since the steps are so similar to [**18**], we will give the first step of the inductive definition, and the properties needed for induction. Details are the same as in [**18**].

Let $\{\alpha_i\}$ and $\{\beta_i\}$ be sequences of positive constants decreasing to 0 and such that $\alpha_1 < \epsilon$ chosen so that $B_{\alpha_1}(0) \cap M \subset B_{\alpha_1}(0) \cap \Omega_2$. Let $\tilde{E}_{11} = \Omega_2 \setminus B_{\alpha_1}(0)$. Then, $|(\tilde{E}_{11})| > 0$. Let $\tilde{F}_{11}$ be a measurable subset of Ω_1 with measure strictly less than $|\Omega_1|$. By Lemma 1.21, there exists $k_1 \in \mathbb{N}$ and $\ell_1 \in \mathcal{L}$ such that

$$|a^{k_1}\tilde{F}_{11} \cap (\tilde{E}_{11} - \ell_1)| \geq \frac{1}{2}|\tilde{E}_{11}| \tag{1.2}$$

Let $G_{11} := a^{k_1}\tilde{F}_{11} \cap (\tilde{E}_{11} - \ell_1)$, let $E_{11} := G_{11} + \ell_1$, and let

$$F_{11} := \tilde{F}_{11} \cap a^{-k_1}(E_{11} - \ell_1) = a^{-k_1}G_{11} \tag{1.3}$$

Then $F_{11} \subset \tilde{F}_{11} \subset \Omega_1$ and $|\Omega_1 \setminus \tilde{F}_{11}| \geq |\Omega_1 \setminus F_{11}| > 0$. Also, $E_{11} \subset \tilde{E}_{11}$, and

$$|E_{11}| = |G_{11}| \geq \frac{1}{2}|\tilde{E}_{11}| \tag{1.4}$$

Also, $G_{11} = a^{k_1}F_{11}$. Now choose $F_{12} \subset \Omega_1$, disjoint from F_{11}, such that $\Omega_1 \setminus (F_{11} \cup F_{12})$ has positive measure less than β_1. Choose m_1 such that $a^{-m_1}F_{12}$ is contained in $N_1 = B_{\alpha_1/2}(0)$ and is disjoint from G_{11}. (This is possible since G_{11} is bounded away from 0.) Set

$$G_{12} := E_{12} := a^{-m_1}F_{12}. \tag{1.5}$$

The first step is complete.

Proceed inductively, obtaining disjoint families of positive measure $\{E_{ij}\}$ in Ω_2, $\{F_{ij}\}$ in Ω_1 and $\{G_{ij}\}$ such that for $i = 1, 2, \ldots$ and $j = 1, 2$ we have

(1) $G_{i1} + \ell_i = E_{i1}$;
(2) $G_{i2} = E_{i2}$;
(3) $a^{-k_1}G_{i1} = F_{i1}$;
(4) $a^{m_i}G_{i2} = F_{i2}$;
(5) $|\Omega_1 \setminus (F_{11} \cup F_{12} \cup \cdots \cup F_{i1} \cup F_{i2})| < \beta_i$, and
(6) $|E_{i1}| \geq \frac{1}{2}|\Omega_2 \setminus (E_{11} \cup E_{12} \cup \cdots \cup E_{i-1,1} \cup E_{i-1,2})| - \frac{1}{2}|N_i|$, where N_i is a ball centered at 0 with radius less than α_i.

Since $\beta_i \to 0$, item 5 implies that $F \setminus (\bigcup F_{ij})$ is a null set, and since $\alpha_i \to 0$, item 6 implies that $(E \setminus (\bigcup E_{ij}))$ is a null set. Let

$$F = \bigcup \{ G_{ij} \mid i = 1, 2, \ldots, j = 1, 2 \} \tag{1.6}$$

then, G is congruent to Ω_2 by items 1 and 2, and the a dilates of G form a partition of M, as desired. $\square$

For sets $\mathcal{D} \subset \mathrm{GL}(n, \mathbb{R})$ which are invariant under an expansive matrix, Corollary 1.19 is nice in that it reduces the question of existence of wavelet sets to the the question of existence of tiling sets for dilations and translations separately. It is still in some sense unsatisfactory, because it relies on the existence of objects external to the sets $(\mathcal{D}, \mathcal{T})$ under consideration. From the point of view of characterizing sets $(\mathcal{D}, \mathcal{T})$ for which wavelet sets exist, something more is needed. We will present in section 4 some progress on this question when $\mathcal{D}$ is a countable subgroup of $\mathrm{GL}(n, \mathbb{R})$.

2. Admissible groups and frames

We will now turn to the applications of those results to frames and wavelets in $\mathbb{R}^n$. But first we recall some results about the continuous wavelet transform.

Recall that translations and dilations on the real line form the so-called $(ax+b)$-group. Assume in general that we have a group G acting on a topological space. Assume that μ is a Radon measure on X and that the measure $g \cdot \mu \colon E \mapsto \mu(g^{-1} \cdot E)$ is absolutely continuous with respect to μ. Then μ is *quasi-invariant*, i.e, there exists a measurable function $j \colon G \times X \to \mathbb{R}^+$ such that

$$\int_X f(g \cdot x)\, d\mu(x) = \int_X j(g, x) f(x)\, d\mu(x)$$

for all $f \in L^1(X)$. Then, we can define a unitary representation of G on $L^2(X)$ by

$$[\pi(g)f](x) = j(g^{-1}, x)^{-1/2} f(g^{-1}x)\,.$$

For a fixed $\psi \in L^2(G)$ define the transform $W_\psi \colon L^2(X) \to C(G)$ by

$$W_\psi(f)(g) := (f, \pi(g)\psi)$$

and notice that W_ψ intertwines the representation π and the *left regular representation*, i.e.,

$$\begin{aligned} W_\psi(\pi(g)f)(x) &= (\pi(g)f, \pi(x)\psi) \\ &= (f, \pi(g^{-1} \cdot x)\psi) \\ &= W_\psi(f)(g^{-1} \cdot x)\,. \end{aligned}$$

Notice that $W_\psi(f)(g) \leq \|f\|_2 \|\psi\|_2$, and hence $W_\psi(f)$ is bounded. One of the important question now is:

QUESTION 6. Given the group G acting on X find a discrete subset $\mathcal{D} \subseteq G$ and a function ψ such that $\{\, \pi(g^{-1})\psi \mid g \in D \,\}$ is a frame for $L^2(X)$.

For a general discussion we refer to the fundamental work of Feichtinger and Gröchenig [**23, 24**].

DEFINITION 2.1. Let $\mathcal{H}$ be a separable Hilbert space, and let $\mathbb{J}$ be a finite our countable infinite index set. A sequence $\{f_j\}_{j\in\mathbb{J}}$ in $\mathcal{H}$ is called a *frame* if there exists constants $0 < A \leq B < \infty$ such that for all $x \in \mathcal{H}$ we have.

$$A\|x\|^2 \leq \sum_{j\in\mathbb{J}} |(x, f_j)|^2 \leq B\|x\|^2\,.$$

$\{f_j\}_{j\in\mathbb{J}}$ is a *tight frame* if we can choose $A = B$ and a *normalized tight frame* or *Parseval frame* if we can choose $A = B = 1$.

EXAMPLE 2.2. Let $\mathcal{H}$ be a Hilbert space and $\mathcal{K} \subset \mathcal{H}$ a closed subspace. Assume that $\{u_n\}$ is a orthonormal basis of $\mathcal{H}$. Let $\mathrm{pr}\colon \mathcal{H} \to \mathcal{K}$ be the orthogonal projection. Define $f_j = \mathrm{pr}(u_j)$. Then $\{f_j\}$ is a Parseval frame for $\mathcal{K}$. In fact it is easy to see that every Parseval frame can be constructed in this way. In particular we can apply this to the situation where $(\Omega, \mathcal{T})$ is a spectral pair and $M \subset \Omega$ is measurable with $|M| > 0$. Then $\{|\Omega|^{-1/2} e_\lambda\}_{\lambda\in\mathcal{T}}$ is an orthonormal basis for $L^2(\Omega)$ and hence $\{f_\lambda := |\Omega|^{-1/2} e_\lambda|_M\}_{\lambda\in\mathcal{T}}$ is a Parseval frame for $L^2(M)$.

EXAMPLE 2.3. For the $(ax+b)$-group $j(a,x) = |a|^{-1}$ is independent of x and we get:

$$\begin{aligned} W_\psi(f)(a,b) &= (f, \pi(a,b)\psi) \\ &= (f, T_b D_a \psi) \\ &= |a|^{-1/2} \int_{\mathbb{R}} f(x)\overline{\psi((x-b)/a)}\, dx\,. \end{aligned}$$

Here $T_b\colon L^2(\mathbb{R}) \to L^2(\mathbb{R})$ stands for the unitary isomorphism corresponding to translation $T_b f(x) = f(x-b)$ and $D_a\colon L^2(\mathbb{R}) \to L^2(\mathbb{R})$ is the unitary map corresponding to dilation $D_a f(x) = |a|^{-1/2} f(x/a)$, $a \neq 0$.

The discrete wavelet transform is obtained by sampling the wavelet transform $W_\psi(f)$ at points gotten by replacing the full $(ax+b)$-group by a discrete subset generated by translation by integers and dilations of the form $a = 2^n$:

$$\begin{aligned} W_\psi^d(f)(2^{-n}, -2^{-n}m) &= (f \mid \pi((2^n,m)^{-1})\psi) \\ &= 2^{n/2} \int_{\mathbb{R}} f(x)\overline{\psi(2^n x + m)}\, dx \\ &= (f, D_{2^{-n}} T_{-m}\psi)\,. \end{aligned}$$

Hence, the corresponding frame is

$$\{\, \pi((2^n,m)^{-1})\psi \mid n,m \in \mathbb{Z}\,\} = \{\, D_{2^n} T_m \psi \mid n,m \in \mathbb{Z}\,\}\,. \tag{2.1}$$

The inverse refers here to the inverse in the $(ax+b)$-group.

As in the last example, it is well known, that the short time Fourier transform, and several other well known integral transforms have a common explanation in this way in the language of representation theory. This observation is the basis for the generalization of the continuous wavelet transform to higher dimensions and more general settings, and was already made by A. Grossmann, J. Morlet, and T. Paul in 1985, see [**34, 35**]. In [**34**] the connection to square integrable representations and the relation to the fundamental paper of M. Duflo and C. C. Moore [**20**] was already pointed out. Several natural questions arise now, in particular to describe the image of the transform W_ψ and how that depends on ψ. But we will not go into that here, but refer to [**2, 3, 6, 8, 19, 21, 24, 25, 28, 30, 31, 32, 33, 34, 35, 39, 46, 51, 59**] for discussion. Here, we will concentrate on the connection to frames, wavelets and wavelet sets.

Denote by $\mathrm{Aff}(\mathbb{R}^n)$ the group of invertible affine linear transformations on $\mathbb{R}^n$. Thus $\mathrm{Aff}(\mathbb{R}^n)$ consists of pairs (x,h) with $h \in \mathrm{GL}(n,\mathbb{R})$ and $x \in \mathbb{R}^n$. The action of $(x,h) \in \mathrm{Aff}(\mathbb{R}^n)$ on $\mathbb{R}^n$ is given by

$$(x,h)(v) = h(v) + x\,.$$

The product of group elements is the composition of maps. Thus

$$(x,a)(y,b) = (a(y)+x, ab)$$

the identity element is $e = (0, \mathrm{id})$ and the inverse of $(x,a) \in \mathrm{Aff}(\mathbb{R}^n)$ is given by

$$(x,a)^{-1} = (-a^{-1}(x), a^{-1})\,. \tag{2.2}$$

Thus $\mathrm{Aff}(\mathbb{R}^n)$ is the semidirect product of the abelian group $\mathbb{R}^n$ and the group $\mathrm{GL}(n,\mathbb{R})$; $\mathrm{Aff}(\mathbb{R}^n) = \mathbb{R}^n \times_s \mathrm{GL}(n,\mathbb{R})$. Let $H \subseteq \mathrm{GL}(n,\mathbb{R})$ be a closed subgroup. (In fact it is not necessary to assume that H is closed, but for simplicity we will assume that.) Define

$$\mathbb{R}^n \times_s H := \{\, (x,a) \in \mathrm{Aff}(\mathbb{R}^n) \mid a \in H,\ x \in \mathbb{R}^n \,\}.$$

Then $H \times_s \mathbb{R}^n$ is a closed subgroup of $\mathrm{Aff}(\mathbb{R}^n)$, and hence a Lie group. *From now on H will always denote a closed subgroup of* $\mathrm{GL}(n,\mathbb{R})$.

Define a unitary representation of $\mathbb{R}^n \times_s H$ on $L^2(\mathbb{R}^n)$ by

$$[\pi(x,a)f](v) := |\det(a)|^{-1/2} f((x,a)^{-1}(v)) = |\det(a)|^{-1/2} f(a^{-1}(v-x))\,. \tag{2.3}$$

Write $\psi_{x,a}$ for $\pi(x,a)\psi$. Because of the role played by the Fourier transform, we will also need another action of H on $\mathbb{R}^n$ given by $a \cdot \omega := (a^{-1})^T(\omega)$. We denote by $\hat{\pi}(x,a)$ the unitary action on $L^2(\mathbb{R}^n)$ given by

$$\hat{\pi}(x,a)f(v) = \sqrt{|\det(a)|}e^{-2\pi i(x|v)} f(a^{-1}\cdot v) = \sqrt{|\det(a)|}e^{-2\pi i(x|v)} f(a^T(v))\,. \tag{2.4}$$

REMARK 2.4. Some authors use the semidirect product $H \times_s \mathbb{R}^n$ instead of $R^n \times_s H$. Thus first the translation and then the linear map is applied, i.e., $(a,x)(v) = a(v+x)$. In this notation the product becomes $(a,x)(b,y) = (ab, ab(y) + a(x))$, the inverse of (a,x) is $(a,x)^{-1} = (a^{-1}, -ax)$, and the wavelet representation is

$$\widetilde{\pi}(a,x)f(v) = |\det a|^{-1/2} f(a^{-1}v - x)\,. \tag{2.5}$$

Furthermore, instead of using the transposed action, we could just as well represent $\mathbb{R}^n$ in the frequency picture as row vectors and use the right action.

The Fourier transform intertwines the representations π and $\hat{\pi}$ [**21**], Lemma 3.1, i.e.,

$$\widehat{\pi(x,a)f}(\omega) = \hat{\pi}(x,a)\hat{f}(\omega)\,, \quad f \in L^2(\mathbb{R}^n)\,. \tag{2.6}$$

Denote by $d\mu_H$ a left invariant measure on H. A left invariant measure on G is then given by $d\mu_G(x,a) = |\det(a)|^{-1} d\mu_H(a)dx$. Let $f, \psi \in L^2(\mathbb{R}^n)$. A simple calculation shows that

$$\int_G |(f \mid \pi(x,a)\psi)|^2\, d\mu_G(x,a) = \int_{\mathbb{R}^n} |\hat{f}(\omega)|^2 \int_H |\hat{\psi}(a^{-1}\cdot\omega)|^2\, d\mu_H(a)d\omega \tag{2.7}$$

$$= \int_{\mathbb{R}^n} |\hat{f}(\omega)|^2 \int_H |\hat{\psi}(a^T(\omega))|^2\, d\mu_H(a)d\omega \tag{2.8}$$

There are several ways to read this. First let $M \subseteq \mathbb{R}^n$ be measurable and invariant under the action $H \times \mathbb{R}^n \ni (h,v) \mapsto h \cdot v := (h^{-1})^T(v) \in \mathbb{R}^n$. Then we denote by $L^2_M(\mathbb{R}^n)$ the space of function $f \in L^2(\mathbb{R}^n)$ such that $\hat{f}(\xi) = 0$ for almost all $\xi \notin M$. Notice that $L^2_M(\mathbb{R}^n)$ is a closed invariant subspace, and that the orthogonal projection onto $L^2_M(\mathbb{R}^n)$ is given by $f \mapsto (\hat{f}\chi_M)^\vee$. The first result is now:

THEOREM 2.5. *Let $M \subseteq \mathbb{R}^n$ be measurable of positive measure, and invariant under the action $(a, v) \mapsto a \cdot v$. Then the wavelet transform*

$$W_\psi \colon f \mapsto (f \mid \pi(x,a)\psi)_{L^2(\mathbb{R}^n)} = |\det a|^{-1/2} \int f(y)\overline{\psi(a^{-1}(y-x))}\, dy$$

is a partial isometry $W_\psi \colon L^2_M(\mathbb{R}^n) \to L^2(G)$ if and only if

$$\Delta_\psi(\omega) := \int_H |\hat{\psi}(a^T\omega)|^2 \, d\mu_H(a) = 1 \tag{2.9}$$

for almost all $\omega \in M$.

Motivated by [**46, 59**] we define

DEFINITION 2.6 (Laugesen, Weaver, Weiss, and Wilson). Let $M \subseteq \mathbb{R}^n$, be measurable, invariant, and $|M| > 0$. Let $\psi \in L^2(\mathbb{R}^n)$ then ψ is said to be *a (normalized) admissible (H, M)-wavelet* if for almost all $\omega \in M$ we have

$$\int_H |\hat{\psi}(a^T\omega)|^2 \, d\mu_H(a) = 1 \, .$$

We say that the pair (H, M) is *admissible* if a (H, M)-admissible wavelet ψ exists. If $M = \mathbb{R}^n$ then we say that H is *admissible* and that ψ is a (normalized) *wavelet* function.

Assume that $\psi \in L^2_M(\mathbb{R}^n)$ is a normalized admissible wavelet. Then

$$G \ni (x,a) \mapsto F(x,a) := W_\psi f(x,a)\psi_{x,a} \in L^2_M(\mathbb{R}^n)$$

is well defined and if $g \in L^2_M(\mathbb{R}^n)$ then

$$\begin{aligned}
\int_G (F(x,a) \mid g)_{L^2(\mathbb{R}^n)} \, d\mu_G(x,a) &= \int_G W_\psi f(x,a) \left(\int_{\mathbb{R}^n} \psi_{x,a}(y)\overline{g(y)} \, dy \right) d\mu_G(x,a) \\
&= \int_G W_\psi f(x,a)\overline{W_\psi g(x,a)} \, d\mu_G(x,a) \\
&= (W_\psi f \mid W_\psi g)_{L^2(G)} \\
&= (f \mid g)_{L^2(\mathbb{R}^n)} \, .
\end{aligned}$$

Hence

LEMMA 2.7. *Assume that $\psi \in L^2_M(\mathbb{R}^n)$ satisfies $\int_H |\hat{\psi}(a^T\omega)|^2 \, d\mu_H(a) = 1$ for almost all $\omega \in M$. Then, as a weak integral,*

$$f = \int W_\psi f(x,a)\psi_{x,a} \, d\mu_G(x,a) \tag{2.10}$$

for all $f \in L^2(\mathbb{R}^n)$.

QUESTION 7 (Laugesen, Weaver, Weiss, and Wilson). Give a characterization of admissible subgroups of $\mathrm{GL}(n, \mathbb{R})$.

It is easy to derive one necessary condition for admissibility. For $\omega \in \mathbb{R}^n$ let

$$H^\omega = \{ h \in H \mid h \cdot \omega = \omega \} = \{ h \in H \mid h^T(\omega) = \omega \} \tag{2.11}$$

be the stabilizer of ω. Then admissibility implies that H^ω is compact for almost all $\omega \in \mathbb{R}^n$. This condition is not sufficient and by now, there is *no* complete characterization of admissible group. The best result is the following, due to Laugesen, Weaver, Weiss, and Wilson [**46**]:

THEOREM 2.8 (Laugesen, Weaver, Weiss, and Wilson, [**46**]). *Let H be a closed subgroup of* $\mathrm{GL}(n,\mathbb{R})$. *For $\omega \in \mathbb{R}^n$ and $\epsilon > 0$ let*

$$H_\epsilon^\omega := \{ h \in H \mid \|a \cdot \omega - \omega\| \leq \epsilon \}$$

be the ϵ-stabilizer of ω. If either

(1) $G = \mathbb{R}^n \times_s H$ *is non-unimodular, or*

(2) $\{\omega \in \mathbb{R}^n \mid H^\omega$ *is non-compact*$\}$ *has positive Lebesgue measure*

holds, then H is not admissible. If both (1) *and*

(3) $\{ \omega \in \mathbb{R}^n \mid H_\epsilon^\omega$ *is non-compact for all* $\epsilon > 0 \}$ *has positive Lebesgue measure*

fail, then H is admissible.

REMARK 2.9. If M is homogeneous under H^T. Then (H, M) is admissible if and only if H^ω is compact for one—and hence all—$\omega \in M$, cf. Lemma 3.1.

Various versions of the following discrete version of the wavelet transform are well known in the literature. Recall that H denotes a closed subgroup of $\mathrm{GL}(n,\mathbb{R})$.

THEOREM 2.10. *Let $M \subseteq \mathbb{R}^n$ be measurable with positive measure and such that $|\overline{M} \setminus M| = 0$. Assume that there exist a countable, discrete subset $\Gamma \subseteq \mathrm{GL}(n,\mathbb{R})$ and a measureable set $\mathbb{F} \subset M$ such that the following holds:*

(1) $|M \setminus \bigcup_{\gamma\in\Gamma} \gamma^T\mathbb{F}| = 0$;

(2) $\sup_{\gamma\in\Gamma} \#\{ \sigma \in \Gamma \mid |\sigma^T\mathbb{F} \cap \gamma^T\mathbb{F}| \neq 0 \} = L < \infty$;

(3) *there exists a set $\mathcal{T} \subset \mathbb{R}^n$, such that*

$$\{ e_t|_\mathbb{F} \mid t \in \mathcal{T} \}$$

is a frame for $L^2(M)$ with frame constants $0 < A \leq B$.

Let $\psi = \chi_\mathbb{F}^\vee$. Then

$$\{ \pi((t,\gamma)^{-1})\psi \mid t \in \mathcal{T},\, \gamma \in \Gamma \}$$

is a frame for $L^2_M(\mathbb{R}^n)$ with frame constants A and LB. In particular if $L = 1$ and $\{ e_t|_\mathbb{F} \mid t \in \mathcal{T} \}$ is a tight frame, then $\{ \pi((t,\gamma)^{-1})\psi \mid t \in \mathcal{T},\, \gamma \in \Gamma \}$ is a tight frame, with same frame constant.

PROOF. First recall that $(t,\gamma)^{-1} = (-\gamma^{-1}, \gamma^{-1})$ by equation (2.2). Hence by (2.4) and (2.6) it follows that (with $\gamma^\# = (\gamma^{-1})^T$):

$$\begin{aligned}
(f, \pi((t,\gamma)^{-1})\psi) &= (\hat{f}, \widehat{\pi((t,\gamma)^{-1})\psi}) \\
&= |\det(\gamma)|^{-1/2} \int \hat{f}(\lambda) e^{-2\pi(\gamma^{-1}t,\lambda)} \chi_\mathbb{F}(\gamma \cdot \lambda)\, d\lambda \\
&= \sqrt{|\det(\gamma)|} \int_\mathbb{F} \hat{f}(\gamma^T\lambda) e^{-2\pi(t,\lambda)}\, d\lambda \\
&= ([D_{\gamma^\#}\hat{f}]|_\mathbb{F}, e_t|_\mathbb{F}) \,.
\end{aligned}$$

As $\{ e_t|_\mathbb{F} \mid t \in \mathcal{T} \}$ is a frame for $L^2(\mathbb{F})$ with frame bounds A and B, it follows that

$$A\|[D_{\gamma^\#}\hat{f}]|_\mathbb{F}\|^2 \leq \sum_{t\in\mathcal{T}} |([D_{\gamma^\#}\hat{f}]|_\mathbb{F}, e_t)|^2 \leq B\|[D_{\gamma^\#}\hat{f}]|_\mathbb{F}\|^2$$

or

$$A\|\hat{f}\chi_{\gamma^T\mathbb{F}}\|^2 \leq \sum_{t\in\mathcal{T}} |(f, \pi((t,\gamma)^{-1}))\psi)|^2 \leq B\|\hat{f}\chi_{\gamma^T\mathbb{F}}\|^2 \,.$$

But by assumptions (1) and (2), it follows that

$$|\hat{f}|^2 \le \sum_{\gamma \in \Gamma} |\hat{f}|^2 \chi_{\gamma^T \mathbb{F}} \le L |\hat{f}|^2$$

and the statement of the theorem follows. □

Few remarks about this theorem are at place here:

REMARK 2.11. If $\mathbb{F}$ is bounded and satisfies (1) and (2) then we there are always infinitely many $\mathcal{T}$ satisfying (2). Just, cf. [**8**], p. 605, take any parallelpiped

$$R = \{ \sum_{j=1}^{n} t_j u_j \mid a_j \le t_j \le b_j,\, j = 1, \dots, n \}$$

such that $\mathbb{F} \subseteq R$. Then $u_1, \dots, u_n$ is a basis for $\mathbb{R}^n$. Let $v_1, \dots, v_n$ be the dual basis, and define

$$\mathcal{T} := \{ \sum_{j=1}^{n} \frac{k_j}{b_j - a_j} v_j \mid k = (k_1, \dots, k_n) \in \mathbb{Z}^n \} .$$

Then $\{ |R|^{-1/2} e_t|_R \mid t \in \mathcal{T} \}$ is an orthonormal basis for $L^2(R)$. By Example 2.2 it follows that $\{ |R|^{-1/2} e_t|_{\mathbb{F}} \mid t \in \mathcal{T} \}$ is a tight frame for $L^2(\mathbb{F})$.

REMARK 2.12. The function ψ in the above Theorem is non smooth, as its Fourier transform is an indicator function and hence not even continuous. But instead of ψ we can choice a smooth, compactly supported function h such that there exists constants $0 \le a \le b$ such that $a \le |h|_{\mathbb{F}} \le$ and such that the support of h satisfies condition (2) (possibly with another L). Then a simple modification of the above proof shows that $\{ \pi((t,\gamma)^{-1} h^{\vee}) \mid (t,\gamma) \in \mathcal{T} \times \Gamma \}$ is a frame, see [**8**], Theorem 3.

In the above theorem we did not assume that M is invariant under a subgroup of $\mathrm{GL}(n,\mathbb{R})$ or that Γ has any special structure. But in applications, or for constructing examples, we will later use that M is invariant under a group H and then use the structure of H to construct both Γ and $\mathbb{F}$. The following theorem is the prime tool for that. We will need the following, which is a part of Lemma 3.3 in [**23**]:

LEMMA 2.13 (Feichtinger and Gröchenig). *Let H be a locally compact, Hausdorff, topological group. Let $\Gamma \subset G$. Then the following conditions are equivalent:*

(1) *There exists a relatively compact open subset $V \subset H$, $e \in H$, such that $\gamma V \cap \sigma V = \emptyset$ if $\gamma, \sigma \in \Gamma$, $\gamma \ne \sigma$;*
(2) *Given any relatively compact subset W in H such that $W^o \ne \emptyset$, then*

$$\sup_{\gamma \in \Gamma} \#\{ \sigma \in \Gamma \mid \gamma W \cap \sigma W \ne \emptyset \} < \infty .$$

THEOREM 2.14. *Assume that M is H-homogeneous under the action $(h,x) \mapsto h \cdot x$, and that (H,M) is admissible. Suppose there exists a relatively compact measurable set $\mathbb{F}_H \subseteq H$ with non-empty interior, and a discrete set $\Gamma \subseteq H$ such that the following holds:*

(1) *$\mu_H(H \setminus \bigcup_{\gamma \in \Gamma} \gamma \mathbb{F}_H) = 0$;*
(2) *$\sup_{\gamma \in \Gamma} \#\{ \sigma \in \Gamma \mid \sigma \mathbb{F}_H \cap \gamma \mathbb{F}_H \ne \emptyset \} < \infty$.*

Let $\omega \in M$ and set $\mathbb{F} = \mathbb{F}_H \cdot \omega$. Then the pair $(\Gamma^{-1}, \mathbb{F})$ satisfies the conditions (1), (2), *and* (3) *in Theorem* 2.10.

PROOF. As M is homogeneous it follows that $M \simeq H/H^\omega$, where $\simeq$ stands for diffeomorphic. Hence $\mathbb{F}$ is measurable with finite measure. Then notice that

$$\begin{aligned}\bigcup_{\gamma\in\Gamma^{-1}} \gamma^T \mathbb{F} &= \bigcup_{\gamma\in\Gamma} \gamma \cdot \mathbb{F}_H \cdot \omega \\ &= \bigcup_{\gamma\in\Gamma} (g\mathbb{F}_H) \cdot \omega \\ &= H \cdot \omega \\ &= M\,.\end{aligned}$$

By Lemma 2.13 there exists an open, relatively compact set $V \subset H$, with $e \in V$, such that

$$\gamma V \cap \sigma V = \emptyset\,, \qquad \gamma \neq \sigma\,. \tag{2.12}$$

As H^ω is compact, it follows that $K := \mathbb{F}_H H^\omega$ is relatively compact. Hence, by (2.12) and Lemma 2.13, we get:

$$\sup_{\gamma\in\Gamma} \#\{\,\sigma \in \Gamma \mid \gamma K \cap \sigma K \neq \emptyset\,\} =: L < \infty\,. \tag{2.13}$$

By Lemma 1.4 in [**51**] we have

$$\begin{aligned}\sup_{\gamma\in\Gamma^{-1}} \#\{\,\sigma \in \Gamma^{-1} \mid \gamma^T\mathbb{F} \cap \sigma^T\mathbb{F} \neq \emptyset\,\} \\ = \sup_{\gamma\in\Gamma} \#\{\,\sigma \in \Gamma \mid \gamma K \cap \sigma K \neq \emptyset\,\} =: L < \infty\end{aligned} \tag{2.14}$$

and condition (2) in Theorem 2.10. As $\overline{\mathbb{F}_H}$ is compact, it follows that $\overline{\mathbb{F}}$ is compact. In particular $\mathbb{F}$ is bounded and the existence of the set $\mathcal{T}$ in (3) follows by Remark 2.11. □

COROLLARY 2.15. *Assume that M is homogeneous and that (H, M) is admissible. If there exists a co-compact subgroup $\Gamma \subseteq H$, then there exists a set $\mathbb{F} \subseteq M$ satisfying the conditions in Theorem* 2.10.

3. The FO condition and frames

Theorem 2.14 combined with Theorem 2.10 gives us a tool to construct frames in subsets of $\mathbb{R}^n$. The problem becomes to find groups such that $(H, \mathbb{R}^n)$ is admissible and such that we can apply Theorem 2.14. In this section we discuss a special class of such group introduced in [**21**] and [**51**]. Those are admissible groups with finitely many open orbits. They groups are closely related to the so-called *prehomogeneous vector spaces of parabolic type* [**9**]. We say that the pair $(H.M)$, $H \subseteq \mathrm{GL}(n, \mathbb{R})$, $\emptyset \neq M \subseteq \mathbb{R}^n$, satisfies the condition *FO* if M is H^T-invariant, and there exists finitely many H^T-invariant open and disjoint sets $U_1, \ldots, U_k \subseteq M$, such that each U_j is H^T-invariant and homogeneous, and $\left|M \setminus \bigcup_{j=1}^k U_j\right| = 0$. We start with a simple lemma:

LEMMA 3.1. *Assume that the pair (H, M) satisfies the condition FO. Then (H, M) is admissible if and only if the stabilizer H^ω is compact for $\omega \in U_j$, $j = 1, \ldots, k$.*

PROOF. It is clear, that if (H, M) is admissible, then H^ω is compact for each $\omega \in \cup_j U_j$. For the other direction, fix $\omega_j \in U_j$. For $j = 1, \dots, k$ let $g_j \in C_c(U_j)$, $g_j \geq 0$, $g \neq 0$. Then the function

$$H \ni a \mapsto g_j(a^T(\omega_j)) \in \mathbb{C}$$

has compact support and $\int_H g_j(a^T(\omega_j))\, d\mu_H(a) > 0$. Let $\omega \in U_j$. Choose $h \in H$ such that $\omega = h^T(\omega_j)$. This is possible because U_j is homogeneous under H^T. Then

$$\begin{aligned} \int_H g_j(a^T(\omega))\, d\mu_H(a) &= \int_H g_j(a^T h^T(\omega_j))\, d\mu_H(a) \\ &= \int_H g_j((ha)^T(\omega_j))\, d\mu_H(a) \\ &= \int_H g_j(a^T(\omega_j))\, d\mu_H(a)\ . \end{aligned}$$

Hence $\Delta_j = \int_H g_j(a^T(\omega))\, d\mu_H(a) > 0$ is independent of $\omega \in U_j$. Define $\varphi\colon \mathbb{R}^n \to \mathbb{C}$ by

$$\varphi(\omega) := \begin{cases} g_j(\omega)/\Delta_j & \text{if } \omega \in U_j, \\ 0 & \text{if } \omega \notin \bigcup_{j=1}^k U_j. \end{cases}$$

Then $\varphi \in C_c(\mathbb{R}^n)$, so in particular $\varphi \in L^2(\mathbb{R}^n)$. Define $\psi := \varphi^\vee$. Then ψ satisfies the admissibility condition (2.9). Hence, H is admissible. □

QUESTION 8. Classify the pairs (H, M) satisfying the condition FO.

We will discuss the construction of frames for $L^2_M(\mathbb{R}^n)$ for some special pairs (H, M) satisfying the condition OF. Groups of the form $H = ANR$ as in the following Lemma showed up naturally in [**21**], also most of the known examples for construction of wavelets are in fact of this form.

LEMMA 3.2. *Let $H \subset \mathrm{GL}(n, \mathbb{R})$ be a closed subgroup such that H can be written as $H = ANR = NAR = RAN$ with R compact, A simply connected abelian, and such that the map*

$$N \times A \times R \ni (n, a, r) \mapsto nar \in H$$

is a diffeomorphism. Assume furthermore that R and A commute, and that R and A normalize N. Finally assume that there exists a co-compact discrete subgroup $\Gamma_N \subset N$. Let $\Gamma_A \subset A$ be a co-compact subgroup in A. Choose bounded measurable subsets $\mathbb{F}_A \subset A$, and $\mathbb{F}_N \subset N$ such that $N = \Gamma_N \mathbb{F}_N$, and $A = \Gamma_A \mathbb{F}$ and such that the union is disjoint. Let $\Gamma = \Gamma_A \Gamma_N$ and $\mathbb{F}_H = \mathbb{F}_N \mathbb{F}_A R \subset H$. Then we have

$$H = \bigcup_{\gamma \in \Gamma} \gamma \mathbb{F}_H$$

and the union is disjoint. Furthermore we can choose $\mathbb{F}_H$ such that $\mathbb{F}_H^{\mathrm{o}} \neq \emptyset$.

PROOF. We have

$$\begin{aligned}
\bigcup_{\gamma} \gamma \mathbb{F}_N \mathbb{F}_A R &= \Gamma_A \Gamma_N (\mathbb{F}_N) \mathbb{F}_A R \\
&= \Gamma_A N \mathbb{F}_A R \qquad \text{because}\, \Gamma_N \mathbb{F}_N = N \\
&= \bigcup_{\gamma \in \Gamma_A} (\gamma N \gamma^{-1}) \gamma \mathbb{F}_A R \\
&= \bigcup_{\gamma \in \Gamma_A} N \gamma \mathbb{F}_A R \qquad \text{because}\, A\, \text{normalizes}\, N \\
&= N \Gamma_A \mathbb{F}_A R \\
&= NAR \\
&= H\,.
\end{aligned}$$

Assume now that

$$\gamma_A \gamma_N f_N f_A r = \sigma_A \sigma_N g_N g_A s$$

for some $\gamma_A, \sigma_A \in \Gamma_A$, $\gamma_N, \sigma_N \in \Gamma_N$, $f_A, g_A \in \mathbb{F}_A$, $f_N, g_N \in \mathbb{F}_N$, and $r, s \in R$. Then, as $A \times N \times R \ni (a, n, r) \mapsto anr \in H$ is a diffeomorphism, it follows that $r = s$. Hence $\gamma_A \gamma_N f_N f_A = \sigma_A \sigma_N g_N g_A$. But then—as A normalizes N—

$$\begin{aligned}
\gamma_N f_N &= (\gamma_A^{-1} \sigma_A) \sigma_N g_N (g_A f_A^{-1}) \\
&= (\gamma_A^{-1} \sigma_A g_A f_A^{-1}) \left((g_A f_A)^{-1} \sigma_N g_N (g_A f_A^{-1}) \right) .
\end{aligned}$$

Hence $\gamma_A^{-1} \sigma_A g_A f_A^{-1} = 1$ or

$$\gamma_A f_A = \sigma_A g_A\,.$$

As the union $\Gamma_A \mathbb{F}_A$ is disjoint, it follows that $\gamma_A = \sigma_A$ and $f_A = g_A$. But then the above implies that

$$\gamma_N f_N = \sigma_N g_N\,.$$

But then—again because the union is disjoint—it follows that $\gamma_N = \sigma_N$ and $f_N = g_N$. □

Our first application of this lemma is to give a simple proof of the main result, Theorem 4.2, of [**51**], without using the results of [**8**]. We will reformulate those results so as to include sampling on irregular grids, see also [**4**].

Assume now that $H = ANR$ satisfies the conditions in Lemma 3.2. Let $\Gamma = \Gamma_A \Gamma_N$ and $\mathbb{F}_H = \mathbb{F}_N \mathbb{F}_A R$ be as in that Lemma. Suppose that $M \subseteq \mathbb{R}^n$ is H invariant and such that there are finitely many open orbits $U_1, \ldots, U_k \subseteq M$ such that $|M \setminus \bigcup_{j=1}^k U_j| = 0$. Finally we assume that for each $\omega_j \in U_j$ the stabilizer of ω_j in H under the action $(h, \omega) \mapsto (h^{-1})^T(\omega)$ is contained in R and hence compact. Let $\mathbb{F}_j = F_H \cdot \omega_j$ and $\mathbb{F} = \bigcup_{j=1}^k \mathbb{F}_j$. Then the Lemma 3.2 implies that we have a multiplicative tiling of M as

$$M = \bigcup_{\gamma \in \Gamma} \gamma \cdot \mathbb{F} = \bigcup_{\gamma \in \Gamma^{-1}} \gamma^T(\mathbb{F}) \tag{3.1}$$

(up to set of measure zero).

THEOREM 3.3. *Let the notation be as above. Suppose that* $\{e_t|_{\mathbb{F}}\}_{t \in \mathcal{T}}$ *is a frame for* $L^2(\mathbb{F})$. *Let* $\psi = \chi_{\mathbb{F}}^{\vee}$. *Then the sequence* $\{\pi((t,\gamma)^{-1})\psi\}_{(t,\gamma) \in \mathcal{T} \times \Gamma}$ *is a frame for* $L^2(M)$ *with the same frame bounds. In particular there exists a constant* $T > 0$

such that $\{T\pi((t,\gamma)^{-1})\psi\}_{(t,\gamma)\in\mathcal{T}\times\Gamma}$ *is a Parseval frame for* $L^2_M(\mathbb{R}^n)$ *if and only if* $\{Te_t|_{\mathbb{F}}\}_{t\in\mathcal{T}}$ *is a Parseval frame for* $L^2(\mathbb{F})$.

PROOF. This follows from Theorem 2.14. □

There are several ways to state different versions of the above theorem. In particular one can have different groups $H_j = A_jN_jR_j$, with compact stabilizers, such that each of them has finitely many open orbits, $U_{j,1},\ldots,U_{j,k_j}$ such that $\mathbb{R}^n = \bigcup_{j,l}\Gamma_jU_{j,l}$ a disjoint union. One can also construct a set $\mathbb{F}_H$ that is not necessarily R-invariant (cf. Example 3.8), and finally, one can allow more than three group. But we will not state all those obvious generalizations, but only notice the following construction from [**21**]. We refer to the Appendix for more details. In [**21**] the authors started with a prehomogeneous vector space (L,V) of parabolic type, see [**9**] for details. Then L has finitely many open orbits in V, but in general the stabilizer of a point is not compact. To deal with that, the authors constructed for each orbits U_j a subgroup $H_j = A_jN_jR_j$ such that U_j is up to measure zero a disjoint union of open H_j orbits $U_{j,i}$. It turns out, that it is not necessary to pick a different group for each orbit, the same group $H = H_j$ works for all the orbits.

THEOREM 3.4. *Let $H = ANR$ be one of the group constructed in* [**21**]. *Then H is admissible.*

PROOF. This follows from Theorem 3.6.3 and Corollary 3.6.4 in [**9**]. □

REMARK 3.5. The statement in [**9**] is in fact stronger than the above remark. In most cases the group AN has finitely many open orbits. This group acts freely and is therefore admissible. The only exception is the so-called Type III spaces, where the group ANR has one orbit and is admissible.

EXAMPLE 3.6 ($\mathbb{R}^+\mathrm{SO}(n)$). Take $A = \mathbb{R}^+\mathrm{id}$, $R = \mathrm{SO}(n)$, the group of orientation preserving rotations in $\mathbb{R}^n$, and $N = \{\mathrm{id}\}$. Then $H = \mathbb{R}^+\mathrm{SO}(n)$ is the group of dilations and orientation preserving rotations. Notice that $g^{-1} = g^T$ if $g \in \mathrm{SO}(n)$ and therefore $g\cdot\omega = g(\omega)$. The group H has two orbits $\{0\}$ and $\mathbb{R}^n\setminus\{0\}$. The stabilizer of $e_1 = (1,0,\ldots,0)^T$ is isomorphic to $\mathrm{SO}(n-1)$. In particular the stabilizer group is compact. It follows that $\mathbb{R}^+\mathrm{SO}(n)$ is admissible. In fact any function with compact support in $\mathbb{R}^n\setminus\{0\}$ is, up to normalization, the Fourier transform of an admissible wavelet.

EXAMPLE 3.7 (Diagonal matrices). Let H be the group of diagonal matrices $H = \{\, d(\lambda_1,\ldots,\lambda_n) \mid \lambda_j \neq 0\,\}$. Thus $A = \{\, d(\lambda_1,\ldots,\lambda_n) \mid \forall j : \lambda_j > 0\,\}$ and $R = \{\, d(\epsilon_1,\ldots,\epsilon_n) \mid \epsilon_j = \pm 1\,\}$. The group N is trivial. Then H has one open and dense orbit

$$U = \{\, (x_1,\ldots,x_n)^T \mid (j=1,\ldots,n)\, x_j \neq 0\,\} = H\cdot(1,\ldots,1)^T\,.$$

The stabilizer of $(1,\ldots,1)^T$ is trivial and hence compact. It follows that H is admissible. We can also replace H by the connected group A. Then we have 2^n open orbits parameterized by $\epsilon \in \{-1,1\}^2$

$$U_\epsilon = \{\, (x_1,\ldots,x_j)^T \mid \mathrm{sign}(x_j) = \epsilon_j\,\} = H\cdot(\epsilon_1,\ldots,\epsilon_n)\,.$$

The stabilizers are still compact and hence H is admissible.

EXAMPLE 3.8 (Some two-dimensional examples). In this example we discuss some 2-dimensional applications. Most of those examples can be found in several other places in the literature, but we would like to show how all of them fits into our general construction.

Upper triangular matrices: Let first H be the group of upper triangular 2×2-matrices of determinant 1,

$$H = \left\{ \begin{pmatrix} a & t \\ 0 & 1/a \end{pmatrix} \;\middle|\; a \neq 0,\, t \in \mathbb{R} \right\}.$$

Here A is the group of diagonal matrices with $a > 0$, N is the group up upper triangular matrices with 1 on the main diagonal and $R = \{\pm \mathrm{id}\}$. Then we have one open orbit of full measure given by

$$U = \{ (x, y)^T \mid y \neq 0 \} = H \cdot e_2$$

where $e_2 = (0, 1)^T$. The stabilizer of e_2 is trivial which implies that H is admissible.

Now take $R = \{\mathrm{id}\}$. Then AN has two open orbits

$$U_1 = \{ (x, y)^T \mid y > 0 \} \quad \text{and} \quad U_2 = \{ (x, y)^T \mid y < 0 \}.$$

Take any $\lambda > 1$ and $b > 0$. Then we can take

$$\Gamma_N := \left\{ \begin{pmatrix} 1 & kb \\ 0 & 1 \end{pmatrix} \;\middle|\; k \in \mathbb{Z} \right\} \quad \text{and} \quad \Gamma_A := \left\{ \begin{pmatrix} \lambda^n & 0 \\ 0 & \lambda^{-n} \end{pmatrix} \;\middle|\; n \in \mathbb{Z} \right\}.$$

Take

$$\mathbb{F}_N = \left\{ \begin{pmatrix} 1 & kb \\ 0 & 1 \end{pmatrix} \;\middle|\; k \in \mathbb{Z} \right\} \quad \text{and} \quad \Gamma_A := \left\{ \begin{pmatrix} \mu & 0 \\ 0 & 1/\mu \end{pmatrix} \;\middle|\; \mu \in (1/\lambda, 1) \right\}.$$

Taking $\omega = (0, 1)^T$ we get

$$\mathbb{F} = \{ (x, y)^T \in \mathbb{R}^2 \mid x = t/y \text{ for some } t \in (0, b) \text{ and some } y \in (1, \lambda) \}.$$

There are several choices for a spectral set Ω containing $\mathbb{F}$. One of them is

$$\Omega = \{ (x, y)^T \mid 0 < x < b,\, \frac{1}{\lambda} < y < 1 \}.$$

To see how the elements in Γ acts, lets us see how the generators of Γ_N and Γ_A acts:

$$\begin{pmatrix} 1 & b \\ 0 & 1 \end{pmatrix} \mathbb{F} = \{ (x, y)^T \in \mathbb{R}^2 \mid x = ty,\, t \in (b, 2b),\, y \in (1/\lambda, 1) \}$$

In particular, the line segment that bound $\mathbb{F}$ are moved into:

$$\begin{aligned} \{ (x, 1)^T \mid 0 < x < b \} &\mapsto \{ (x, 1) \mid b < x < 2b \}, \\ \{ (x, 1/\lambda)^T \mid 0 < x < b/\lambda \} &\mapsto \{ (x, 1/\lambda) \mid b/\lambda < x < 2b/\lambda \}, \\ \{ (0, y)^T \mid \frac{1}{\lambda} < y < 1 \} &\mapsto \{ (x, y) \mid x = by,\, \frac{1}{\lambda} < y < 1 \}, \\ \{ (x, y)^T \mid x = by,\, \frac{1}{\lambda} < y < 1 \} &\mapsto \{ (x, y) \mid x = 2by,\, \frac{1}{\lambda} < y < 1 \}. \end{aligned}$$

Similarly

$$\begin{pmatrix} \lambda & 0 \\ 0 & 1/\lambda \end{pmatrix} \mathbb{F} = \{ (x, y)^T \in \mathbb{R}^2 \mid x = sy \\ \text{for some } s \in (0, \lambda^2 b) \text{ and some } y \in (\lambda^{-2}, \lambda^{-1}) \}.$$

Dilations and rotations: In this case we let $A = \mathbb{R}^+ \,\mathrm{id}$, the group of dilations,

$$N = \mathrm{SO}_o(2) = \left\{ R_\theta = \begin{pmatrix} \cos(2\pi t) & \sin(2\pi t) \\ -\sin(2\pi t) & \cos(2\pi t) \end{pmatrix} \;\middle|\; 0 \le t \le 1 \right\},$$

the group of orientation preserving rotations, and $R = \{\mathrm{id}\}$. (Notice, that we could have interchanged to role of R and N, as N is compact in this case.) Then we have one open orbit $U = \mathbb{R}^2 \setminus \{(0,0)^T\}$. Fix $\lambda > 1$ and $s \in \mathbb{N}$ and take $\Gamma_A = \{ \lambda^n \,\mathrm{id} \mid n \in Z \}$ and $\Gamma_N = \{ R_{k/s} \mid k = 0, 1, \dots, s-1 \}$. Then Γ_N is a finite subgroup of N and $\Gamma_A \Gamma_N$ is a group. In this case we can take

$$\mathbb{F} = \{ r(\cos(2\pi\theta), \sin(2\pi\theta))^T \mid \lambda^{-1} < r < 1,\, \theta \in (-1/(2s), 1/(2s)) \}.$$

Dilations and hyperbolic rotations: The example $H = \mathbb{R}^+\mathrm{SO}_o(1,n)$ was discussed in details in [**51**]. Take $n = 2$. Then we can take $A = \mathbb{R}^+ \,\mathrm{id}$ and and

$$N = \left\{ \begin{pmatrix} \cosh(t) & \sinh(t) \\ \sinh(t) & \cosh(t) \end{pmatrix} \;\middle|\; t \in \mathbb{R} \right\}.$$

In this case we have four open orbits

$$U_1 = \{ (x,y)^T \mid 0 < |y| < x \}, \qquad U_2 = \{ (x,y)^T \mid 0 < |y| < -x \},$$
$$U_3 = \{ (x,y)^T \mid 0 < |x| < y \}, \qquad U_4 = \{ (x,y)^T \mid 0 < |x| < -y \}.$$

Fix $\lambda > 1$ and $b > 0$. Let $\Gamma_A := \{ \lambda^n \,\mathrm{id} \mid n \in \mathbb{Z} \}$ and

$$\Gamma_N := \left\{ \begin{pmatrix} \cosh(kb) & \sinh(kb) \\ \sinh(kb) & \cosh(kb) \end{pmatrix} \;\middle|\; k \in \mathbb{Z} \right\}.$$

Notice, that in this case Γ is in fact a group as A and N commute. In this case we can take

$$\mathbb{F}_A = \{ \mu \,\mathrm{id} \mid 1 < \mu < \lambda \} \quad \text{and} \quad \mathbb{F}_N = \left\{ \begin{pmatrix} \cosh(t) & \sinh(t) \\ \sinh(t) & \cosh(t) \end{pmatrix} \;\middle|\; t \in (-b/2, b/2) \right\}.$$

The fundamental domain in U_1 is then

$$\mathbb{F} = \{ (x,y)^T \mid 1 < x < \cosh(b/2),\; x^2 - y^2 \in (1, \lambda^2) \}.$$

The boundary of this domain is given by

$$\begin{aligned} \partial\mathbb{F} = \{ (\cosh(t), \sinh(t))^T \mid -b/2 < t < b/2 \} \\ \cup \{ \lambda(\cosh(t), \sinh(t))^T \mid -b/2 < t < b/2 \} \\ \cup \{ s(\cosh(b/2), \pm\sinh(b/2) \mid 1 \le s \le \lambda \}. \end{aligned}$$

We leave it to the reader to determine a spectral pair $(\Omega, \mathcal{T})$ such that $\mathbb{F} \subset \Omega$.

4. Construction of wavelet sets

We apply now our construction in section 1 to discrete subgroups of $\mathrm{GL}(n, \mathbb{R})$. We start by the following reformulation of Theorem 2.8 for discrete groups. Our aim is later to apply it to the discrete subgroup Γ_A from the last section. As before we use the notation $a \cdot x = (a^{-1})^T(x) = a^\#(x)$.

LEMMA 4.1. *Let D be a discrete subgroup of $\mathrm{GL}(n,\mathbb{R})$. If for almost every $x \in \mathbb{R}^n$, there exists an $\epsilon > 0$ such that ϵ-stabilizer D^x_ϵ is finite, then there exists a measurable function h such that*

$$\sum_{d\in D} |h(d^T x)|^2 = 1 \quad a.e. \tag{4.1}$$

We have the following improvement of Lemma 4.1.

PROPOSITION 4.2. *Let D be a discrete subgroup of $\mathrm{GL}(n,\mathbb{R})$. If for almost every $x \in \mathbb{R}^n$, there exists an $\epsilon > 0$ such that D^x_ϵ is finite, then there exists a measurable function g of the form $g = \chi_K$ such that*

$$\sum_{d\in D} |g(d^T x)|^2 = 1 \quad a.e. \tag{4.2}$$

PROOF. We first recall some notation and preliminary results from [**46**]. For an open ball $B \subset \mathbb{R}^n$, we define the orbit density function $f_B \colon \mathbb{R}^n \to [0,\infty]$ by

$$f_B(x) = \mu(\{\, d \in D \mid d^T x \in \overline{B}\,\}), \tag{4.3}$$

where μ is counting measure. Lemma 2.6 of [**46**] asserts that

$$\begin{aligned} \Omega_0 &:= \{\, x \in \mathbb{R}^n \mid D^x_\epsilon \text{ non-compact } \forall \epsilon > 0\,\} \\ &= \{\, x \in \mathbb{R}^n \mid f_B(x) = \infty,\ \forall B : B \cap \mathcal{O}_x \neq \emptyset\,\} \end{aligned} \tag{4.4}$$

Now, let $\mathcal{B} = \{B_j\}$, $j \in \mathbb{N}$, be an enumeration of the balls in $\mathbb{R}^n$ having rational centers and positive rational radii. Let $f_j = f_{B_j}$. We claim that

$$\mathbb{R}^n = \bigcup_{j\geq 1} \{\, x \in \mathbb{R}^n \mid f_j(x) = 1\,\} \cup \Omega_0 \cup N, \tag{4.5}$$

where

$$N := \bigcup_{d\in D, d\neq \mathrm{id}} \{\, x \in \mathbb{R}^n \mid d^T x = x\,\}. \tag{4.6}$$

To see this, suppose that $x \notin (\Omega_0 \cup N)$. Then, there exists an open ball B such that $B \cap \mathcal{O}_x \neq \emptyset$, and $f_B(x) < \infty$. Since $B \cap \mathcal{O}_x \neq \emptyset$, there is a $d_0 \in D$ such that $d_0^T x \in B$. Therefore, there is a j such that $d_0^T x \in B_j \subset B$; in particular $B_j \cap \mathcal{O}_x \neq \emptyset$ and $\infty > f_j(x) > 0$. Now, write $\{\, d \in D \mid d^T x \in \overline{B_j}\,\} = \{d_0^T, \dots, d_k^T\}$. Since $x \notin N$, the $d_i^T x = d_j^T x$ only if $i = j$. Hence, there is an open set O such that $\mu(\{\, d \in D \mid d^T x \in O\,\} = 1$. Choose j such that $d_0^T x \in B_j \subset O$ so that $f_j(x) = 1$.

Continuing along the lines of [**46**], let

$$\Omega_1 = \{\, x \in \mathbb{R}^n \mid f_1(x) = 1\,\}$$

and

$$\Omega_j = \{\, x \in \mathbb{R}^n \mid f_j(x) = 1\,\} \setminus (\Omega_1 \cup \cdots \cup \Omega_{j-1})$$

The sets $\{\Omega_j\}_{j\geq 1}$ form a disjoint collection of Borel sets such that $\mathbb{R}^n \setminus (\bigcup_{j=1}^\infty \Omega_j)$ has measure 0 (it is a subset of $\Omega \cup N$). Let us define

$$g(x) = \sum_{j=1}^\infty \chi_{\Omega_j}(x)\chi_{\overline{B_j}}(x) \tag{4.7}$$

and $g(x) = 0$ for $x \notin (\bigcup_{j\geq 1} \Omega_j)$. Note that

$$g(x) = \chi_K\,, \quad K = \bigcup_{j=1}^{\infty} (\Omega_j \cap \overline{B_j}), \tag{4.8}$$

so all that is needed to complete the proof, is to show $\{\, d^T K \mid d \in D \,\}$ is a tiling of $\mathbb{R}^n$, equivalently, $\sum_{d\in D} g(d^T x) = 1$ a.e. This is a special case of the argument in [**46**], which we outline now.

First, note that if $x \in \mathbb{R}^n$ such that $f_j(x) = 1$ for some smallest j, then there is a unique $d \in D$ such that $dx \in \overline{B_j}$. Since f_j is constant on orbits, $d^T x \in \Omega_j \cap \overline{B_j}$ and $d^T x \notin (\Omega_1 \cup \cdots \cup \Omega_{j-1})$. Therefore, $x \in d \cdot K$ and $\bigcup_{d\in D} d^T K = \mathbb{R}^n$ up to a set of measure 0.

For disjointness, since $d^T \Omega_j = \Omega_j$, it suffices to check that $(\Omega_j \cap \overline{B_j}) \cap (\Omega_j \cap \overline{B_j}) a$ has measure 0 for all d not the identity id. If $x \in (\Omega_j \cap \overline{B_j})$, then $f_j(x) = 1$. If, in addition, $x = d^T \omega$ for some $\omega \in (\Omega_j \cap \overline{B_j})$, then $d^{-1}x \in \overline{B_j}$ which means that $d \cdot x = x$ and $d = \text{id}$ since $f_j(x) = 1$. □

THEOREM 4.3. *Let D be a discrete subgroup of* $\mathrm{GL}(n, \mathbb{R})$ *that contains an expansive matrix, and $\mathcal{L} \subset \mathbb{R}^n$ a full rank lattice. If for almost every $x \in \mathbb{R}^n$, there exists an $\epsilon > 0$ such that D^x_ϵ is finite, then, there exists a $(D, \mathcal{L})$-wavelet set.*

PROOF. By Proposition 4.2, there exists a function $g = \chi_K$ such that equation 4.2 holds. Therefore, $\{\, d^T K \mid d \in D \,\}$ tiles $\mathbb{R}^n$. Thus, by Theorem 1.17 there exists a $(D, \mathcal{L})$-wavelet set. □

Note that there are examples of discrete subgroups of $\mathrm{GL}(n, \mathbb{R})$ that are not generated by a single element, cf. Example 3.8.

QUESTION 9. If $\mathcal{D}$ is a subgroup of $\mathrm{GL}(n, \mathbb{R})$ and $\mathcal{L}$ is a full rank lattice such that there exists a $(\mathcal{D}, \mathcal{T})$-wavelet set, then for almost every x does there exist an $\epsilon > 0$ such that the ϵ-stabilizer D^x_ϵ is finite?

One final comment is that in all of the above considerations, the set D is assumed to be invariant under multiplication by an expansive matrix. Removing this condition seems to be very hard. Indeed, even when the set $D = \{\, a^j \mid j \in \mathbb{Z} \,\}$, it is not clear what happens when a is not an expansive matrix. In this case, the interplay between dilations and translations becomes crucial in understanding when there exists a wavelet set. For example, let $a = \begin{pmatrix} 2 & 0 \\ 0 & 2/3 \end{pmatrix}$, $D = \{\, a^j \mid j \in \mathbb{Z} \}$, and $\mathcal{T} = \mathbb{Z}^2$. It is easy to see that there is a set of finite measure Ω such that $\{a^j \Omega\}$ tiles $\mathbb{R}^2$. However, there exist full rank lattices $\mathcal{L}_1$ and $\mathcal{L}_2$ such that there are no $(D, \mathcal{L}_1)$-wavelet sets, yet there are $(D, \mathcal{L}_2)$-wavelet sets [**54**]. Hence, in the non-expansive case, it is not enough to simply prove the existence of sets that tile via dilations and translations separately.

We will now apply this to the discrete subgroup $\Gamma_A \subset A$ from the last section, where A is as in Theorem 3.4.

THEOREM 4.4. *Let $H = ANR$ be one of the group constructed in* [**21**]. *Then the following holds:*

(1) *The group A contains a discrete co-compact subgroup $\Gamma_A \subset A$ such that $E_A = \{\, d \in \Gamma_A \mid d$ is expansive$\,\}$ is a non-trivial subsemigroup. In particular there exists an expansive matrix a such that Γ_A is a invariant.*

(2) *Let* Γ_A *and* Γ_A^+ *be as in the Appendix and let* $\Gamma = \Gamma_A\Gamma_N$. *Then* $\Gamma\Gamma_A^+ \subset \Gamma$.

PROOF. (1) follows from Lemma 5.3 in the appendix. For (2) we recall that Γ_A^+ is contained in the center of ANR. Hence

$$\begin{aligned}\Gamma\Gamma_A^+ &= \Gamma_A\Gamma_N\Gamma_A^+ \\ &= \Gamma_A\Gamma_A^+\Gamma_N \\ &\subseteq \Gamma_A\Gamma_N = \Gamma .\end{aligned}$$

□

We have now proved, using Theorem 4.3 the following theorem:

THEOREM 4.5. *Let the notation be as in Theorem 4.4. Let* $\mathcal{L}$ *be a full rank lattice in* $\mathbb{R}^n$. *Then there exists a* $(\Gamma_A, \mathcal{L})$*-wavelet set and a* $(\Gamma, \mathcal{L})$*-wavelet set.*

REMARK 4.6. Theorem 4.4 gives several examples of non-groups of dilations for which wavelet sets exist. Unfortunately from the point of view of characterizing sets $(\mathcal{D}, \mathcal{T})$ for which wavelet sets exist, if one starts with the set $\mathcal{D}$, one still has to rely on the existence of an object external to the set $\mathcal{D}$ for the existence of wavelet sets. It would also be interesting to remove the condition that $\mathcal{L}$ is a lattice.

5. Symmetric cones

In this section we discuss the important example of homogeneous cones in $\mathbb{R}^n$. Those cones show up in several places in analysis. As an example one can take Hardy spaces of holomorphic function on tube type domains $\mathbb{R}^n + i \oplus \Omega$ [**55**]. An excellent reference for harmonic analysis on symmetric cones is the book by J. Faraut and A. Koranyi [**22**]. A nonempty open subset $\Omega \subset \mathbb{R}^n$ is called an *open (convex) cone* if Ω is convex and $\mathbb{R}^+\Omega \subseteq \Omega$. Let Ω be an open cone, define the *dual cone* Ω^* by

$$\Omega^* := \{\, v \in \mathbb{R}^n \mid \forall u \in \overline{\Omega} \setminus \{0\} : (v, u) > 0 \,\} .$$

If Ω^* is nonempty, then Ω^* is a open cone. Ω is *self-dual* if $\Omega = \Omega^*$. Let

$$\mathrm{GL}(\Omega) = \{\, g \in \mathrm{GL}(n, \mathbb{R}) \mid g(\Omega) = \Omega \,\} .$$

Then Ω is *homogeneous* if $\mathrm{GL}(\Omega)$ acts transitively on Ω. From now on we assume that Ω is a self-dual homogeneous cone. Let $g \in \mathrm{GL}(\Omega)$ and $u \in \overline{\Omega} \setminus \{0\}$. Then $g(u) \in \overline{\Omega} \setminus \{0\}$. Hence if $v \in \Omega = \Omega^*$, then

$$(g^T(v), u) = (v, g(u)) > 0 .$$

Thus $g^T(v) \in \Omega^* = \Omega$. It follows that $\mathrm{GL}(\Omega)$ is invariant under transposition, and hence reductive. Let $e \in \Omega$. Then

$$K = \mathrm{GL}(\Omega)^e = \{\, g \in \mathrm{GL}(\Omega) \mid g(e) = e \,\} .$$

Let $\theta(g) = (g^{-1})^T$. Then it is always possible to choice e such that $K = \{\, g \in \mathrm{GL}(\Omega) \mid \theta(g) = g \,\} = \mathrm{SO}(n) \cap \mathrm{GL}(n, \mathbb{R})$. Define the *Lie algebra* of $\mathrm{GL}(\Omega)$ by

$$\mathfrak{gl}(\Omega) := \{\, X \in M(n, \mathbb{R}) \mid \forall t \in \mathbb{R} : e^{tX} \in \mathrm{GL}(\Omega) \,\} .$$

Then $\mathfrak{gl}(\Omega)$ is invariant under the Lie algebra automorphism $\dot\theta(X) = -X^T$. Let

$$\mathfrak{k} = \{\, X \in \mathfrak{gl}(\Omega) \mid \dot\theta(X) = X \,\}$$

and

$$\mathfrak{s} = \{\, X \in \mathfrak{gl}(\Omega) \mid \dot\theta(X) = -X \,\} = \mathrm{Symm}(n, \mathbb{R}) \cap \mathfrak{gl}(\Omega)$$

where $\mathrm{Symm}(n,\mathbb{R})$ stand for the space of symmetric matrices. Let $\mathfrak{a}$ be a maximal subspace in $\mathfrak{s}$ such that $[X,Y] = XY - YX = 0$ for all $X, Y \in \mathfrak{a}$. Notice that $(X,Y) = \mathrm{Tr}(XY^T)$ is an inner product on $\mathfrak{gl}(\Omega)$ and that, with respect to this inner product, $\mathrm{ad}(X)\colon \mathfrak{gl}(\Omega) \to \mathfrak{gl}(\Omega)$, $Y \mapsto [X,Y]$ satisfies

$$\mathrm{ad}(X)^T = \mathrm{ad}(X^T)\ .$$

Hence the algebra $\{\, \mathrm{ad}(X) \mid X \in \mathfrak{a}\,\}$ is a commuting family of self adjoint operator on the finite dimensional vector space $\mathfrak{gl}(\Omega)$. Hence there exists a basis $\{X_j\}_j$ of $\mathfrak{gl}(\Omega)$ consisting of joint eigenvectors of $\{\, \mathrm{ad}(X) \mid X \in \mathfrak{a}\,\}$. Let $\mathfrak{z}(\mathfrak{a})$ be the zero eigenspace, i.e., the maximal subspace of $\mathfrak{gl}(\Omega)$ commuting with all $X \in \mathfrak{a}$. Then there exists a finite subset $\Delta \subset \mathfrak{a}^* \setminus \{0\}$ such that with

$$\mathfrak{gl}(\Omega)^\alpha = \{\, Y \in \mathfrak{gl}(\Omega) \mid \forall X \in \mathfrak{a} : \mathrm{ad}(X)Y = \alpha(X)Y\,\}$$

we have

$$\mathfrak{gl}(\Omega) = \mathfrak{z}(\mathfrak{a}) \oplus \bigoplus_{\alpha\in\Delta} \mathfrak{gl}(\Omega)^\alpha\ .$$

Notice that if $\alpha \in \Delta$ then $-\alpha \in \Delta$. In fact, if

$$X \in \mathfrak{gl}(\Omega)^\alpha \Longrightarrow X^T \in \mathfrak{gl}(\Omega)^{-\alpha}\ . \tag{5.1}$$

Let $\mathfrak{a}' = \{\, X \in \mathfrak{a} \mid \forall \alpha \in \Delta : \alpha(X) \neq 0\,\}$. Then $\mathfrak{a}'$ is open and dense in $\mathfrak{a}$. In particular $\mathfrak{a}' \neq \{0\}$. Fix $Z \in \mathfrak{a}'$ and let $\Delta^+ = \{\, \alpha \in \Delta \mid \alpha(Z) > 0\,\}$. Then $\Delta = \Delta^+ \cup -\Delta^+$, and if $\alpha, \beta \in \Delta^+$ are such that $\alpha + \beta \in \Delta$, then $\alpha + \beta \in \Delta^+$. Let

$$\mathfrak{n} = \bigoplus_{\alpha\in\Delta^+} \mathfrak{gl}(\Omega)^\alpha.$$

Then $\mathfrak{n}$ is a nilpotent Lie algebra (as $[\mathfrak{gl}(\Omega)^\alpha, \mathfrak{gl}(\Omega)^\beta] \subset \mathfrak{gl}(\Omega)^{\alpha+\beta}$) and $[\mathfrak{a},\mathfrak{n}] \subseteq \mathfrak{n}$. In particular it follows that $\mathfrak{q} = \mathfrak{a} \oplus \mathfrak{n}$ is a solvable Lie algebra. Notice that the algebra $\mathfrak{z}(\mathfrak{a})$ is invariant under transposition. Hence $\mathfrak{z}(\mathfrak{a}) = \mathfrak{z}(\mathfrak{a}) \cap \mathfrak{k} \oplus \mathfrak{a}$. Because of (5.1) it therefore follows that

$$\mathfrak{gl}(\Omega) = \mathfrak{k} \oplus \mathfrak{a} \oplus \mathfrak{n}\ .$$

This decomposition is called *the Iwasawa decomposition of* $\mathfrak{gl}(\Omega)$. Let $A = \{\, e^X \mid X \in \mathfrak{a}\,\}$ and $N = \{\, e^Y \mid Y \in \mathfrak{n}\,\}$. Then A and N are Lie groups, A is abelian, and $aNa^{-1} = N$ for all $a \in A$. It follows that $Q := AN = NA$ is a Lie group with N a normal subgroup. Furthermore we have the following *Iwasawa decomposition of* $\mathrm{GL}(\Omega)$:

LEMMA 5.1 (The Iwasawa decomposition). *The map*

$$A \times N \times K \ni (a,n,k) \mapsto ank \in \mathrm{GL}(\Omega)$$

is an analytic diffeomorphism.

We note that the one dimensional group $Z = \mathbb{R}^+\mathrm{id}$ is a subgroup of $\mathrm{GL}(\Omega)$ and in fact $Z \subset A$. If $a(\lambda) = \lambda\mathrm{id} \in Z$, with $\lambda > 1$, then $a(\lambda)$ is expansive. In particular it follows that the set E of expansive matrices in A is a nonempty subsemigroup of A. Let $X_0 = \mathrm{id}, X_1, \ldots, X_r$ be a basis of $\mathfrak{a}$ and let

$$\Gamma_A = \{\, \exp(n_0X_0 + \cdots + n_rX_r) \mid n_j \in \mathbb{Z}\,\}\ .$$

Then A/Γ_A is compact. Furthermore there exists a discrete subgroup $\Gamma_N \subset N$ such that N/Γ_N is compact.

Let now $\mathcal{D} = \Gamma$ and $d = \exp(2X_0)$. Then d is expansive and $d\mathcal{D} = \mathcal{D}d \subset \mathcal{D}$, because d is central in $\mathrm{GL}(\Omega)$. It follows that the results from the previous sections are applicable in this case.

Appendix: Prehomogeneous vector spaces

One way to find admissible groups with finitely many open orbits is to start with *prehomogeneous vector spaces.* Those are pairs (H, V) where H is a reductive Lie group, say $H^T = H$, and V is a finite dimensional vector space, such that H has finitely many open orbits in V. There is no full classification of those spaces at the moment, but a subclass, the *prehomogeneous vector spaces of parabolic type* has been classified. We refer to [**9**] Section 2.11, for detailed discussion and references. The problem, from the point of view of our work is, that the compact stabilizer condition does not hold in general, but as shown in [**21**] one can always replace H by a subgroup of the form ANR as before, such that ANR is admissible. Notice that, by using either ANR or $A^TN^TR^T$, which satisfies the same conditions, we can consider either the standard action on $\mathbb{R}^n$ or the action $(a, x) \mapsto (a^{-1})^T(x)$. We will use the second action in what follows.

Let $H = H^T$ be a reductive Lie group acting on $V = \mathbb{R}^n$. Then H can be written as $H = LC$ where $C = C^T$ is a vector group, isomorphic to a abelian subalgebra $\mathfrak{c}$ of $\mathfrak{gl}(n, \mathbb{R}) = M(n, \mathbb{R})$. The isomorphism is simply given by the matrix exponential function

$$X \mapsto \exp(X) = e^X = \sum_{j=0}^{\infty} \frac{X^j}{j!} \,.$$

The vector space V is graded in the sense that there exists a subset $\Delta \subset \mathfrak{c}^*$ such that

$$V = \bigoplus_{\alpha\in\Delta} V_\alpha \tag{5.2}$$

where

$$V_\alpha = \{ v \in V \mid (\forall \in \mathfrak{c}) \, : \, X \cdot v = \alpha(X)v \} \,. \tag{5.3}$$

If $c = \exp(X) \in C$ and $\lambda \in \mathfrak{c}^*$, then we write $c^\lambda = e^{\lambda(X)}$. In particular $c \cdot v = c^\alpha v$ for all $v \in V_\alpha$.

Denote by pr_α the projection onto V_α along $\bigoplus_{\beta\neq\alpha} V_\beta$.

LEMMA 5.2. *We have $H \cdot V_\alpha \subset V_\alpha$ for all $\alpha \in \Delta$. Furthermore if $v \in V$ and $H \cdot v$ is open, then $\mathrm{pr}_\alpha(v) \neq 0$ for all $\alpha \in \Delta$.*

PROOF. Let $c \in C$, $h \in H$ and $v \in V_\alpha$. As C is central in C it follows that $c \cdot (h \cdot v) = h \cdot (c \cdot v) = c^\alpha h \cdot v$. □

The set Δ has the properties that $0 \notin \Delta$, if $\alpha \in \Delta$, then $-\alpha \notin \Delta$, and finally there exists $\alpha_1, \ldots, \alpha_k \in \Delta$ such that if $\alpha \in \Delta$, then there are $n_1, \ldots, n_r \in \mathbb{N}_0$ such that

$$\alpha = n_1\alpha_1 + \cdots + n_r\alpha_r \,. \tag{5.4}$$

For $\alpha \in \Delta$ let $\mathcal{N}_\alpha = \{ X \in \mathfrak{c} \mid \alpha(X) = 0 \}$. Then $\bigcup_{\alpha\in\Delta} \mathcal{N}_\alpha$ is a finite union of hyperplanes and hence the complement is open and dense in $\mathfrak{c}$. Let $\mathfrak{c}^+$ be a connected component of the complement of $\bigcup \mathcal{N}_\alpha$. Because of (5.4) we can choose $\mathfrak{c}^+$ such that

$$\forall X \in \mathfrak{c}^* \forall \alpha \in \Delta : \alpha(X) > 0 . \tag{5.5}$$

Notice that $\mathfrak{c}^+$ is convex, $\mathfrak{c}^+ + \mathfrak{c}^+ \subset \mathfrak{c}^+$ and $\mathbb{R}^+ \mathfrak{c}^+ \subset \mathfrak{c}^+$.

LEMMA 5.3. *The group A contains a non-trivial abelian semigroup of expanding matrices.*

PROOF. Let $C^+ := \exp(\mathfrak{c}^+)$. Suppose that $a, b \in C^+$. Choose $X, Y \in \mathfrak{c}^+$ such that $a = \exp(X)$ and $b = \exp(Y)$. Then $ab = \exp(X + Y) \in C^+$. Thus C^+ is a semigroup. Let $a = \exp(X)$ be as above. Let $v = \sum_\alpha v_\alpha \in V$ with $v_\alpha \in$, then

$$\exp(X) \cdot v = \sum_\alpha e^{\alpha(X)} v_\alpha$$

and $e^{\alpha(X)} > 1$ because $\alpha(X) > 0$ for all α. □

Choice a basis $X_1, \ldots, X_r$ of $\mathfrak{a}$ such that the vectors $X_1, \ldots, X_k$ and the vectors $X_{k+1}, \ldots, X_r$ form a basis for the orthogonal complement of $\mathfrak{c}$ in $\mathfrak{a}$. Here we use the inner product $(X, Y) = \operatorname{Tr}(XY^T)$. As $\mathfrak{c}^+$ is an open cone in $\mathfrak{c}$ we can choose $X_j \in \mathfrak{c}^+$, $j = 1, \ldots, k$. Let

$$\Gamma_A := \{ \exp(\sum_{j=1}^{r} n_j X_j) \mid \forall j \, : \, n_j \in \mathbb{Z} \} .$$

Then Γ_A is a co-compact, discrete subgroup of A and every element of

$$\Gamma_A^+ := \{ \exp(n_1 X_1 + \cdots + n_k X_k) \mid \forall j \, : \, n_j \geq 0 \} \setminus \{\mathrm{id}\}$$

is expansive. As

$$\begin{aligned} \exp(n_1 X_1 + \cdots + n_r X_r) &\exp(m_1 X_1 + \cdots + m_r X_r) \\ &= \exp((n_1 + m_1) X_1 + \cdots + (n_r + m_r) X_r) \end{aligned}$$

it follows that Γ_A^+ is a subsemigroup of Γ_A such that $\Gamma_A \Gamma_A^+ \subset \Gamma_A$. Thus Γ_A is γ invariant for all $\gamma \in \Gamma_A^+$. We have therefore shown the following:

LEMMA 5.4. *There exists a co-compact discrete subgroup Γ of A and a subsemigroup Γ_A^+ such that each element of Γ_A^+ is expansive and $\Gamma_A \Gamma_A^+ \subset \Gamma_A$.*

References

[1] S.T. Al, J.-P. Antoine, and J.-P. Gazeau: Square integrability of group representations on homogeneous space, I. Reproducing triples and frames. *Ann. Inst. Henri Poincaré* **55** (1991), 829–856

[2] ———: *Coherent states, wavelets and their generalizations.* New York, Springer, 2000

[3] A. Aldroubi, D. Larson, W.-S. Tang, and E. Weber: Geometric Aspects of Frame Representations of Abelian Groups. Preprint, 2002

[4] A. Aldroubi, C. Cabrelli, U. Molter: Wavelet on irregular grids with arbitrary dilation matrices, and frame atome for $L^2(R^d)$. Preprint, 2003

[5] P. Aniello, G. Cassinelli, E. De Vito, and A. Levrero: Wavelet transforms and discrete frames associated to semidirect prodicts. *J. Math. Phys.* **39** (1998), 3965–3973

[6] L. Baggett, A. Carey, W. Moran, and P. Ohring: General Existence Theorem for Orthonormal Wavelet. An Abstract Approach. *Publ. Res. Inst. Mth. Sci. (1)* **31** (1995), 95–111

[7] L.W. Baggett, K.D. Merrill: Abstract harmonic analysis and wavelets in $\mathbb{R}^n$. The functional and harmonic analysis of wavelets and frames. In (San Antonio, TX, 1999), 17–27, Contemp. Math., 247, Amer. Math. Soc., Providence, RI, 1999.

[8] D. Bernier, K. F. Taylor: Wavelets from square-integrable reepresentations. *SIAM J. Math. Anal.* **27** (1996), 594–608

[9] N. Bopp, H. Rubenthaler, *Local Zeta functions attached to the minimal spherical series for a class of symmetric spaces.* Preprint, Strasbourg, 2002

[10] A. Borel, Harish-Chandra: Arithmetic subgroups of algebraic groups. *Ann. Math.* **75** (1962) 485–535

[11] ______: Combined MSF multiwavelets. *J. Fourier Anal. Appl.* **8** (2002), no. 2, 201–210.

[12] ______: On a problem of Daubechies. *Constr. Approx.* (2003)**19**, 179–190.

[13] M. Bownik and D. Speegle: The wavelet dimension function for real dilations and dilations admitting non-MSF wavelets. *Approximation theory, X (St. Louis, MO, 2001),* 63–85, Innov. Appl. Math., Vanderbilt Univ. Press, Nashville, TN, 2002.

[14] C.K. Chui, X. Shi: Orthonormal wavelets and tight frames with arbitrary real dilations. *Appl. Comput. Harmon. Anal.* **9** (2000), no. 3, 243–264.

[15] X. Dai, Y. Diao, Q. Gu and D. Han: The existence of subspace wavelet sets. Preprint

[16] X, Dai and D.R. Larson, *Wandering vectors for unitary systems and orthogonal wavelets.* Mem. Amer. Math. Soc. **134**, no 640 (1998)

[17] X. Dai, D.R. Larson and D. Speegle: Wavelet sets in $\mathbb{R}^n$, *J. Fourier Anal. Appl.* **3** (1997), no. 4, 451–456.

[18] X. Dai, D.R. Larson and D. Speegle: Wavelet sets in $\mathbb{R}^n$ II, *Wavelets, multiwavelets, and their applications (San Diego, CA, 1997),* 15–40, Contemp. Math., 216, *Amer. Math. Soc.,* Providence, RI, 1998.

[19] I. Daubechies, A. Grossmann, and Y. Meyer: Painless nonorthogonal expansions. *J. Math. Phys.* **27** (1986), 1271–1283

[20] M. Duflo, C.C. Moore: On the regular representation of a nonunimodular locally compact group. *J. Funct. Anal.* **21** (1976), 209–243

[21] R. Fabec, G. Ólafsson: The Continuous Wavelet Transform and Symmetric Spaces. *Acta Applicandae Math.* **77** (2003), 41–69

[22] J. Faraut, and A. Koranyi: Analysis on symmetric cones. Oxford University Press, Oxford, 1994

[23] H.G. Feichtinger, and K. Gröchenig: A unified approach to atomic decompositions via integrable group representations. In: *Proceedings, Conference on Functions, Spaces and Applications. Lund 1986* Lecture Notes in Mathematics **1302** (1986), 52–73, Springer-Verlag, New York/Berlin 1987

[24] ______: Banach spaces related to integrable group representations and their atomic decomposition. *J. Funct. Anal.* **86** (1989), 307–340

[25] ______: Gabor wavelets and the Heisenberg group: Gabor expansions and short time Fourier transform from the group theoretical point of view. In C. K. Chui (ed): *Wavelets: A tutorial in theory and applications*, 359–398. Academic Press, Boston, 1992

[26] H.G. Feichtinger, and W. Kozek: Quantization of TF lattice-invariant operators on elementary LCA groups. In: H. G. Feichtinger, and T. Strohmer (ed): *Gabor Analysis and Algorithms* p. 233-266, Birkhäuser, Boston, 1998,

[27] B. Fuglede: Commuting sel-adjoint partial differential operators and a group theoretical problem. *J. Func. Anal.* **16** (1974), 101–121

[28] H. Führ: Wavelet frames and admissibility in higher dimensions. *J. Math. Phys.* **37** (1996), 6353–6366

[29] ______: Continuous wavelet transforms with Abelian dilation groups. *J. Math. Phys.* **39** (1998), 3974–3986

[30] ______: Admissible vectors for the regular representation. To appear in *Proceedings of the AMS*

[31] H. Führ, M. Mayer: Continuous wavelet transforms from semidirect products: Cyclic representations and Plancherel measure. To appear in *J. Fourier Anal. Appl.*

[32] K. Gröchenig: Aspects of Gabor analysis on locally compact abelian groups. In: H. G. Feichtinger, and T. Strohmer (ed): *Gabor Analysis and Alogrithms* p. 211–231, Birkhäuser, Boston, 1998,

[33] ______: Foundations of Time-Frequency Analysis. Birkhäuser, Boston, 2001

[34] A. Grossmann, J. Morlet, and T. Paul: Transforms associated to square integrable group representations I. General results. *J. Math. Phys.* **26** (1985), 2473–2479

[35] ———: Transforms associated to square integrable group representations II. Examples. *Ann. Inst. Henri Poincaré: Phys. Theor.* **45** (1986), 293–309

[36] Y-H. Ha, H. Kang, J. Lee, J. Seo: Unimodular wavelets for L^2 and the Hardy space H^2. *Michigan Math. J.* **41**, (1994), 345–361.

[37] D. Han, and D. Larson: Frames, Bases, and Group Representations. *Mem. Amer. Mth. Soc.* **147** (2000)

[38] D. Han, and Y. Wang: Lattice Tiling and the Weyl-Heisenberg frames. *Geom. Funct. Anal.* **11** (2001), 741–758

[39] C.E. Heil, and D.F. Walnut: Continuous and discrete wavelet transform. *SIAMRev.* **31** (1989), 628–666

[40] E. Hernandez and G. Weiss: A first course on wavelets. With a foreword by Yves Meyer. Studies in Advanced Mathematics. *CRC Press, Boca Raton, FL,* 1996

[41] A. Iosevich, N.H. Katz, and T. Tao: Convex bodies with a point of curvature do not have Fourier bases. *Amer. J. Math.* **123** (2001), 115–120

[42] P. Jorgensen, and S. Pedersen: Orthogonal harmonic analysis of fractal measures. *Elec. Res. Announc. Amer. Math. Soc.* **4** (1998), 35–42

[43] ———: Spectral pairs in Cartesian coordinates *J. Fourier Anal. Appl.* **5** (1999), 285–302

[44] J.C. Lagarias, and Y. Wang: Spectral sets and factorizations of finite abelian groups. *J. Func. Anal.* **145** (1997), 73–98

[45] D.R. Larson: *Von Neumann algebras and wavelets,* Operator algebras and applications (Samos, 1996), 267-312, NATO Adv. Sci. Inst. Ser. C Math. Phys. Sci., 495, Kluwer Acad. Publ., Dordrecht, 1997

[46] R.S. Laugesen, N. Weaver, G.L. Weiss, E.N. Wilson: A characterization of the higher dimensional groups associated with continuous wavelets. *J. Geom. Anal.* **12** (2002), 89–102.

[47] R. Lipsman, J.A. Wolf: The Plancherel formula for parabolic subgroups of the classical groups. *J. Analyse Math.* **34** (1978), 120–161

[48] ———: Canonical semi-invariants and the Plancherel formula for parabolic groups. *Trans. AMS* **269** (1982), 111–131

[49] Macev: On a class of homogeneous spaces. *Izvestiya Akademii Nauk SSSR* Seriya Mathematischeskia **13** (1949), 9–32

[50] G. Ólafsson: Fourier and Poisson transform associated to semisimple symmetric spaces. *Invent. Math.* **90** (1987), 605-629

[51] ———: Continuous action of Lie groups on $\mathbb{R}^n$ and frames. Preprint, 2002

[52] Z. Rzeszotnik: Characterization theories in the theory of wavelets. PhD Dissertation (2001), Washington University in Saint Louis

[53] Z. Rzeszotnik and D. Speegle: On wavelets interpolated from a pair of wavelet sets. *Proc. Amer. Math. Soc.* **130** (2002), no. 10, 2921–2930

[54] D. Speegle: On the existence of wavelets for non-expansive dilations, *Collect. Math.* **54**, 2, (2003), 163–179.

[55] E. Stein and G. Weiss: Introduction to Fourier analysis on Euclidean spaces. Princeton Mathematical Series **32**. Princeton, New Jersey, 1971

[56] T. Tao: Fuglede conjecture is false in 5 and higher dimensions. Preprint 2003, http://arXiv.org/abs/math/0306134

[57] Yang Wang: Wavelets, tiling, and spectral sets. *Duke Math. J.* **114** (2002), no. 1, 43–57.

[58] E. Weber: Frames and Single Wavelets for Unitary Groups. *Canad. J. Math.* **54** (3) (2002), 634–647

[59] G. Weiss and E.N. Wilson: The Mathematical Theory of Wavelets. In: J.S. Byrnes (ed.) *Twenty Century Harmonic Analysis—A Celebration*, 329–366, Kluwer Academic Publisher, 2001

[60] J. Wolf, *Classification and Fourier inversion for parabolic subgroups with square integrable nilradical*, Mem. AMS **225** (1979)

Department of Mathematics, Louisiana State University, Baton Rouge, LA 70803, USA

E-mail address: olafsson@math.lsu.edu

Department of Mathematics, Saint Louis University, St. Louis, MO 63103

E-mail address: speegled@slu.edu

Contemporary Mathematics
Volume **345**, 2004

Orthonormalized coherent states

Alexander M. Powell

ABSTRACT. We examine the uncertainty principle for orthonormal bases and a theorem of Bourgain. We survey recent work in this area and prove a result on the time and frequency localization of orthonormalized Gaussian coherent states which shows the connection between Bourgain's theorem and Gabor analysis.

1. Introduction

Given $f \in L^2(\mathbb{R})$, we formally define the Fourier transform of f by

$$\widehat{f}(\gamma) = \int f(t) e^{-2\pi i t\gamma} dt.$$

The uncertainty principle in harmonic analysis is the general statement that a function and its Fourier transform can not both be "too well localized", [**HJ**], [**FS**], [**B1**]. A classical example of an uncertainty principle is Heisenberg's inequality.

THEOREM 1.1 (Heisenberg Uncertainty Principle). *For every* $f \in L^2(\mathbb{R})$ *and any* $a, b \in \mathbb{R}$,

$$\|(t-a)f\|_{L^2(\mathbb{R})} \, \|(\gamma - b)\widehat{f}\|_{L^2(\mathbb{R})} \geq \frac{1}{4\pi} \|f\|^2_{L^2(\mathbb{R})}. \tag{1.1}$$

Moreover, equality holds in (1.1) *if and only if* $f(t) = Ce^{2\pi i b t} e^{-c(t-a)^2}$ *for some constants* $C \in \mathbb{C}$ *and* $c > 0$.

There is an extensive literature on generalizations of Heisenberg's inequality, e.g., see [**FS**]. Heisenberg's uncertainty principle uses the weights t^2 and γ^2 to measure localization. Throughout this paper we shall also focus on generalizations of the uncertainty principle which use different combinations of weights.

In the next definition, $\widehat{\mathbb{R}} = \mathbb{R}$ denotes the dual group of $\mathbb{R}$. We shall use this notation whenever we wish to emphasize that we are dealing with the frequency domain. For example, if $f \in L^2(\mathbb{R})$ then $\widehat{f} \in L^2(\widehat{\mathbb{R}})$.

2000 *Mathematics Subject Classification.* Primary 42C15; Secondary 42C40.
Key words and phrases. Fourier analysis, Gabor analysis, the uncertainty principle.
The author was supported in part by ONR Grant N000140210398.

DEFINITION 1.2. *Let u and v be nonnegative Borel measurable functions on $\widehat{\mathbb{R}}$ and $\mathbb{R}$, respectively. If $1 < p, p', q, q' < \infty$ with $\frac{1}{p} + \frac{1}{p'} = 1$, $\frac{1}{q} + \frac{1}{q'} = 1$ and if there is a constant $K > 0$ such that*

$$\sup_{s>0} \left(\int_0^{1/s} u(\gamma) d\gamma \right)^{1/q} \left(\int_0^s v(t)^{-p'/p} dt \right)^{1/p'} = K$$

then we say $(u, v) \in F(p, q)$.

DEFINITION 1.3. *Given a nonnegative Borel measurable function v, $L^p_v(\mathbb{R})$ is the set of all Borel measurable functions f for which*

$$\|f\|_{p,v} \equiv \left(\int |f(t)|^p v(t) dt \right)^{1/p} < \infty.$$

The following weighted uncertainty principle [**B1**] is an example of how the classical uncertainty principle can be generalized.

THEOREM 1.4 (Benedetto, Heinig). *Let $1 < p \leq q < \infty$, and let u and v be even weights on $\widehat{\mathbb{R}}$ and $\mathbb{R}$ for which $(u, v) \in F(p, q)$ with constant K (as above). Assume $1/u$ and v are increasing on $(0, \infty)$. Then there is a constant $C(K)$ such that*

$$\|f\|^2_{L^2(\mathbb{R})} \leq 4\pi C(K) \|tf(t)\|_{p,v} \|\gamma \widehat{f}(\gamma)\|_{q', u^{-q'/q}}$$

for all f in the Schwartz class $\mathcal{S}(\mathbb{R})$. Here, p', q' are as in definition 1.2.

This result illustrates the theme of how different pairings of weights can be used to formulate uncertainty principles.

2. Contents

In section 3, we present background on frames and Gabor systems which will be needed later. In section 4, we discuss theorems concerning the uncertainty principle for orthonormal bases. We first consider standard results for the (t^2, γ^2) weight and then look at more recent work involving the (t^p, γ^q) weights. Our main focus will be on a result of Bourgain and a weighted generalization due to John Benedetto and the author.

In section 5, we discuss some issues arising in the proof of Bourgain's theorem. Sections 6 through 8 are devoted to the proof of our main result, theorem 6.1. This result gives localization estimates for orthonormalized Gaussian coherent states. This is closely related to a key step in the proof of Bourgain's theorem and thus gives insight into connections with Gabor systems.

3. Background: frames and Gabor systems

We shall need some standard background material on frames and Gabor systems. [**G1**] and [**C**] provide excellent introductions to these topics.

3.1. Frames.

DEFINITION 3.1. *A sequence $\{f_n : n \in \mathbb{N}\} \subseteq L^2(\mathbb{R})$ is a frame for $L^2(\mathbb{R})$ if there exist constants $0 < A \leq B < \infty$ such that*

$$\forall f \in L^2(\mathbb{R}), \quad A \|f\|^2_{L^2(\mathbb{R})} \leq \sum_{n \in \mathbb{N}} |\langle f, f_n \rangle|^2 \leq B \|f\|^2_{L^2(\mathbb{R})}.$$

Orthonormal bases are the canonical examples of frames. However, in general, frames are neither orthonormal nor bases. We have the following definition for the case when a frame is a basis.

DEFINITION 3.2. *A sequence* $\{f_n : n \in \mathbb{N}\} \subseteq L^2(\mathbb{R})$ *is a Riesz basis for its closed linear span if there exist constants* $0 < A \leq B < \infty$ *such that for every finite sequence* $\{c_n\} \subseteq \mathbb{C}$

$$A\,\|c\|_{l^2(\mathbb{N})} \leq \left\|\sum_{n\in\mathbb{N}} c_n f_n\right\|_{L^2(\mathbb{R})} \leq B\,\|c\|_{l^2(\mathbb{N})}. \tag{3.1}$$

If $\{f_n : n \in \mathbb{N}\}$ *is also a frame for* $L^2(\mathbb{R})$ *then we say* $\{f_n : n \in \mathbb{N}\}$ *is a Riesz basis for* $L^2(\mathbb{R})$.

For example, any finite set of linearly independent functions in $L^2(\mathbb{R})$ is a Riesz basis for its span. The following elementary result illustrates the "minimality" property of Riesz bases.

LEMMA 3.3. *Suppose* $\{f_n : n \in \mathbb{N}\} \subseteq L^2(\mathbb{R})$ *is a Riesz basis for its closed linear span. One can not have*

$$f_N \in \overline{\text{span}}\,\{f_j : j \neq N\}$$

for any $N \in \mathbb{N}$.

PROOF. Without loss of generality suppose $N = 1$, and that

$$f_1 \in \overline{\text{span}}\,\{f_j : j \neq 1\}.$$

Let $0 < A \leq B < \infty$ be the constants in (3.1). By our assumption there exists a sequence $\{c_j\}_{j=2}^M$ such that

$$\left\|f_1 - \sum_{j=2}^{M} c_j f_j\right\|_{L^2(\mathbb{R})} \leq \frac{A}{2}.$$

Combining this with (3.1) and the fact that $\{f_j\}$ is a Riesz basis for its span gives

$$A \leq A\left(1 + \sum_{j=2}^{M} |c_j|^2\right)^{\frac{1}{2}} \leq \left\|f_1 - \sum_{j=2}^{M} c_j f_j\right\|_{L^2(\mathbb{R})} \leq \frac{A}{2}.$$

This is a contradiction, since $A > 0$. □

Frames are widely studied because they can be used to give stable decompositions of functions, e.g. see Theorem 3.2 of [**B1**]. Moreover, the possibility of redundancy in frames makes them more desirable than bases in many applications, e.g. [**B2**]. A large portion of frame theory has focused on the study of very specific classes of frames. Frames given by Gabor systems have attracted much attention.

3.2. Gabor systems.

DEFINITION 3.4. *Given* $f \in L^2(\mathbb{R})$ *and* $a, b > 0$, *the Gabor system*

$$\mathcal{G}(f, a, b) = \{f_{m,n} : m, n \in \mathbb{Z}\}$$

is defined by

$$f_{m,n}(t) = e^{-2\pi i a m t} f(t - bn).$$

In view of the previous section, one is especially interested in finding $f \in L^2(\mathbb{R})$ and $a, b > 0$ for which $\mathcal{G}(f, a, b)$ is a frame for $L^2(\mathbb{R})$. A great deal is known about this problem. The Ron-Shen duality theorem, [**G1**], is particularly illustrative as it highlights the role of the parameters a and b.

THEOREM 3.5 (Ron-Shen Duality). *Let $f \in L^2(\mathbb{R})$ and $a, b > 0$. $\mathcal{G}(f, a, b)$ is a frame for $L^2(\mathbb{R})$ if and only if $\mathcal{G}(f, \frac{1}{b}, \frac{1}{a})$ is a Riesz basis for its closed linear span.*

The next result shows that Gabor systems possess a nice linear independence property.

THEOREM 3.6. *Let $f \in L^2(\mathbb{R})$ and $a, b > 0$. Any finite subset of $\mathcal{G}(f, a, b)$ is linearly independent.*

The case $a = b = 1$ of the above result is first due to Heil, Ramanathan and Topiwala, [**HRT**], and the general (lattice) case was later solved by Linnell, [**L**]. Heil, Ramanathan, and Topiwala conjectured this result remains true even for irregular Gabor systems.

4. Uncertainty principles for orthonormal bases

The introduction discussed how the uncertainty principle constrains the localization properties of an individual function and its Fourier transform. We saw how different weights can be used to measure uncertainty. In this section, we discuss how the uncertainty principle constrains the localization properties of orthonormal bases for $L^2(\mathbb{R})$. Once again, we shall see different combinations of weights playing a role.

4.1. The weights t^2 and γ^2. The Balian-Low theorem is an uncertainty principle for Gabor orthonormal bases. It says that a function which generates a Gabor orthonormal basis can not be too well localized in the following sense.

THEOREM 4.1 (Balian-Low). *If $f \in L^2(\mathbb{R})$ and $\mathcal{G}(f, 1, 1)$ is an orthonormal basis for $L^2(\mathbb{R})$ then either*

$$\int |t|^2 |f(t)|^2 dt = \infty$$

or

$$\int |\gamma|^2 |\widehat{g}(\gamma)|^2 d\gamma = \infty.$$

This result is also true in much greater generality. For example, it holds for Riesz bases and also holds in higher dimensions, [**DJ**], [**BCM**], [**GHHK**].

The weights t^2 and γ^2 play a crucial role here. In contrast to the Balian-Low theorem, it has been shown [**BCGP**] that if weaker weights are used, then one *can* have Gabor orthonormal bases whose generator is well-localized in time and frequency with respect to the weakened weights. To be precise, it was shown

THEOREM 4.2 (Benedetto, Czaja, Gadziński, Powell). *There exists $f \in L^2(\mathbb{R})$ such that $\mathcal{G}(f, 1, 1)$ is an orthonormal basis for $L^2(\mathbb{R})$ and such that for every $d > 2$*

$$\int \frac{1 + |t|^2}{\log^d(|t| + 2)} |f(t)|^2 dt < \infty \tag{4.1}$$

and

$$\int \frac{1 + |\gamma|^2}{\log^d(|\gamma| + 2)} |\widehat{f}(\gamma)|^2 d\gamma < \infty. \tag{4.2}$$

See [**BCGP**] for more information on the range of d in this theorem and the role of the corresponding logarithm. For example, (4.1) still holds with $d > 1$. The question of deriving sharper estimates for (4.2) was left open in [**BCGP**].

In view of theorems 4.1 and 4.2, it is natural to ask what happens for general (i.e., non-Gabor) orthonormal bases, namely, what sort of "uniform localization" can a general orthonormal basis for $L^2(\mathbb{R})$ have with respect to the weights t^2 and γ^2. This question was posed by Balian, [**Bal**], and answered by Bourgain, [**Bou**].

THEOREM 4.3 (Bourgain). *Let $\epsilon > 0$. There exists an orthonormal basis, $\{g_n : n \in \mathbb{N}\}$, for $L^2(\mathbb{R})$ such that*

$$\forall n \in \mathbb{N}, \quad \Delta^2(g_n) \leq \left(\frac{1}{2\sqrt{\pi}} + \epsilon\right)^2$$

and

$$\forall n \in \mathbb{N}, \quad \Delta^2(\widehat{g_n}) \leq \left(\frac{1}{2\sqrt{\pi}} + \epsilon\right)^2.$$

The variance, $\Delta^2(\cdot)$, is defined by

$$\Delta^2(f) \equiv \inf_{a\in\mathbb{R}} \int |t-a|^2 |f(t)|^2 dt.$$

This result uses variances to measure localization; the uniform boundedness of the variances reflects a type of uniform localization of the basis with respect to the t^2 and γ^2 weights.

To put this in perspective, note that there are $\psi \in \mathcal{S}(\mathbb{R})$ which generate wavelet orthonormal bases, $\{\psi_{m,n} : m, n \in \mathbb{Z}\}$, for $L^2(\mathbb{R})$, [**LM**]. Since $\psi \in \mathcal{S}(\mathbb{R})$, each $\Delta^2(\psi_{m,n})$ and $\Delta^2(\widehat{\psi_{m,n}})$ is finite. However, these variances are not uniformly bounded for any wavelet system (although their product may be), e.g. see [**B1**].

Theorems 4.1, 4.2, and 4.3 form a trio of results which gives insight into the boundaries of uncertainty for the t^2 and γ^2 weights.

4.2. The weights t^p and γ^q. Motivated by the theme of using different weights to obtain different uncertainty principles, we now consider the situation for the weights t^p and γ^q, where $\frac{1}{p} + \frac{1}{q} = 1$. This choice of weights arises naturally in the following Balian-Low type theorem of Feichtinger and Gröchenig.

THEOREM 4.4 (Feichtinger, Gröchenig). *Let $\epsilon > 0$ and $\frac{1}{p} + \frac{1}{q} = 1$ with $1 < p, q < \infty$. If $f \in L^2(\mathbb{R})$ and $\mathcal{G}(f, 1, 1)$ is an orthonormal basis for $L^2(\mathbb{R})$ then either*

$$\int |t|^{p+\epsilon} |g(t)|^2 dt = \infty$$

or

$$\int |\gamma|^{q+\epsilon} |\widehat{g}(\gamma)|^2 d\gamma = \infty.$$

The reference for this result is that it follows from theorem 4.4 of [**FG**] and theorem 1 of [**G2**]. As in the case $(p, q) = (2, 2)$, it was proven that Gabor bases are possible if the t^p and γ^q weights are weakened slightly, [**BCGP**]. In particular,

THEOREM 4.5 (Benedetto, Czaja, Gadziński, Powell). *Let $\frac{1}{p} + \frac{1}{q} = 1$ with $1 < p, q < \infty$. There exists $f \in L^2(\mathbb{R})$ such that $\mathcal{G}(f, 1, 1)$ is an orthonormal basis* *for $L^2(\mathbb{R})$ and such that for every $d > 2$*

$$\int \frac{1+|t|^p}{\log^d(|t|+2)}|f(t)|^2 dt < \infty$$

and

$$\int \frac{1+|\gamma|^q}{\log^d(|\gamma|+2)}|\widehat{f}(\gamma)|^2 d\gamma < \infty.$$

In [**BP**], we consider what sort of localization a general orthonormal basis can have with respect to the t^p and γ^q weights. We generalize theorem 4.3 to deal with these weights. We use the following "generalized variances" to measure localization.

DEFINITION 4.6. *Given $f \in L^2(\mathbb{R})$ and $\lambda > 0$ we define the generalized variance of f by*

$$\Delta^2_\lambda(f) = \inf_{a\in\mathbb{R}} \int |t-a|^\lambda |f(t)|^2 dt.$$

In terms of this definition, we proved [**BP**]

THEOREM 4.7 (Benedetto, Powell). *Let $1 < p, q < \infty$ satisfy $\frac{1}{p} + \frac{1}{q} = 1$. Fix $\epsilon > 0$ and $\varphi \in C_c^\infty$ with $\|\varphi\|_{L^2(\mathbb{R})} = 1$. There exists an orthonormal basis, $\{g_n : n \in \mathbb{N}\}$, for $L^2(\mathbb{R})$, such that*

$$\forall n \in \mathbb{N}, \quad \Delta^2_p(g_n) \le \int |t|^p |\varphi(t)|^2 dt + \epsilon$$

and

$$\forall n \in \mathbb{N}, \quad \Delta^2_q(\widehat{g_n}) \le \int |\gamma|^q |\widehat{\varphi}(\gamma)|^2 d\gamma + \epsilon.$$

Theorems 4.4, 4.5, and 4.7 form a trio of results which gives insight into the role of the weights t^p and γ^q in uncertainty principles for orthonormal bases, and which extend the original $(p,q) = (2,2)$ results.

5. On Bourgain's theorem

The proofs of Bourgain's theorem and Theorem 4.7 are both rather technical. Both proofs start by constructing finite, orthonormal, well-localized systems of functions in $L^2(\mathbb{R})$. Since these systems are not complete, an orthonormal basis is built by carefully taking linear combinations of these finite systems with a dense set of functions. The proof of Bourgain's theorem is summarized as:

(1) Take $\{f_n\} \subseteq C_c^\infty(\mathbb{R})$ which is dense in the unit sphere of $L^2(\mathbb{R})$.

(2) For each f_n construct a finite, "well-localized", orthonormal system, S_n.

(3) To construct the basis, take appropriate linear combinations of f_n with the elements of S_n.

Roughly speaking, the sequence $\{f_n\}$ contributes completeness to the basis, while the elements of each S_n control the generalized variances $\Delta^2_p(\cdot)$ and $\Delta^2_q(\cdot)$. Our goal here is to answer the question of why the proofs follow the above blueprint.

QUESTION 5.1. *Why do the proofs of Bourgain's theorem and Theorem 4.7 proceed in the manner mentioned above?*

To answer this, let's begin by taking another look at Bourgain's theorem. A naive alternate idea for constructing a basis of the type in Bourgain's theorem is to start with a complete set of functions which already has the desired uniform localization and to orthonormalize it to obtain a basis.

For example, if $g(t) = 2^{1/4}e^{-\pi t^2}$ then the elements of the Gabor system $\mathcal{G}(g, 1, 1)$ all satisfy

$$\Delta^2(g_{m,n}) = \frac{1}{4\pi} \quad \text{and} \quad \Delta^2(\widehat{g_{m,n}}) = \frac{1}{4\pi}. \tag{5.1}$$

Moreover, $\mathcal{G}(g, 1, 1)$ is complete, but is not a Riesz basis for $L^2(\mathbb{R})$, or even a frame for $L^2(\mathbb{R})$. This follows (e.g., see [**F**], [**G1**], [**BW**]) from

THEOREM 5.2. *Let* $g(t) = 2^{1/4}e^{-\pi t^2}$.

(1) $\mathcal{G}(g, a, b)$ *is a frame for* $L^2(\mathbb{R})$, *if* $ab < 1$.
(2) $\mathcal{G}(g, a, b)$ *is incomplete in* $L^2(\mathbb{R})$, *if* $ab > 1$.
(3) $\mathcal{G}(g, a, b)$ *is complete in* $L^2(\mathbb{R})$, *but is not a frame, if* $ab = 1$. *Moreover,* $\mathcal{G}(g, a, b)$ *remains complete if any element is removed, but is no longer complete if two elements are removed.*

Thus, if one orthonormalizes $\mathcal{G}(g, 1, 1)$ with respect to some indexing of $\mathbb{Z} \times \mathbb{Z}$ then one obtains a new system, $\mathcal{O}(g, 1, 1)$, which is an orthonormal basis for $L^2(\mathbb{R})$. Since all the elements of $\mathcal{G}(g, 1, 1)$ satisfy (5.1) it is plausible that the elements of the orthonormalized system also have uniformly bounded time and frequency variances. However, it is difficult to estimate the variances here because $\mathcal{G}(g, 1, 1)$ is not a Riesz basis. The difficulties arise specifically because it is difficult to estimate the spectra of the Gramian matrices which arise in the Gram-Schmidt orthogonalization process. The relationship between Riesz bases and Gramian matrices sheds some light on this, e.g. see Proposition 5.1.5. in [**G1**].

On the other hand, if one chooses $ab > 1$ then it follows from the Ron-Shen duality theorem that $\mathcal{G}(g, a, b)$ is a Riesz basis for its closed linear span. By theorem 5.2, $\mathcal{G}(g, a, b)$ is incomplete. Thus, if one orthonormalizes $\mathcal{G}(g, a, b)$, the resulting system $\mathcal{O}(g, a, b)$ is not complete. However, since $\mathcal{G}(g, a, b)$ is a Riesz basis for its closed linear span, we show it is possible to estimate the variances of the elements of $\mathcal{O}(g, a, b)$.

In summary,

- $\mathcal{O}(g, 1, 1)$ is an orthonormal basis for $L^2(\mathbb{R})$, but it is difficult to estimate the time and frequency variances of the elements in $\mathcal{O}(g, 1, 1)$.
- If $ab > 1$ then $\mathcal{O}(g, a, b)$ is not complete in $L^2(\mathbb{R})$, but one can derive variance estimates.

This explains why the proof of Bourgain's theorem proceeds as it does. Since the idea outlined in the first bullet is difficult to carry out, Bourgain's proof essentially uses the second idea and adds in completeness by cleverly taking linear combinations with a dense sequence. In particular, the finite, orthonormal, well-localized systems used in the proof of Bourgain's theorem are just finite subsets of a Gabor system which has been orthonormalized.

In Bourgain's theorem, one doesn't work with the Gaussian (as in the above discussion), but instead with a compactly supported function. The compact support makes the orthonormalization easier. The remaining sections focus on orthonormalizing $\mathcal{G}(g, 2, 2)$, where g is the Gaussian. For comparison see [**J**], where orthonormalization calculations for wavelet systems were done.

6. Orthonormalizing coherent states

Consider the indexing of $\mathbb{Z} \times \mathbb{Z}$ which begins

$$(0,0),(1,1),(0,1),(-1,1),(-1,0),(-1,-1),(0,-1),(1,-1),(1,0),$$
$$(2,2),(0,2),(-2,2),(-2,0),(-2,-2),(0,-2),\dots$$

and continues to spiral outwards in this manner. Let $\mathcal{O}(g,2,2)$ be the system which results when $\mathcal{G}(g,2,2)$ is orthonormalized in the above order. We examine the time and frequency localization of the elements in $\mathcal{O}(g,2,2)$. To simplify the exposition, we shall only derive estimates for those elements whose index is of the form $(n,n), n \in \mathbb{N}$.

Let $\varphi_{n,n}$ be the function obtained when $g_{n,n}$ is orthogonalized with respect to the previous elements of $\mathcal{G}(g,2,2)$, indexed as above, namely,

$$\varphi_{n,n}(t) = g_{n,n}(t) - \sum_{j=-(n-1)}^{n-1} \sum_{k=-(n-1)}^{n-1} a_{j,k}^n g_{j,k}(t), \tag{6.1}$$

where the $a_{j,k}^n$ are chosen to ensure that

$$\varphi_{n,n} \text{ is orthogonal to } \{g_{j,k} : -(n-1) \leq j,k \leq (n-1)\}. \tag{6.2}$$

Note that by theorem 3.6, $\varphi_{n,n}$ and the $\{a_{j,k}^n\}$ are uniquely determined. To normalize the $\varphi_{n,n}$, let

$$\psi_{n,n}(t) = \varphi_{n,n}(t) / \|\varphi_{n,n}\|_{L^2(\mathbb{R})}\,.$$

Thus, $\psi_{n,n}$ is the element of $\mathcal{O}(g,2,2)$ with index (n,n). We shall show

THEOREM 6.1. *For every $p \in \mathbb{N}$ there exists a constant C_p such that*

$$\Delta_p^2(\psi_{n,n}) \leq C_p \quad \textit{and} \quad \Delta_p^2(\widehat{\psi_{n,n}}) \leq C_p$$

holds for all $n \in \mathbb{N}$.

The main part of the proof is devoted to estimating the $\{a_{j,k}^n\}$ in (6.1). This is the content of the next section.

7. Estimating the $\{a_{j,k}^n\}$

To estimate the $\{a_{j,k}^n\}$, we begin by using (6.2) to take inner products in (6.1). This yields

$$\langle g_{n,n}, g_{p,q}\rangle = \sum_{j=1-n}^{n-1} \sum_{k=1-n}^{n-1} a_{j,k}^n \langle g_{j,k}, g_{p,q}\rangle, \tag{7.1}$$

for all (p,q) satisfying $-(n-1) \leq p,q \leq n-1$. The following lemma gives explicit values for the inner products in (7.1).

LEMMA 7.1. *Let $h(t) = 2^{1/4}e^{-\pi t^2}$. Fix $a,b>0$ and let $\mathcal{G}(h,a,b) = \{h_{m,n}\}_{m,n\in\mathbb{Z}}$. Then*

$$\widehat{h} = h \quad \textit{and} \quad \|h\|_{L^2(\mathbb{R})} = 1 \tag{7.2}$$

and

$$\langle h_{m,n}, h_{k,l}\rangle = e^{-\frac{\pi}{2}b^2(n-l)^2} e^{-\frac{\pi}{2}a^2(m-k)^2} e^{-\pi i ab(n+l)(m-k)}. \tag{7.3}$$

PROOF. A standard calculation gives (7.2). To see (7.3) we calculate as follows:

$$\begin{aligned}
\langle h_{m,n}, h_{k,l}\rangle &= \int h(t-nb)\overline{h(t-lb)}e^{-2\pi i a t(m-k)}dt \\
&= \sqrt{2}\int e^{-\pi(t-nb)^2}e^{-\pi(t-lb)^2}e^{-2\pi i a t(m-k)}dt \\
&= \sqrt{2}\int e^{-\pi(t^2-2nbt+n^2b^2+t^2-2lbt+l^2b^2)}e^{-2\pi i a t(m-k)}dt \\
&= \sqrt{2}e^{-\pi b^2(n^2+l^2)}\int e^{-2\pi(t^2-b(n+l)t)}e^{-2\pi i a t(m-k)}dt \\
&= \sqrt{2}e^{-\pi b^2(n^2+l^2)}e^{2\pi(\frac{b}{2}(n+l))^2}\int e^{-2\pi(t^2-b(n+l)t+\frac{b^2}{4}(n+l)^2)}e^{-2\pi i a t(m-k)}dt \\
&= \sqrt{2}e^{-\pi b^2(n^2+l^2-\frac{1}{2}(n+l)^2)}\int e^{-2\pi(t-\frac{b}{2}(n+l))^2}e^{-2\pi i a t(m-k)}dt \\
&= \sqrt{2}e^{-\frac{\pi}{2}b^2(n-l)^2}e^{-\pi i a b(n+l)(m-k)}\mathcal{F}\left[e^{-2\pi t^2}\right](a(m-k)) \\
&= \sqrt{2}e^{-\frac{\pi}{2}b^2(n-l)^2}e^{-\pi i a b(n+l)(m-k)}\frac{1}{\sqrt{2}}e^{-\frac{\pi}{2}a^2(m-k)^2} \\
&= e^{-\frac{\pi}{2}b^2(n-l)^2}e^{-\frac{\pi}{2}a^2(m-k)^2}e^{-\pi i a b(n+l)(m-k)},
\end{aligned}$$

where $\mathcal{F}$ denotes the Fourier transform. □

COROLLARY 7.2. *Let $g(t) = 2^{1/4}e^{-\pi t^2}$ and let $\mathcal{G}(g,2,2) = \{g_{m,n}\}_{m,n\in\mathbb{Z}}$. Then*

$$\langle g_{m,n}, g_{k,l}\rangle = e^{-2\pi(n-l)^2}e^{-2\pi(m-k)^2}. \tag{7.4}$$

Using this corollary, we see that (7.1) is equivalent to

$$G\mathfrak{a} = \mathfrak{g}, \tag{7.5}$$

where

$$\mathfrak{g} = \begin{pmatrix} e^{-2\pi(1)^2}v \\ e^{-2\pi(2)^2}v \\ \cdots \\ e^{-2\pi(2n-1)^2}v \end{pmatrix} \text{ with } v = \begin{pmatrix} e^{-2\pi(1)^2} \\ e^{-2\pi(2)^2} \\ \cdots \\ e^{-2\pi(2n-1)^2} \end{pmatrix}$$

and

$$\mathfrak{a} = \begin{pmatrix} y_{n-1} \\ y_{n-2} \\ \cdots \\ y_{1-n} \end{pmatrix} \text{ with } y_j = \begin{pmatrix} a^n_{j,n-1} \\ a^n_{j,n-2} \\ \cdots \\ a^n_{j,1-n} \end{pmatrix}$$

and

$$G = \begin{pmatrix} B_0 & B_1 & B_2 & \cdots & B_{2n-2} \\ B_1 & B_0 & B_1 & \cdots & B_{2n-3} \\ \cdots & \cdots & \cdots & \cdots & \cdots \\ B_{2n-2} & \cdots & B_2 & B_1 & B_0 \end{pmatrix}$$

where

$$B_j = e^{-2\pi j^2}C$$

and

$$C \equiv \begin{pmatrix} 1 & e^{-2\pi(1)^2} & e^{-2\pi(3)^2} & \cdots & e^{-2\pi(2n-2)^2} \\ e^{-2\pi(1)^2} & 1 & e^{-2\pi(1)^2} & \cdots & e^{-2\pi(2n-3)^2} \\ \cdots & \cdots & \cdots & \cdots & \cdots \\ e^{-2\pi(2n-2)^2} & e^{-2\pi(2n-3)^2} & e^{-2\pi(2n-4)^2} & \cdots & 1 \end{pmatrix}.$$

We have suppressed the fact that all these matrices depend on n. We may use the following result to estimate the spectrum of G. It appears, among other places, in [**QSS**].

THEOREM 7.3 (Gershgorin Theorem). *Let $M = (M_{i,j})$ be a $d \times d$ matrix with complex entries. The spectrum of M satisfies*

$$\sigma(M) \subseteq \bigcup_{i=1}^{d} C_i, \qquad C_i = \{z \in \mathbb{C} : |z - M_{i,i}| \leq \sum_{j=1, j\neq i}^{d} |M_{i,j}|\}.$$

We now apply the Gershgorin theorem to our matrix G. Note that

$$\sup_i \sum_{j=1, j\neq i}^{(2n-1)^2} |G_{i,j}| = \left(\sum_{j=-n}^{n} e^{-2\pi j^2} \right)^2 - 1 \leq \left(\sum_{j\in\mathbb{Z}} e^{-2\pi j^2} \right)^2 - 1 \equiv r.$$

One may verify that $0 < r < 1$, and, in fact, numerically one has that $r \approx .0075$. Thus, since G is self-adjoint, $\sigma(G) \subseteq [1 - r, 1 + r]$ is bounded away from 0. Note that while G depends on n, this spectrum bound is independent of n. Likewise, if we let $R = I - G$, then we have $\sigma(R) \subseteq [-r, r]$. This allows us to derive the following block-version of Jaffard's lemma for our matrix G. Although the proof is essentially the same as Jaffard's, [**J**], we nonetheless include it for the sake of completeness.

LEMMA 7.4. *There exist $C, \delta > 0$ independent of n, such that*

$$|G^{-1}(j,k)| \leq Ce^{-\delta|j'-k'|}e^{-\delta|J-K|},$$

where $j = J(2n-1) + j', k = K(2n-1) + k'$, and $0 \leq J, K < 2n-1$ and $0 < j', k' \leq 2n-1$. Recall G is a $(2n-1)^2 \times (2n-1)^2$ matrix.

PROOF. Let $R = I - G$. We have $\sigma(R) \subseteq [-r, r]$, where $0 < r < 1$ is as above. Therefore,

$$|R^m(j,k)| \leq \left(\sum_{l=1}^{(2n-1)^2} |R^m(j,l)|^2 \right)^{\frac{1}{2}} = \|R^m e_j\|_2 \leq \|R^m\| \leq r^m. \tag{7.6}$$

Here, $\|\cdot\|_2$ denotes the l^2 norm of a vector and $\|\cdot\|$ denotes the corresponding matrix norm it induces. Also, $\{e_j\}$ is the canonical basis for l^2. By the definitions of G and R we have

$$|R(j,k)| \leq ce^{-|J-K|}e^{-|j'-k'|}. \tag{7.7}$$

In particular,

$$\forall\, 0 < \delta < 1, \;\; |R(j,k)| \leq ce^{-\delta|J-K|}e^{-\delta|j'-k'|}. \tag{7.8}$$

Next, note that one has the following simple inequality

$$\sum_{j=1}^{n} e^{-|m-j|}e^{-\delta|j-k|} \leq \sum_{j=1}^{\infty} e^{-|m-j|}e^{\delta|m-j|}e^{-\delta|m-k|} \leq \tilde{c}e^{-\delta|m-k|}. \tag{7.9}$$

Combining (7.7), (7.8), and (7.9) gives

$$\begin{aligned}
|R^l(j,k)| &\leq \sum_{m_1=1}^{(2n-1)^2} \cdots \sum_{m_{l-1}=1}^{(2n-1)^2} \sum_{m_l=1}^{(2n-1)^2} |R(j,m_1)| \cdots |R(m_{l-1},m_l)|\,|R(m_l,k)| \\
&\leq \sum_{M_1=0}^{2n-2} \sum_{m'_1=1}^{2n-1} \cdots \sum_{M_{l-1}=0}^{2n-2} \sum_{m'_{l-1}=1}^{2n-1} \sum_{M_l=0}^{2n-2} \sum_{m'_l=1}^{2n-1} \\
&\quad c^l e^{-|J-M_1|}e^{-|j'-m'_1|} \ldots e^{-|M_{l-1}-M_l|}e^{-|m'_{l-1}-m'_l|}e^{-\delta|K-M_l|}e^{-\delta|k'-m'_l|} \\
&\leq \sum_{M_1=0}^{2n-2} \sum_{m'_1=1}^{2n-1} \cdots \sum_{M_{l-1}=0}^{2n-2} \sum_{m'_{l-1}=1}^{2n-1} \\
&\quad \tilde{c}c^l e^{-|J-M_1|}e^{-|j'-m'_1|} \ldots e^{-|M_{l-1}-M_l|}e^{-|m'_{l-1}-m'_l|}e^{-\delta|K-M_{l-1}|}e^{-\delta|k'-m'_{l-1}|} \\
&\ \ \vdots \\
&\leq (c\tilde{c})^l e^{-\delta|J-K|}e^{-\delta|j'-k'|} = C^l e^{-\delta|J-K|}e^{-\delta|j'-k'|}.
\end{aligned}$$

Combining this with (7.6) gives that for all $m \in \mathbb{N}$

$$|R^m(j,k)| \leq \min\{r^m, C^m e^{-\delta|J-K|}e^{-\delta|j'-k'|}\}. \tag{7.10}$$

Take M large enough so that $\rho \equiv Cr^{M-1} < 1$. By (7.10)

$$|R^m(j,k)| \leq \left((r^m)^{M-1}C^m e^{-\delta|J-K|}e^{-\delta|j'-k'|}\right)^{\frac{1}{M}} = \left(\rho^m e^{-\delta|J-K|}e^{-\delta|j'-k'|}\right)^{\frac{1}{M}}.$$

Using $G^{-1} = \sum_{m=0}^{\infty} R^m$ and (7.10), we have

$$|G^{-1}(j,j)| \leq \sum_{m=0}^{\infty} r^m = \frac{1}{1-r}$$

and

$$\begin{aligned}
|G^{-1}(j,k)| &\leq \delta_{j,k} + \sum_{n=1}^{\infty} (\rho^{\frac{1}{M}})^n e^{-\frac{\delta}{M}|j'-k'|}e^{-\frac{\delta}{M}|J-K|} \\
&\leq \left(\frac{\rho^{1/M}}{1-\rho^{1/M}}\right) e^{-\frac{\delta}{M}|j'-k'|}e^{-\frac{\delta}{M}|J-K|}.
\end{aligned}$$

This completes the proof. □

We shall use lemma 7.4 together with (7.5) and the definition of $\mathfrak{g}$ to estimate the $\{a^n_{j,k}\}$. The following lemma will be useful.

LEMMA 7.5. *There exist constants C and α, independent of n, such that*

$$\sum_{l=1}^{2n-1} e^{-\delta|n-j-l|}e^{-2\pi l^2} \leq Ce^{-\alpha|n-j|}$$

holds for all $n \in \mathbb{N}$ and $-(n-1) \leq j \leq n-1$.

PROOF.

$$\begin{aligned}\sum_{l=1}^{2n-1} e^{-\delta|n-j-l|}e^{-2\pi l^2} &\leq \sum_{1\leq l\leq \frac{1}{2}|n-j|} e^{-\delta|n-j-l|}e^{-2\pi l^2} + \sum_{l>\frac{1}{2}|n-j|} e^{-\delta|n-j-l|}e^{-2\pi l^2}\\ &\leq e^{-\frac{1}{2}\delta|n-j|}\sum_{1\leq l\leq \frac{1}{2}|n-j|} e^{-2\pi l^2} + e^{-\frac{1}{2}\pi(n-j)^2}\sum_{l>\frac{1}{2}|n-j|} e^{-\delta|n-j-l|}\\ &\leq e^{-\frac{1}{2}\delta|n-j|}\sum_{l\in\mathbb{Z}} e^{-2\pi l^2} + e^{-\frac{1}{2}\pi(n-j)^2}\sum_{l\in\mathbb{Z}} e^{-\delta|n-j-l|}\\ &\leq Ce^{-\alpha|n-j|}.\end{aligned}$$

The second inequality holds because

$$1 \leq l \leq \frac{1}{2}|n-j| \implies |n-j-l| \geq \frac{1}{2}|n-j|. \qquad \square$$

Using lemmas 7.4 and 7.5 we may now estimate the coefficients $\{a^n_{j,k}\}$.

LEMMA 7.6. *There exist constants $C, \alpha > 0$ such that the coefficients $\{a^n_{j,k}\}$ satisfy*

$$|a^n_{j,k}| \leq Ce^{-\alpha|n-j|}e^{-\alpha|n-k|}.$$

Recall that $\{a^n_{j,k}\}$ depends on n. The above constants are independent of n.

PROOF. Using (7.5), we have

$$\mathfrak{a} = G^{-1}\mathfrak{g}.$$

Using lemma 7.4 and the definitions of $\mathfrak{g}$ and $\mathfrak{a}$, we have

$$\begin{aligned}|a^n_{n-j,n-k}| &\leq \left|\sum_{l=1}^{(2n-1)^2} G^{-1}((j-1)(2n-1)+k, l)\mathfrak{g}_l\right|\\ &= \left|\sum_{l'=1}^{2n-1}\sum_{L=1}^{2n-1} G^{-1}((j-1)(2n-1)+k, (L-1)(2n-1)+l')e^{-2\pi(l')^2}e^{-2\pi(L)^2}\right|\\ &\leq \sum_{l'=1}^{2n-1}\sum_{L=1}^{2n-1} e^{-\delta|j-L|}e^{-\delta|k-l'|}e^{-\delta|l'|}e^{-\delta|L|}\\ &\leq Ce^{-\alpha|j|}e^{-\alpha|k|}\end{aligned}$$

for each $-(n-1) \leq j, k \leq n-1$. Thus,

$$|a^n_{j,k}| \leq Ce^{-\alpha|n-j|}e^{-\alpha|n-k|}. \qquad \square$$

8. Localization estimates

First, note that by theorem 5.2 the $L^2(\mathbb{R})$ norms of the $\{\varphi_{n,n}\}_{n\in\mathbb{N}}$ stay uniformly bounded away from 0.

LEMMA 8.1. *There exists $D > 0$ such that*

$$\|\varphi_{n,n}\|^2_{L^2(\mathbb{R})} > D$$

holds for all $n \in \mathbb{N}$.

PROOF. We proceed by contradiction. Suppose there exists a subsequence of $\{\varphi_{n,n}\}_{n=0}^{\infty}$ satisfying

$$\lim_{m\to\infty} \|\varphi_{n_m,n_m}\|_{L^2(\mathbb{R})}^2 = 0.$$

By (6.1),

$$\left\| g_{n_m,n_m} - \sum_{j=1-n_m}^{n_m-1} \sum_{k=1-n_m}^{n_m-1} a_{j,k}^{n_m} g_{j,k} \right\|_{L^2(\mathbb{R})} \longrightarrow 0, \qquad \text{as } m \to \infty.$$

We may use the translation and modulation invariance of Gabor systems to convert this from a statement about g_{n_m,n_m} to a statement about $g_{0,0}$. In particular, we have that for every $\epsilon > 0$ there exists a finite $\{c_{j,k}\} \subseteq \mathbb{C}$ such that

$$\left\| g_{0,0} - \sum_{j<0} \sum_{k<0} c_{j,k} g_{j,k} \right\|_{L^2(\mathbb{R})} < \epsilon.$$

This implies that

$$g_{0,0} \in \overline{\operatorname{span}}\ \{g_{j,k} : j, k < 0\}. \tag{8.1}$$

Recall that $\mathcal{G}(g, 2, 2)$ is a Riesz basis for its closed linear span by theorem 5.2 and the Ron-Shen duality theorem. Therefore, (8.1) is a contradiction by lemma 3.3. □

We need one final lemma before we can prove theorem 6.1.

LEMMA 8.2. *For every $M \in \mathbb{N}$ there exists $C = C_M > 0$ such that*

$$\left| \int |t - 2n|^M g_{j,k}(t) \overline{g_{l,m}(t)} dt \right| \leq C e^{-2\pi(k-m)^2} |2n - (k+m)|^M$$

holds for all $n \in \mathbb{N}$ and all $j, l \in \mathbb{Z}$ and all $k, m \in \mathbb{Z}$ such that $k + m \neq 2n$.

PROOF. Since $g_{0,k}(t) g_{0,m}(t) = \sqrt{2} e^{-2\pi(k-m)^2} e^{-2\pi[t-(k+m)]^2}$, it suffices to examine

$$\int |t - 2n|^M e^{-2\pi[t-(k+m)]^2} dt.$$

Using

$$|t - x|^M \leq K_M \sum_{k=0}^{M} |t|^k |x|^{M-k}$$

we have

$$\begin{aligned} \int |t - 2n|^M e^{-2\pi[t-(k+m)]^2} dt &\leq K_M \sum_{l=0}^{M} |2n - (k+m)|^{M-l} \int |t|^l e^{-2\pi t^2} dt \\ &\leq C_M |2n - (k+m)|^M \end{aligned}$$

and the lemma follows. □

We are now in position to estimate the localization of the $\{\psi_{n,n}\}$. Recall that

$$\psi_{n,n}(t) = \varphi_{n,n}(t) / \|\varphi_{n,n}\|_{L^2(\mathbb{R})} \,.$$

THEOREM 8.3. *Fix $p \in \mathbb{N}$. There exists a constant C_p such that*

$$\int |t-2n|^p |\psi_{n,n}(t)|^2 dt \leq C_p$$

and

$$\int |\gamma-2n|^p |\widehat{\psi_{n,n}}(\gamma)|^2 d\gamma \leq C_p$$

hold for all $n \in \mathbb{N}$.

PROOF.

$$\begin{aligned}
\int |t-2n|^p |\psi_{n,n}(t)|^2 dt &= \frac{1}{\|\varphi_{n,n}\|^2_{L^2(\mathbb{R})}} \int |t-2n|^p |\varphi_{n,n}(t)|^2 dt \\
&= \frac{1}{\|\varphi_{n,n}\|^2_{L^2(\mathbb{R})}} \left(\int |t-2n|^p \left| g_{n,n}(t) - \sum_{j=-(n-1)}^{n-1} \sum_{k=-(n-1)}^{n-1} a^n_{j,k} g_{j,k}(t) \right|^2 dt \right) \\
&\leq \frac{1}{D} \left(S_1+S_2+S_3\right),
\end{aligned}$$

where

$$\begin{aligned}
S_1 &= \int |t-2n|^p |g_{n,n}(t)|^2 dt = \int |t|^p |g(t)|^2 dt, \\
S_2 &= \sum_{j=-(n-1)}^{n-1} \sum_{k=-(n-1)}^{n-1} a^n_{j,k} \int |t-2n|^p g_{j,k}(t) \overline{g_{n,n}(t)} dt \\
&\quad + \sum_{j=-(n-1)}^{n-1} \sum_{k=-(n-1)}^{n-1} \overline{a^n_{j,k}} \int |t-2n|^p \overline{g_{j,k}(t)} g_{n,n}(t) dt, \\
S_3 &= \sum_{j=-(n-1)}^{n-1} \sum_{k=-(n-1)}^{n-1} \sum_{l=-(n-1)}^{n-1} \sum_{m=-(n-1)}^{n-1} a^n_{j,k} \overline{a^n_{l,m}} \int |t-2n|^p g_{j,k}(t) \overline{g_{l,m}(t)} dt.
\end{aligned}$$

It is clear that S_1 is uniformly bounded in n. To see that S_2 is bounded independently of n we use lemmas 7.6 and 8.2 in the following way:

$$\begin{aligned}
|S_2| &\leq 2C \sum_{j=1-n}^{n-1} \sum_{k=1-n}^{n-1} \left(e^{-\alpha|n-j|} e^{-\alpha|n-k|}\right) \left(e^{-2\pi(j-n)^2} |2n-(n+j)|^p\right) \\
&= 2C \left(\sum_{k=1-n}^{n-1} e^{-\alpha|n-k|} \right) \left(\sum_{j=1-n}^{n-1} e^{-\alpha|n-j|} e^{-2\pi(n-j)^2} |n-j|^p \right) \\
&\leq 2C \left(\sum_{k=1-n}^{n-1} e^{-\alpha|n-k|} \right) \left(\sum_{j=1-n}^{n-1} e^{-\alpha'|n-j|} \right) \\
&\leq 2C \left(\sum_{k=1}^{\infty} e^{-\alpha|k|} \right) \left(\sum_{j=1}^{\infty} e^{-\alpha'|j|} \right).
\end{aligned}$$

Likewise, for S_3 we have

$$|S_3| \leq C \sum_{j,k=1-n}^{n-1} \sum_{l,m=1-n}^{n-1} e^{-\alpha|j-n|}e^{-\alpha|k-n|}e^{-\alpha|l-n|}e^{-\alpha|m-n|}e^{-2\pi(j-l)^2}\,|2n-(j+l)|^p$$
$$= C \sum_{j,k=1}^{2n-1} \sum_{l,m=1}^{2n-1} e^{-\alpha|j|}e^{-\alpha|k|}e^{-\alpha|l|}e^{-\alpha|m|}e^{-2\pi(j-l)^2}\,|j+l|^p$$
$$\leq C \left(\sum_{j,l=1}^{\infty} e^{-\alpha|j|}e^{-\alpha|l|}e^{-2\pi(j-l)^2}|j+l|^p \right) \left(\sum_{k,m=1}^{\infty} e^{-\alpha|k|}e^{-\alpha|m|} \right).$$

Thus, we see that for any $p > 0$ there is C_p, independent of n, such that

$$\int |t-2n|^p|\psi_{n,n}(t)|^2 dt < C_p, \quad \forall n \in \mathbb{N}.$$

For the other inequality, recall that $g_{m,n}$ was defined with $a = b = 2$ and note that

$$\widehat{g_{m,n}}(\gamma) = e^{-2\pi i n b(\gamma+am)}\widehat{g}(\gamma+am) = (\widehat{g})_{n,-m}(\gamma) = g_{n,-m}(\gamma).$$

Therefore, the same calculations as above yield the uniform boundedness of

$$\int |\gamma-2n|^p|\widehat{\psi_{n,n}}(\gamma)|^2 d\gamma.$$

□

Acknowledgements. The author wishes to thank John Benedetto and Anwar Saleh for valuable discussions on the material.

References

[Bal] R. Balian, Un precipe d'incertitude fort en théorie du signal ou en méchanique quantique, *C.R. Acad. Sci. Paris*, 292 (20): 1357–1362, 1981.

[B1] J. J. Benedetto, Frame decompositions, sampling, and uncertainty principle inequalities, in *Wavelets: Mathematics and Applications*, Stud. Adv. Math., CRC, Boca Raton, FL, 1994.

[B2] J. J. Benedetto, Noise reduction in terms of the theory of frames, in *Signal and image representation in combined spaces*, Wavelet Anal. Appl., Academic Press, San Diego, CA, 1998.

[BCGP] J. J. Benedetto, W. Czaja, P. Gadziński, A. M. Powell, The Balian-Low theorem and regularity of Gabor systems, *J. Geom. Anal.*, 13 (2): 217–232, 2003.

[BCM] J. J. Benedetto, W. Czaja and A. Maltsev, The Balian-Low theorem for the symplectic form on $\mathbb{R}^{2d}$, *J. Math. Phys.*, 44 (4): 1735-1750, 2003.

[BP] J. J. Benedetto and A. M. Powell, *A (p,q) version of Bourgain's theorem*, preprint.

[BW] J. J. Benedetto and D. F. Walnut, Gabor frames for L^2 and related spaces, in *Wavelets: Mathematics and Applications*, Stud. Adv. Math., CRC, Boca Raton, FL, 1994.

[Bou] J. Bourgain, A remark on the uncertainty principle for Hilbertian basis, *J. Funct. Anal.*, 79 (1): 136–143, 1988.

[C] O. Christensen, Frames, Riesz bases, and discrete Gabor/wavelet expansions, *Bull. Amer. Math. Soc. (N.S.)*, 38 (3): 273–291, 2001.

[DJ] I. Daubechies and A. J. E. M. Janssen, Two theorems on lattice expansions, *IEEE Trans. Inform. Theory*, 39 (1): 3–6, 1993.

[F] G. B. Folland *Harmonic Analysis in Phase Space*, Annals of Mathematics Studies, Princeton University Press, Princeton, NJ, 1989.

[FG] H. Feichtinger and K. Gröchenig, Gabor frames and time-frequency analysis of distributions, *J. Func. Anal.*, 146: 464-495, 1997.

[FS] G. B. Folland and A. Sitaram, The uncertainty principle: a mathematical survey, *J. Fourier Anal. Appl.*, 3 (3): 207–238, 1997.

[G1] K. Gröchenig, *Foundations of Time-Frequency Analysis*, Applied and Numerical Harmonic Analysis, Birkäuser, Boston, MA, 2001.

[G2] K. Gröchenig, An uncertainty principle related to the Poisson summation formula, *Studia Math.*, 121 (1):87–104.

[GHHK] K. Gröchenig, D. Han, C. Heil and G. Kutyniok, The Balian-Low theorem for symplectic lattices in higher dimensions, *Appl. Comput. Harmon. Anal.*, 13 (2): 169–176, 2002.

[HJ] V. Havin and B. Jöricke, *The Uncertainty Principle in Harmonic Analysis*, Ergebnisse der Mathematik und ihrer Grezgebiete, 3. Folge, Band 28, Springer Verlag, Berlin, 1994.

[HRT] C. Heil, J. Ramanathan and R. Topiwala, Linear independence of time-frequency translates, *Proc. Amer. Math. Soc.*, 124 (9): 2787–2795, 1996.

[J] S. Jaffard, Propriétés des matrices "bien localisées" près de leur diagonale et quelques applications, *Ann. Inst. H. Poincaré Anal. Non Linéaire*, 7 (5):461–476, 1990.

[L] P. Linnell, von Neumann algebras and linear independence of translates, *Proc. Amer. Math. Soc.*, 127 (11):3269–3277, 1999.

[LM] P. G. Lemarié and Y. Meyer, Ondelettes et bases hilbertiennes, *Rev. Mat. Iberoamericana*, 2 (1-2):1–18, 1986.

[QSS] A. Quarteroni, R. Sacco and F. Saleri, *Numerical Mathematics*, Texts in Applied Mathematics, Springer-Verlag, New York, 2000.

Program in Applied and Computational Mathematics, Princeton University, Fine Hall, Washington Road, Princeton, NJ 08540

E-mail address: apowell@math.princeton.edu

Contemporary Mathematics
Volume **345**, 2004

Localization of stability and p-frames in the Fourier domain

Qiyu Sun

ABSTRACT. In this paper, we introduce and study the localization of stability and p-frame properties of a finitely generated shift-invariant system in the Fourier domain, and then provide more information to that shift invariant system. Especially for a shift-invariant system generated by finitely many compactly supported functions, we show that it has p-frame property at almost all frequencies, and that it either has stability property at almost all frequencies or does not have stability property at all frequencies.

1. Introduction

In this paper, we are interested in the localization of stability and p-frame properties of the shift invariant system generated by finitely many functions in the Fourier domain. The localization of stable shifts and p-frames in the Fourier domain provides us more information to that shift invariant system, especially when the system is generated by compactly supported functions. For instance, we show that a vector-valued compactly supported distribution either has stable shifts at almost all frequencies or does not have stable shifts at all frequencies, and that a vector-valued compactly supported bounded function generates a p-frame at almost all frequencies.

Let $\ell^p, 1 \le p \le \infty$, be the space of all p-summable sequences on $\mathbb{Z}^d$, and let $\|\cdot\|_{\ell^p}$ denote the usual ℓ^p norm. For a linear space X, we denote its direct sum of r copies by $X^{(r)}$. For compactly supported bounded functions $f_1, \ldots, f_r$ on $\mathbb{R}^d$, we define the *semi-convolution* $F*'$ on $(\ell^p)^{(r)}, 1 \le p \le \infty$, by

$$F*' : (\ell^p)^{(r)} \ni D := \{D(k)\} \longmapsto \sum_{k\in\mathbb{Z}^d} D(k)^T F(\cdot - k) := F *' D,$$

and denote the range of the semi-convolution $F*'$ on $(\ell^p)^{(r)}$ by

$$V_p(F) := \{F *' D : D \in (\ell^p)^{(r)}\},$$

where $F = (f_1, \ldots, f_r)^T$. The space $V_p(F)$ is the set of all linear combinations of $f_i(\cdot - k), 1 \le i \le r, k \in \mathbb{Z}^d$, using ℓ^p coefficients. Thus it is a shift invariant space generated by the shift invariant system

$$\mathcal{F} := \{f_i(\cdot - k) : 1 \le i \le r, k \in \mathbb{Z}^d\}.$$

Here we say that a linear space V of functions on $\mathbb{R}^d$ is *shift invariant* if $f \in V$ implies that $f(\cdot - k) \in V$ for all $k \in \mathbb{Z}^d$. The shift invariant space $V_p(F)$ is used in wavelet analysis for the case that the generator is a scaling function or a mother wavelet ([**7, 9, 19**]), and is also used as the model space for sampling, where the sinc function is the generator (see for instance [**1, 5**]).

Before starting to discuss localization in the Fourier domain, we recall the definition of Fourier series. For any summable sequence $D = \{D(k)\}$, its Fourier series is defined by $\mathcal{F}(D)(\xi) := \sum_{k\in\mathbb{Z}^d} D(k)e^{-ik\xi}$. The above definition of Fourier series can be extended to any sequence D with polynomial growth, that is, $|D(k)| \le p(k), k \in \mathbb{Z}^d$, for some polynomial p. In this situation, the Fourier series is a periodic tempered distribution, which we still denote by $\mathcal{F}(D)$. For any measurable set E, we let $\ell^p_E, 1 \le p \le \infty$, be the set of all ℓ^p sequences whose Fourier series are supported in $E + 2\pi\mathbb{Z}^d$.

The semi-convolution $F*'$ from $(\ell^\infty)^{(r)}$ to $V_\infty(F)$ is the first operator we want to localize in the Fourier domain. We say that F has *stable shifts* if the semi-convolution $F*'$ is one-to-one on $(\ell^\infty)^{(r)}$. For the case that F has compact support, it was known that F has stable shifts if and only if the rank of the $r \times \mathbb{Z}^d$ matrix $(\widehat{F}(\xi+2k\pi))_{k\in\mathbb{Z}^d}$ is r for all $\xi \in \mathbb{R}^d$ ([**16, 21**]). Here we define the *Fourier transform* $\hat{f}$ of an integrable function f by $\hat{f}(\xi) = \int_{\mathbb{R}^d} e^{-ix\xi} f(x)dx$, and understand the one of a tempered distribution as usual. Moreover, by the Possion formula, it follows that if the $r \times \mathbb{Z}^d$ matrix $(\widehat{F}(\xi_0 + 2k\pi))_{k\in\mathbb{Z}^d}$ has rank strictly less than r, then $F*'(vE_{\xi_0}) = 0$, where $E_{\xi_0} := \{e^{ik\xi_0}\} \in \ell^\infty$ and v is a nonzero vector in $\mathbb{C}^r$ so chosen that $v^T\widehat{F}(\xi_0 + 2k\pi) = 0$ for all $k \in \mathbb{Z}^d$. We observe that the Fourier series of E_{ξ_0} is supported in $\xi_0 + 2\pi\mathbb{Z}^d$. On the other hand, if the $r \times \mathbb{Z}^d$ matrix $(\widehat{F}(\xi_0 + 2k\pi))_{k\in\mathbb{Z}^d}$ is of full rank, then $F*'D \not\equiv 0$ for any nonzero sequence $D \in (\ell^\infty)^{(r)}$ whose Fourier transform is supported in $B(\xi_0, \delta) + 2\pi\mathbb{Z}^d$ for some small $\delta > 0$, and hence the semi-convolution $F*'$ is one-to-one on $(\ell^\infty_{B(\xi_0,\delta)})^{(r)}$ for some $\delta > 0$. That property is called *stable shifts at the frequency* ξ_0 in this paper. Note that F has stable shifts if and only if it has stable shifts at any frequency (Corollary 2.3). We may then consider the stable shifts at some frequency as the localization of the stable shifts in the Fourier domain. In Section 2, we discuss the stable shifts of F at a frequency when F is a tempered distribution having ℓ^1 decay, while the class of tempered distributions with ℓ^1 decay contains all integrable functions, compactly supported distributions, and also refinable distributions with smooth symbols (see Section 2 for more details about that function space). In Theorem 2.2, we characterize the stable shifts at a frequency and hence generalize the corresponding result for the stable shifts in [**24**]. In Theorem 2.4, we discuss the stable shifts of F at a frequency when F is a compactly supported distribution, and we show that a compactly supported distribution F either does not have stable shifts at all frequencies, or has stable shifts at almost all frequencies (Corollary 2.5).

Let $L^p, 1 \le p \le \infty$, be the space of all p-integrable functions on $\mathbb{R}^d$, and let $\|\cdot\|_p$ be the usual L^p norm. The semi-convolution operator $F*'$ from $(\ell^p)^{(r)}$ to $V_p(F), 1 \le p \le \infty$, is the second operator we want to localize in the Fourier domain. We say that F has ℓ^p *stable shifts* if $F*'$ is an isomorphism between $(\ell^p)^{(r)}$ and $V_p(F)$, that is, there exists a positive constant C so that

$$C^{-1}\|D\|_{\ell^p} \le \|F *' D\|_p \le C\|D\|_{\ell^p} \quad \forall\, D \in (\ell^p)^{(r)}. \tag{1.1}$$

The concept of ℓ^p stable shifts plays an important role in the approximation by shift invariant spaces, the regularity of scaling functions, and the convergence of cascade algorithms (see for instance [**13, 14, 25, 27**] and references therein). Comparing the definitions of the stable shifts and the ℓ^p stable shifts of F, we see that (a) the semi-convolution $F*'$ is required to be a bounded operator from $(\ell^p)^{(r)}$ to L^p in the definition of the ℓ^p stable shifts of F, but it is only required to be well defined in the distributional sense in the definition of the stable shifts of F, and (b) the semi-convolution $F*'$ is assumed to be one-to-one and to have bounded inverse in the definition of the ℓ^p stable shifts of F, but it is assumed to be one-to-one only in the definition of the stable shifts of F. For a compactly supported L^p function F, the semi-convolution is a bounded operator from $(\ell^p)^{(r)}$ to L^p because

$$\begin{aligned}\|F *' D\|_p &\leq \|F\|_p \|D\|_p \times \sup_{x\in\mathbb{R}^d} \Big(\sum_{k\in\mathbb{Z}^d} \chi_{[-K,K]^d}(x-k)\Big)^{(p-1)/p} \\ &\leq (2K+2)^{d(p-1)/p} \|F\|_p \|D\|_p,\end{aligned}$$

where K is so chosen that F is supported in $[-K,K]^d$ and χ_E is the characteristic function on a measurable set E (see (3.1) for a more general result). Moreover, it is shown that the ℓ^p stable shifts of F and the stable shifts of F are equivalent to each other ([**15**]). We say that F has *ℓ^p stable shifts at the frequency* $\xi_0 \in \mathbb{R}^d$ if the semi-convolution $F*'$ is an isomorphism between $(\ell^p_{B(\xi_0,\delta)})^{(r)}$ and its image for some $\delta > 0$, that is, (1.1) holds for all sequences $D \in (\ell^p_{B(\xi_0,\delta)})^{(r)}$. Note that $F \in (\mathcal{L}^p)^{(r)}$ has ℓ^p stable shifts if and only if it has ℓ^p stable shifts at any frequency (Corollary 3.2). Then we may consider the ℓ^p stable shifts at a certain frequency as the localization of the ℓ^p stable shifts in the Fourier domain. In Theorem 3.1 of Section 3, we generalize the equivalence between the ℓ^p stable shifts and the stable shifts in [**15**] to the equivalence of their localization in the Fourier domain for those functions F in $(\mathcal{L}^p)^{(r)}$, which contains all compactly supported L^p functions.

We see from the definition of the ℓ^p stable shifts that $F = (f_1, \ldots, f_r)^T$ has stable shifts if and only if $\{f_i(\cdot - k) : 1 \leq i \leq r, k \in \mathbb{Z}^d\}$ is a Riesz basis of $V_p(F)$. This observation inspires us to introduce the concept of a p-frame of a finitely generated shift invariant space in [**3**]. We say that F generates a *p-frame* if $\{f_i(\cdot - k) : 1 \leq i \leq r, k \in \mathbb{Z}^d\}$ is a p-frame for $V_p(F)$, that is, there exist positive constants A and B so that

$$A\|f\|_p \leq \sum_{i=1}^{r} \|\{\langle f, f_i(\cdot - k)\rangle\}\|_{\ell^p} \leq B\|f\|_p \quad \forall\, f \in V_p(F), \tag{1.2}$$

which, in turn, is equivalent to the operator T from $V_p(F)$ to $(\ell^p)^{(r)}$,

$$T : V_p(F) \ni f \to \big\{(\langle f, f_1(\cdot - k)\rangle, \ldots, \langle f, f_r(\cdot - k)\rangle)^T\big\} \in (\ell^p)^{(r)},$$

is an isomorphism on $V_p(F)$. The operator T is called the *analysis operator* T. The third operator we want to localize in the Fourier domain is the analysis operator T. For a shift invariant space V and a measurable set E, we define

$$V_E = \big\{f \in V : \hat{f} \text{ is supported in } E + 2\pi\mathbb{Z}^d\big\}.$$

If $V = V_p(F)$ for some generator F and $p \in [1,\infty]$, we use $V_{p,E}(F)$ to denote V_E. We say that F generates a *p-frame at the frequency* $\xi_0 \in \mathbb{R}^d$ if (1.2) holds for all $f \in V_{p,B(\xi_0,\delta)}(F)$, where $\delta > 0$. As we show later that for those functions F with certain decay at infinity, it generates a p-frame if and only if it generates a p-frame at any frequency (Corollary 4.4). Then we may consider the p-frame property at some frequency as the localization of the p-frame property in the Fourier domain. In [**3**], it is shown that a function F in $(W(L^\infty, \ell^1))^{(r)}$, (a function in that space is locally bounded and globally L^1, see Section 4 for the precise definition), generates a p-frame if and only if the rank of the $r \times \mathbb{Z}^d$ matrix $(\widehat{F}(\xi+2k\pi))_{k\in\mathbb{Z}^d}$ is independent of $\xi \in \mathbb{R}^d$ (Proposition 4.1). In Section 4, we establish a corresponding result for the p-frame at any frequency (Theorem 4.2). Applying Theorem 4.2, we show that any compactly supported bounded function F generates a p-frame for $V_p(F)$ at almost all frequencies (Theorem 4.5).

2. Stable shifts at certain frequency

In this section, we consider the localization of the stable shifts of finitely generated shift invariant system in Fourier domain.

First let us recall a class of function space from which the generators of the shift invariant system are chosen. Let $\mathcal{S}$ to be the space of all Schwartz functions, and let $\ll \cdot, \cdot \gg$ be the action between a tempered distribution and a Schwartz function. We say that a tempered distribution f has ℓ^1 *decay* if $\ll f(\cdot - x), h \gg$ is continuous about x for any $h \in \mathcal{S}$, and if there exist positive constants C and k_0 independent of $h \in \mathcal{S}$ such that

$$\sum_{k\in\mathbb{Z}^d} |\ll f(\cdot - k), h\gg| \le C \sum_{|\alpha|\le k_0} \|D^\alpha h(\cdot)(1+|\cdot|)^{k_0}\|_\infty$$

for any $h \in \mathcal{S}$ ([**24**]). The class of tempered distributions with ℓ^1 decay contains most of the functions (tempered distributions) we are interested in for the study of stable shifts. In fact, any integrable function f on $\mathbb{R}^d$ is a tempered distribution with ℓ^1 decay because for any $h \in \mathcal{S}$,

$$\begin{aligned}
\sum_{k\in\mathbb{Z}^d} |\ll f(\cdot - k), h\gg| &= \sum_{k\in\mathbb{Z}^d} |\ll f, h(\cdot + k)\gg| \\
&\le \int_{\mathbb{R}} |f(x)| \times \Big(\sum_{k\in\mathbb{Z}^d} |h(x+k)|\Big) dx \\
&\le \|h(1+|\cdot|)^{d+1}\|_\infty \times \int_{\mathbb{R}^d} |f(x)| \Big(\sum_{k\in\mathbb{Z}^d} (1+|x+k|)^{-d-1}\Big) dx \\
&\le \|h(1+|\cdot|)^{d+1}\|_\infty \Big(1 + \sum_{0\ne k\in\mathbb{Z}^d} |k|^{-d-1}\Big) \|f\|_1.
\end{aligned}$$

The class of tempered distributions with ℓ^1 decay also contains all compactly supported distributions and globally supported refinable distributions with smooth symbol ([**24**]), but it does not contain the sinc function $\mathrm{sinc}(x) := \frac{\sin \pi x}{\pi x}$ on the line because

$$\begin{aligned}
\sum_{k\in\mathbb{Z}} |\ll f(\cdot - k), h\gg| &= \frac{1}{2\pi} \sum_{k\in\mathbb{Z}^d} \Big| \int_{-\pi}^{\pi} e^{-i(k+1/2)\xi} d\xi \Big| \\
&= \sum_{k\in\mathbb{Z}} \Big| \frac{\sin(k+1/2)\pi}{(k+1/2)\pi} \Big| = +\infty
\end{aligned}$$

for any Schwartz function h whose Fourier transform satisfies $\hat{h}(\xi) = e^{-i\xi/2}$ for $|\xi| \le \pi$.

For $F = (f_1, \dots, f_r)^T$ having ℓ^1 decay, $\{\ll F(\cdot - k), h\gg\} \in \ell^1$ for any $h \in \mathcal{S}$, and hence the sum $\sum_{k\in\mathbb{Z}^d} D(k)^T \ll F(\cdot - k), h\gg$ is well defined for any bounded sequence $D = \{D(k)\}$. So we may define the *semi-convolution* $F *' D$ of a vector-valued tempered distribution F and a bounded sequence $D = \{D(k)\}$ by

$$\ll F *' D, h\gg := \sum_{k\in\mathbb{Z}^d} D(k)^T \ll F(\cdot - k), h\gg \quad \text{for any } h \in \mathcal{S}. \tag{2.1}$$

Thus the semi-convolution $F *' D$ is a tempered distribution ([**24**]).

We define the stable shifts of a tempered distribution $F = (f_1, \dots, f_r)^T$ with ℓ^1 decay and its localization in Fourier domain as follows.

DEFINITION 2.1. Let $F = (f_1, \dots, f_r)^T$ be a vector-valued tempered distribution with ℓ^1 decay.

- We say that F has *stable shifts* if $F*'$ is one-to-one on $(\ell^\infty)^{(r)}$, that is, the only bounded sequence D such that $F *' D = 0$ is the zero sequence ([**24**]).
- We say that F has *stable shifts at frequency* $\xi_0 \in \mathbb{R}^d$ if there exists $\delta > 0$ so that the only sequence $D \in (\ell^\infty_{B(\xi_0,\delta)})^{(r)}$ so that $F *' D = 0$ is the zero sequence.

For the stable shifts, there is a long list of publications on the characterizations and applications, especially for compactly supported distributions and refinable distributions, (see, for instance, [**12, 16, 21, 23, 26**] for compactly supported distributions, [**8, 10, 11, 18, 22, 28**] for compactly supported refinable distributions, [**15**] for globally supported functions in $\mathcal{L}^p$, and [**24**] for tempered distributions with ℓ^1 decay).

We say that a function f is a C^∞ *function with* ℓ^1 *decay* if

$$\sum_{k\in\mathbb{Z}^d} \|D^n f\|_{L^\infty(k+[0,1)^d)} < \infty \text{ for any } n \in (\mathbb{Z}_+)^d.$$

We may understand that C^∞ functions with ℓ^1 decay are functions locally C^∞ and globally ℓ^1. Thus a Schwartz function is a C^∞ function with ℓ^1 decay, and so is a linear combination of the integer shifts of a Schwartz function using ℓ^1 coefficients. In [**24**], the author established the following result for the stable shifts of a tempered distribution with ℓ^1 decay (The original result is slightly different from the one stated below).

PROPOSITION 2.1. *Let $F = (f_1, \dots, f_r)^T$ be a tempered distribution with ℓ^1 decay. Then the following statements are equivalent:*

(i) *F has stable shifts.*
(ii) *The matrix $(\widehat{F}(\xi + 2k\pi))_{k\in\mathbb{Z}^d}$ is of full rank for any $\xi \in \mathbb{R}^d$.*
(iii) *There exist C^∞ functions $h_1, \dots, h_r$ with ℓ^1 decay so that*

$$d_i(k) = \ll F *' D, h_i(\cdot - k)\gg, \ 1 \le i \le r, \ k \in \mathbb{Z}^d$$

for all bounded sequences $D = \{(d_1(k), \dots, d_r(k))^T\}$.

We remark that the equivalence between the first and second statement is established in [**21**] for a compactly supported function F with $r = 1$ and in [**16**] with $r \ge 1$, and in [**15**] for an integrable function F.

In this section, we establish a corresponding version of Proposition 2.1 for the stable shifts at a frequency.

THEOREM 2.2. *Let $\xi_0 \in \mathbb{R}^d$, and let $F = (f_1, \ldots, f_r)^T$ be a tempered distribution with ℓ^1 decay. Then the following three statements are equivalent to each other.*

(i) *F has stable shifts at the frequency $\xi_0 \in \mathbb{R}^d$.*
(ii) *The $r \times \mathbb{Z}^d$ matrix $(\widehat{F}(\xi_0 + 2k\pi))_{k\in\mathbb{Z}^d}$ is of full rank.*
(iii) *There exist C^∞ functions $h_1, \ldots, h_r$ with ℓ^1 decay and a positive constant δ so that*

$$d_i(k) = \ll F *' D, h_i(\cdot - k)\gg, \ 1 \le i \le r, \ k \in \mathbb{Z}^d \tag{2.2}$$

for any bounded sequences $D := \{(d_1(k), \ldots, d_r(k))^T\}$ whose Fourier series $\mathcal{F}(D)$ are supported in $B(\xi_0, \delta) + 2\pi\mathbb{Z}^d$.

We may understand the above theorem as the equivalence among stable shifts at a certain frequency, full dimension of the Fourier fiber bundle in the neighborhood of that frequency, and the inverse process of the semi-convolution at that frequency. Combining Theorem 2.2 and Proposition 2.1, we have:

COROLLARY 2.3. *A tempered distribution with ℓ^1 decay has stable shifts if and only if it has stable shifts at any frequency.*

Then Theorem 2.2 can be thought as the localization version of Proposition 2.1 in the Fourier domain.

For a compactly supported distribution $F = (f_1, \ldots, f_r)^T$, we say that F has *finitely linearly independent shifts* if there does not exist a nonzero sequence D with finite support so that $F *' D \equiv 0$ ([**28**]). In other words, there does not exist a linear dependent finite subsystem of $\mathcal{F} := \{f_i(\cdot - k) : 1 \le i \le r, k \in \mathbb{Z}^d\}$. In the Fourier domain, we may interpret the finitely linearly independent shifts as the nonexistence of a nonzero vector $Q(\xi) = (q_1(\xi), \ldots, q_r(\xi))^T$ so that all entries of $Q(\xi)$ are trigonometric polynomials and that $Q(\xi)^T\widehat{F}(\xi) \equiv 0$. Applying Theorem 2.2, we have the following result about the stable shifts of a compactly supported distribution at a frequency.

THEOREM 2.4. *Let $F = (f_1, \ldots, f_r)^T$ be a compactly supported nonzero distribution. Then:*

(i) *F has finitely linearly dependent shifts if and only if F does not have stable shifts at all frequencies.*
(ii) *F has finitely linearly independent shifts if and only if F has stable shifts at almost all frequencies.*

As an easy consequence of Theorem 2.4, we have:

COROLLARY 2.5. *Let $F = (f_1, \ldots, f_r)^T$ be a compactly supported nonzero distribution. Then either F does not have stable shifts at all frequencies, or F has stable shifts at almost all frequencies. Especially if $r = 1$ then F has stable shifts at almost all frequencies.*

2.1. Proof of Theorem 2.2. We recall some properties of a tempered distribution with ℓ^1 decay in [**24**].

LEMMA 2.6. *Let F be a vector-valued tempered distribution on $\mathbb{R}^d$ with ℓ^1 decay. Then:*

(i) *$\widehat{F}$ is continuous.*

(ii) $\widehat{F}$ *has polynomial increase at infinity, i.e., there exists a polynomial* $Q(\xi)$ *such that* $|\widehat{F}(\xi)| \le Q(\xi)$ *for all* $\xi \in \mathbb{R}^d$.

(iii) *For any* $g \in \mathcal{S}$, *the following Poisson summation formula holds*

$$\sum_{k\in\mathbb{Z}^d} \widehat{F}(\xi+2k\pi)\overline{\hat{g}(\xi+2k\pi)} = \sum_{j\in\mathbb{Z}^d} \ll F(\cdot+j), g\gg e^{-ij\xi}.$$

PROOF OF THEOREM 2.2. The implication (iii)$\Longrightarrow$(i) is obvious. Then it suffices to prove (i) $\Longrightarrow$ (ii)$\Longrightarrow$ (iii). First we prove (i) $\Longrightarrow$ (ii). Suppose, on the contrary, that the rank of $r \times \mathbb{Z}^d$ matrix $(\widehat{F}(\xi_0+2k\pi))_{k\in\mathbb{Z}^d}$ is strictly less than r. Then there exists a (complex-valued) nonzero vector $v = (v_1, \ldots, v_r)^T$ so that the function $g := v^T F$ satisfies $\hat{g}(\xi_0+2k\pi) = 0$ for all $k \in \mathbb{Z}^d$. The sequence $E_{\xi_0} = (e^{ij\xi_0})_{j\in\mathbb{Z}^d}$ belongs to ℓ^∞, has its Fourier series supported in $\xi_0 + 2\pi\mathbb{Z}^d$, and satisfies $g *' E_{\xi_0} = 0$. This contradicts the assumption (i).

Next we prove (ii)$\Longrightarrow$(iii). By (ii), there exist $k_1, \ldots, k_r \in \mathbb{Z}^d$ such that

$$A := \big(\widehat{F}(\xi_0+2k_1\pi), \ldots, \widehat{F}(\xi_0+2k_r\pi)\big) \quad \text{is nonsingular.} \tag{2.3}$$

Let $e_i, 1 \le i \le r$, be vectors with i-th component one and other component zero. By (2.3), $Av_i = e_i$ for some $v_i = (v_i(k_1), \ldots, v_i(k_r))^T \in \mathbb{C}^r, 1 \le i \le r$. Let w_i be a C^∞ function so that w_i is supported in $\bigcup_{i'=1}^{l} B(\xi_0, \delta) + 2k_{i'}\pi$ and $w_i(\xi) = v_i(k_{i'})$ for all $\xi \in B(\xi_0, \delta/2) + 2k_{i'}\pi$, where $\delta > 0$ is a sufficiently small positive number chosen later. Recall that $\widehat{f_i}, 1 \le i \le r$, are continuous on $\mathbb{R}^d$ by Lemma 2.6. Then for sufficiently small $\delta > 0$, the matrix

$$A(\xi) := \Big(\sum_{k\in\mathbb{Z}^d} \widehat{f_i}(\xi+2k\pi)\overline{w_{i'}(\xi+2k\pi)}\Big)_{1\le i,i'\le r}$$

has its entries in the Wiener class, and is nonsingular in a neighborhood of $B(\xi_0, \delta) + 2\pi\mathbb{Z}^d$. Hence there exists another matrix $B(\xi)$ whose entries belong to the Wiener class so that

$$A(\xi)B(\xi) = I_r \tag{2.4}$$

on a smaller neighborhood of $\xi + 2\pi\mathbb{Z}^d$, say $B(\xi_0, \delta_1) + 2\pi\mathbb{Z}^d$ for some $0 < \delta_1 < \delta/2$. Define $h_1, \ldots, h_r$ by

$$\big(\widehat{h}_1(\xi), \ldots, \widehat{h}_r(\xi)\big)^T = \overline{B(\xi)}^T (w_1(\xi), \ldots, w_r(\xi))^T. \tag{2.5}$$

Then $h_1, \ldots, h_r$ are linear combinations of the integer shifts of some Schwartz functions using ℓ^1 sequences, and hence are C^∞ functions with ℓ^1 decay. By (2.4) and (2.5),

$$\sum_{k\in\mathbb{Z}^d} \widehat{f_i}(\xi+2k\pi)\overline{\widehat{h}_{i'}(\xi+2k\pi)} = \delta_{ii'}, \ 1 \le i, i' \le r \tag{2.6}$$

for all $\xi \in B(\xi_0, \delta_1) + 2\pi\mathbb{Z}^d$, where $\delta_{ii'}$ is the usual Kronecker symbol. Let $D = \{(d_1(k), \ldots, d_r(k))^T\}$ be any bounded sequence with its Fourier series $\mathcal{F}(D)$ supported in $B(\xi_0, \delta_1) + 2\pi\mathbb{Z}^d$. Multiplying $\mathcal{F}(D_i)$ at both sides of the equation (2.6) and then summing up for i from 1 to r, we obtain

$$\mathcal{F}(D_{i'}) = \sum_{k\in\mathbb{Z}^d} \widehat{F *' D}(\cdot+2k\pi)\overline{\widehat{h}_{i'}(\cdot+2k\pi)}, \ 1 \le i' \le r,$$

where $D_i = \{d_i(k)\}, 1 \le i \le r$. Then the assertion (iii) follows. □

2.2. Proof of Theorem 2.4. To prove Theorem 2.4, we need a characterization of finitely linearly dependent shifts of a compactly supported distribution, which is essentially given in [**28**] for the one dimensional case.

LEMMA 2.7. *Let $F = (f_1, \ldots, f_r)^T$ be a compactly supported distribution. Then F has finitely linearly dependent shifts if and only if the matrix $(\widehat{F}(\xi + 2k\pi))_{k\in\mathbb{Z}^d}$ has rank less than or equal to $r-1$ for all $\xi \in \mathbb{R}^d$.*

To prove Lemma 2.7, we recall a result in [**2**].

LEMMA 2.8. *Let $F = (f_1, \ldots, f_r)^T$ be a compactly supported distribution. Then there exists a compactly supported distribution $G = (g_1, \ldots, g_s)^T$ so that G has stable shifts, and such that $f_1, \ldots, f_r$ are finite linear combinations of $\{g_i(\cdot - k) : 1 \le i \le s, k \in \mathbb{Z}^d\}$.*

PROOF OF LEMMA 2.7. First the sufficiency. Let $Q(\xi) = (q_1(\xi), \ldots, q_r(\xi))^T$ be a nonzero vector with polynomial entries so that $Q(\xi)^T \widehat{F}(\xi) = 0$ for all $\xi \in \mathbb{R}^d$. Therefore

$$Q(\xi)^T A(\xi) = 0 \quad \forall\, \xi \in \mathbb{R}^d, \tag{2.7}$$

where $A(\xi) = (\widehat{F}(\xi+2k\pi))_{k\in\mathbb{Z}^d}$. By our assumption, $Q(\xi) \ne 0$ for almost all $\xi \in \mathbb{R}^d$, which together with (2.7) implies that the matrix $A(\xi)$ has rank less than or equals to $r-1$ for almost all $\xi \in \mathbb{R}^d$. On the other hand, $\widehat{F}$ is an analytic function by the assumption on F, and hence either the rank of $(\widehat{F}(\xi + 2k\pi))_{k\in\mathbb{Z}^d}$ is strictly less than r for all $\xi \in \mathbb{R}^d$, or is equal to r for almost all $\xi \in \mathbb{R}^d$. Then the assertion follows.

Then the necessity. Let G be the function as in Lemma 2.8, and $P(\xi)$ be the trigonometric polynomial matrix so chosen that $\widehat{F}(\xi) = P(\xi)\widehat{G}(\xi)$. By the stable shifts of G, $(\widehat{G}(\xi + 2k\pi))_{k\in\mathbb{Z}^d}$ has rank s for all $\xi \in \mathbb{R}^d$ ([**16, 21**]). This together with our assumption on F implies that the matrix $P(\xi)$ has rank less than or equal to $r-1$ for all $\xi \in \mathbb{R}^d$. Thus there exists a nonzero trigonometric polynomial vector $Q(\xi) = (q_1(\xi), \ldots, q_r(\xi))^T$ so that $Q(\xi)^T P(\xi) = 0$ for all $\xi \in \mathbb{R}^d$. Hence $Q(\xi)^T \widehat{F}(\xi) = 0$ for all $\xi \in \mathbb{R}^d$ and the assertion follows. □

PROOF OF THEOREM 2.4. By the assumption on F, the Fourier transform $\widehat{F}(\xi)$ of F is an analytic function. Hence either the rank of $(\widehat{F}(\xi + 2k\pi))_{k\in\mathbb{Z}^d}$ is strictly less than r for all $\xi \in \mathbb{R}^d$, or is equal to r for almost all $\xi \in \mathbb{R}^d$. The assertion then follows from Theorem 2.2 and Lemma 2.7. □

3. ℓ^p stable shifts at certain frequencies

In this section, we consider ℓ^p stable shifts at some frequency for the finitely generated shift invariant system, and show the equivalence between the stable shifts at some frequency and ℓ^p stable shifts at that frequency when the generators have a certain decay property at infinity.

The generators of the finitely generated shift invariant system are chosen from the function space $\mathcal{L}^p, 1 \le p \le \infty$, which contains all functions f with finite $\|f\|_{\mathcal{L}^p}$, where $\|f\|_{\mathcal{L}^p} = \big\| \sum_{j\in\mathbb{Z}^d} |f(\cdot + j)| \big\|_{L^p([0,1)^d)}$ ([**15**]). Here $L^p(K), 1 \le p \le \infty$, is the space of all p-integrable functions on a measurable set K, and $\|\cdot\|_{L^p(K)}$ is the usual $L^p(K)$ norm. Clearly for $1 \le p \le \infty$, a compactly supported L^p belongs to $\mathcal{L}^p$, and an $\mathcal{L}^p$ function is integrable. Thus a function in $\mathcal{L}^p$ is a tempered distribution with ℓ^1 decay.

For any $F = (f_1, \dots, f_r)^T \in (\mathcal{L}^p)^{(r)}$, we have

$$\begin{aligned}\|F *' D\|_\infty &\le \sup_{x\in\mathbb{R}^d} \sum_{j\in\mathbb{Z}^d} |D(j)||F(x-j)| \\ &\le \|D\|_{\ell^\infty} \sup_{x\in\mathbb{R}^d} \sum_{j\in\mathbb{Z}^d} |F(x-j)| = \|D\|_{\ell^\infty}\|F\|_{\mathcal{L}^\infty}\end{aligned}$$

when $p = \infty$, and

$$\begin{aligned}\|F *' D\|_p^p &\le \int_{\mathbb{R}^d} \Big(\sum_{j\in\mathbb{Z}^d} |D(j)||F(x-j)|\Big)^p dx \\ &\le \int_{\mathbb{R}^d} \Big(\sum_{j\in\mathbb{Z}^d} |D(j)|^p|F(x-j)|\Big) \times \Big(\sum_{j\in\mathbb{Z}^d} |F(x-j)|\Big)^{p-1} dx \\ &= \|D\|_{\ell^p}^p \int_{[0,1]^d} \Big(\sum_{j\in\mathbb{Z}^d} |F(x-j)|\Big)^p dx = \|D\|_{\ell^p}^p \|F\|_{\mathcal{L}^p}^p\end{aligned}$$

when $1 \le p < \infty$. Therefore

$$\|F *' D\|_p \le \|D\|_{\ell^p}\|F\|_{\mathcal{L}^p} \quad \text{for all } D \in (\ell^p)^{(r)} \tag{3.1}$$

([**15**]). Thus we may consider the semi-convolution $F*'$ as a bounded operator from $(\ell^p)^{(r)}$ to $L^p, 1 \le p \le \infty$.

For any function $F = (f_1, \dots, f_r)^T \in (\mathcal{L}^p)^{(r)}$ and $1 \le p \le \infty$, it is known that F has ℓ^p stable shifts if and only if F has stable shifts ([**15**]). In this section, we show the equivalence between the ℓ^p stable shifts at a certain frequency and the stable shifts at that frequency.

THEOREM 3.1. *Let $F = (f_1, \dots, f_r)^T \in (\mathcal{L}^p)^{(r)}, 1 \le p \le \infty$, and $\xi_0 \in \mathbb{R}^d$. Then F has ℓ^p stable shifts at the frequency ξ_0 if and only if F has stable shifts at the frequency ξ_0.*

Combining Corollary 2.3 and Theorem 3.1, we have

COROLLARY 3.2. *A function in $(\mathcal{L}^p)^{(r)}$ has ℓ^p stable shifts if and only if it has ℓ^p stable shifts at any frequency.*

To prove Theorem 3.1, we recall a lemma in [**3**].

LEMMA 3.3. *Let $f \in \mathcal{L}^p$ for some $1 \le p < \infty$. Assume that $\sum_{j\in\mathbb{Z}^d} f(\cdot - j) = 0$. Then for any Schwartz function h on $\mathbb{R}^d$,*

$$\lim_{\delta\to 0} \delta^d \Big\| \sum_{j\in\mathbb{Z}^d} h(\delta j) f(\cdot - j)\Big\|_{\mathcal{L}^p} = 0.$$

Note that the Fourier transform of

$$\delta^d \sum_{j\in\mathbb{Z}^d} h(\delta j) f(\cdot - j)$$

is

$$\hat{f}(\xi) \times \sum_{k\in\mathbb{Z}^d} \hat{h}(\delta^{-1}(\xi + 2k\pi)),$$

which can be considered as a periodic smooth cutoff of $\hat{f}$ at the origin. Then the result in Lemma 3.3 can be understood as: if $\hat{f}(2k\pi) = 0$ for all $k \in \mathbb{Z}^d$, then the $\mathcal{L}^p$ norm of the periodic smooth cutoff of a function $f \in \mathcal{L}^p$ at the δ neighborhood of the origin tends to zero as δ tends to zero. A periodic smooth cutoff is crucial when we consider the shift invariant problem in $L^p, 1 \le p \le \infty$, instead of in L^2, where a bounded periodic cutoff is usually used. The limit result in Lemma 3.3 is not true for $p = \infty$ and a counterexample is given in [**3**].

PROOF OF THEOREM 3.1. First we prove the necessity, which is essentially given in [**15**]. Suppose, on the contrary, that F does not have stable shifts at the frequency ξ_0. Then the matrix $(\widehat{F}(\xi_0 + 2k\pi))_{k\in\mathbb{Z}^d}$ is not of full rank by Theorem 2.2, and hence there exists a (complex-valued) nonzero vector $v = (v_1, \ldots, v_r)^T$ so that the function $g := v^T F$ satisfies

$$\hat{g}(\xi_0 + 2k\pi) = 0 \quad \forall\, k \in \mathbb{Z}^d. \tag{3.2}$$

If $g \equiv 0$, then the proof is done since $\sum_{i=1}^r f_i *' D_i = 0$ for sequences $D_i, 1 \le i \le r$, chosen so that $D_i = v_i D$ for some $D \in \ell^p$, which contradicts the stable assumption on F. Now we suppose $g \not\equiv 0$. Set $E_{\xi_0} = (e^{-ij\xi_0})_{j\in\mathbb{Z}^d} \in \ell^\infty$. For $p = \infty$, it follows from (3.2) that

$$g *' E_{\xi_0} = 0, \tag{3.3}$$

which contradicts the stable shifts at the frequency ξ_0 since the Fourier series of E_{ξ_0} is supported in $\xi_0 + 2\pi\mathbb{Z}^d$. Let h_0 be a Schwartz function so chosen that $\widehat{h_0}$ is supported in $B(0,1)$ and $\widehat{h_0}(x) = 1$ for all $x \in B(0, 1/2)$. For $1 \le p < \infty$, it follows from (3.3) and Lemma 3.3 that

$$\lim_{\delta\to 0} \delta^d \Big\| \sum_{j\in\mathbb{Z}^d} h_0(\delta j) e^{-i(\cdot - j)\xi_0} g(\cdot - j)\Big\|_{\mathcal{L}^p} = 0. \tag{3.4}$$

Note that the Fourier transform of

$$\delta^d \sum_{j\in\mathbb{Z}^d} h_0(\delta j) e^{-i(\cdot - j)\xi_0} g(\cdot - j)$$

is

$$\hat{g}(\xi + \xi_0) \sum_{k\in\mathbb{Z}^d} \widehat{h_0}(\delta^{-1}(\xi + 2k\pi)),$$

which equals $\hat{g}(\xi + \xi_0)$ on $B(0, \delta/2)$. This together with (3.4) shows that for any $\epsilon > 0$ there exists $\delta_0 > 0$ so that

$$\|g *' D\|_p \le \delta^d \Big\| \sum_{j\in\mathbb{Z}^d} h_0(\delta j) e^{-i(\cdot - j)\xi_0} g(\cdot - j)\Big\|_{\mathcal{L}^p} \times \|D\|_{\ell^p} \le \epsilon \|D\|_{\ell^p}$$

for any ℓ^p sequence D whose Fourier series is supported in $B(\xi_0, \delta/2) + 2\pi\mathbb{Z}^d$ and $\delta \in (0, \delta_0)$. This is a contradiction since $\epsilon > 0$ can be chosen arbitrarily.

Now we prove the sufficiency. By Theorem 2.2, there exist C^∞ functions $h_1, \ldots, h_r$ with ℓ^1 decay and $\delta > 0$ so that

$$d_i(k) = \ll F *' D, h_i(\cdot - k) \gg, \; 1 \le i \le r, \; k \in \mathbb{Z}^d, \tag{3.5}$$

where $D = \{(d_1(k), \ldots, d_r(k))^T\} \in (\ell^p)^{(r)}$ and $\mathcal{F}(D)$ is supported in $B(\xi_0, \delta) + 2\pi\mathbb{Z}^d$. Therefore

$$\|D\|_{\ell^\infty} \leq \sum_{i=1}^{r} \|F *' D\|_\infty \|h_i\|_1 \leq C_0 \|F *' D\|_\infty$$

for $p = \infty$, and

$$\|D\|_{\ell^p}^p \leq C_1 \sum_{i=1}^{r} \|h_i\|_1^{p-1} \sum_{k\in\mathbb{Z}^d} \int_{\mathbb{R}^d} |F *' D(x)|^p |h_i(x-k)| dx \leq C_2 \|F * D\|_p^p$$

for $1 \leq p < \infty$, where C_0, C_1, C_2 are positive constants independent of $D \in \ell^p$. Hence the sufficiency follows. □

4. p-frame at certain frequency

In this section, we consider localization of a p-frame in the finitely generated shift invariant system in the Fourier domain. The generators of the shift invariant system are chosen from the function space $W(L^p, \ell^q), 1 \leq p, q \leq \infty$, which contains all functions f whose norm $\|f\|_{W(L^p,\ell^q)}$ is finite, where

$$\|f\|_{W(L^p,\ell^q)} := \|\{\|f\|_{L^p(k+[0,1)^d)}\}\|_{\ell^q}.$$

For any $1 \leq p, q \leq \infty$, we have

$$W(L^\infty, \ell^q) \subset W(L^p, \ell^q)$$

since $\|f\|_{L^p(k+[0,1)^d)} \leq \|f\|_{L^\infty(k+[0,1)^d)}$ for all $k \in \mathbb{Z}^d$, and

$$W(L^p, \ell^1) \subset \mathcal{L}^p \subset W(L^p, \ell^p) = L^p$$

because

$$\|f\|_{\mathcal{L}^p} = \Big\| \sum_{k\in\mathbb{Z}^d} |f(k+\cdot)| \Big\|_{L^p([0,1)^d)} \leq \sum_{k\in\mathbb{Z}^d} \|f(k+\cdot)\|_{L^p([0,1)^d)} = \|f\|_{W(L^p,\ell^1)}.$$

Let $F = (f_1, \ldots, f_r)^T \in (W(L^{p/(p-1)}, \ell^1))^{(r)}$. Then for any $f \in L^p$,

$$\sum_{i=1}^{r} \|\{\langle f, f_i(\cdot - k)\rangle\}\|_{\ell^p} \leq \sum_{l\in\mathbb{Z}^d} \sum_{i=1}^{r} \|\{\langle f, f_{i,l}(\cdot - k)\rangle\}\|_{\ell^p}$$

$$\leq \sum_{i=1}^{r} \sum_{l\in\mathbb{Z}^d} \|f\|_p \|f_{i,l}\|_{p/(p-1)} \leq \sum_{i=1}^{r} \|f_i\|_{W(L^{p/(p-1)},\ell^1)} \|f\|_p, \tag{4.1}$$

where $f_{i,l} = f_i \chi_{l+[0,1)^d}, l \in \mathbb{Z}^d$. Thus $\{f_i(\cdot - k) : 1 \leq i \leq r, k \in \mathbb{Z}^d\}$ is the Bessel sequence for the shift invariant space $V_p(F)$ when $F = (f_1, \ldots, f_r)^T \in (\mathcal{L}^p)^{(r)} \cap (W(L^{p/(p-1)}, \ell^1))^{(r)}$, and the analysis operator T is a bounded operator from L^p to $(\ell^p)^{(r)}$. So it is good to assume that the generators of the shift invariant system belong to $\mathcal{L}^p \cap W(L^{p/(p-1)}, \ell^1)$ when we consider the corresponding p-frame property.

When the generators are assumed to be in the class $W(L^\infty, \ell^1)$, which is a subspace of $\mathcal{L}^p$ and also of $W(L^{p/(p-1)}, \ell^1)$ for $1 \leq p \leq \infty$, the following characterization of the p-frame property was established in [**3**].

PROPOSITION 4.1. *Let $F = (f_1, \ldots, f_r)^T \in (W(L^\infty, \ell^1))^{(r)}$ and $1 \leq p \leq \infty$. Then the following statements are equivalent to each other.*

(i) *$V_p(F)$ is a closed subspace of L^p.*

(ii) *F generates a p-frame for $V_p(F)$.*
(iii) *The rank of the $r \times \mathbb{Z}^d$ matrix $(\widehat{F}(\xi+2k\pi))_{k\in\mathbb{Z}^d}$ is independent of $\xi \in \mathbb{R}^d$.*
(iv) *There exist sequences $\{a_{ii'}(k)\} \in \ell^1, 1 \le i, i' \le r$, so that*

$$f = \sum_{i,i'=1}^{r} \sum_{k,k'\in\mathbb{Z}^d} a_{ii'}(k-k')\langle f, f_{i'}(\cdot - k')\rangle f_i(\cdot - k)$$

for all $f \in V_p(F)$.

For any $F = (f_1, \ldots, f_r)^T \in (W(L^\infty, \ell^1))^{(r)}$, it was pointed out in [**3**] that if F has ℓ^p stable shifts, then it generates a p-frame for $V_p(F), 1 \le p \le \infty$. In fact, when F has ℓ^p stable shifts, the shift invariant space $V_p(F)$ is closed in L^p since it is isomorphic to the sequence space ℓ^p, and hence F generates a p-frame for $V_p(F), 1 \le p \le \infty$, by Proposition 4.1. The above assertion is known for $p = 2$ under the weak assumption $\sum_{k\in\mathbb{Z}^d} |\widehat{F}(\cdot + 2k\pi)|^2 \in L^\infty$ in [**6**]. So the frame property of the shift invariant system is a generalization of the stability (Riesz) property. The p-frame was introduced in [**3**] as a class of Banach frames related to the finitely generated shift invariant system, which preserves most of the frame properties in Hilbert space (see [**4, 20**] for more properties of p-frames).

In this section, we establish the corresponding version of Proposition 4.1 for a p-frame at a certain frequency.

THEOREM 4.2. *Let $\xi_0 \in \mathbb{R}^d$, and let*

$$F = (f_1, \ldots, f_r)^T \in (\mathcal{L}^p)^{(r)} \cap (W(L^{p/(p-1)}, \ell^1))^{(r)} \qquad \text{for } 1 \le p < \infty$$

and

$$F \in (W(L^\infty, \ell^1))^{(r)} \qquad \text{for } p = \infty.$$

Then the following statements are equivalent:

(i) *F generates a p-frame for $V_p(F)$ at the frequency ξ_0.*
(ii) *The rank of the $r \times \mathbb{Z}^d$ matrix $(\widehat{F}(\xi + 2k\pi))_{k\in\mathbb{Z}^d}$ is independent of ξ at a small neighborhood of ξ_0.*
(iii) *There exist a positive constant δ and sequences $\{a_{ii'}(k)\} \in \ell^1, 1 \le i, i' \le r$, so that*

$$f = \sum_{i,i'=1}^{r} \sum_{k,k'\in\mathbb{Z}^d} a_{ii'}(k-k')\langle f, f_{i'}(\cdot - k')\rangle f_i(\cdot - k) \tag{4.2}$$

for all $f \in V_{p,B(\xi_0,\delta)}(F)$.

and

THEOREM 4.3. *Let $\xi_0 \in \mathbb{R}^d$, and let $F = (f_1, \ldots, f_r)^T \in (\mathcal{L}^p)^{(r)}$ for $1 \le p < \infty$ and $F \in (W(L^\infty, \ell^1))^{(r)}$ for $p = \infty$. Then the space $V_{p,B(\xi_0,\delta)}(F)$ is a closed subspace of L^p for sufficiently small $\delta > 0$ if and only if the rank of the $r \times \mathbb{Z}^d$ matrix $(\widehat{F}(\xi + 2k\pi))_{k\in\mathbb{Z}^d}$ is independent of ξ at a small neighborhood of ξ_0.*

We may understand the above two theorems as the equivalence among p-frame at certain frequency, the constant dimension of Fourier fiber bundle at a neighborhood of that frequency, inverse process of the semi-convolution at that frequency, and closedness of the shift invariant space at that frequency.

Combining Theorem 4.2 and Proposition 4.1, we have

COROLLARY 4.4. *A function in* $(\mathcal{L}^p)^{(r)} \cap (W(L^{p/(p-1)}, \ell^1))^{(r)}$ *generates a p-frame if and only if it generates a p-frame at all frequencies.*

Then Theorems 4.2 and 4.3 can be considered as the localization version of Proposition 4.1 in the Fourier domain. Moreover, we see that the conditions on the generators in Theorem 4.2 are weaker than the ones in Proposition 4.1. For instance, a compactly supported function $F \in L^{\max(p,p/(p-1))}$ satisfies the conditions in Theorem 4.2, but does not satisfy the conditions in Proposition 4.1.

Note that the Fourier transform of the function

$$h = \chi_{[0,1]} - \chi_{[1,2]}$$

is

$$(e^{-2i\xi} - 2e^{-i\xi} + 1)/(i\xi).$$

Then h generates a p-frame at any frequency $\xi_0 \in \mathbb{R}\backslash(2\pi\mathbb{Z})$ by Theorem 4.2. Moreover, applying Theorem 4.2 yields the following interesting result for compactly supported bounded function.

THEOREM 4.5. *Let* $1 \le p \le \infty$, $f_1, \dots, f_r$ *be compactly supported and bounded, and set* $F = (f_1, \dots, f_r)^T$. *Then* F *generates a p-frame at almost all frequencies.*

4.1. Proof of Theorem 4.2. To prove Theorem 4.2, we need two lemmas.

LEMMA 4.6. *Let* $\xi_0 \in \mathbb{R}^d$, $F = (f_1, \dots, f_r)^T$ *be a tempered distribution with* ℓ^1 *decay. Then there exists a nonsingular matrix* $P_{\xi_0}(\xi)$ *so that all entries are periodic functions in the Wiener class, and the functions* F_{1,ξ_0} *and* F_{2,ξ_0} *defined by*

$$P_{\xi_0}(\xi)\widehat{F}(\xi) = \begin{pmatrix} \widehat{F}_{1,\xi_0}(\xi) \\ \widehat{F}_{2,\xi_0}(\xi) \end{pmatrix} \tag{4.3}$$

satisfy (i) F_{1,ξ_0} *has stable shifts at the frequency* ξ_0, *and* (ii) $\widehat{F}_{2,\xi_0}(\xi_0 + 2k\pi) = 0$ *for all* $k \in \mathbb{Z}^d$. *Furthermore,* F_{2,ξ_0} *can be chosen so that* $\widehat{F}_{2,\xi_0}(\xi + 2k\pi) = 0$ *for all* $\xi \in B(\xi_0, \delta_0)$ *and* $k \in \mathbb{Z}^d$ *if the rank of* $(\widehat{F}(\xi + 2k\pi))_{k\in\mathbb{Z}^d}$ *is a constant at a neighborhood of* ξ_0, *where* $\delta_0 > 0$.

The proof of the above lemma under the assumption that $F \in (W(L^\infty, \ell^1))^{(r)}$ was given in [**3**, Lemma 2]. We omit the details of the proof here since it can be done similarly.

LEMMA 4.7. *Let* $f \in W(L^p, \ell^1)$ *for some* $1 \le p \le \infty$. *Assume that* $\sum_{j\in\mathbb{Z}^d} f(\cdot - j) = 0$. *Then for any function* h *on* $\mathbb{R}^d$ *satisfying*

$$|h(x)| \le C(1 + |x|)^{-d-1} \quad \text{for all } x \in \mathbb{R}^d, \tag{4.4}$$

and

$$|h(x) - h(y)| \le C|x - y|\,(1 + \min(|x|, |y|))^{-d-1} \quad \text{for all } x, y \in \mathbb{R}^d, \tag{4.5}$$

we have

$$\lim_{\delta\to 0} \delta^d \Big\| \sum_{j\in\mathbb{Z}^d} h(\delta j) f(\cdot - j) \Big\|_{W(L^p,\ell^1)} = 0. \tag{4.6}$$

PROOF. Given a positive number ϵ, there exists $N_0 \ge 2$ such that

$$\sum_{|k|\ge N_0} \|f\|_{L^p(k+[0,1)^d)} \le \epsilon. \tag{4.7}$$

Set

$$f_1(x) = f(x)\chi_{O_{N_0}}(x) + \sum_{|k|\geq N_0} f(x+k)\chi_{[0,1]^d}(x),$$

where $O_{N_0} = \bigcup_{|k|<N_0}(k+[0,1]^d)$. Then

$$\|f_1 - f\|_{W(L^p,\ell^1)} \leq 2 \sum_{|k|\geq N_0} \|f\|_{L^p(k+[0,1)^d)} \leq 2\epsilon \tag{4.8}$$

by (4.7), and

$$\sum_{k\in\mathbb{Z}^d} f_1(x-k) = \sum_{k\in\mathbb{Z}^d} f(x-k) = 0 \quad \forall\, x\in\mathbb{R}^d \tag{4.9}$$

by the definition of f_1. By (4.4) and (4.8), we have

$$\begin{aligned} &\Big\|\delta^d \sum_{k\in\mathbb{Z}^d} h(\delta k)(f(\cdot - k) - f_1(\cdot - k))\Big\|_{W(L^p,\ell^1)} \\ &\qquad \leq \delta^d \sum_{k\in\mathbb{Z}^d} |h(\delta k)|\, \|f - f_1\|_{W(L^p,\ell^1)} \leq C\epsilon, \end{aligned} \tag{4.10}$$

where C is a positive constant independent of $\delta\in(0,1)$. From (4.5) and (4.9) it follows that

$$\begin{aligned} &\sum_{k\in\mathbb{Z}^d} \delta^d \Big\| \sum_{j\in\mathbb{Z}^d} h(\delta j) f_1(\cdot - j)\Big\|_{L^p(k+[0,1)^d)} \\ &\quad \leq \delta^d \sum_{k\in\mathbb{Z}^d} \Big\| \sum_{j\in\mathbb{Z}^d} (h(\delta j) - h(\delta k)) f_1(\cdot - j)\Big\|_{L^p(k+[0,1)^d)} \\ &\quad \leq C_1(N_0)\delta^{d+1} \\ &\qquad \times \sum_{k\in\mathbb{Z}^d} \Big\| \sum_{|j-k|\leq N_0+1} \frac{|f_1(\cdot - j)|}{(1+\delta\min(|j|,|k|))^{d+1}} \Big\|_{L^p(k+[0,1)^d)} \\ &\quad \leq C_2(N_0)\delta \|f_1\|_{W(L^p,\ell^1)} \leq C_3(N_0)\delta(\|f\|_{W(L^p,\ell^1)} + 2\epsilon), \end{aligned} \tag{4.11}$$

where $C_1(N_0), C_2(N_0), C_3(N_0)$ are positive constants depending only on N_0, d and the constant C in (4.5). Therefore the estimate (4.6) follows from (4.10) and (4.11) when δ is chosen sufficiently small. □

Proof of Theorem 4.2. We divide the proof into the following steps: (i) $\Longrightarrow$ (ii) $\Longrightarrow$ (iii) $\Longrightarrow$ (i). The implication of (i) $\Longrightarrow$ (ii) can be done by using the same technique in [**3**, Section 5.6] except the estimate (5.18) and the limit following the estimate (5.18) there being replaced by the estimate (4.1) and the limit in Lemma 4.7. We omit the details of the proof here.

Now we prove (ii) $\Longrightarrow$ (iii). Let functions F_{1,ξ_0} and F_{2,ξ_0}, matrix $P_{\xi_0}(\xi)$, and the positive number δ_0 be as in Lemma 4.6. Then the rank of $(\widehat{F}_{1,\xi_0}(\xi+2k\pi))_{k\in\mathbb{Z}^d}$ is k_0

for all $\xi \in B(\xi_0, \delta_1)$ for some $\delta_1 \le \delta_0$. Note that for $f \in \mathcal{L}^p$ and $g \in W(L^{p/(p-1)}, \ell^1)$,

$$\begin{aligned}\sum_{k\in\mathbb{Z}^d}\Big|\int_{\mathbb{R}^d} f(x-k)g(x)dx\Big| & \\ &\le \sum_{l\in\mathbb{Z}^d}\int_{[0,1)^d}|g(x-l)|\times\Big(\sum_{k\in\mathbb{Z}^d}|f(x-k)|\Big)dx \\ &\le \sum_{l\in\mathbb{Z}^d}\|g\|_{L^{p/(p-1)}(l+[0,1)^d)}\|f\|_{\mathcal{L}^p} \le \|f\|_{\mathcal{L}^p}\|g\|_{W(L^{p/(p-1)},\ell^1)}.\end{aligned}\tag{4.12}$$

Then the matrix $A(\xi) := \sum_{k\in\mathbb{Z}^d}\widehat{F}_{1,\xi_0}(\xi+2k\pi)\overline{\widehat{F}_{1,\xi_0}(\xi+2k\pi)}^T$ has all entries in the Wiener class. From the rank properties of the matrix $(\widehat{F}(\xi+2k\pi))_{k\in\mathbb{Z}^d}$ it follows that $A(\xi)$ is nonsingular in $B(\xi_0,\delta_1)+2\pi\mathbb{Z}^d$, which implies that there exists an inverse $B(\xi)$ of $A(\xi)$ in $B(\xi_0,\delta_1/2)+2\pi\mathbb{Z}^d$ so that its entries are still in the Wiener class. For any $f \in V_{p,B(\xi_0,\delta_1/2)}(F_{1,\xi_0})$, $\hat{f}(\xi) = \mathcal{F}(D)(\xi)^T\widehat{F}_{1,\xi_0}(\xi)$ for some sequence $D \in (\ell^p)^{(r)}$ with $\mathcal{F}(D)$ supported in $B(\xi_0,\delta_1/2)+2\pi\mathbb{Z}^d$. Thus the Fourier transforms of the sequences $W_i := \{\langle f, f_{i,\xi_0}(\cdot-k)\rangle\}, 1 \le i \le k_0$, satisfies

$$(\mathcal{F}(W_1)(\xi),\ldots,\mathcal{F}(W_{k_0})(\xi)) = \mathcal{F}(D)(\xi)^T A(\xi).$$

Therefore

$$\mathcal{F}(D)(\xi)^T = (\mathcal{F}(W_1)(\xi),\ldots,\mathcal{F}(W_{k_0})(\xi))B(\xi).$$

Substituting the above equation into $\hat{f}(\xi) = \mathcal{F}(D)(\xi)^T\widehat{F}_{1,\xi_0}(\xi)$ and using $\widehat{F}_{1,\xi_0}(\xi) = C_1(\xi)\widehat{F}(\xi)$ for some $k_0 \times r$ matrix $C_1(\xi)$ with entries in the Wiener class, we obtain

$$f = \sum_{i,i'=1}^{r}\sum_{k,k'\in\mathbb{Z}^d} a_{ii'}(k-k')\langle f, f_{i'}(\cdot-k')\rangle f_i(\cdot-k)$$

for any $f \in V_p(F)$ with its Fourier transform supported in $B(\xi_0,\delta_1/2)+2\pi\mathbb{Z}^d$, where $\{a_{ii'}(k)\} \in \ell^1, 1 \le i,i' \le r$. This proves the assertion (iii).

Finally we prove (iii) $\Longrightarrow$ (i). By (4.2), we have

$$\begin{aligned}\|f\|_p &\le \sum_{i=1}^{r}\Big\|\Big\{\sum_{i'=1}^{r}\sum_{k'\in\mathbb{Z}^d} a_{ii'}(k-k')\langle f, f_{i'}(\cdot-k')\rangle\Big\}\Big\|_{\ell^p} \\ &\le \sum_{i,i'=1}^{r}\|\{a_{ii'}(k)\}\|_{\ell^1}\|\{\langle f, f_i(\cdot-k)\rangle\}\|_{\ell^p} \le C\|\{\langle f, f_i(\cdot-k)\rangle\}\|_{\ell^p}\end{aligned}$$

for some positive constant C independent of f. Hence the implication (iii) $\Longrightarrow$ (i) follows. □

4.2. Proof of Theorem 4.3.

PROOF. First the sufficiency. By the continuity of $\widehat{F}$, the rank of $(\widehat{F}(\xi+2k\pi))_{k\in\mathbb{Z}^d}$ is at least that of $(\widehat{F}(\xi_0+2k\pi))_{k\in\mathbb{Z}^d}$ for all ξ in a small neighborhood of ξ_0. Therefore if (ii) does not hold, then the set of all $\xi \in B(\xi_0,\delta)$ such that the rank of $(\widehat{F}(\xi+2k\pi))_{k\in\mathbb{Z}^d}$ is strictly larger than the one of $(\widehat{F}(\xi_0+2k\pi))_{k\in\mathbb{Z}^d}$ is not empty for any positive δ. Using the same technique in [**3**, Section 5.2], the implication of (i) $\Longrightarrow$ (ii) follows.

Then the necessity. Let functions F_{1,ξ_0} and F_{2,ξ_0}, matrix $P_{\xi_0}(\xi)$ and the positive number δ_0 be as chosen in Lemma 4.6. Then $\widehat{F}_{2,\xi_0}(\xi)=0$ for all $\xi\in B(\xi_0,\delta)+2\pi\mathbb{Z}^d$, which together with the nonsingularity of the matrix P_{ξ_0} imply that

$$V_{p,B(\xi_0,\delta)}(F)=V_{p,B(\xi_0,\delta)}(F_{1,\xi_0})$$

for all $\delta\in(0,\delta_0)$. Recall that F_{1,ξ_0} has stable shifts at the frequency ξ_0, and hence it has ℓ^p stable shifts at the frequency ξ_0 by Theorem 3.1. Therefore there exists $\delta_1\in(0,\delta_0)$ so that $V_{p,B(\xi_0,\delta)}(F_{1,\xi_0})$ is a closed subspace of L^p for all $\delta<\delta_1$, and hence the assertion (i) follows. □

4.3. Proof of Theorem 4.5.

Proof. For a compactly supported bounded function $F=(f_1,\ldots,f_r)^T$, let k_0 be maximum of the rank of the $r\times\mathbb{Z}^d$ matrix $(\widehat{F}(\xi+2k\pi))_{k\in\mathbb{Z}^d}, \xi\in\mathbb{R}^d$. Then there exist $1\le i_1<i_2<\cdots<i_{k_0}\le r$ and $j_1,\ldots,j_{k_0}\in\mathbb{Z}^d$ so that the matrix $\big(\widehat{f}_{i_s}(\xi_0+2j_t\pi)\big)_{1\le s,t\le k_0}$ has nonzero determinant for some $\xi_0\in\mathbb{R}^d$. Recall that $\widehat{F}$ is an analytic function. Then the determinant of the above matrix is a nonzero analytic function, which implies that the set E of all $\xi\in\mathbb{R}^d$ with nonzero determinant is an open set and the complement $\mathbb{R}^d\backslash E$ has zero Lebesgue measure. Thus F generates a frame for $V_p(F)$ at any frequency in E. The assertion follows. □

Acknowledgment The research of the author was partially supported by the Texas Higher Education Coordinating Board Award ARP-003652-0071-2001. The author would like to thank the referee for his(her) suggestions.

References

[1] A. Aldroubi and K. Gröchenig, Nonuniform sampling and reconstruction in shift-invariant spaces. *SIAM Rev.*, 43(2001), 585–620.

[2] A. Aldroubi and Q. Sun, Locally finite dimensional shift-invariant spaces in $\mathbb{R}^d$, *Proc. Amer. Math. Soc.*, 130(2002), 2641–2654.

[3] A. Aldroubi, Q. Sun, and W.-S. Tang, p-frames and shift-invariant spaces of L^p. *J. Fourier Anal. Appl.*, 7(2001), 1–21.

[4] A. Aldroubi, Q. Sun and W.-S. Tang, Connection between p-frames and p-Riesz bases in locally finite SIS of $L^p(\mathbb{R}^d)$, In *Proceeding of SPIE 4119: Wavelet Applications in Signal and Image Processing VIII* (A. Aldroubi, A. F. Laine and M. A. Unser eds.), SPIE, 2000, pp. 668–674.

[5] A. Aldroubi, Q. Sun, and W.-S. Tang, Non-uniform average sampling and reconstruction in multiply generated shift-invariant spaces, *Constr. Approx.*, to appear.

[6] J. Benedetto and S. Li, The theory of multiresolution analysis frames and applications to filter banks, *Appl. Comput. Harmon. Anal.*, 5(1998), 398–427.

[7] C. K. Chui, *An Introduction to Wavelets*, Academic Press, 1992.

[8] A. Cohen and Q. Sun, An arithmetic characterization of the conjugate quadrature filters associated to orthonormal wavelet bases, *SIAM J. Math. Anal.*, 24(1993), 1355–1360.

[9] I. Daubechies, *Ten Lectures on Wavelets*, SIAM, Philadelphia 1992.

[10] T. A. Hogan, Stability and independence of the shifts of finitely many refinable functions, *J. Fourier Anal. Appl.*, 3(1997), 757–774.

[11] T. A. Hogan, Stability and independence for multivariate refinable distributions, *J. Approx. Th.*, 98(1999), 248–270.

[12] T. A. Hogan and R. Q. Jia, Dependency relations among the shifts of a multivariate refinable distribution, *Constr. Approx.*, 17(2001), 19–37.

[13] G. Jetter and G. Plonka, A survey on L^2 approximation orders from shift-invariant spaces, In *Multivariate Approximation and Applications* (N. Dyn, D. Leviatan, D. Levin, A. Pinkus, eds.), Cambridge University Press, 2001, pp. 73–111.

[14] R. Q. Jia, Subdivision schemes in L^p spaces, *Adv. Comput. Math.*, 3(1995), 309–341.

[15] R. Q. Jia and C. A. Micchelli, Using the refinement equation for the construction of pre-wavelets II: powers of two, In *Curve and Surfaces* (P. J. Laurent, A. Le Méhauté and L. L. Schumaker eds.), Academic Press, New York 1991, pp. 209–246.

[16] R. Q. Jia and C. A. Micchelli, On linear independence of integer translates of a finite number of functions, *Proc. Edinburgh Math. Soc.*, 36(1992), 69–75.

[17] R. Q. Jia and J.-Z. Wang, Stability and linear independence associated with wavelet decompositions, *Proc. Amer. Math. Soc.*, 117(1993), 1115–1124.

[18] W. Lawton, Necessary and sufficient conditions for constructing orthonormal wavelet bases, *J. Math. Phys.*, 32(1991), 57–61.

[19] S. Mallat, *A Wavelet Tour of Signal Processing*, Academic Press, 1998.

[20] O. Christensen and D. T. Stoeva, p-frames in separable Banach spaces, *Adv. Comput. Math.*, 18(2003), 117–126.

[21] A. Ron, A necessary and sufficient condition for the linear independence of the integer translates of a compactly supported distribution, *Constr. Approx.*, 5(1989), 297–308.

[22] Z. Shen, Refinable function vectors, *SIAM J. Math. Anal.*, 29(1998), 239–250.

[23] Q. Sun, Sequence spaces and stability of integer translates, *Zeitschrift Analysis und ihre Anwendungen*, 12(1993), 567–584.

[24] Q. Sun, Stability of the shifts of global supported distributions, *J. Math. Anal. and Appl.*, 261(2001), 113–125.

[25] Q. Sun, Convergence of cascade algorithm and smoothness of refinable distributions, *Chinese. Ann. Math.*, to appear.

[26] Q. Sun, Linear dependence of the shifts of vector-valued compactly supported distributions, submitted.

[27] L. F. Villemoes, Wavelet analysis of refinement equations, *SIAM J. Math. Anal.*, 25(1994), 1433–1460.

[28] J.-Z. Wang, Stability and linear independence associated with scaling vectors, *SIAM J. Math. Anal.*, 29(1998), 1140–1156.

Sun, Department of Mathematics, University of Houston, Houston, TX77204, USA

Contemporary Mathematics
Volume 345, 2004

Orthonormal wavelets arising from HDAFs

Jiansheng Yang, Lixin Shen, Manos Papadakis, Ioannis Kakadiaris,
Donald J. Kouri, and David K. Hoffman

ABSTRACT. Hermite Distributed Approximating Functionals (HDAFs) were introduced by Hoffman and Kouri for the numerical computation of the solutions of certain types of PDEs. These functions and their Fourier transforms are both dominated by Gaussians. In this paper we modify HDAFs in order to construct a family of continuous orthonormal wavelets. These wavelets can have any prescribed number of vanishing moments. The main part of the paper is devoted to the study of the smoothness of these wavelets.

1. Introduction and main results

The theory of Distributed Approximating Functionals (DAF's) was introduced in the work of Hoffman, Kouri, and their collaborators (e.g., see [**HNSK91**], [**HK93**]). DAFs are infinitely differential functions very localized in the spatial domain, which, depending on a certain parameter, form sequences of functions converging to the Dirac's delta function in the sense of distributions. This property made DAFs useful for the computation of numerical solutions of certain types of PDEs. In an effort to understand their efficiency for computational applications, the rigorous study of the mathematical foundations of DAFs was intitiated in [**CG99**] and continued in [**WWH01**]. The first DAFs that were created were the Hermite DAFs (HDAFs, see [**HNSK91**]) of whom their definition we include in the next paragraph.

The present paper is primarily devoted to the study of the smoothness of the Modified Hermite DAF orthonormal scaling functions and of their corresponding orthonormal wavelets, which were introduced in [**KPS2, KPS1**]. A Hermite DAF (HDAF) of order $N = 1, 2, 3, \ldots,$ is defined by the following equation:

$$h_{N,\sigma}(x) = \frac{1}{\sqrt{2\pi}\sigma} e^{-(x^2/(2\sigma^2))} \sum_{n=0}^{N} \left(-\frac{1}{4}\right)^n \frac{1}{n!} H_{2n}\left(\frac{x}{\sqrt{2}\sigma}\right),$$

2000 *Mathematics Subject Classification.* Primary 42C40; Secondary 65T60.

Key words and phrases. Hermite Distributed Approximating Functionals, wavelet.

This research was supported in part by the following grants: Texas Higher Eduaction Coordinating board ARP 003652-0071-2001, NSF DMS-0070376, NSF CHE-0074311, and R.A. Welch Foundation E-0608. Additional financial support was also provided by the Texas Learning and Computation Center of the University of Houston.

where σ is a positive constant and H_{2n} is the Hermite polynomial of degree $2n$. The Hermite polynomial H_n can be expressed by Rodrigues' Formula $H_n(x) = (-1)^n e^{x^2} \frac{d^n}{dx^n} e^{-x^2}$ (see also [**AS72**]). The Fourier transform of the HDAF is routinely shown to be

$$\widehat{h}_{N,\sigma}(\xi) = e^{-(\xi^2\sigma^2/2)} \sum_{n=0}^{N} \frac{(\xi^2\sigma^2)^n}{2^n n!}, \tag{1.1}$$

where the Fourier transform of an L^1-function f is defined by

$$\widehat{f}(\xi) := \int_{-\infty}^{\infty} f(x) e^{-ix\xi} dx, \quad \xi \in \mathbb{R}.$$

It is clear that $\lim_{N\to\infty} \widehat{h}_{N,\sigma}(\xi) = 1$ and $\lim_{\sigma\to 0} \widehat{h}_{N,\sigma}(\xi) = 1$ for all $\xi \in \mathbb{R}$. These two properties justify why HDAFs form sequences of functions converging to the Dirac's delta function in the sense of distributions (see [**CG99, WWH01**]). We will refrain from repeating the definition of multiresolution analysis (MRA), since it is widely known. We only point out that we use the definition of an MRA given in [**HW96**] and that an *orthonormal scaling function* is a square-integrable function ϕ satisfying, first, a two-scale relation, i.e.,

$$\widehat{\phi}(2\xi) = m_0(\xi)\widehat{\phi}(\xi) \quad \text{a.e. in } \mathbb{R}, \tag{1.2}$$

where m_0 is 2π-periodic and belongs to $L^2([-\pi,\pi))$; second, the integer translates of ϕ form an orthonormal system. The closed linear span of this system is the core subspace V_0 of the MRA $\{V_j\}_j, j \in \mathbb{Z}$. The 2π-periodic function m_0 associated with ϕ is called the *low pass filter* associated with ϕ. The wavelet ψ corresponding to ϕ is given by

$$\widehat{\psi}(\xi) = e^{-i\xi/2}\overline{m_0(\xi/2+\pi)}\widehat{\phi}(\xi/2).$$

The most widely known and utilized wavelets are these with compact support in the time domain, because the wavelet transforms induced by them are implemented by finite impulse response (FIR) filter banks. However, wavelets associated with low pass filters, which are not trigonometric polynomials are implemented by infinite impulse response (IIR) filter banks, and such filterbanks are less popular than their FIR counterparts. However, IIR filters with sharp frequency selectivity may be more desirable for certain types of applications, especially if they are symmetric as well. The low pass filters associated with the scaling functions we study in this paper have both these two properties. Next, we present the construction of the MHDAF orthonormal scaling functions. Let N be a non-negative integer and $\sigma > 0$. We define $m_{N,\sigma}$, which we also extend 2π-periodically over the real line, by the following equation:

$$m_{N,\sigma}(\xi) := \begin{cases} \sqrt{1-\widehat{h}^2_{N,\sigma}(\xi+\pi)}, & -\pi \le \xi < -\frac{\pi}{2}, \\ \widehat{h}_{N,\sigma}(\xi), & -\frac{\pi}{2} \le \xi \le \frac{\pi}{2}, \\ \sqrt{1-\widehat{h}^2_{N,\sigma}(\xi-\pi)}, & \frac{\pi}{2} < \xi < \pi. \end{cases}$$

We also impose the following condition:

$$\widehat{h}_{N,\sigma}\left(\frac{\pi}{2}\right) = \frac{\sqrt{2}}{2}. \tag{1.3}$$

The definition of $m_{N,\sigma}$ and (1.3) readily imply

$$|m_{N,\sigma}(\xi)|^2 + |m_{N,\sigma}(\xi+\pi)|^2 = 1, \qquad \xi \in \mathbb{R}. \tag{1.4}$$

Since, $\widehat{h}_{N,\sigma}(0) = 1$ we obviously have $m_{N,\sigma}(0) = 1$ and $m_{N,\sigma}(\pm\pi) = 0$. It is also easy to see that $m_{N,\sigma}$ is a 2π-periodic continuous function.

On the other hand, $\widehat{h}_{N,\sigma}$ is decreasing in the interval $[0, +\infty)$. Therefore

$$\inf_{|\xi|\le \pi/2} |m_{N,\sigma}(\xi)| \ge m_{N,\sigma}(\pi/2) = \frac{\sqrt{2}}{2} > 0.$$

If we combine this fact with (1.4), we obtain that ϕ, defined by

$$\widehat{\phi}(\xi) = \prod_{j=1}^{\infty} m_{N,\sigma}(\xi/2^j), \quad \xi \in \mathbb{R}\,,$$

is an orthonormal scaling function (see [**Mal89**] or Corollary 4.14 in [**HW96**]), which we call a *Modified-HDAF (MHDAF) scaling function.* Apparently $m_{N,\sigma}$ is the low pass filter corresponding to ϕ. We call $m_{N,\sigma}$ the *Modified-HDAF low pass filter of order N.* Since every $m_{N,\sigma}$ is even and takes on non-negative values only, the Fourier transforms of MHDAF scaling functions are even and non-negative. Thus the MHDAF scaling functions are themselves even in the time domain.

MHDAF low pass filters have an infinite number of nonzero Fourier coefficients, and the magnitude of the k-th coefficient is majorized by $O(|k|^{-3})$ ([**KPS2, KPS1**]). Thus MHDAF scaling functions are not compactly supported. This poses some technical constraints in the study of their smoothness properties, since it leaves us with one choice only: the study of the decay properties of their Fourier transforms. There is a considerable literature devoted to the study of this subject (e.g., [**CR95, HE95, V94**]). The reader may also find a in [**S99**] a brief and rather comprehensive account on the various smoothness measures of refinable tempered distributions based on the decay of their Fourier transform along with a generalization of Proposition 3.5 of [**CR95**] (Theorem 1.2). However, the conditions guaranteeing a certain degree of smoothness for scaling or refinable functions are not always easy to check, as the reader will soon realize (see also [**FS01**]). Remarkably enough, Theorem 1.4 reveals that, the Fourier domain techniques proposed in [**CR95**] for the study of the smoothness of compactly supported scaling functions (primarily Propositions 3.3 and 3.5) can also be applied in the case of the MHDAF scaling functions.

The investigation of the smoothness of the MHDAF scaling functions begins with the study of the relationship between σ and N, the two parameters that determine an HDAF. Recall that (1.3) implicitly determines σ in terms of N, which is a non-negative integer. The unique σ satisfying (1.3) will be denoted by $\sigma(N)$. The estimation of the smoothness of the MHDAF scaling function is given by Theorem 1.5. However, before obtaining this result we need several intermediate results, namely Theorems 1.1, 1.2 and 1.4. We prove these theorems in the following sections.

THEOREM 1.1. *Set* $\gamma_N := \dfrac{\pi^2\sigma^2(N)}{8}$. *Then the following are true:*

(1) $\gamma_N < 1.08$ *for* $N = 1$, *and* $\gamma_N < N$ *for* $N \ge 2$;

(2) $\gamma_N > \dfrac{N+2}{3}$ *for all* $N \ge 1$;

In order to improve the readability of the paper we introduce the following notation; $\sigma(N)$, $\widehat{h}_{N,\sigma(N)}$ and $m_{N,\sigma(N)}$ will be denoted by σ, $\widehat{h}_N$ and m_N, respectively.

THEOREM 1.2. *The following are true:*

(1) *For every N, the low pass filter m_N has a zero of order of $N+1$ at each $\xi = (2k+1)\pi$, with $k \in \mathbb{Z}$.*
(2) *If N is odd, then m_N is C^∞ at $(2k+1)\pi$, for all $k \in \mathbb{Z}$; if N is even, then m_N is C^N at $(2k+1)\pi$, for all $k \in \mathbb{Z}$.*
(3) *m_N is C^1 at every $k\pi + \pi/2$, $k \in \mathbb{Z}$. At all other points m_N is C^∞.*

For $\alpha = n+\beta$, where n is a non-negative integer and $0 < \beta \leq 1$, we define C^α to be the set of all bounded functions f which are n times continuously differentiable and their n-th order derivative $f^{(n)}$ is Hölder continuous of order β, i.e.,

$$|f^{(n)}(x+h) - f^{(n)}(x)| \leq C|h|^\beta$$

for every $x, h \in \mathbb{R}$.

LEMMA 1.3. *([**HW96**]) Let $f \in \mathbb{L}^2(\mathbb{R})$ and $\alpha > 0$. Suppose*

$$\int_{2^j\pi}^{2^{j+1}\pi} |\hat{f}(\xi)| d\xi \leq C2^{-\alpha j}, \quad j = 0, 1, 2, \ldots.$$

Then the following are true: If α is not an integer, then $f \in C^\alpha$, otherwise $f \in C^{\alpha-\epsilon}$, for every $0 < \epsilon < \alpha$.

The previous technical result is Lemma 3.22 in [**HW96**]. Using Lemma 1.3 we obtain that if $|\hat{f}(\xi)| \leq C(1+|\xi|)^{-1-\alpha}$, then f belongs to C^α, if α is not an integer; otherwise $f \in C^{\alpha-\epsilon}$, for every ϵ in $(0, \alpha)$.

For each $N \geq 0$ we define the function

$$L_N(\xi) := \frac{m_N(\xi)}{|\cos^{N+1}(\xi/2)|}.$$

Theorem 1.2 implies that L_N is well-defined everywhere on $\mathbb{R}$. The following theorem is the key result we need in order to prove Theorem 1.5.

THEOREM 1.4. *The following are true:*

(1) *$L_N(\xi)$ is an increasing function on the interval $[0, \pi]$;*
(2) *$|L_N(\xi)L_N(2\xi)|$ is a decreasing function on the interval $[\frac{2\pi}{3}, \pi]$.*

From the previous theorem, we have

$$\begin{array}{ll} |L_N(\xi)| \leq |L_N(\frac{2\pi}{3})| & \text{for} \quad |\xi| \leq \frac{2\pi}{3}, \\ |L_N(\xi)L_N(2\xi)| \leq |L_N(\frac{2\pi}{3})|^2 & \text{for} \quad \frac{2\pi}{3} \leq |\xi| \leq \pi. \end{array}$$

This allows us to apply Propositions 3.3 and 3.5 in [**CR95**] in order to derive Theorem 1.5 below, which is the main result of this paper. Although these two results from [**CR95**] require that L_N is C^∞, it is enough to use the C^1 continuity of L_N to prove them.

THEOREM 1.5. *For every $N = 1, 2, \ldots$ we have*

$$0 \leq \widehat{\phi}(\xi) \leq C(1+|\xi|)^{-N-1+b_N}$$

with $b_N := \log|L_N(2\pi/3)|/\log 2$. Therefore, $\phi \in C^{N-b_N-\epsilon}(\mathbb{R})$, for every $\epsilon > 0$. Furthemore, $\lim_{N\to+\infty}(N - b_N) = +\infty$.

Now, from $\lim_{N\to+\infty}(N-b_N) = +\infty$, *we infer that, the smoothness of the Modified HDAF scaling functions increases, as N increases.*

2. Estimating $\sigma(N)$

In this section, we prove Theorem 1.1. We begin with a lemma, which will be frequently used in Sections 2 and 4. The proof of the lemma is straightforward, so it will be omitted.

LEMMA 2.1. *Let a, b and c be arbitrary real numbers such that $(a+c-b)(a+c) > 0$ and $bc \geq 0$. Then*

$$\frac{a-b}{a+c-b} \leq \frac{a}{a+c}. \tag{2.1}$$

PROOF OF THEOREM 1.1. Define

$$g_N(\gamma) := e^{-\gamma} \sum_{n=0}^{N} \frac{\gamma^n}{n!} = \frac{\sum_{n=0}^{N} (\gamma^n/n!)}{\sum_{n=0}^{+\infty} (\gamma^n/n!)}, \qquad \gamma \in \mathbb{R}.$$

Then, $g'_N(\gamma) = -e^{-\gamma}\frac{\gamma^N}{N!} \leq 0$, so g_N is decreasing on $[0, +\infty)$. On the other hand, $g_N(\gamma_N) = \frac{\sqrt{2}}{2}$. Therefore, in order to prove the first part of Theorem 1.1, it is enough to show $g_1(1.08) < \frac{\sqrt{2}}{2}$ and $g_N(N) < \frac{\sqrt{2}}{2}$ for every $N \geq 2$. Likewise, in order to establish the second part of Theorem 1.1, it is enough to show $g_N\left(\frac{N+2}{3}\right) > \frac{\sqrt{2}}{2}$, for all $N \geq 1$.

(1): By direct computations one can verify that $g_1(1.08)$, and $g_N(N)$ for $N = 2, 3, 4$ are less than $\frac{\sqrt{2}}{2}$, hence $\gamma_1 < 1.08$ and $\gamma_N < N$ for $N = 2, 3, 4$.

Let $N \geq 5$. Now, set

$$I_1 := \sum_{n=0}^{N} \frac{N^n}{n!} = \left\{ \sum_{k=1}^{N-1} \left(\frac{N-1}{N} \cdots \frac{N-k}{N} \right) + 2 \right\} \frac{N^{N-1}}{(N-1)!} \tag{2.2}$$

and

$$I_2 := \sum_{n=N+1}^{\infty} \frac{N^n}{n!} = \left\{ \sum_{k=1}^{\infty} \left(\frac{N}{N+1} \cdots \frac{N}{N+k} \right) \right\} \frac{N^{N-1}}{(N-1)!}.$$

On the other hand, it is easy to see

$$\begin{aligned} \sum_{k=1}^{\infty} \left(\frac{N}{N+1} \cdots \frac{N}{N+k} \right) &> \sum_{k=1}^{N-1} \left(\frac{N}{N+1} \cdots \frac{N}{N+k} \right) \\ &> \sum_{k=1}^{N-1} \left(\frac{N-1}{N} \cdots \frac{N-k}{N} \right). \end{aligned}$$

Moreover, setting $a = N$, $b = N-5$ and $c = n$, where $n = 1, 2, \ldots, k$, Lemma 2.1 implies

$$\begin{aligned} \sum_{k=1}^{\infty} \left(\frac{N}{N+1} \cdots \frac{N}{N+k} \right) &> \sum_{k=1}^{\infty} \left(\frac{5}{5+1} \cdots \frac{5}{5+k} \right) \\ &> \sum_{k=1}^{4} \left(\frac{5}{5+1} \cdots \frac{5}{5+k} \right) > 2. \end{aligned}$$

Combining the last two inequalities with the definitions of I_1 and I_2, we obtain

$$I_1 < 2I_2.$$

Now, using lemma 2.1 for $a = I_1 + (2I_2 - I_1)$, $b = 2I_2 - I_1$ and $c = I_2$ we derive

$$g_N(N) = \frac{I_1}{I_1 + I_2} \leq \frac{I_1 + (2I_2 - I_1)}{I_1 + I_2 + (2I_2 - I_1)} = \frac{2}{3} < \frac{\sqrt{2}}{2}.$$

The final conclusion comes from the fact that γ_N is decreasing on the positive half-axis.

(2): For all $N \geq 1$ define

$$II_1 := \sum_{n=0}^{N} \frac{(N+2/3)^n}{n!} \quad \text{and} \quad II_2 := \sum_{n=N+1}^{\infty} \frac{(N+2/3)^n}{n!}.$$

Then

$$II_1 \geq \frac{(N+2/3)^{N+1}}{(N+1)!} \left(\frac{N+1}{(N+2/3)} + \frac{N+1}{(N+2/3)} \times \frac{N}{(N+2/3)} \right) \geq 4 \cdot \frac{(N+2/3)^{N+1}}{(N+1)!}$$

and

$$II_2 \leq \frac{(N+2/3)^{N+1}}{(N+1)!} \sum_{k=0}^{\infty} \left(\frac{1}{3}\right)^k = \frac{3}{2} \frac{(N+2/3)^{N+1}}{(N+1)!}.$$

Applying Lemma 2.1 again, with $a = II_1$, $b = II_1 - 4\frac{(N+2/3)^{N+1}}{(N+1)!}$ and $c = II_2$ we obtain,

$$g_N\left(\frac{N+2}{3}\right) = \frac{II_1}{II_1 + II_2} > \frac{4}{4 + \frac{3}{2}} > \frac{\sqrt{2}}{2},$$

which implies $\gamma_N > \frac{N+2}{3}$. □

3. The vanishing moments of the MHDAF orthonormal wavelets

In this section we prove Theorem 1.2.

PROOF OF THEOREM 1.2. By the symmetry and the periodicity of $m_N(\xi)$, we only have to check the continuity and the order of the zero of m_N at $\xi = \pi$.

Using (1.4) and $m_N(\xi + \pi) = m_N(\xi - \pi)$, which is true for all ξ, we have:

$$m_N(\xi) = \sqrt{1 - \widehat{h}_N^2(\xi - \pi)}$$

in $(\pi - \delta, \pi + \delta)$, where $0 < \delta < \frac{\pi}{2}$. For notational convenience, let us define

$$d(\xi) := \sqrt{1 - \widehat{h}_N^2(\xi)}.$$

The conclusion of Theorem 1.2 will follow, once we have established the properties of d at the origin.

Since,

$$\begin{aligned} 1 - \widehat{h}_N^2(\xi) &= (1 + \widehat{h}_N(\xi))(1 - \widehat{h}_N(\xi)) \\ &= (1 + \widehat{h}_N(\xi)) e^{-(\xi^2\sigma^2/2)} \sum_{n=N+1}^{\infty} \frac{(\xi^2\sigma^2)^n}{2^n n!} \\ &= (1 + \widehat{h}_N(\xi)) \frac{\sigma^{2(N+1)}\xi^{2(N+1)} e^{-(\xi^2\sigma^2/2)}}{2^{N+1}} \sum_{n=N+1}^{\infty} \frac{(\xi^2\sigma^2)^{n-N-1}}{2^{n-N-1} n!}, \end{aligned}$$

$d(\xi)$ can be factored as

$$d(\xi) = |\xi|^{N+1} g(\xi), \tag{3.1}$$

where

$$g(\xi) = \sqrt{(1 + \widehat{h}_N(\xi))\frac{\sigma^{2(N+1)}e^{-(\xi^2\sigma^2/2)}}{2^{N+1}} \sum_{n=N+1}^{\infty} \frac{(\xi^2\sigma^2)^{n-N-1}}{2^{n-N-1}n!}}$$

Clearly, $g(\xi) \in C^\infty$. Hence, if N is odd, then $d(\xi) \in C^\infty$, and the order of the zero of m_N at $\xi = \pi$ is equal to $N + 1$.

If N is even, then d is only C^N at 0, but the order of the zero of m_N at $\xi = \pi$ is still equal to $N + 1$. We omit the proof of the last item in Theorem 1.2, because it is straightforward. □

4. The smoothness of M-HDAF scaling function

This section begins with the proof of the first part of Theorem 1.4. Then, we give some technical lemmas, useful only in the proof the second part of Theorem 1.4. The proofs of the second part of Theorem 1.4 and of the last assertion of Theorem 1.5 conclude the present section.

Let us first introduce some auxilliary functions.

Let $\gamma > 0$. Define,

$$\begin{aligned}
\mathcal{Q}_{N,1}(\gamma) &:= \left(\frac{\gamma^{N+1}}{(N+1)!}\right)^{-1} \sum_{n=N+1}^{\infty} \frac{\gamma^n}{n!}, \\
\mathcal{Q}_{N,2}(\gamma) &:= \left(\frac{\gamma^{N}}{N!}\right)^{-1} \sum_{n=0}^{N} \frac{\gamma^n}{n!}, \\
\mathcal{Q}_{N,3}(\gamma) &:= \left(\sum_{n=0}^{N} \frac{\gamma^n}{n!}\right)^{-1} \sum_{n=N+1}^{\infty} \frac{\gamma^n}{n!}, \\
\mathcal{Q}_{N,4}(\gamma) &:= \frac{1}{\mathcal{Q}_{N,1}(\gamma)} - \frac{2}{(2+\mathcal{Q}_{N,3}(\gamma))\mathcal{Q}_{N,1}(\gamma)}.
\end{aligned}$$

Clearly,

$$\mathcal{Q}_{N,1}(\gamma) = 1 + \sum_{n=1}^{\infty} \frac{\gamma^n}{(N+2)\cdots(N+n+1)}; \tag{4.1}$$

$$\mathcal{Q}_{N,2}(\gamma) = \sum_{n=0}^{N} \frac{N!}{(N-n)!\gamma^n}; \tag{4.2}$$

$$\mathcal{Q}_{N,3}(\gamma) = \frac{\gamma}{N+1} \cdot \frac{\mathcal{Q}_{N,1}(\gamma)}{\mathcal{Q}_{N,2}(\gamma)}. \tag{4.3}$$

LEMMA 4.1. *The following are true for every* $N \geq 1$*:*

(1) $1 - \dfrac{1}{\mathcal{Q}_{N,1}(\gamma)} \leq \dfrac{\gamma}{N+2}$;

(2) $\mathcal{Q}_{N,2}$ *is decreasing on* $[0,\infty)$*, and* $\mathcal{Q}_{N,2}(N) > \mathcal{Q}_{N-1,2}(N-1)$.

(3) $\mathcal{Q}_{N,4}(\gamma) \leq \dfrac{\mathcal{Q}_{N,3}(\gamma)}{2}$*, and* $\mathcal{Q}_{N,4}$ *is increasing on* $(0, N+1]$.

PROOF. (1): Using (4.1), we have

$$1 - \frac{1}{\mathcal{Q}_{N,1}(\gamma)} = \frac{(\gamma/(N+2)) \cdot (1 + \sum_{n=1}^{\infty}(\gamma^n/((N+3)\cdots(N+2+n))))}{1 + \sum_{n=1}^{\infty}(\gamma^n/((N+2)\cdots(N+1+n)))} \leq \frac{\gamma}{N+2}.$$

(2): Since

$$\mathcal{Q}'_{N,2}(\gamma) = \frac{\gamma^{N-1}}{(N-1)!}\left(\sum_{n=1}^{N} \frac{\gamma^n}{(n-1)!}\left(\frac{1}{N} - \frac{1}{n}\right) - 1\right)\frac{1}{(\gamma^N/N!)^2} < 0,$$

we conclude that $\mathcal{Q}_{N,2}$ is decreasing on $[0,\infty)$. But,

$$\frac{N^N}{N!} = \frac{N^{N-1}}{(N-1)!} > \frac{(N-1)^{N-2}}{(N-2)!}$$

and

$$\frac{N-1}{N}\cdots\frac{N-k}{N} > \frac{(N-1)-1}{N-1}\cdots\frac{(N-1)-k}{N-1}$$

for every $1 \leq k \leq N-1$. Then, using (2.2), we have

$$\mathcal{Q}_{N,2}(N) > \mathcal{Q}_{N-1,2}(N-1).$$

(3): Eq. (4.1) implies $\mathcal{Q}_{N,1}(\gamma) \geq 1$. This inequality and $2 + \mathcal{Q}_{N,3}(\gamma) \geq 2$, for all $\gamma \geq 0$, imply

$$\mathcal{Q}_{N,4}(\gamma) = \frac{1}{\mathcal{Q}_{N,1}(\gamma)} \cdot \frac{\mathcal{Q}_{N,3}(\gamma)}{2 + \mathcal{Q}_{N,3}(\gamma)} \leq \frac{\mathcal{Q}_{N,3}(\gamma)}{2}.$$

On the other hand, it is easy to check

$$\mathcal{Q}_{N,4}(\gamma) = \frac{(\gamma^{N+1}/(N+1)!)}{e^\gamma + \sum_{n=0}^{N}(\gamma^n/n!)}.$$

Then, we clearly have,

$$\mathcal{Q}'_{N,4}(\gamma) = \frac{\dfrac{\gamma^N}{N!}\left(e^\gamma + \displaystyle\sum_{n=0}^{N}\frac{\gamma^n}{n!}\right) - \dfrac{\gamma^{N+1}}{(N+1)!}\left(e^\gamma + \displaystyle\sum_{n=0}^{N-1}\frac{\gamma^n}{n!}\right)}{\left(e^\gamma + \displaystyle\sum_{n=0}^{N}\frac{\gamma^n}{n!}\right)^2} > 0$$

for every $\gamma \in (0, N+1]$, which implies that $\mathcal{Q}_{N,4}$ is an increasing function on the interval $(0, N+1]$. □

PROOF OF ITEM 1 OF THEOREM 1.4. In order to establish that L_N is increasing on the interval $[0,\pi]$, it is sufficient to prove $L'_N \geq 0$ on each one of the intervals $[0,\pi/2]$ and $[\pi/2,\pi]$.

First let $\xi \in [0,\pi/2]$. Then

$$L_N(\xi) = \frac{\widehat{h}_N(\xi)}{\cos^{N+1}(\xi/2)}.$$

In order to show $L'_N(\xi) \geq 0$, for every $\xi \in [0,\pi/2]$, it is enough to prove

$$\frac{\left(\widehat{h}_N(\xi)\right)'}{\widehat{h}_N(\xi)} \geq \frac{\left(\cos^{N+1}(\xi/2)\right)'}{\cos^{N+1}(\xi/2)},$$

equivalently,

$$\frac{1}{2}\tan(\xi/2) \geq \frac{\sigma^2\xi}{N+1}(\mathcal{Q}_{N,2}(\sigma^2\xi^2/2))^{-1}.$$

First, let $N \leq 4$. Then by item 1 of Theorem 1.1, we have

$$\frac{\sigma^2\xi^2}{2} \leq \frac{\sigma^2\pi^2}{8} = \max(1.08, N),$$

so routine calculations imply

$$\frac{\sigma^2\xi}{N+1}\left(\mathcal{Q}_{N,2}\left(\frac{\sigma^2\xi^2}{2}\right)\right)^{-1} \leq \frac{\xi}{4} \leq \frac{1}{2}\tan(\xi/2),$$

for $N = 1, 2, 3, 4$.

On the other hand, if $N \geq 5$, then item 1 of Theorem 1.1, gives $\sigma^2 \leq \frac{8N}{\pi^2}$. Then,

$$\frac{\sigma^2}{N+1} < \frac{8}{\pi^2}$$

and item 2 of the previous lemma implies

$$\left(\mathcal{Q}_{N,2}\left(\frac{\sigma^2\xi^2}{2}\right)\right)^{-1} \leq (\mathcal{Q}_{N,2}(N))^{-1} \leq (\mathcal{Q}_{N-1,2}(N-1))^{-1} \leq \cdots \leq (\mathcal{Q}_{5,2}(5))^{-1} < \frac{2}{7}.$$

Hence

$$\frac{\sigma^2\xi}{N+1}\left(\mathcal{Q}_{N,2}\left(\frac{\sigma^2\xi^2}{2}\right)\right)^{-1} \leq \frac{16}{7\pi^2}\xi \leq \frac{2.3\xi}{\pi^2} \leq \frac{\xi}{4} \leq \frac{1}{2}\tan(\xi/2)\ ,$$

which establishes that, for $N \geq 5$, L_N is increasing on $[0, \pi/2]$.

Now, let $\xi \in [\frac{\pi}{2}, \pi]$. Define

$$\tilde{L}_N(\xi) := L_N(\pi - \xi) = \frac{\sqrt{1 - \widehat{h}_N^2(\xi)}}{\cos^{N+1}(\frac{\pi-\xi}{2})} = \frac{\sqrt{1 - \widehat{h}_N^2(\xi)}}{\sin^{N+1}(\xi/2)}.$$

We must now prove that $\tilde{L}_N$ is decreasing on $[0, \frac{\pi}{2}]$. To accomplish this goal it is enough to show

$$\frac{\left(\sqrt{1 - \widehat{h}_N^2(\xi)}\right)'}{\sqrt{1 - \widehat{h}_N^2(\xi)}} \leq \frac{(\sin^{N+1}(\xi/2))'}{\sin^{N+1}(\xi/2)}$$

or equivalently,

$$\frac{1}{\tan\xi/2} \geq \frac{1}{(\xi/2)\mathcal{Q}_{N,1}(\sigma^2\xi^2/2)}. \tag{4.4}$$

But, (4.1) implies $\mathcal{Q}_{N,1}(\frac{\sigma^2\xi^2}{2}) \geq 1 + \frac{\frac{\sigma^2\xi^2}{2}}{N+2}$, so (4.4) is true if the following inequality is true as well:

$$\tan(\xi/2) \leq \frac{\xi}{2} + \frac{(\sigma^2\xi^3/4)}{N+2}. \tag{4.5}$$

Now, item 2 of Theorem 1.1 implies $\sigma^2 > \frac{8}{\pi^2} \cdot \frac{N+2}{3}$. Therefore, to obtain (4.5), it is enough to prove

$$\tan\xi \leq \xi + \frac{16\xi^3}{3\pi^2}, \tag{4.6}$$

for every $\xi \in [0, \pi/4]$. So, let

$$g(\xi) := \tan\xi - \xi - \frac{16\xi^3}{3\pi^2}.$$

Then

$$\begin{aligned} g'(\xi) &= \sec^2\xi - 1 - \frac{16\xi^2}{\pi^2} \\ &= \tan^2\xi - \frac{16\xi^2}{\pi^2} \\ &= \xi^2\left(\frac{\tan^2\xi}{\xi^2} - \frac{16}{\pi^2}\right) \\ &\leq \xi^2\left(\frac{\tan^2\xi}{\xi^2} - \frac{16}{\pi^2}\right)|_{\xi=\frac{\pi}{4}} = 0. \end{aligned}$$

so, $g(\xi) \leq g(0) = 0$, for every $\xi \in [0, \pi/4]$. This completes the proof of (4.5) and so of the first part of Theorem 1.4. □

The proof of second part of Theorem 1.4 is more complicated and it requires several intermediate results, which we will state and prove in the sequel.

LEMMA 4.2. *Let $x \in (0, \pi/4]$. Then, the following are true:*

(1) $\dfrac{1}{2(\frac{x}{2} + \frac{2}{3!}(\frac{x}{2})^3)} - \dfrac{1}{2\tan\frac{x}{2}} \geq \dfrac{x^3}{135}$;

(2) $\dfrac{1}{x} - \dfrac{1}{2(\frac{x}{2} + \frac{2}{3!}(\frac{x}{2})^3)} \geq \dfrac{x}{13}$.

PROOF. Since the proof of item 2 is easy, we will only give the proof of item 1. We first have to prove

$$\tan x \geq x + \frac{1}{3!}x^3 + \frac{16}{5!}x^5\,, \quad 0 < x \leq \pi/8\,. \tag{4.7}$$

In order to prove the latter inequality we have to use Taylor's expansion of the tangent function ([**AS72**]):

$$\tan x = \sum_{n=1}^{\infty} \frac{(-4)^n(1-4^n)B_{2n}}{(2n)!}x^{2n-1}\,,$$

where B_n is a Bernoulli number. Since,

$$B_{2n} = \frac{(-1)^{n-1}2(2n)!}{(2\pi)^{2n}}\sum_{p=1}^{\infty} p^{-2n} \quad \text{for} \quad n = 1, 2, \ldots.$$

we conclude

$$\frac{(-4)^n(1-4^n)B_{2n}}{(2n)!} > 0 \qquad n = 1, 2, \ldots.$$

This establishes (4.7). Using (4.7) we obtain

$$\begin{aligned} \frac{1}{2(\frac{x}{2} + \frac{2}{3!}(\frac{x}{2})^3)} - \frac{1}{2\tan\frac{x}{2}} &\geq \frac{1}{2(\frac{x}{2} + \frac{2}{3!}(\frac{x}{2})^3)} - \frac{1}{2(\frac{x}{2} + \frac{2}{3!}(\frac{x}{2})^3 + \frac{16}{5!}(\frac{x}{2})^5)} \\ &\geq \frac{\frac{x^3}{120}}{(1 + \frac{2}{3!}(\frac{\pi}{8})^2)(1 + \frac{2}{3!}(\frac{\pi}{8})^2 + \frac{16}{5!}(\frac{\pi}{8})^4)} \\ &\geq \frac{x^3}{135}\,. \end{aligned}$$

□

PROPOSITION 4.3. For every $N \geq 1$ and $x \in (0, \frac{\pi}{4}]$ we have

$$\frac{1}{(x/2)\cdot(2+\mathcal{Q}_{N,3}(\sigma^2x^2/2))\mathcal{Q}_{N,1}(\sigma^2x^2/2)} - \frac{1}{2\tan(x/2)} + \tan x - \frac{(4\sigma^2/(N+1))x}{\mathcal{Q}_{N,2}(2\sigma^2x^2)} \geq 0. \tag{4.8}$$

PROOF. First, we discuss the case $N \leq 3$. Applying item 1 of Theorem 1.1 we obtain

$$\frac{\sigma^2x^2}{2} \leq \frac{\gamma_N}{4}, \quad \text{for every} \quad x \in [0, \pi/4],$$

where $\gamma_N = \frac{\pi^2\sigma^2}{8}$. Eqs. (4.1) and (4.2) imply

$$\mathcal{Q}_{N,1}\left(\frac{\sigma^2x^2}{2}\right) \leq \sum_{n=0}^{\infty}\left(\frac{\gamma_N}{4(N+2)}\right)^n, \quad \mathcal{Q}_{N,2}\left(\frac{\sigma^2x^2}{2}\right) \geq \sum_{n=0}^{N}\frac{N!}{(N-n)!(\frac{\gamma_N}{4})^n}.$$

Now, let $N = 1$. According to Theorem 1.1 $\gamma_1 < 1.08$, so $\sigma^2 \leq \frac{8\times 1.08}{\pi^2}$. Therefore,

$$\mathcal{Q}_{1,1}\left(\frac{\sigma^2x^2}{2}\right) \leq \frac{10}{9}, \quad \mathcal{Q}_{1,2}\left(\frac{\sigma^2x^2}{2}\right) \geq \frac{127}{27}.$$

Now, applying (4.3) we get

$$\mathcal{Q}_{1,3}\left(\frac{\sigma^2x^2}{2}\right) \leq \frac{16.2}{127}\cdot\frac{4x^2}{\pi^2}. \tag{4.9}$$

Furthermore, item 1 of Lemma 4.1 gives us

$$1 - \frac{1}{\mathcal{Q}_{1,1}(\sigma^2x^2/2)} \leq \frac{\sigma^2x^2}{6} \leq \frac{4.32}{3\pi^2}x^2. \tag{4.10}$$

Using the definition of $\mathcal{Q}_{1,4}$ one can easily verify

$$\frac{1}{(x/2)\cdot(2+\mathcal{Q}_{1,3}(\sigma^2x^2/2))\mathcal{Q}_{1,1}(\sigma^2x^2/2)} - \frac{1}{2\tan(x/2)} = \left(\frac{1}{x} - \frac{1}{2\tan(x/2)}\right) - \left(\frac{1}{x} - \frac{1}{x\mathcal{Q}_{1,1}(\sigma^2x^2/2)}\right) - \frac{\mathcal{Q}_{1,4}(\sigma^2x^2/2)}{x}. \tag{4.11}$$

Using the Taylor series expansion of the tangent and the fact that for $0 \leq x < \pi/2$ we can easily see $\tan\frac{x}{2} \geq \frac{x}{2} + \frac{2}{3!}(\frac{x}{2})^3$. Combining the previous inequality with item 2 of Lemma 4.2 we conclude

$$\frac{1}{x} - \frac{1}{2\tan(x/2)} \geq \frac{x}{13}. \tag{4.12}$$

On the other hand, (4.10) implies

$$\frac{1}{x} - \frac{1}{x\mathcal{Q}_{1,1}(\sigma^2x^2/2)} \leq \frac{4.32x}{3\pi^2}. \tag{4.13}$$

Item 3 of Lemma 4.1 and (4.9) finally imply

$$\frac{\mathcal{Q}_{1,4}(\sigma^2x^2/2)}{x} \leq \frac{8.1}{127}\cdot\frac{4x}{\pi^2}. \tag{4.14}$$

Combining inequalities (4.12), (4.13), (4.14) and (4.11) we obtain

$$\begin{aligned}\frac{1}{(x/2)\cdot(2+\mathcal{Q}_{1,3}(\sigma^2x^2/2))\mathcal{Q}_{1,1}(\sigma^2x^2/2)} &- \frac{1}{2\tan(x/2)}\\ &\geq \frac{x}{13} - \frac{1.08}{3}\cdot\frac{4x}{\pi^2} - \frac{8.1}{127}\cdot\frac{4x}{\pi^2}\\ &\geq \frac{x}{13} - (0.4238)\cdot\frac{4x}{\pi^2}.\end{aligned}$$

Hence, the left-hand-side of (4.8), which we denote by LHS, satisfies

$$LHS \quad\geq\quad \tan x + \frac{x}{13} - \left(0.4238 + \frac{4.32}{\mathcal{Q}_{1,2}(2\sigma^2x^2)}\right)\frac{4x}{\pi^2},$$

so (4.8) will be valid, if the right-hand-side of previous inequality is non-negative for every $x \in (0, \frac{\pi}{4}]$. Let us first, assume $x \in (0, \frac{\pi}{6}]$. Since $\sigma^2 \leq \frac{8\times 1.08}{\pi^2}$ and $\mathcal{Q}_{1,2}$ is decreasing we obtain $\mathcal{Q}_{1,2}(2\sigma^2x^2) \geq \mathcal{Q}_{1,2}(\frac{4.32}{9}) > 3$. Therefore,

$$\begin{aligned}\tan x + \frac{x}{13} - \left(0.4238 + \frac{4.32}{\mathcal{Q}_{1,2}(2\sigma^2x^2)}\right)\frac{4x}{\pi^2} &\geq \tan x + \frac{x}{13} - \left(0.4238 + \frac{4.32}{3}\right)\frac{4x}{\pi^2}\\ &\geq x - \frac{8x}{\pi^2} \geq 0;\end{aligned}$$

Now, let $x \in [\frac{\pi}{6}, \frac{\pi}{4}]$. Arguing as in the previous case we obtain $\mathcal{Q}_{1,2}(2\sigma^2x^2) \geq \mathcal{Q}_{1,2}(1.08) = \frac{52}{27}$. Combining these inequalities with $\tan x + \frac{x}{13} \geq \frac{7x}{6}$, we conclude

$$\tan x + \frac{x}{13} - \left(0.4238 + \frac{4.32}{\mathcal{Q}_{1,2}(2\sigma^2x^2)}\right)\frac{4x}{\pi^2} \geq \frac{7x}{6} - \left(0.4238 + \frac{4.32}{52/27}\right)\frac{4x}{\pi^2} \geq 0.$$

The cases $N = 2$ and $N = 3$ are similar to the case $N = 1$, so we will omit their proofs.

Now, let $N \geq 4$. The item 1 of Theorem 1.1 implies

$$\frac{\sigma^2x^2}{2} \leq \frac{\gamma_N}{4} \leq \frac{N}{4}, \tag{4.15}$$

for every $x \in [0, \pi/4]$, where $\gamma_N = \frac{\pi^2\sigma^2}{8}$. Then, using the previous inequalities and (4.1) once again, we get

$$\mathcal{Q}_{N,1}\left(\frac{\sigma^2x^2}{2}\right) \leq \sum_{n=0}^{\infty}\left(\frac{N}{4(N+2)}\right)^n \leq \sum_{n=0}^{\infty}\left(\frac{1}{4}\right)^n = \frac{4}{3}.$$

Applying the definition of $\mathcal{Q}_{N,2}$ and using (4.15), we have

$$\begin{aligned}\mathcal{Q}_{N,2}\left(\frac{\sigma^2x^2}{2}\right) &\geq \left(\frac{(\sigma^2x^2/2)^N}{N!}\right)^{-1}\sum_{n=0}^{N-1}\frac{(\sigma^2x^2/2)^n}{n!}\\ &\geq \frac{N}{(\sigma^2x^2/2)}\left(1 + \sum_{n=1}^{N-1}\frac{(N-1)\cdots(N-n)}{(N/4)^n}\right).\end{aligned}$$

Now, observe

$$\sum_{n=1}^{N-1}\frac{(N-1)\cdots(N-n)}{(N/4)^n} > \sum_{n=1}^{3}\frac{(N-1)\cdots(N-n)}{(N/4)^n}.$$

Letting $n = 1, 2, 3$ and a, b, c in Lemma 2.1 be equal to $N-n$, $N-4$, n, respectively we obtain $(N-n)/N > (4-n)/4$. So

$$\frac{(N-1)\cdots(N-n)}{(N/4)^n} > (4-1)\cdots(4-n), \quad n = 1, 2, 3.$$

Therefore,

$$\mathcal{Q}_{N,2}\left(\frac{\sigma^2 x^2}{2}\right) \geq \frac{N}{(\sigma^2x^2/2)}\left(1 + \sum_{n=1}^{3}(4-1)\cdots(4-n)\right) \geq \frac{16N}{(\sigma^2x^2/2)}.$$

Thus, using also (4.3), we conclude

$$\mathcal{Q}_{N,3}\left(\frac{\sigma^2 x^2}{2}\right) \leq \frac{\sigma^2}{N+1}\cdot\frac{\sigma^2}{N}\cdot\frac{x^4}{48} \leq \left(\frac{8}{\pi^2}\right)^2\left(\frac{x^4}{48}\right) \leq \frac{x^4}{72}. \tag{4.16}$$

Next, we will establish the following two inequalities, which clearly imply (4.8).

$$\frac{1}{(x/2)(2+\mathcal{Q}_{N,3}(\sigma^2x^2/2))\mathcal{Q}_{N,1}(\sigma^2x^2/2)} - \frac{1}{2\tan(x/2)} \geq \frac{1}{x\mathcal{Q}_{N,1}(\sigma^2x^2/2)} - \frac{1}{2((x/2)+(2/3!)(x/2)^3)} \tag{4.17}$$

and

$$\frac{1}{x\mathcal{Q}_{N,1}(\sigma^2x^2/2)} - \frac{1}{2((x/2)+(2/3!)(x/2)^3)} + \tan x - \frac{\frac{4\sigma^2}{N+1}\cdot x}{\mathcal{Q}_{N,2}(2\sigma^2x^2)} \geq 0. \tag{4.18}$$

We begin with the proof of (4.17). The definition of $\mathcal{Q}_{N,4}$ readily implies

$$\frac{1}{x\mathcal{Q}_{N,1}(\sigma^2x^2/2)} - \frac{1}{(x/2)\cdot(2+\mathcal{Q}_{N,3}(\sigma^2x^2/2))\mathcal{Q}_{N,1}(\sigma^2x^2/2)} = \frac{\mathcal{Q}_{N,4}(\sigma^2x^2/2)}{x}.$$

But,

$$\frac{\mathcal{Q}_{N,4}(\sigma^2x^2/2)}{x} \leq \frac{\mathcal{Q}_{N,3}(\sigma^2x^2/2)}{2x},$$

due to item 3 of Lemma 4.1. Combining the previous inequality with (4.16) we get

$$\frac{1}{x\mathcal{Q}_{N,1}(\sigma^2x^2/2)} - \frac{1}{(x/2)\cdot(2+\mathcal{Q}_{N,3}(\sigma^2x^2/2))\mathcal{Q}_{N,1}(\sigma^2x^2/2)} \leq \frac{x^3}{144}, \tag{4.19}$$

which together with item 1 of Lemma 4.2 imply (4.17).

Let us now prove (4.18). Item 1 of lemma 4.1 and item 2 of 4.2 imply

$$\begin{aligned}
&\frac{1}{x\mathcal{Q}_{N,1}(\sigma^2x^2/2)} - \frac{1}{2((x/2)+(2/3!)(x/2))^3)} \\
&= \left(\frac{1}{x\mathcal{Q}_{N,1}(\sigma^2x^2/2)} - \frac{1}{x}\right) + \left(\frac{1}{x} - \frac{1}{2((x/2)+(2/3!)(x/2)^3)}\right) \\
&\geq \frac{x}{13} - \frac{\sigma^2 x}{2(N+2)} \geq \frac{x}{13} - \frac{N}{N+2}\cdot\frac{4x}{\pi^2}.
\end{aligned}$$

Using the fact that $\gamma_N < N$ and the previous inequality, we obtain

$$\begin{aligned}
&\frac{1}{x\mathcal{Q}_{N,1}(\sigma^2x^2/2)} - \frac{1}{2((x/2)+(2/3!)(x/2)^3)} + \tan x - \frac{\frac{4\sigma^2}{N+1}\cdot x}{\mathcal{Q}_{N,2}(2\sigma^2x^2)} \\
&\qquad \geq \tan x + \frac{x}{13} - \left(\frac{N}{N+2} + \frac{(8N/(N+1))}{\mathcal{Q}_{N,2}(2\sigma^2x^2)}\right)\frac{4x}{\pi^2},
\end{aligned}$$

for all $0 \leq x \leq \pi/4$. So, the conclusion of proposition 4.3 will be established once we have that the RHS of previous inequality is non-negative, for all $0 \leq x \leq \pi/4$ and $N \geq 4$.

In order to prove this fact, let us first assume $0 \leq x \leq \frac{\pi}{6}$. Using the fact that $\mathcal{Q}_{N,2}$ is decreasing we derive

$$\mathcal{Q}_{N,2}(2\sigma^2 x^2) \geq \mathcal{Q}_{N,2}\left(\frac{4N}{9}\right) \geq \sum_{n=0}^{4} \frac{4!}{(4-n)!(16/9)^n} > 8.$$

Thus,

$$\tan x + \frac{x}{13} - \left(\frac{N}{N+2} + \frac{(8N/(N+1))}{\mathcal{Q}_{N,2}(2\sigma^2 x^2)}\right)\frac{4x}{\pi^2} \geq x - \frac{8x}{\pi^2} \geq 0\,.$$

Now, let $\frac{\pi}{6} \leq x \leq \frac{\pi}{4}$. Since, $\tan x + \frac{x}{13} \geq \frac{7x}{6}$, in order to prove

$$\tan x + \frac{x}{13} - \left(\frac{N}{N+2} + \frac{(8N/(N+1))}{\mathcal{Q}_{N,2}(2\sigma^2 x^2)}\right)\frac{4x}{\pi^2} \geq 0\,, \tag{4.20}$$

it is enough to establish

$$\left(\frac{N}{N+2} + \frac{(8N/(N+1))}{\mathcal{Q}_{N,2}(2\sigma^2 x^2)}\right)\frac{4x}{\pi^2} \leq \frac{7x}{6} \tag{4.21}$$

for all $N \geq 4$ and $x \in (\frac{\pi}{6}, \frac{\pi}{4}]$.

Since for every $x \in (\frac{\pi}{6}, \frac{\pi}{4}]$ we have $2\sigma^2 x^2 \leq N$, the fact that $\mathcal{Q}_{N,2}$ is decreasing, implies $\mathcal{Q}_{N,2}(2\sigma^2 x^2) \geq \mathcal{Q}_{N,2}(N)$. Direct calculations show that for $N = 4, 5, 6, 7, 8, 9$, $\mathcal{Q}_{N,2}(N)$ is greater than 3.2, 3.5, 3.77, 4, 4.24, and 4.45 respectively. This establishes (4.21) for $N = 4, 5, 6, 7, 8, 9$ and $x \in (\frac{\pi}{6}, \frac{\pi}{4}]$.

Now, let $N \geq 10$. Using item 2 of Lemma 4.1 we have

$$\mathcal{Q}_{N,2}(2\sigma^2 x^2) \geq \mathcal{Q}_{N,2}(N) \geq \mathcal{Q}_{10,2}(10) > 4.65,$$

which implies

$$\left(\frac{N}{N+2} + \frac{(8N/(N+1))}{\mathcal{Q}_{N,2}(2\sigma^2 x^2)}\right)\frac{4x}{\pi^2} \leq \left(1 + \frac{8}{4.65}\right)\frac{4x}{\pi^2} \leq \frac{7x}{6}.$$

Therefore, (4.20), and thus, (4.18) hold for every $x \in (\frac{\pi}{6}, \frac{\pi}{4}]$ and $N \geq 4$. This concludes the proof of the Proposition 4.3. □

Before stating the next lemma we need to introduce some notation which will be helpful in its proof. If x is an arbitrary real number, we denote by $\lfloor x \rfloor$ and $\lceil x \rceil$ the greatest integer less than or equal to x and the smallest integer greater than or equal to x, respectively.

LEMMA 4.4. *The function*

$$g(y) = \frac{1}{1 + (y/(N+2))} - \frac{1}{\mathcal{Q}_{N,1}(y)}$$

is increasing on the interval $(0, N+1]$.

Before we give the proof of Lemma 4.4, we first show the plots of the function g for $N = 1$, 10, and 100, respectively, in Figure 1.

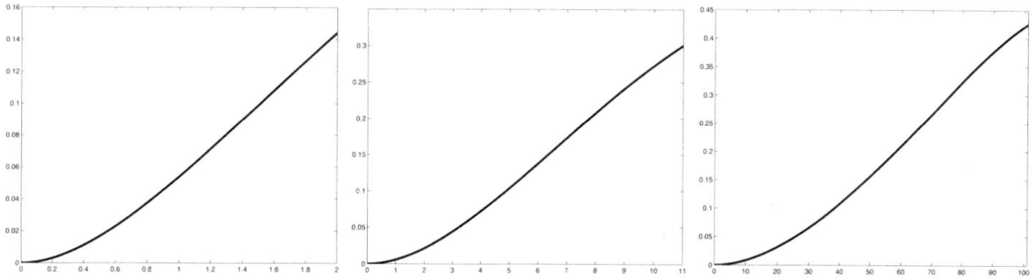

FIGURE 1. The plots of the function g in Lemma 4.4 (from left to right) with $N = 1$, 10, and 100, respectively.

PROOF OF LEMMA 4.4. The first derivative of g is

$$g'(y) = \frac{\sum_{n=0}^{\infty}((n+1)(N+1)!y^n/(N+2+n)!)}{\left(\sum_{n=0}^{\infty}((N+1)!y^n/(N+1+n)!)\right)^2} - \frac{(1/(N+2))}{(1+(y/(N+2)))^2}$$
$$= \frac{(1/(N+2))I(y)}{\left(\sum_{n=0}^{\infty}((N+1)!y^n/(N+1+n)!)\right)^2 (1+(y/(N+2)))^2},$$

where

$$I(y) := \left(1 + \frac{y}{N+2}\right)^2 \sum_{n=0}^{\infty} \frac{(n+1)(N+2)!y^n}{(N+2+n)!} - \left(\sum_{n=0}^{\infty} \frac{(N+1)!y^n}{(N+1+n)!}\right)^2. \tag{4.22}$$

Our goal will be to show $I(y) \geq 0$ for all $y \in (0, N+1]$.

First, note that the series involved in the definition of I are bounded by geometric series, because $0 < y \leq N+1$. This implies that the series in the right-hand side of eq. (4.22) converge uniformly. Now,

$$\left(\sum_{n=0}^{\infty} \frac{(N+1)!y^n}{(N+1+n)!}\right)^2$$
$$= \left(1 + \sum_{n=1}^{\infty} \frac{(N+1)!y^n}{(N+1+n)!}\right)\left(1 + \sum_{l=1}^{\infty} \frac{(N+1)!y^l}{(N+1+l)!}\right)$$
$$= 1 + 2\sum_{n=1}^{\infty} \frac{(N+1)!y^n}{(N+1+n)!} + \left(\sum_{n=1}^{\infty} \frac{(N+1)!y^n}{(N+1+n)!}\right)\left(\sum_{l=1}^{\infty} \frac{(N+1)!y^l}{(N+1+l)!}\right)$$
$$= 1 + 2\sum_{n=1}^{\infty} \frac{1}{(N+2)\cdots(N+1+n)} y^n$$
$$+ \sum_{n=1}^{\infty}\sum_{l=1}^{\infty} \frac{1}{(N+2)\cdots(N+1+n)} \cdot \frac{1}{(N+2)\cdots(N+1+l)} y^{n+l}$$
$$= 1 + \frac{2}{N+2}y + \frac{2}{(N+2)(N+3)}y^2 + \frac{2}{(N+2)(N+3)(N+4)}y^3$$
$$+ 2\sum_{n=4}^{\infty} \frac{1}{(N+2)\cdots(N+1+n)}y^n + \frac{1}{(N+2)^2}y^2 + \frac{2}{(N+2)^2(N+3)}y^3$$
$$+ \sum_{n=4}^{\infty}\left[\sum_{l=1}^{n-1} \frac{1}{(N+2)\cdots(N+1+n-l)} \cdot \frac{1}{(N+2)\cdots(N+1+l)}\right] y^n$$

$$
\begin{aligned}
&= 1 + \frac{2}{N+2}y + \left(\frac{2}{(N+2)(N+3)} + \frac{1}{(N+2)^2}\right)y^2 \\
&\quad + \left(\frac{2}{(N+2)(N+3)(N+4)} + \frac{2}{(N+2)^2(N+3)}\right)y^3 \\
&\quad + \sum_{k=2}^{\infty}\left(2\sum_{i=k+1}^{2k-1}\frac{1}{(N+2)\cdots(N+i+1)}\cdot\frac{1}{(N+2)\cdots(N+2k-i+1)}\right. \\
&\qquad\qquad \left. + \left(\frac{1}{(N+2)\cdots(N+k+1)}\right)^2\right)y^{2k} \\
&\quad + \sum_{k=2}^{\infty}\left(2\sum_{i=k+1}^{2k}\frac{1}{(N+2)\cdots(N+i+1)}\right. \\
&\qquad\qquad \left. \cdot\frac{1}{(N+2)\cdots(N+2k+1-i+1)}\right)y^{2k+1}.
\end{aligned}
$$

This expansion of the term $\left(\sum_{n=0}^{\infty}\frac{(N+1)!y^n}{(N+1+n)!}\right)^2$ now yields

$$
\begin{aligned}
I(y) &= \frac{2}{N+3}y + \left(\frac{3}{(N+3)(N+4)} + \frac{2}{(N+2)(N+3)}\right)y^2 \\
&\quad + \left(\frac{4}{(N+3)(N+4)(N+5)} + \frac{4}{(N+2)(N+3)(N+4)}\right)y^3 \\
&\quad + \sum_{n=4}^{\infty}\left(\frac{n+1}{(N+3)\cdots(N+n+2)} + \frac{2n-2}{(N+2)\cdots(N+n+1)}\right. \\
&\qquad\qquad \left. + \frac{n-1}{(N+2)^2(N+3)\cdots(N+n)}\right)y^n \\
&\quad - \sum_{k=2}^{\infty}\left(2\sum_{i=k+1}^{2k-1}\frac{1}{(N+2)\cdots(N+i+1)}\cdot\frac{1}{(N+2)\cdots(N+2k-i+1)}\right. \\
&\qquad\qquad \left. + \left(\frac{1}{(N+2)\cdots(N+k+1)}\right)^2\right)y^{2k} \\
&\quad - \sum_{k=2}^{\infty}\left(2\sum_{i=k+1}^{2k}\frac{1}{(N+2)\cdots(N+i+1)}\right. \\
&\qquad\qquad \left. \cdot\frac{1}{(N+2)\cdots(N+2k+1-i+1)}\right)y^{2k+1}.
\end{aligned}
$$

Using $0 < y \leq N+1$, one can easily verify the following inequalities

$$
\frac{y^i}{(N+2)\cdots(N+i+1)}\cdot\frac{y^{2k-i}}{(N+2)\cdots(N+2k-i+1)} \leq \frac{y^{i-1}}{(N+3)\cdots(N+i+1)};
$$

$$
\left(\frac{y^k}{(N+2)\cdots(N+k+1)}\right)^2 \leq \frac{y^{k-1}}{(N+3)\cdots(N+k+1)};
$$

$$\frac{y^i}{(N+2)\cdots(N+i+1)} \cdot \frac{y^{2k+1-i}}{(N+2)\cdots(N+2k+1-i+1)} \leq \frac{y^{i-1}}{(N+3)\cdots(N+i+1)}.$$

The previous inequalities imply

$$\begin{aligned}
&\left(2\sum_{i=k+1}^{2k-1} \frac{1}{(N+2)\cdots(N+i+1)} \cdot \frac{1}{(N+2)\cdots(N+2k-i+1)} \right.\\
&\qquad \left. + \left(\frac{1}{(N+2)\cdots(N+k+1)}\right)^2\right) y^{2k} \\
&\quad = 2\sum_{i=k+1}^{2k-1} \frac{y^i}{(N+2)\cdots(N+i+1)} \cdot \frac{y^{2k-i}}{(N+2)\cdots(N+2k-i+1)} \\
&\qquad + \left(\frac{y^k}{(N+2)\cdots(N+k+1)}\right)^2 \\
&\quad \leq 2\sum_{i=k+1}^{2k-1} \frac{y^{i-1}}{(N+3)\cdots(N+i+1)} + \frac{y^{k-1}}{(N+3)\cdots(N+k+1)}.
\end{aligned}$$

Using again $0 < y \leq N+1$ we have,

$$\begin{aligned}
&\left(2\sum_{i=k+1}^{2k} \frac{1}{(N+2)\cdots(N+i+1)} \cdot \frac{1}{(N+2)\cdots(N+2k+1-i+1)}\right) y^{2k+1} \\
&\qquad \leq 2\sum_{i=k+1}^{2k} \frac{y^{i-1}}{(N+2)\cdots(N+i+1)} < 2\sum_{i=k+1}^{2k} \frac{y^{i-1}}{(N+3)\cdots(N+i+1)}.
\end{aligned}$$

Hence

$$\begin{aligned}
&\sum_{k=2}^{\infty}\left(2\sum_{i=k+1}^{2k-1} \frac{1}{(N+2)\cdots(N+i+1)} \cdot \frac{1}{(N+2)\cdots(N+2k-i+1)} \right.\\
&\qquad \left. + \left(\frac{1}{(N+2)\cdots(N+k+1)}\right)^2\right) y^{2k} \\
&\quad \leq \sum_{k=2}^{\infty}\left(2\sum_{i=k+1}^{2k-1} \frac{y^{i-1}}{(N+3)\cdots(N+i+1)} + \frac{y^{k-1}}{(N+3)\cdots(N+k+1)}\right) \\
&\quad = \sum_{k=2}^{\infty} \frac{y^{k-1}}{(N+3)\cdots(N+k+1)} + \sum_{i=3}^{\infty}\sum_{k=\lceil \frac{i+1}{2}\rceil}^{i-1} \frac{2y^{i-1}}{(N+3)\cdots(N+i+1)} \\
&\quad = \sum_{k=2}^{\infty} \frac{y^{k-1}}{(N+3)\cdots(N+k+1)} + \sum_{i=3}^{\infty} \frac{2(\lfloor \frac{i+1}{2}\rfloor - 1)y^{i-1}}{(N+3)\cdots(N+i+1)} \\
&\quad = \frac{y}{N+3} + \sum_{k=3}^{\infty} \frac{(2\lfloor \frac{k+1}{2}\rfloor - 1)y^{k-1}}{(N+3)\cdots(N+k+1)}
\end{aligned}$$

and

$$\sum_{k=2}^{\infty}\left(2\sum_{i=k+1}^{2k}\frac{1}{(N+2)\cdots(N+i+1)}\cdot\frac{1}{(N+2)\cdots(N+2k+1-i+1)}\right)y^{2k+1}$$
$$\leq\sum_{k=3}^{\infty}\frac{2\lfloor\frac{k}{2}\rfloor y^{k-1}}{(N+3)\cdots(N+k+1)}.$$

By adding the last two inequalities we obtain

$$\begin{aligned}
&\sum_{k=2}^{\infty}\left(2\sum_{i=k+1}^{2k-1}\frac{1}{(N+2)\cdots(N+i+1)}\cdot\frac{1}{(N+2)\cdots(N+2k-i+1)}\right.\\
&\qquad\left.+\left(\frac{1}{(N+2)\cdots(N+k+1)}\right)^2\right)y^{2k}\\
&\quad+\sum_{k=2}^{\infty}\left(2\sum_{i=k+1}^{2k}\frac{1}{(N+2)\cdots(N+i+1)}\cdot\frac{1}{(N+2)\cdots(N+2k+1-i+1)}\right)y^{2k+1}\\
&\quad\leq\frac{y}{N+3}+\sum_{k=3}^{\infty}\frac{(2(\lfloor\frac{k}{2}\rfloor+\lfloor\frac{k+1}{2}\rfloor)-1)y^{k-1}}{(N+3)\cdots(N+k+1)}\\
&\quad=\frac{y}{N+3}+\sum_{k=3}^{\infty}\frac{(2k-1))y^{k-1}}{(N+3)\cdots(N+k+1)}\\
&\quad=\frac{y}{N+3}+\sum_{k=2}^{\infty}\frac{(2k+1)y^{k}}{(N+3)\cdots(N+k+2)}\\
&\quad=\frac{y}{N+3}+\frac{5y^2}{(N+3)(N+4)}+\frac{7y^3}{(N+3)(N+4)(N+5)}\\
&\qquad+\sum_{n=4}^{\infty}\frac{(2n+1)y^n}{(N+3)\cdots(N+n+2)}.
\end{aligned}$$

Furthermore, it can easily be verified that

$$\begin{aligned}
&\sum_{n=4}^{\infty}\left(\frac{n+1}{(N+3)\cdots(N+n+2)}+\frac{2n-2}{(N+2)\cdots(N+n+1)}\right.\\
&\qquad\left.+\frac{n-1}{(N+2)^2(N+3)\cdots(N+n)}\right)y^n\\
&\quad\geq\sum_{n=4}^{\infty}\frac{(4n-2)y^n}{(N+3)\cdots(N+n+2)}\\
&\quad\geq\sum_{n=4}^{\infty}\frac{(2n+1)y^n}{(N+3)\cdots(N+n+2)}.
\end{aligned}$$

Combining the previous inequalities we readily obtain

$$I(y)\geq 0,\quad y\in(0,N+1],$$

so, $g'(y)\geq 0$, for all $y\in(0,N+1]$. □

LEMMA 4.5. *For every* $x \in [\frac{\pi}{4}, \frac{\pi}{3}]$, *define*

$$h_x(a) = \frac{1}{x(1+ax^2)} - \frac{2}{(\pi - 2x)(1 + a(\pi - 2x)^2)}.$$

Then, h_x *is an increasing function on the interval* $[\frac{4}{3\pi^2}, \frac{4}{\pi^2}]$.

PROOF. Routine calculations yield

$$\begin{aligned} h'_x(a) &= -\frac{x}{(1+ax^2)^2} + \frac{2(\pi-2x)}{(1+a(\pi-2x)^2)^2} \\ &= \frac{2xa(\pi-2x)(2x-(\pi-2x)) + 2(\pi-2x)}{(1+ax^2)^2(1+a(\pi-2x)^2)^2} \\ &\quad + \frac{-x + a^2x(\pi-2x)(2x^3-(\pi-2x)^3)}{(1+ax^2)^2(1+a(\pi-2x)^2)^2} \\ &\geq \frac{2(\pi-2x) - x + a^2x(\pi-2x)(2x^3-(\pi-2x)^3)}{(1+ax^2)^2(1+a(\pi-2x)^2)^2}. \end{aligned}$$

First, observe that if $x \in [0, \frac{\pi}{3}]$, then $2(\pi - 2x) - x \geq \pi/3$, and if $x \in [\frac{\pi}{2+2^{1/3}}, \frac{\pi}{3}]$, then $2x^3 - (\pi - 2x)^3 \geq 0$, and thus

$$2(\pi - 2x) - x + a^2x(\pi - 2x)(2x^3 - (\pi - 2x)^3) > 0.$$

So, regardless of the values of a, $h'_x(a)$ is positive for every $x \in [\frac{\pi}{2+2^{1/3}}, \frac{\pi}{3}]$.

Now, assume $x \in [\frac{\pi}{4}, \frac{\pi}{2+2^{1/3}}]$. Then, $2x^3 - (\pi - 2x)^3 < 0$, so, if $a \in [\frac{4}{3\pi^2}, \frac{4}{\pi^2}]$, then

$$2(\pi - 2x) - x + a^2x(\pi - 2x)(2x^3 - (\pi - 2x)^3) \geq \omega(x),$$

where $\omega(x) := 2(\pi - 2x) - x + \frac{16}{\pi^4}x(\pi - 2x)(2x^3 - (\pi - 2x)^3)$. The proof of the lemma will be completed, once we establish $\omega(x) > 0$, for every x in $[\frac{\pi}{4}, \frac{\pi}{2+2^{1/3}})$.

First, note that $\omega(\pi/4) > 0$. On the other hand, $\omega'(x) \geq \frac{8}{\pi^2}[x^2+(2x-5)^2]$. But the quadratic polynomial $x^2 + (2x-5)^2$ is decreasing on $(-\infty, 2)$, thus ω' restricted on $[\frac{\pi}{4}, \frac{\pi}{2+2^{1/3}}]$ attains its minimum value at $x = \frac{\pi}{2+2^{1/3}}$. Since, $\frac{\pi}{2+2^{1/3}} < 1$, we obtain

$$\omega'\left(\frac{\pi}{2+2^{1/3}}\right) > \omega'(1) \geq 5\left(\frac{16}{\pi^2} - 1\right) > 0.$$

This establishes that ω is increasing in the interval $[\frac{\pi}{4}, \frac{\pi}{2+2^{1/3}}]$, and since $\omega(\pi/4) > 0$, we finally conclude that $h'_x(a) > 0$, for all $x \in [\frac{\pi}{4}, \frac{\pi}{2+2^{1/3}}]$ and $a \in [\frac{1}{3}(\frac{4}{\pi^2}), \frac{4}{\pi^2}]$. Thus, h_x is an increasing function on the interval $[\frac{1}{3}(\frac{4}{\pi^2}), \frac{4}{\pi^2}]$. □

LEMMA 4.6. *Let*

$$p(y) = \frac{1}{y^3} - \frac{1}{\sin^2 y \tan y} = \frac{(\sin^2 y \tan y / y^3) - 1}{\sin^2 y \tan y}$$

and

$$q(y) = \frac{\sin^2 y \tan y}{y^3} - 1.$$

Then, for every $y \in [\frac{\pi}{8}, \frac{\pi}{4}]$ *we have* $p(y) \geq 0$ *and that* q *is increasing.*

PROOF. Obviously,

$$p(y) = \frac{q(y)}{\sin^2 y \tan y}.$$

It is also easy to verify

$$q'(y) = \frac{2y\sin^2 y + y\tan^2 y - 3\sin^2 y \tan y}{y^4} = \frac{\tan^2 y(2y\cos^2 y + y - \frac{3}{2}\sin 2y)}{y^4}.$$

Now, let

$$r(y) = 2y\cos^2 y + y - \frac{3}{2}\sin 2y.$$

Then,

$$r'(y) = 2\cos^2 y - 4y\cos y \sin y + 1 - 3\cos 2y = 2 - 2\cos 2y - 2y\sin 2y,$$

$$r''(y) = 4\sin 2y - 2\sin 2y - 4y\cos 2y = 2\cos 2y(\tan 2y - 2y).$$

Clearly, $r''(y) \geq 0$ for all $y \in [0, \frac{\pi}{4}]$ and $r(0) = r'(0) = 0$. Thus, $r'(y) > 0$ and so $r(y) > 0$, for every $y \in [0, \frac{\pi}{4}]$. Therefore,

$$q'(y) \geq 0, \quad y \in \left(0, \frac{\pi}{4}\right].$$

Since $\lim_{y\to 0+} q(y) = 0$ we have $q(y) > 0$ for every $y \in [0, \frac{\pi}{4}]$, so $p(y) > 0$ for all $y \in [\frac{\pi}{8}, \frac{\pi}{4}]$. □

LEMMA 4.7. *The function*

$$D(x) = \left(\frac{1}{x} - \frac{1}{2\tan(x/2)}\right) - \frac{(4x/(3\pi^2))}{1 + (4x^2/(3\pi^2))}$$
$$+ 2 \cdot \frac{(4(\pi - 2x)/(3\pi^2))}{1 + (4(\pi - 2x)^2/(3\pi^2))} - 2\left(\frac{1}{(\pi - 2x)} - \frac{1}{2\tan((\pi - 2x)/2)}\right)$$

is non-negative on the interval $x \in [\frac{\pi}{4}, \frac{\pi}{3}]$.

PROOF. Elementary computations give

$$D\left(\frac{\pi}{4}\right) > 0.013, \quad D\left(\frac{\pi}{3}\right) > 0.034.$$

In view of these numerical results, the conclusion the lemma will be established, once we show $D''(x) \leq 0$ for $x \in [\frac{\pi}{4}, \frac{\pi}{3}]$, because, then D is concave downward.

Routine calculations give

$$D'(x) = -\frac{(4/(3\pi^2)) \cdot (1 - (4x^2/(3\pi^2)))}{(1 + (4x^2/(3\pi^2)))^2} - 4 \cdot \frac{(4/(3\pi^2)) \cdot (1 - (4(\pi - 2x)^2/(3\pi^2)))}{(1 + (4(\pi - 2x)^2/(3\pi^2)))^2}$$
$$+ \left(-\frac{1}{x^2} + \frac{1}{4\sin^2(x/2)}\right) + 4 \cdot \left(-\frac{1}{(\pi - 2x)^2} + \frac{1}{4\sin^2((\pi - 2x)/2)}\right)$$

and

$$\begin{aligned}
D''(x) = & \frac{2(4/(3\pi^2))^2 x(3-(4x^2/(3\pi^2)))}{(1+(4x^2/(3\pi^2)))^3} \\
& - 8\left(\frac{2(4/(3\pi^2))^2(\pi-2x)\cdot(3-(4(\pi-2x)^2/(3\pi^2)))}{(1+(4(\pi-2x)^2/(3\pi^2)))^3}\right) \\
& + \left(\frac{2}{x^3} - \frac{1}{4\sin^2(x/2)\tan(x/2)}\right) \\
& - 8\left(\frac{2}{(\pi-2x)^3} - \frac{1}{4\sin^2((\pi-2x)/2)\tan((\pi-2x)/2)}\right) \\
= & \frac{(32/(9\pi^4))x(3-(4x^2/(3\pi^2)))}{(1+(4x^2/(3\pi^2)))^3} \\
& - 8\left(\frac{(32/(9\pi^4))(\pi-2x)\cdot(3-(4(\pi-2x)^2/(3\pi^2)))}{(1+(4(\pi-2x)^2/(3\pi^2)))^3}\right) \\
& + \frac{1}{4}\left(\frac{1}{(x/2)^3} - \frac{1}{\sin^2(x/2)\tan(x/2)}\right) \\
& - 2\left(\frac{1}{(\pi-2x/2)^3} - \frac{1}{\sin^2((\pi-2x)/2)\tan((\pi-2x)/2)}\right).
\end{aligned}$$

For $\frac{\pi}{4} \le x \le \frac{\pi}{3}$, we have $\frac{\pi}{3} \le \pi - 2x \le \frac{\pi}{2}$. A routine first-order derivative test shows that $a(y) := y(3 - \frac{4y^2}{3\pi^2})$ is increasing on $[\frac{\pi}{4}, \frac{\pi}{2}]$. Applying this result both for the restriction of a on the interval $[\frac{\pi}{4}, \frac{\pi}{2}]$ and the composite function $x \to a(\pi - 2x)$, where $\frac{\pi}{4} \le x \le \frac{\pi}{3}$, we obtain

$$\begin{aligned}
& \frac{(32/(9\pi^4))x(3-(4x^2/(3\pi^2)))}{(1+(4x^2/(3\pi^2)))^3} - 8\left(\frac{(32/(9\pi^4))(\pi-2x)\cdot(3-(4(\pi-2x)^2/(3\pi^2)))}{(1+(4(\pi-2x)^2/(3\pi^2)))^3}\right) \\
& \le \frac{(32/(9\pi^4))(\pi/3)\cdot(3-(4/(3\pi^2))(\pi^2/9))}{(1+(4/(3\pi^2))(\pi^2/16))^3} \\
& \quad - 8\left(\frac{(32/(9\pi^4))(\pi/3)\cdot(3-(4/(3\pi^2))(\pi^2/9))}{(1+(4/(3\pi^2))(\pi^2/4))^3}\right) \\
& < -0.2\,.
\end{aligned}$$

Moreover, Lemma 4.6 implies

$$-2\left(\frac{1}{(\pi-2x/2)^3} - \frac{1}{\sin^2((\pi-2x)/2)\tan((\pi-2x)/2)}\right) \le 0$$

and

$$\begin{aligned}
\frac{1}{4}\left(\frac{1}{(x/2)^3} - \frac{1}{\sin^2(x/2)\tan(x/2)}\right) &= \frac{\frac{\sin^2(x/2)\tan(x/2)}{(x/2)^3} - 1}{4\sin^2(x/2)\tan(x/2)} \\
&\le \frac{\frac{\sin^2(\pi/6)\tan(\pi/6)}{(\pi/6)^3} - 1}{4\sin^2(\pi/8)\tan(\pi/8)} < 0.03\,.
\end{aligned}$$

Combining all the previous inequalities we conclude

$$D''(x) \le -0.2 + 0.03 + 0 < 0\,.$$

□

Eq. (3.1) in the previous section implies that L_N is continuous on $[0,\pi]$. Therefore, in order to prove Theorem 1.4 it is enough to restrict our discussion on the interval $(0,\pi)$.

PROOF OF ITEM 2 OF THEOREM 1.4. Since L_N is even and

$$|L_N(\xi)L_N(2\xi)| = |L_N(\xi)L_N(2\pi-2\xi)| = L_N(\xi)L_N(2\pi-2\xi)\ .$$

Changing the variable from $\pi-\xi$ to x, we assert that $L_N(\xi)L_N(2\xi)$ is decreasing on $[\frac{2\pi}{3},\pi]$ if and only if $L_N(\pi-x)L_N(2x)$ is increasing on $[0,\frac{\pi}{3}]$, equivalently

$$(\ln(L_N(\pi-x)L_N(2x)))' \geq 0,$$

for every $0\leq x\leq\frac{\pi}{3}$.

Since, $\frac{2\pi}{3}\leq\pi-x\leq\pi$ for every $x\in[0,\frac{\pi}{3}]$, we get

$$L_N(\pi-x) = \frac{\sqrt{1-\widehat{h}^2_{N,\sigma}(x)}}{\sin^{N+1}(x/2)}.$$

We will give the proof of above inequality on the intervals $(0,\frac{\pi}{4})$ and $[\frac{\pi}{4},\frac{\pi}{3}]$, separately.

First, let $x\in(0,\frac{\pi}{4})$. Then, $0<2x<\frac{\pi}{2}$. Therefore,

$$L_N(2x) = \frac{\widehat{h}_{N,\sigma}(2x)}{\cos^{N+1}(x)}.$$

So, if $x\in(0,\frac{\pi}{4})$, routine calculations give

$$\frac{[\ln(L_N(\pi-x)L_N(2x))]'}{N+1} = \frac{1}{(x/2)\cdot(2+\mathcal{Q}_{N,3}(\sigma^2x^2/2))\cdot\mathcal{Q}_{N,1}(\sigma^2x^2/2)} - \frac{1}{2\tan(x/2)} + \tan x - \frac{(4\sigma^2/(N+1))x}{\mathcal{Q}_{N,2}(2\sigma^2x^2)}.$$

A direct application of Proposition 4.3, implies

$$\frac{(\ln L_N(\pi-x)L_N(2x))'}{N+1} \geq 0.$$

Therefore, $L_N(\pi-x)L_N(2x)$ is increasing on $[0,\frac{\pi}{4}]$.

We now focus on $x\in[\frac{\pi}{4},\frac{\pi}{3})$. Obviously, all such x satisfy $\frac{\pi}{2}\leq 2x<\frac{2\pi}{3}$. Therefore,

$$L_N(2x)L_N(\pi-x) = \frac{\sqrt{1-\widehat{h}^2_{N,\sigma}(\pi-2x)}}{\sin^{N+1}(\pi-2x/2)}\frac{\sqrt{1-\widehat{h}^2_{N,\sigma}(x)}}{\sin^{N+1}(x/2)}.$$

Thus,

$$\begin{aligned}
&\frac{(\ln L_N(\pi-x)L_N(2x))'}{N+1}\\
&= \frac{1}{(x/2)\cdot(2+\mathcal{Q}_{N,3}(\sigma^2x^2/2))\cdot\mathcal{Q}_{N,1}(\sigma^2x^2/2)} - \frac{1}{2\tan(x/2)}\\
&\quad - \frac{2}{(\pi-2x/2)\cdot(2+\mathcal{Q}_{N,3}(\sigma^2(\pi-2x)^2/2))\cdot\mathcal{Q}_{N,1}(\sigma^2(\pi-2x)^2/2)}\\
&\quad + \frac{1}{\tan(\pi-2x/2)}.
\end{aligned}$$

We denote the right-hand-side of above equation as I. Using the definition of $\mathcal{Q}_{N,4}$, we rewrite I:

$$I = I_1 + I_2 + I_3 ,$$

where

$$I_1 = \frac{2}{\pi - 2x}\mathcal{Q}_{N,4}\left(\frac{\sigma^2(\pi-2x)^2}{2}\right) - \frac{1}{x}\mathcal{Q}_{N,4}\left(\frac{\sigma^2 x^2}{2}\right),$$

$$I_2 = \frac{1}{x\mathcal{Q}_{N,1}(\sigma^2x^2/2)} - \frac{2}{(\pi-2x)\mathcal{Q}_{N,1}(\sigma^2(\pi-2x)^2/2)},$$

$$I_3 = -\frac{1}{2\tan(x/2)} + \frac{1}{\tan(\pi - 2x/2)}.$$

Now, observe that for every $x \in [\frac{\pi}{4}, \frac{\pi}{3}]$ we have,

$$x \leq \pi - 2x \leq \pi/2 ,$$

and thus,

$$\frac{1}{x} \leq \frac{2}{\pi - 2x}. \tag{4.23}$$

Moreover, due to Theorem 1.1, we obtain $\frac{\sigma^2x^2}{2} < \gamma_N < N+1$ and $\frac{\sigma^2(\pi-2x)^2}{2} < \gamma_N < N+1$, for all $N \geq 1$ and $\frac{\pi}{4} \leq x \leq \frac{\pi}{3}$; so, by (4.23) and item 3 of Lemma 4.1 we conclude $I_1 \geq 0$.

An elementary calculation establishes

$$\begin{aligned} I_2 = {} & \frac{1}{x\mathcal{Q}_{N,1}(\sigma^2x^2/2)} - \frac{1}{x(1+(\sigma^2x^2/(2(N+2))))} + \frac{1}{x(1+(\sigma^2x^2/(2(N+2))))} \\ & - \frac{2}{(\pi-2x)\mathcal{Q}_{N,1}(\sigma^2(\pi-2x)^2/2)} + \frac{2}{(\pi-2x)(1+(\sigma^2(\pi-2x)^2/(2(N+2))))} \\ & - \frac{2}{(\pi-2x)(1+(\sigma^2(\pi-2x)^2/(2(N+2))))} . \end{aligned}$$

Using g defined in Lemma 4.4, we obtain

$$\begin{aligned} I_2 = {} & \frac{2}{\pi-2x}g\left(\frac{\sigma^2(\pi-2x)^2}{2}\right) - \frac{1}{x}g\left(\frac{\sigma^2x^2}{2}\right) \\ & + \frac{1}{x(1+(\sigma^2x^2/(2(N+2))))} - \frac{2}{(\pi-2x)(1+(\sigma^2(\pi-2x)^2/(2(N+2))))}. \end{aligned}$$

Combining $\frac{\sigma^2x^2}{2} < N+1$ and $\frac{\sigma^2(\pi-2x)^2}{2} < N+1$, (4.23) and the fact that g is increasing on $[0, N+1)$, for all $N \geq 1$,

$$I_2 \geq \frac{1}{x(1+(\sigma^2x^2/(2(N+2))))} - \frac{2}{(\pi-2x)(1+(\sigma^2(\pi-2x)^2/(2(N+2))))} . \tag{4.24}$$

Using once again $\gamma_N < N+1$, for every $N \geq 1$, and item 2 of Theorem 1.1 we have

$$\frac{4}{3\pi^2} \leq \frac{\sigma^2}{2(N+2)} \leq \frac{4}{\pi^2}, \quad N \geq 1.$$

Setting $a = \frac{\sigma^2}{2(N+2)}$ and then applying Lemma 4.5, inequality (4.24) implies,

$$I_2 \geq \frac{1}{x(1+(4x^2/(3\pi^2)))} - \frac{2}{(\pi-2x)(1+(4(\pi-2x)^2/(3\pi^2)))}.$$

Combining the previous inequality with $I_1 \geq 0$, we have

$$\begin{aligned} I &\geq \frac{1}{x(1+(4x^2/(3\pi^2)))} - \frac{1}{2\tan(x/2)} \\ &\quad + \frac{1}{\tan((\pi-2x)/2).} - \frac{2}{(\pi-2x)(1+(4(\pi-2x)^2/(3\pi^2)))} \\ &= \left(\frac{1}{x} - \frac{1}{2\tan(x/2)}\right) - \frac{(4x/(3\pi^2))}{1+(4x^2/(3\pi^2))} + 2\cdot\frac{(4(\pi-2x)/(3\pi^2))}{1+(4(\pi-2x)^2/(3\pi^2))} \\ &\quad - 2\left(\frac{1}{(\pi-2x)} - \frac{1}{2\tan((\pi-2x)/2)}\right), \end{aligned}$$

for $\frac{\pi}{4} \leq x \leq \frac{\pi}{3}$. Finally, Lemma 4.7 readily implies,

$$I \geq 0,$$

which completes the proof of second part of Theorem 1.4. □

We will conclude the present section with the proof of Theorem 1.5.

PROOF OF THEOREM 1.5. According to the remarks preceding the statement of Theorem 1.5 the only conclusion that remains to be proved is $\lim_{N\to\infty}(N-b_N) = \infty$.

The definition of b_N implies

$$b_N = \log_2\sqrt{1-\widehat{h}^2_{N,\sigma}\left(\frac{\pi}{3}\right)} + N + 1\,, \tag{4.25}$$

for every $N \geq 1$.

In order to establish $\lim_{N\to+\infty}(N-b_N) = +\infty$ it suffices to prove

$$\lim_{N\to+\infty} \widehat{h}_{N,\sigma}(\pi/3) = 1 \tag{4.26}$$

Using g_N defined in the proof of Theorem 1.1 we infer

$$\widehat{h}_{N,\sigma}\left(\frac{\pi}{3}\right) = g_N\left(\frac{4}{9}\gamma_N\right)\,, \tag{4.27}$$

where γ_N is defined so that $\widehat{h}_{N,\sigma}(\frac{\pi}{2}) = g_N(\gamma_N) = \frac{\sqrt{2}}{2}$. Recall that g_N is decreasing (refer again to the proof of Theorem 1.1). By Theorem 1.1 $\gamma_N < N$ for all $N \geq 2$. Thus,

$$g_N\left(\frac{4}{9}N\right) < g_N\left(\frac{4}{9}\gamma_N\right) < 1\,.$$

In order to prove (4.26), it is enough to show $\lim_{N\to+\infty} g_N(\frac{4}{9}N) = 1$. In fact, a more general result is true: $\lim_{N\to+\infty} g_N(xN) = 1$, for $0 < x < 1$. Let us now establish this result.

Fix $0 < x < 1$. Using the definition of g_N we have

$$\begin{aligned} 1 - g_N(xN) &= e^{-Nx}\sum_{n=N+1}^{\infty}\frac{(Nx)^n}{n!} \\ &= e^{-Nx}\frac{(Nx)^{N+1}}{(N+1)!}\left(1+\sum_{k=1}^{\infty}\frac{(Nx)^k}{(N+2)\cdots(N+1+k)}\right). \end{aligned}$$

On the other hand the series $\sum_{k=1}^{\infty} \frac{(Nx)^k}{(N+2)\cdots(N+1+k)}$ is dominated by the geometric series $\sum_{k=1}^{\infty} x^k$, which converges since $0 < x < 1$. Therefore,

$$0 \leq 1 - g_N(xN) < e^{-Nx} \frac{(Nx)^{N+1}}{(N+1)!} (1-x)^{-1} .$$

Using Stirling's formula ($n! > e^{-n}n^n$, for all $n \in \mathbb{N}$) for $n = N+1$, we now obtain

$$e^{-Nx} \frac{(Nx)^{N+1}}{(N+1)!} (1-x)^{-1} < e^{-Nx} \frac{(Nx)^{N+1} e^{N+1}}{(N+1)^{N+1}} (1-x)^{-1} ,$$

which finally gives

$$0 \leq 1 - g_N(xN) < (xe^{1-x})^N ex(1-x)^{-1} .$$

Observing that the function $x \to xe^{1-x}$ is increasing on $[0,1]$ we have $xe^{1-x} < 1$, for every $0 < x < 1$, which establishes $\lim_{N \to +\infty}(1 - g_N(xN)) = 0$. □

5. Acknowledgments

This research was supported in part by the following grants: Texas Higher Eduaction Coordinating board ARP 003652-0071-2001, NSF DMS-0070376, NSF CHE-0074311 and R.A. Welch Foundation E-0608. Partial finacial support was also provided by the Texas Learning and Computation Center of the University of Houston and by a University of Houston GEAR 2002 grant. We finally wish to express our most sincere thanks to Professor Q. Sun, currently with the University of Houston's Mathematics Department, for his valuable suggestions that helped us to improve the original manuscript.

References

[AS72] M. Abramowitz and I.A. Stegun (Eds.), *Handbook of Mathematical Functions*, Dover Publications Inc., NY 1972.

[CG99] C. Chandler and A. G. Gibson, *Uniform approximation of functions with discrete approximation functionals*, J. Approximation Theory **100** (1999), 233–250.

[CR95] A. Cohen and R. D. Ryan, *Wavelets and Multirate Signal Processing*, Chapman & Hall, 1995.

[FS01] A. Fan, Q. Sun, *Regularity of Butterworth refinable functions*, Asian J. Math. **5**(3), pp. 433-440, 2001.

[HK93] D. K. Hoffman and D. J. Kouri, *Distributed approximating function theory for an arbitrary number of particles in a coordinate system-independent formalism*, J. Phys. Chem. **97** (1993), 4984–4988.

[HNSK91] D. K. Hoffman, N. Nayar, O. A. Sharafeddin, and D. J. Kouri, *On an analytic banded approximation for the discretized free propagator*, J. Phys. Chem. **95** (1991), 8299–8305.

[HW96] E. Hernández and G. Weiss, *A First Course on Wavelets*, CRC Press, 1996.

[HE95] , L. Herve, *Construction et regularite des fonctions d' echelle*, SIAM J. Math. Anal., **26**, pp. 1361-1385, 1995.

[KPS1] I.A. Kakadiaris, M. Papadakis, L. Shen, D.J. Kouri, and D.J. Hoffman, *g-HDAF multiresolution deformable models for shape modeling and reconstruction*, British Machine Vision Conference (Cardiff, UK), Sep 2002.

[KPS2] ———, *m-HDAF multiresolution deformable models*, The 14th International Conference on Digital Signal Processing (Santorini, Greece), Jul 2002.

[Mal89] S.G. Mallat, *Multiresolution approximations and wavelet orthonormal bases of $l^2(\vec{r})$*, Trans. Amer. Math. Soc. **315** (1989), 69–87.

[S99] Q. Sun, *Sobolev exponent estimates and asymptotic regularity of the M-band Daubechies' scaling functions*, Constr. Approx., **15**, pp. 441-465, 1999.

[V94] L.F. Villemoes, *Wavelet analysis of refinement equations*, SIAM J. Math. Anal., **25**, pp. 1433-1466, 1994.

[WWH01] G. Wei, H. Wang, D. K. Hoffman, D.J. Kouri, I. A. Kakadiaris, and M. Papadakis, *On the mathematical properties of distributed approximating functionals*, J. Phys. Chem. **30** (2001), no. 1, 83–107.

School of Mathematical Sciences, Peking University, Beijing 100871, P. R. China
E-mail address: yjs@math.pku.edu.cn

Department of Mathematics, West Virginia University, Morgantown, WV 26506
E-mail address: lshen@math.wvu.edu

Department of Mathematics, University of Houston, Houston, TX 77204-3008
E-mail address: mpapadak@math.uh.edu

Department of Computer Science, University of Houston, Houston, TX 77204-3008
E-mail address: ioannisk@uh.edu

Department of Physics, University of Houston, Houston, TX 77204
E-mail address: Kouri@uh.edu

Department of Chemistry, Iowa State University, Ames, Iowa 50011
E-mail address: hoffman@ameslab.gov

Titles in This Series

345 **Christopher Heil, Palle E. T. Jorgensen, and David R. Larson, Editors,** Wavelets, frames and operator theory, 2004

344 **Ricardo Baeza, John S. Hsia, Bill Jacob, and Alexander Prestel, Editors,** Algebraic and arithmetic theory of quadratic forms, 2004

343 **N. Sthanumoorthy and Kailash C. Misra, Editors,** Kac-Moody Lie algebras and related topics, 2004

342 **János Pach, Editor,** Towards a theory of geometric graphs, 2004

341 **Hugo Arizmendi, Carlos Bosch, and Lourdes Palacios, Editors,** Topological algebras and their applications, 2004

340 **Rafael del Río and Carlos Villegas-Blas, Editors,** Spectral theory of Schrödinger operators, 2004

339 **Peter Kuchment, Editor,** Waves in periodic and random media, 2003

338 **Pascal Auscher, Thierry Coulhon, and Alexander Grigor'yan, Editors,** Heat kernels and analysis on manifolds, graphs, and metric spaces, 2003

337 **Krishan L. Duggal and Ramesh Sharma, Editors,** Recent advances in Riemannian and Lorentzian geometries, 2003

336 **José González-Barrios, Jorge A. León, and Ana Meda, Editors,** Stochastic models, 2003

335 **Geoffrey L. Price, B. Mitchell Baker, Palle E.T. Jorgensen, and Paul S. Muhly, Editors,** Advances in quantum dynamics, 2003

334 **Ron Goldman and Rimvydas Krasauskas, Editors,** Topics in algebraic geometry and geometric modeling, 2003

333 **Giovanni Alessandrini and Gunther Uhlmann, Editors,** Inverse problems: Theory and applications, 2003

332 **John Bland, Kang-Tae Kim, and Steven G. Krantz, Editors,** Explorations in complex and Riemannian geometry, 2003

331 **Luchezar L. Avramov, Marc Chardin, Marcel Morales, and Claudia Polini, Editors,** Commutative algebra: Interactions with algebraic geometry, 2003

330 **S. Y. Cheng, C.-W. Shu, and T. Tang, Editors,** Recent advances in scientific computing and partial differential equations, 2003

329 **Zhangxin Chen, Roland Glowinski, and Kaitai Li, Editors,** Current trends in scientific computing, 2003

328 **Krzysztof Jarosz, Editor,** Function spaces, 2003

327 **Yulia Karpeshina, Günter Stolz, Rudi Weikard, and Yanni Zeng, Editors,** Advances in differential equations and mathematical physics, 2003

326 **Kenneth D. T-R McLaughlin and Xin Zhou, Editors,** Recent developments in integrable systems and Riemann-Hilbert problems, 2003

325 **Seok-Jin Kang and Kyu-Hwan Lee, Editors,** Combinatorial and geometric representation theory, 2003

324 **Caroline Grant Melles, Jean-Paul Brasselet, Gary Kennedy, Kristin Lauter, and Lee McEwan, Editors,** Topics in algebraic and noncommutative geometry, 2003

323 **Vadim Olshevsky, Editor,** Fast algorithms for structured matrices: theory and applications, 2003

322 **S. Dale Cutkosky, Dan Edidin, Zhenbo Qin, and Qi Zhang, Editors,** Vector bundles and representation theory, 2003

321 **Anna Kamińska, Editor,** Trends in Banach spaces and operator theory, 2003

320 **William Beckner, Alexander Nagel, Andreas Seeger, and Hart F. Smith, Editors,** Harmonic analysis at Mount Holyoke, 2003

319 **W. H. Schikhof, C. Perez-Garcia, and A. Escassut, Editors,** Ultrametric functional analysis, 2003